Transition Metals in Supramolecular Chemistry

Editorial Board

Transition Metals in Supramolecular Chemistry

Perspectives in Supramolecular Chemistry
Volume 5

Edited by Jean-Pierre Sauvage
Université Louis Pasteur, France

JOHN WILEY & SONS
Chichester • New York • Weinheim • Brisbane • Singapore • Toronto

Baffins Lane, Chichester,
West Sussex PO19 1UD, England

National 01243 779777
International (+44) 1243 779777

e-mail (for orders and customer service enquiries): cs-books@wiley.co.uk
Visit our Home Page on http://www.wiley.co.uk
or http://www.wiley.com

Other Wiley Editorial Offices

John Wiley & Sons, Inc., 605 Third Avenue,
New York, NY 10158-0012, USA

WILEY-VCH Verlag GmbH, Pappelallee 3,
D-69469 Weinheim, Germany

Jacaranda Wiley Ltd, 33 Park Road Milton,
Queensland 4064, Australia

John Wiley Sons (Asia) Pte Ltd, Clementi Loop #02-01,
Jim Xing Distripark, Singapore 129809

John Wiley & Sons (Canada) Ltd, 22 Worcester Road,
Rexdale, Ontario M9W 1L1, Canada

Library of Congress Cataloging-in-Publication Data

Transition metals in supramolecular chemistry / edited by Jean-Pierre Sauvage.
p. cm. – (Perspectives in supramolecular chemistry ; v. 5)
Includes bibliographical references and index.
ISBN 0471-97620-2
1. Transition metal complexes. 2. Macromolecules. I. Sauvage, Jean-Pierre. II. Series.
QD474.T69 1999
546′.6–dc21 98-54677
CIP

British Library Cataloguing in Publication Data

A catalogue record for this book is available from the British Library

ISBN 0 471 97620 2

Typeset in 10/12pt Times by Techset Composition Ltd, Salisbury, Wiltshire
Printed and bound in Great Britain by Biddles Ltd, Guildford and King's Lynn
This book is printed on acid-free paper responsibly manufactured from sustainable forestry, in which at least two trees are planted for each one used for paper production.

Contents

Contributors

Duncan W. Bruce, School of Chemistry, University of Exeter, Stocker Road, Exeter, EX4 4QD, UK

Jean-Claude Chambron, Faculté de Chimie, Université Louis Pasteur, 67000 Strasbourg, France

Simon Collinson, School of Chemistry, University of Exeter, Stocker Road, Exeter, EX4 4QD, UK

Bernhard Demleitner, Institut für Organische Chemie der Universität Erlangen-Nürnberg, Henkestrasse 42, D-91054 Erlangen, Germany

Luigi Fabbrizzi, Dip. Chimica Generale, Università di Pavia, Via Taramelli 12, I-27100 Pavia, Italy

Maurizio Licchelli, Dip. Chimica Generale, Università di Pavia, Via Taramelli 12, I-27100 Pavia, Italy

Kimoon Kim, National Creative Research Initiative, Center for Smart Supramolecules and Department of Chemistry, Pohang University of Science and Technology, San 31, Hyojadong, Pohang 790–784, South Korea

Piersandro Pallavicini, Dip. Chimica Generale, Università di Pavia, Via Taramelli 12, I-27100 Pavia, Italy

Luisa Parodi, Dip. Chimica Generale, Università di Pavia, Via Taramelli 12, I-27100 Pavia, Italy

Christophe Provent, Département de Chimie Minérale, Analytique et Appliquée, Université de Genève, 30 quai Ernest Ansermet, CH 1211 Geneva 4, Switzerland

José Antonio Real, Departament de Quimica Inorgànica, Facultat de Quimica, Universitat de València, 46100 Burjassot, Spain

Rolf W. Saalfrank, Institut für Organische Chemie der Universität Erlangen-Nürnberg, Henkestrasse 42, D-91054 Erlangen, Germany

Angelo Taglietti, L Dip. Chimica Generale, Università di Pavia, Via Taramelli 12, I-27100 Pavia, Italy

Alan F. Williams, Département de Chimie Minérale, Analytique et Appliquée, Université de Genève, 30 quai Ernest Ansermet, CH 1211 Geneva 4, Switzerland

Richard E. P. Winpenny, Department of Chemistry, University of Edinburgh, West Mains Road, Edinburgh, EH9 3JJ, Scotland

Preface

The pioneering work of Pedersen, Lehn and Cram on various cyclic structures acting as hosts, and their interactions with cationic species, is considered as the start of modern supramolecular chemistry—the chemistry of weak forces and noncovalent interactions. Clearly, 30 years ago transition metals and their complexes were not regarded as important components in such structures, and the fields of 'host–guest' recognition and coordination chemistry were very distinct with almost nothing to share. Things have dramatically changed! It suffices to wander through the eight following contributions to realize that transition metal complexes are nowadays used almost routinely to build large multicomponent architectures. These often display new and exciting properties on the way to molecular devices for specific functions.

Transition metals are utilized to construct fascinating structures such as 'rotaxanes' (a rotaxane is a ring threaded by an open-chain fragment bearing two bulky groups at its ends so as to prevent dethreading), or beautiful and novel multicomponent assemblies such as helices, grids and related high-nuclearity complexes. Another very active field of research, related to supramolecular sciences, is concerned with liquid crystals incorporating more and more sophisticated transition metal complexes and thus displaying increasingly well-defined and specific properties related to both their mesogen nature and the presence of transition metal centres.

Molecular magnetism constitutes another promising facet of modern coordination chemistry and the field has produced in recent years both magnificent solid-state structures and properties. The use of transition metal binding sites associated with various luminophores, for instance has also led to the development of promising molecular switches.

If the present volume covers a relatively broad field of coordination and supramolecular chemistry, I am conscious that other important related areas of research are not represented, in particular those related to biology. Nevertheless, this book gives an excellent and contemporary view of the 'abiotic' side, and we hope that it will contribute to stimulate ideas and interactions between researchers from various disciplines.

Jean-Pierre Sauvage
Strasbourg, May 27, 1998

Chapter 1

Ligand and Metal Control of Self-Assembly in Supramolecular Chemistry

ROLF W. SAALFRANK AND BERNHARD DEMLEITNER

University Erlangen-Nürnberg, Germany

1 INTRODUCTION

Self-assembly [1–16] is ubiquitous in nature, where the process is used to create complex functional biological structures with precision. Biologically self-assembled structures are built up by a modular approach from simple, ordinary subunits, thus minimizing the amount of information required for a specific ensemble. Among the most notable biologically self-assembled structures is the *tobacco mosaic virus* (TMV) [17]. However, chemistry is not limited to systems similar to those found in nature; chemists are free to create unknown unnatural species and to invent novel processes. Driven by the quest for new host molecules, self-assembly has been recognized as a powerful methodology for the construction of supramolecular systems. The conceptual basis underlying self-assembled transition metal complexes makes structures generally analogous to the well-established organic-based molecular host–guest systems accessible. This analogy has proven highly successful for the controlled self-assembly of numerous clusters with predesigned molecular architectures. While much of the present work in supramolecular chemistry remains focused on the development of organic molecular recognition agents, there is an increasing interest in hosts that contain transition metals. Consequently, the predictable nature of coordination chemistry has been used successfully to generate the metallatopomers **4–6** of the coronates **1**, {2}-cryptates **2**, and {3}-cryptates **3**

Transition Metals in Supramolecular Chemistry, edited by J. P. Sauvage.

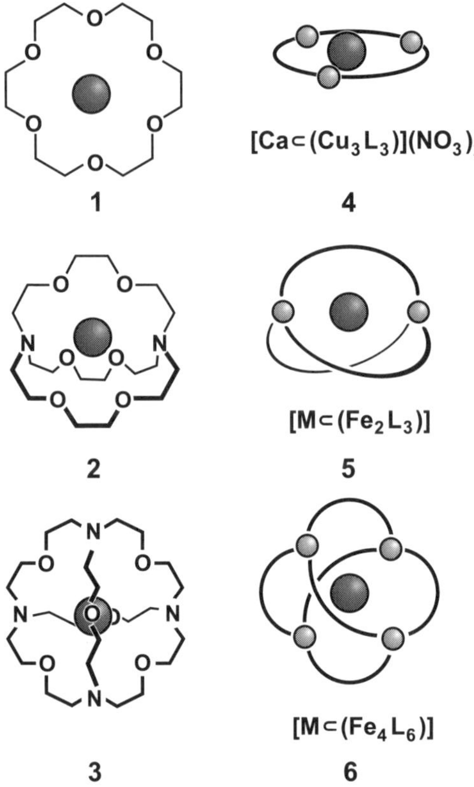

Figure 1 Pictorial representation: metallatopomers **4–6** of coronates **1**, {2}-cryptates **2**, and {3}-cryptates **3**.

efficiently by simply mixing the component ligands and metal ions in solution (Figure 1) [18,19].

These compounds enjoy a number of advantages over their organic counterparts, in particular, one-pot reactions, high yields, spectroscopic, electronic, and magnetic properties, which are inaccessible with organic species [20,21]. Furthermore, the use of transition metal coordination has been explored by Lehn and co-workers and many others. The new strategy was successfully used for the construction of molecular racks [1,22], ladders [1,11,23], grids [1,11,24,25], squares [7,26,27], cylinders [11,28], molecular boxes [29], catenanes [13,15,30], rotaxanes [31], knots [16,30,32], dendrimers [11,33], double, triple [12,13], and circular [34,35] helicates. Self-assembly of Werner-type complexes features the participation of weak, reversible, non-covalent bonding interactions, which facilitates error checking and self-correction. Over the years, the number of laboratories around the world, attracted by the interdisciplinary character of supramolecular chemistry, has increased considerably. We do not attempt to cover the literature exhaustively, but more recent notable advances in this area have been given prominence here and are detailed below.

2 METALLACORONATES, CIRCULAR HELICATES, {2}-METALLACRYPTATES AND {3}-METALLACRYPTATES

2.1 Metallacoronands and Metallacoronates

Pecoraro *et al.* demonstrate [20] in an excellent comprehensive review article that cyclic structures generally analogous to classical organocrown ethers are accessible by simply substituting an ethane bridge with a transition metal ion. The metallacrown analogy has proven highly successful for the controlled preparation of homonuclear, heteronuclear, and mixed-valent assemblies of moderate nuclearity (3–12 metal atoms) with predesigned molecular architectures. Real metallacrown analogy [36] is nicely illustrated (Figure 2) by the trinuclear vanadium complex **7** [37,38], prepared from VCl_3 with salicylhydroxamic acid and three equivalents of sodium methoxide in methanol. A series of related inorganic metallacrown copper(II) complexes have also been prepared and characterized by X-ray diffraction [39]. The tetranuclear copper(II) complex **8** [38,40], prepared with β-alaninehydroxamic acid, displays an almost planar conformation of the macrocycle. Schematic diagrams of [9]crown-3 (9-C-3) **9** and [12]crown-4 (12-C-4) **10** are compared with the analogous cores of the metallacrowns [9]metallacrown-3 (9-MC-3) **7** and [12]metallacrown-4 (12-MC-4) **8**.

Salicylhydroxamic acid (H_3shi), which contains both hydroximate and phenolate donors, and many hydroximate-based ligands form tri- and tetranuclear metallacrowns with moderate-to-high valent metal ions [e.g. Fe^{II}, Mn^{III}, Ga^{III}] [20,21,39]. The shi^{3-} ligand combines a five- and a six-membered chelate ring and the next ligand is required to bind to this building block at a 90° angle (Figure 3). With rare exceptions, the molecules so far prepared can be realized in a one-step procedure with yields of crystalline product between 50 and 95%. The schematic diagrams of (9-MC-3) **7**, (12-MC-4) **11** and (15-MC-5) **12** demonstrate (Figures 2 and 3), that stable metallacrowns are not limited to a (9-MC-3) or (12-MC-4) motif. A logical concept for the predesigned synthesis of a planar (15-MC-5) structure is to alter the angle to 108° using picoline hydroxamic acid, which forms two five-membered chelate rings as shown for **12**.

2.1.1 Trinuclear metallacoronands and metallacoronates

Crown ethers selectively complex alkaline ions [6,41], and the complexation of different sized cations leads to coronates [5,42] of various structures. The structural analogy between crown ethers and their topologically equivalent metallacrown ethers

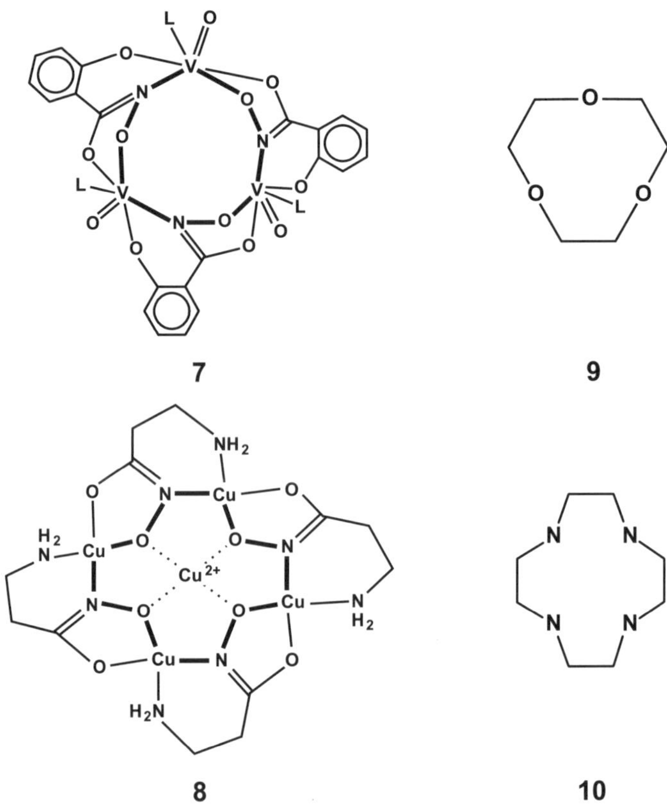

Figure 2 Schematic diagrams of (9-C-3) **9** and (12-C-4) **10** compared with (9-MC-3) **7** and (12-MC-4) **8**.

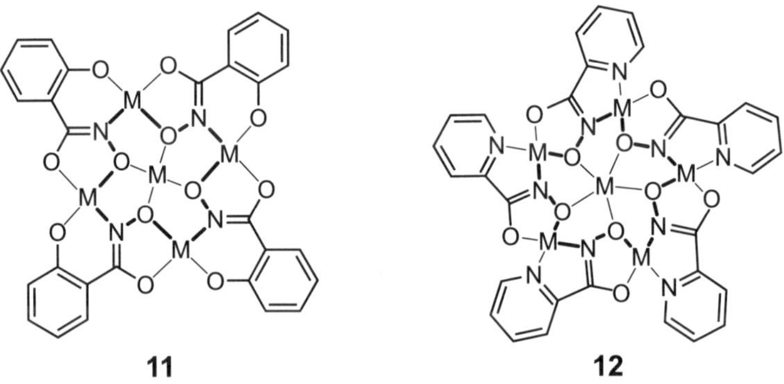

Figure 3 Schematic diagrams of (12-MC-4) **11** and (15-MC-5) **12**.

(MC) [2,43–45] should lead to novel supramolecular architectures of well-defined host–guest compounds. Since the ionic radii of alkaline and alkaline earth metal cations differ significantly in size, whereas the diameter of a given metallacrown essentially does not change, the inclusion of small cations such as Na^+ or Ca^{2+} should yield a metallacoronate [19,45–47] with 1 : 1 stoichiometry (M^+: MC = 1 : 1). On the contrary, encapsulation of the larger K^+ or $NH_4{}^+$ ions should lead to the formation of metallacrown ether sandwich complexes (M^+: MC = 1 : 2). Reaction of diethyl ketipinate H_2L **13** with copper(II) acetate in the presence of calcium nitrate affords green microcrystals **14** after crystallization from tetrahydrofuran/diethyl ether (Scheme 1).

According to the X-ray crystallographic analysis, metallacoronate **14** is a neutral trinuclear metal cluster of the general composition $[Ca \subset (Cu_3L_3)(NO_3)_2]\cdot THF\cdot H_2O$. The copper atoms are linked across the triangular corners by bis(bidentate) diethyl ketipinate dianions L^{2-}. Thereby each copper(II) ion is primarily coordinated in a square-planar fashion to four oxygen atoms. Additional coordination of water, tetrahydrofuran and nitrate, respectively, leads to a square-pyramidal environment of each copper(II) center by O donors. In the core of the resulting [15]metallacrown-6, a calcium ion is encapsulated, and for charge compensation, two nitrate ions are coordinated axially to calcium (Figure 4). Formal replacement of the three copper(II) centers in the calcium free [15]metallacrown-6 system of **14** by ethane bridges leads to the topologically equivalent [18]crown-6.

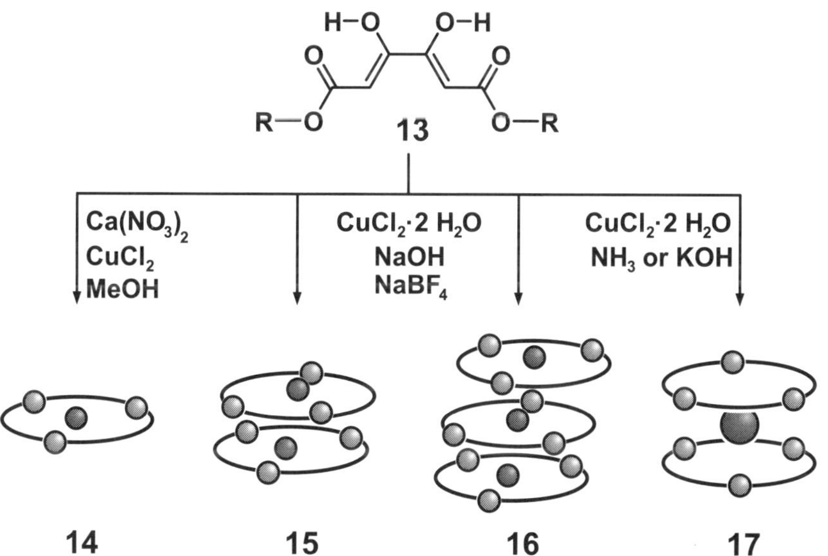

Scheme 1 Formation of metallacoronate **14**, double-decker **15**, triple-decker **16**, and sandwich **17**.

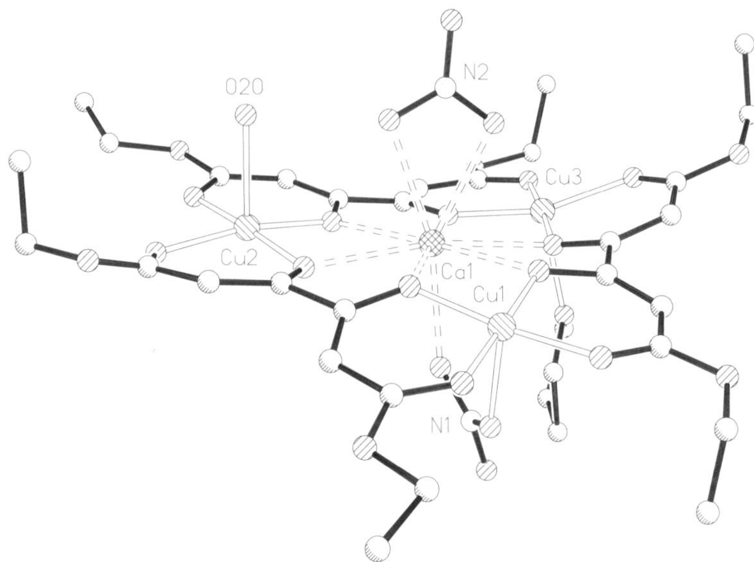

Figure 4 X-ray crystallographic structure of trinuclear metallacoronate **14**.

Double deprotonation of diethyl ketipinate H_2L **13** by sodium hydroxide and reaction of the formed dianion with a methanolic solution of $NaBF_4$ and copper(II) chloride dihydrate yields a mixture of two different crystalline metallacoronates **15** and **16** (Scheme 1), whose molecular structures were determined by X-ray crystallographic analyses. The metallacoronate **15** is a dimer of two $[Na \subset (Cu_3L_3)BF_4]$ building blocks. The monomer is composed of a $[Cu_3L_3]$ metallacrown with a sodium ion located in the core which is η^6-coordinated to all oxygen atoms in the inner ring. The counterion of the $[Na \subset (Cu_3L_3)]^+$ cation is a BF_4^- ion. The eightfold donor solvation, the Na^+ ions are striving for, leads to dimerization. The linkage of the monomers to give double-decker $\{[Na \subset (Cu_3L_3)BF_4]\cdot THF\cdot H_2O\}_2$ **15** is accomplished via η^1-coordination of the two Na^+ ions to one ring oxygen of each neighboring coronate. This leads to a minor mutual deformation of the monomeric building blocks (Figure 5).

However, similar suitable coordination around copper(II) and sodium is also achieved by the formation of triple-decker metallacoronate $[Na \subset (Cu_3L_3)BF_4]_3\cdot 2THF$ **16** (Figure 6; 3D-1). In order to accomplish eightfold coordination for the sodium ions, aggregation of the $[Na \subset (Cu_3L_3)]^+$ monomers furnishes double- and triple-decker metallacoronates **15** and **16**. The features of stacking are governed by the coordination of water molecules. The coordination of water leads to the formation of dimeric complex **15**. In **15** both sites of the stack are totally coordinatively blocked by solvent molecules or tetrafluoroborate counterions. However, without coordination of water, the most suitable ligation around copper(II) and sodium is achieved by the formation of triple-decker metallacoronate **16**. A

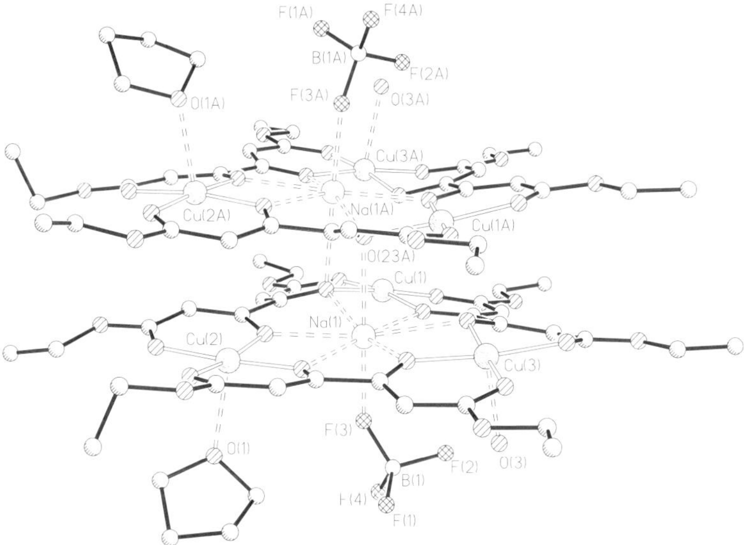

Figure 5 X-ray crystallographic structure of double-decker **15**.

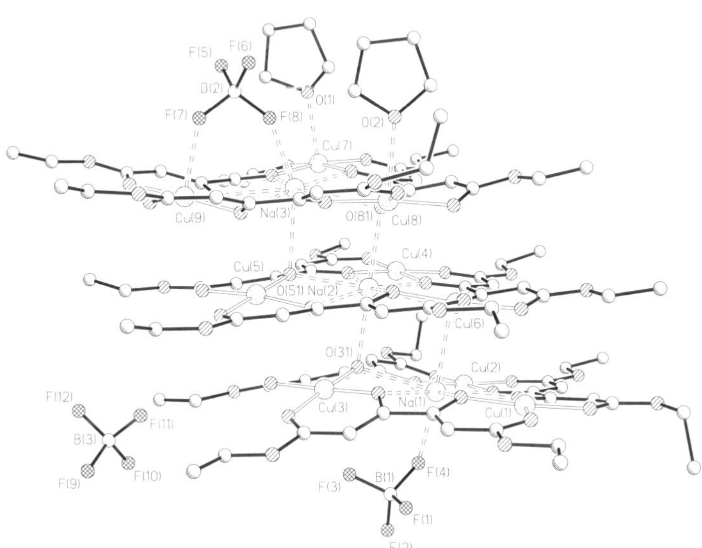

Figure 6 X-ray crystallographic structure of triple-decker **16**.

common feature of **15** and **16** is their neutral, 15-membered building block. Formal replacement of the three copper(II) centers in the [15]metallacrown-6 fragments of **15** or **16** by ethane bridges leads to the topologically equivalent crown ether [18]crown-6.

Whereas, encapsulation of the small cations Na^+ and Ca^{2+} leads to metallacoronates of 1 : 1 stoichiometry, double deprotonation of ketipinate H_2L **13** with potassium hydroxide and reaction with copper(II) chloride dihydrate affords metallacrown ether sandwich complex **17** with 2 : 1 stoichiometry (Scheme 1). The X-ray crystallographic analysis reveals that $[K \subset (Cu_3L_3)_2OMe]\cdot 6HOMe$ **17** is constructed of two neutral [15]metallacrown-6 building blocks which are turned relative to each other by 60°. The sandwich-like linkage results via a crystallographically disordered potassium ion in the center between the two metallacrowns. The counterion of the cation $\{[K \subset (Cu_3L_3)_2]\cdot 6HOMe\}^+$ of **17** is a methoxide ion, which is bound to the potassium ion. Coordinative saturation of the six copper centers results from methanol molecules. Thus, each Cu^{2+} ion is surrounded by O donors (Figure 7; 3D-2) in a tetragonal pyramidal geometry.

2.1.2 Hexa-, octa-, and decanuclear metallacoronands and metallacoronates

For a better understanding of electron transfer processes and of magnetic properties of polynuclear iron [48,49], manganese [50], and nickel complexes [51], further studies on compounds of this type are necessary. The same is true for metallacrown ethers with enclosed cations [38,43,44,51–53]. With respect to potential applications, iron(III) compounds certainly play a central role. Polyiron-oxo species are present in aqueous solution, but the growth of polyiron complexes cannot be controlled, and in the absence of additional ligands beyond oxo, hydroxo and aquo ligands, the final product of hydrolysis in water is ferrihydrite [53]. The situation is different in the presence of additional ligands. In fact, methanolysis of simple iron(III) salts in the presence of 1,3-diketonates has proved to be an excellent route to Fe_2, Fe_3, Fe_4, Fe_6, and Fe_{10} clusters [54].

Several different types of Fe_6 clusters have been reported, which were classified as planar, twisted boat, chair, parallel triangles, octahedral and fused clusters [55]. The $[Fe_6(\mu_2\text{-}OMe)_{12}(dbm)_6]$ ring, where Hdbm is dibenzoylmethane, is a neutral species, but in the solid state it crystallizes with NaCl to give **18** (Figure 8). In fact, the sodium ion is trapped in the center of the iron ring, which acts like a crown ether complexing the alkaline earth ion [53].

Recently, reaction of triethanolamine with sodium hydride and addition of iron(III) chloride in THF (Scheme 2) afforded yellow crystals of **19**, in which the iron-to-ligand ratio was 1 : 1 and the ratio of iron to sodium chloride was 6 : 1.

For the unequivocal characterization of the molecular structure of **19**, an X-ray crystallographic structure analysis was carried out [46]. According to this analysis, **19** is present in the crystal as a cyclic iron(III) complex with a [12]metallacrown-6

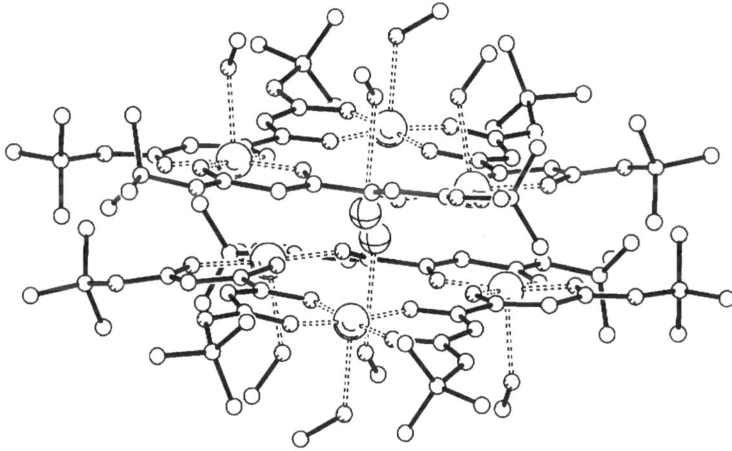

Figure 7 X-ray crystallographic structure of sandwich $[K \subset (Cu_3L_3)_2OMe]\cdot 6HOMe$ **17**, with K^+ and MeO^- disordered.

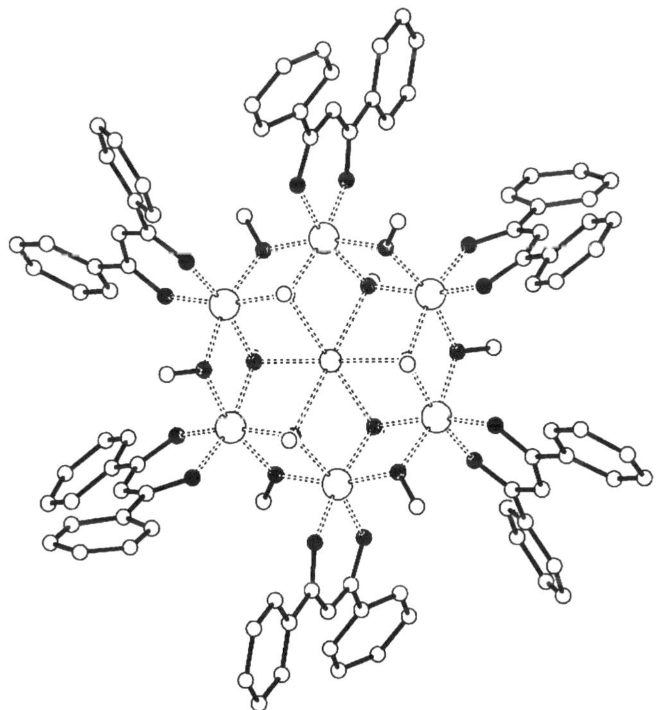

Figure 8 X-ray crystallographic structure of cation $[Na \subset Fe_6(\mu_2\text{-}OMe)_{12}(dbm)_6]^+$ of **18**.

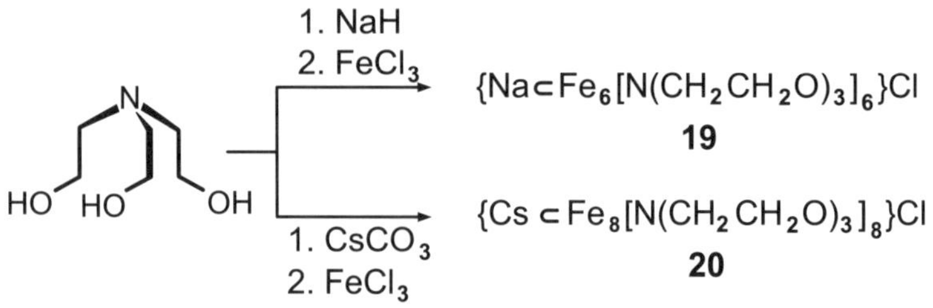

1. NaH
2. $FeCl_3$

$\{Na \subset Fe_6[N(CH_2CH_2O)_3]_6\}Cl$
19

HO HO OH

1. $CsCO_3$
2. $FeCl_3$

$\{Cs \subset Fe_8[N(CH_2CH_2O)_3]_8\}Cl$
20

Scheme 2 Formation of six- and eight-membered iron coronates **19** and **20**.

structure, in which a sodium ion is encapsulated in the center, and a chloride is the counterion (Figure 9). The six crystallographically equivalent iron atoms of the centrosymmetric cation $\{Na \subset Fe_6[N(CH_2CH_2O)_3]_6\}^+$ of **19** are located in the corners of a regular hexagon. The diameter of the hexagon, defined as the distance of two opposite iron atoms, is 6.431 Å. The distorted octahedral coordination sphere of the iron atoms is composed of one nitrogen donor, one μ_1- and two μ_2- and two μ_3-oxygen donors. Consequently, triethanolamine acts as a tetradentate ligand and links three iron(III) ions.

Three sets of six oxygen atoms are located in the corners of three pairs of regular triangles, which are rotated 60° relative to each other (Figure 10, left). The trigonal faces are located parallel and equidistant in pairs with one plane above and one

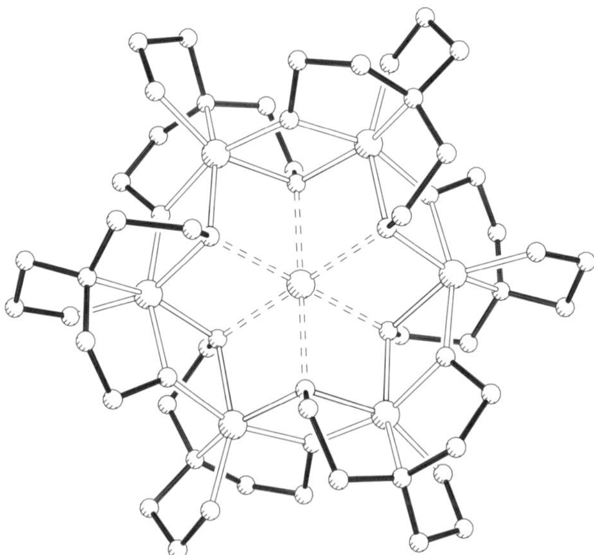

Figure 9 X-ray crystallographic structure of cation $\{Na \subset Fe_6[N(CH_2CH_2O)_3]_6\}^+$ of **19**.

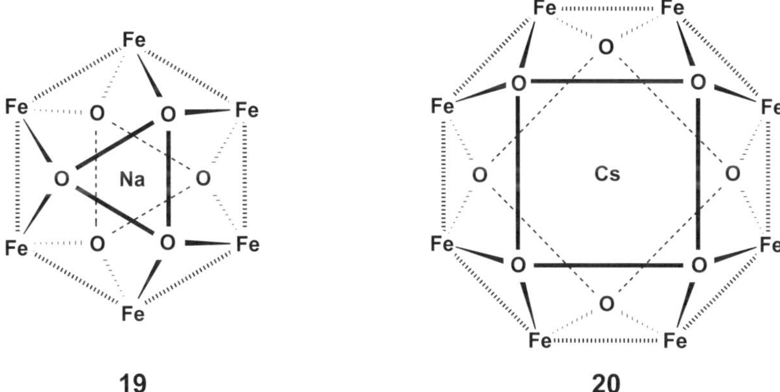

Figure 10 Left: schematic representation of the central Fe_6O_6 backbone with one set of six oxygens of **19**. Right: schematic representation of the central Fe_8O_8 backbone with one set of eight oxygens of **20**.

below the hexagonal plane generated by the iron atoms. These six oxygen atoms are related by an S_6 axis. The sodium ion is located in the center of the cation of **19** and has a distorted octahedral coordination sphere from the six μ_3-O atoms. The diameter (1.98 Å) of the cavity marked by opposite μ_3-O atoms nearly corresponds to double the ionic radius of sodium (2.04 Å).

According to Lehn *et al.* [34], in the case of the template mediated self-construction of a supramolecular system, amidst a set of possibilities, a combination of building blocks is realized, which leads to the best receptor for the substrate. Therefore, it is possible that the six-membered cyclic structure $\{Na \subset Fe_6[N(CH_2CH_2O)_3]_6\}Cl$ **19** is exclusively selected from all the imaginable iron triethanolamine oligomers, when sodium ions are present. Thus, in the presence of cations with ionic radii that differ from that of sodium, variant structures are expected.

When triethanolamine was allowed to react with cesium carbonate and iron(III) chloride (Scheme 2), an X-ray structure analysis of the reaction product **20** revealed a cyclic iron(III) complex with a [16]metallacrown-8 structure, in which the cesium was located in the center of the ring and a chloride ion functioned as counterion (Figure 11; 3D-3) [46]. The eight iron atoms of the almost centrosymmetric cation $\{Cs \subset Fe_8[N(CH_2CH_2O)_3]_8\}^+$ of **20** are located in the corners of a nearly regular octagon. The diameter of the octagon, defined as the mean distance between opposite iron atoms, is 8.224 Å. The distorted octahedral coordination sphere of the iron atoms is composed of one nitrogen donor, one μ_1-, two μ_2- and two μ_3-oxygen donors.

Three sets of eight out of a total of 24 oxygen atoms are located on the corners of three pairs of square faces that are rotated 45° relative to each other (Figure 10,

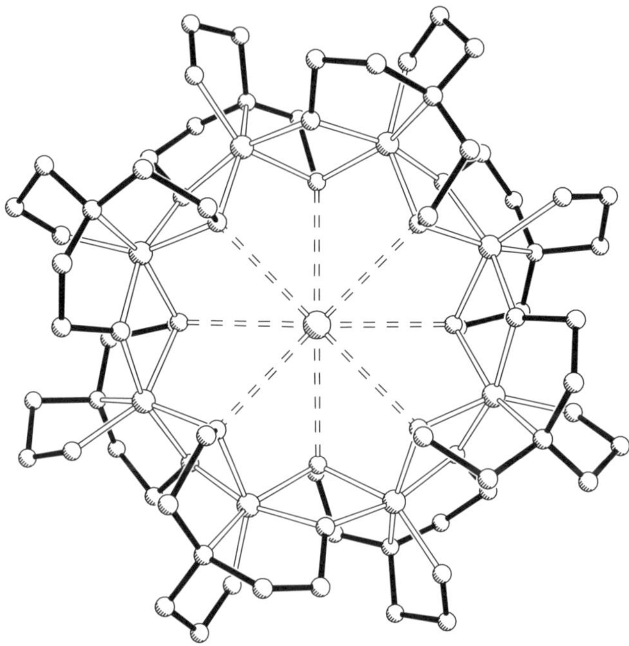

Figure 11 X-ray crystallographic structure of cation $\{Cs \subset Fe_8[N(CH_2CH_2O)_3]_8\}^+$ of **20**.

right). The square faces are located parallel and almost equidistant in pairs and one plane lies above and one below the octagon generated by the iron atoms. These structural elements share a common S_8 axis. The cesium ion lies directly above the center of the cation of **20** and is shifted about 0.5 Å towards the chloride counterion. The eight μ_3-O atoms form a quadratic antiprism coordination sphere around the cesium center. Furthermore, a common feature of complexes **19** and **20** is that the μ_1-O atoms do not participate in the formation of the hexa- and octanuclear structures. They solely function as ligands and for the coordinative saturation of the iron atoms. However, other donors such as chloride ions could also be candidates for this function. Therefore, *N*-methyl-diethanolamine was allowed to react with calcium hydride and iron(III) chloride. According to the X-ray crystallographic analysis, product **21** is present in the crystal as an unoccupied neutral iron(III) complex with a [12]metallacrown-6-structure (Figure 12).

The six iron atoms of the approximately centrosymmetric neutral molecule $\{Fe_6[H_3C-N(CH_2CH_2O)_2]_6Cl_6\}$ **21** are located in the corners of an almost regular hexagon. The diameter of the hexagon, defined as the mean distance between two opposite iron atoms, is 6.361 Å. The distorted octahedral coordination sphere of the iron atoms is composed of one nitrogen donor, one chloride ion, and four μ_2-oxygen donors.

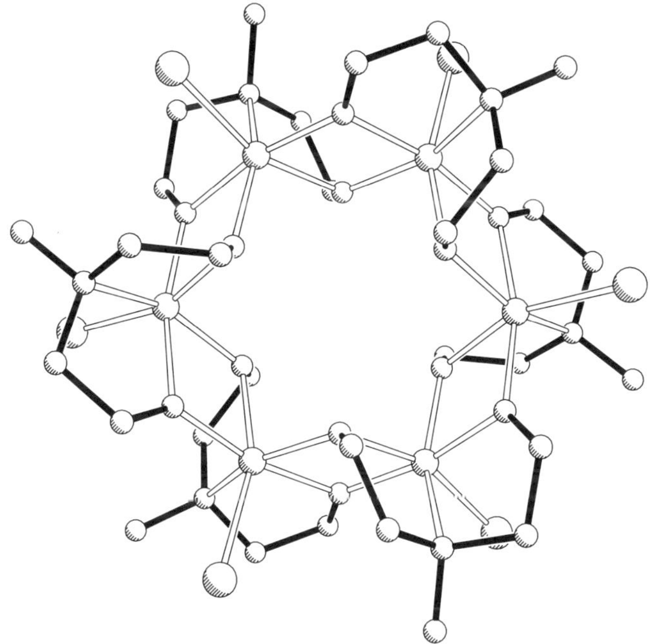

Figure 12 X-ray crystallographic structure of $\{Fe_6[H_3C{-}N(CH_2CH_2O)_2]_6Cl_6\}$ **21**.

Most of the oligonuclear cyclic chelate complexes were discovered by serendipity. This is also true for $\{Zn \subset [Zn(hmp)_2]_6\}Cl_2$ **22** which was generated from $Zn[N(SiMe_3)_2]_2$ and 2-hydroxymethylpyridine (Hhmp) in the presence of $CHCl_3$ or CH_2Cl_2 [56]. The pictogram of **22** (Figure 13) indicates the high symmetry of the complex dication composed of a central Zn^{2+} ion surrounded in a disk-like fashion by six ZnO_2N_2 fragments.

Interestingly, the reaction between $MnCl_2{\cdot}4H_2O$, 2-hydroxymethylpyridine (Hhmp) and NEt_4MnO_4 in acetonitrile gives the mixed-valence manganese cluster topology $\{Mn \subset [Mn_6(OH)_3Cl_3(hmp)_9]\}Cl[MnCl_4]$ **23**, in which three Mn^{II} and three Mn^{III} cations comprise a Mn_6 hexagon [57]. The central Mn^{II} ion is held by three μ_3-OH^- and three μ_3-$O_{hmp}{}^-$ ions, the latter bridging two peripheral and the central manganese atoms. The remaining μ_2-$O_{hmp}{}^-$ ions bridge the manganese atoms of the hexagon. The complexed dication of **23** has virtual C_3 symmetry. The Cl^- counterion is hydrogen bonded to the μ_3-OH^- groups, but the $NEt_4{}^+$ and $MnCl_4{}^{2-}$ ions are well separated from the nearly planar Mn_7 unit (Figure 14).

In some cases the conditions are even more complex. For instance, it has been shown, that two metallacoronates, containing 17 and 19 iron(III) ions, respectively, crystallize together in the same unit cell and have very similar structures and formulas [53,58]. That of trication $\{Fe \subset [Fe_{16}(\mu_3\text{-}O)_4(\mu_3\text{-}OH)_6$

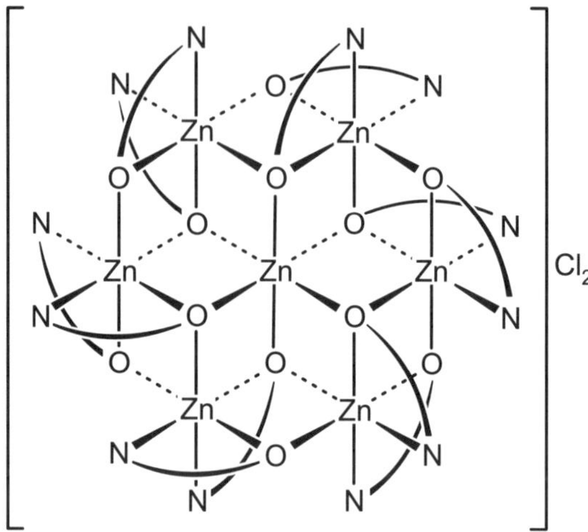

Figure 13 Pictorial representation of $\{Zn \subset [Zn(hmp)_2]_6\}Cl_2$ **22**.

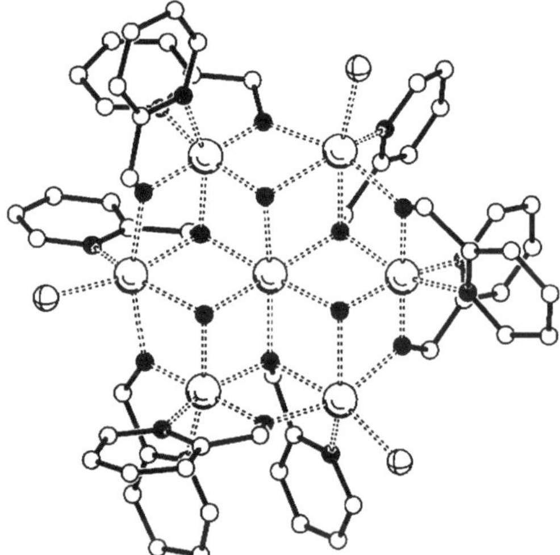

Figure 14 X-ray crystallographic structure of cation $\{Mn \subset [Mn_6(OH)_3Cl_3(hmp)_9]\}^{2+}$ of **23**.

$(\mu_2\text{-}OH)_{10}(heidi)_8(H_2O)_{12}]\}^{3+}$ **24** $[H_3heidi = HO(CH_2)_2N(CH_2CO_2H)_2]$ is shown in Figure 15. It is interesting to note that the central core of **24**, which is composed of seven iron ions is very similar to that observed for the cation $[Na \subset Fe_6(\mu_2\text{-}OMe)_{12}(dbm)_6]^+$ of **18** [53].

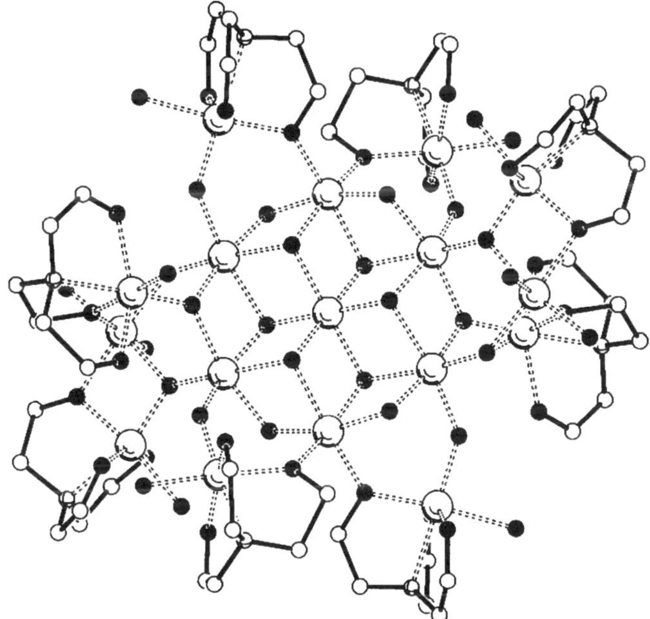

Figure 15 X-ray crystallographic structure of trication $\{Fe \subset [Fe_{16}(\mu_3\text{-}OH)_4(\mu_3\text{-}OH)_6\text{-}(\mu_2\text{-}OH)_{10}(heidi)_8(H_2O)_{12}]\}^{3+}$ of **24**.

The ring size does not necessarily determine the diameter of the hole of a metallacoronate, as was shown by the cyclic structure $[Fe(OMe)_2(O_2CCH_2Cl)]_{10}$ **25** (Figure 16), better known as *ferric wheel* [11,48,53]. The *wheel* is prepared in methanol solution from the reaction between $[Fe_3O(O_2CCH_2Cl)_6(H_2O)_3](NO_3)$ and $Fe(NO_3)_3{\cdot}9H_2O$. Four μ_2-OMe donors and two monochloroacetate ions link three Fe(III) ions each out of the 10 Fe(III) ions of **25** and is unoccupied in the center.

Similarily, toroidal inclusion complex $\{(NH_4) \subset [Co(OMe)_2(O_2CMe)]_8[PF_6]\}$ is isolated from the reaction between cobalt(III) acetate and methanol in the presence of NH_4PF_6 [59].

Recently a hexadecanuclear polyolatometalate of copper(II) and multideprotonated D-sorbitol was synthesized and characterized by X-ray diffraction [60]. However, the hitherto largest cyclic Fe^{III} cluster contains 18 iron(III) ions in the ring. The octadecairon(III) complex $[Fe(OH)(XDK)Fe_2OMe)_4(O_2CMe)_2]_6$ **26** [where $H_2XDK = m$-xylilene diamine bis(Kemp's triacid imide)] [61] was prepared in the presence of tetraalkylammonium carboxylate salts from slightly alkaline methanolic solutions of the diiron(III) complex $[Fe_2O(XDK)(HOMe)_5\text{-}(H_2O)](NO_3)_2{\cdot}4H_2O$. The composition of the crystalline product, a double salt having the formula **26**$\cdot 6Et_4N(NO_3)\cdot 15HOMe\cdot 6Et_2O\cdot 24H_2O$, was determined by microanalysis and a single-crystal X-ray diffraction analysis (Figure 17). The

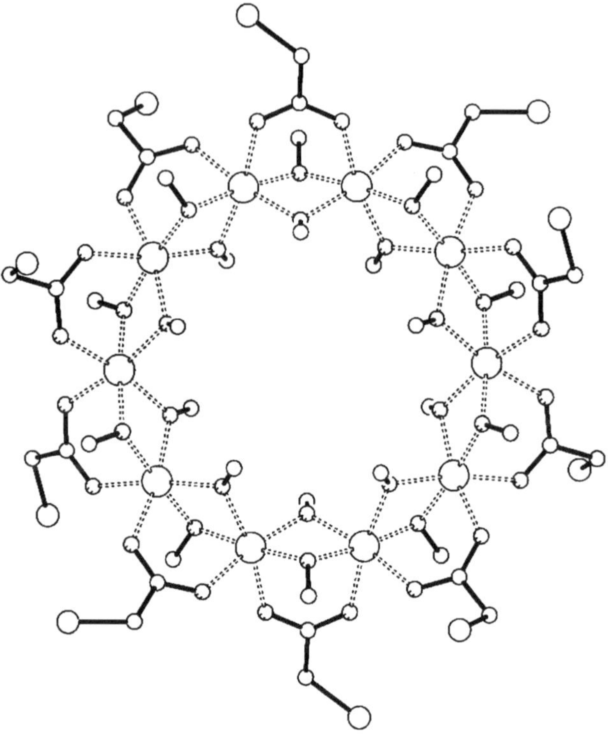

Figure 16 X-ray crystallographic structure of $[Fe(OMe)_2(O_2CCH_2Cl)]_{10}$ **25**.

molecular 18-wheeler has idealized D_{3d} symmetry. The repeating unit comprises a (μ-hydroxo)bis(μ-carboxylato)-diiron(III) moiety linked by an acetate and two methoxide ions to a third iron atom. Each iron atom in **26** has distorted octahedral symmetry. The hydrophobic environment afforded by the XDK^{2-} ligands on the inner surface of the wheel is incompatible with its occupancy by cations, as seen in the case of smaller metallacoronates lacking this feature [2,19,43–47,51–53].

The study of molecules possessing unusual large spin values in their ground state is an area of current interest [50,57,61,62]. It has recently become apparent that a relatively high ground-state spin value is one of the necessary requirements for molecules exhibiting single-molecule magnetism. The synthesis of new high-spin molecules is thus of interest.

Oxalate (ox) is widely used to prepare molecular-based magnetic materials. The structures of $(NBu_4)[Mn^{II}Cr^{III}(ox)_3]$ and $(PPh_4)[Mn^{II}Cr^{III}(ox)_3]$ have been determined and shown to be lamellar and isostructural. In the layer planes, each Mn^{II} ion is bound to three Cr^{III} ions (and vice versa) via oxalato bridges generating a graphite-like pattern. The repeating unit $[Mn^{II}Cr^{III}(ox)_3]^-$ **27** is reminiscent of a heteronuclear metallacrown ether and is represented schematically in Figure 18 [63].

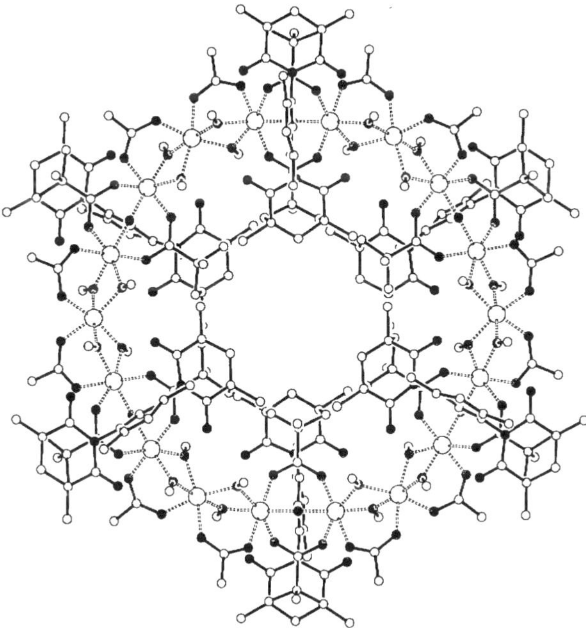

Figure 17 X-ray crystallographic structure of $[Fe(OH)(XDK)Fe_2(OMe)_4(O_2CMe)_2]_6$ **26**.

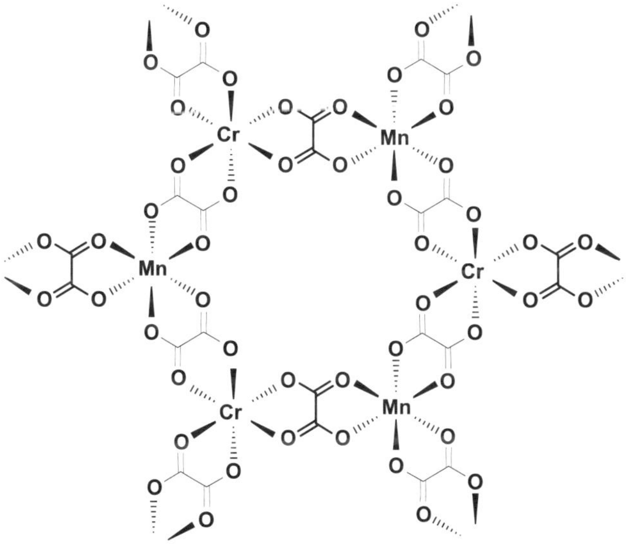

Figure 18 Schematic representation of the repeat unit **27** of $^{2}_{\infty}[Mn^{II}Cr^{III}(ox)_3]^-$ without the bulky cations being located between adjacent layers.

2.2 Circular Helicates

Inorganic double or triple helices are formed by two or three ligand strands wrapped around linearly disposed metal ions [13]. Among cyclic transition metal complexes, circular helicates $[n]^m$**cH** {$[n]^m$**cH** is a general notation characterizing circular helicates (**cH**) with $n =$ number of metal ions and $m =$ helicity ($m = 2$ for a double helix)} have specific features and may be considered as toroidal helices [34]. There are two different kinds of circular helical systems. Some structures self-assemble from the metal ions and the ligands only in the presence of an anion, which could act as a template [34,35,64–67], whereas, in other cases, the circular helicates self-assemble from the metal ions and the ligands alone [68–70].

2.2.1 Anion-centered circular helicates

The interest in polynuclear supramolecular iron complexes mainly stems from the importance of oxo-centered polyiron aggregates as model compounds for iron-oxo proteins. In principle, μ_3-oxo-centered complexes related to $[Fe_3O(O_2CR)_6(H_2O)_3]$ **28** [71] (Figure 19, left) should also be accessible with other double negatively charged ligands. Double deprotonation of H_2L **29** gives rise to two dianionic rotamers $(L^A)^{2-}$ and $(L^B)^{2-}$, which with Fe^{III} or Zn^{II} ions selectively furnish two different complexes $[Fe_3OL^A{}_3]$ **30** [35] and $[Zn_8O_2L^B{}_6]$ **31** (Scheme 3) [64].

According to the X-ray crystallographic analysis, **30** is present in the crystal as a neutral, trinuclear, iron chelate complex (Figure 19, right; 3D-4). The core of

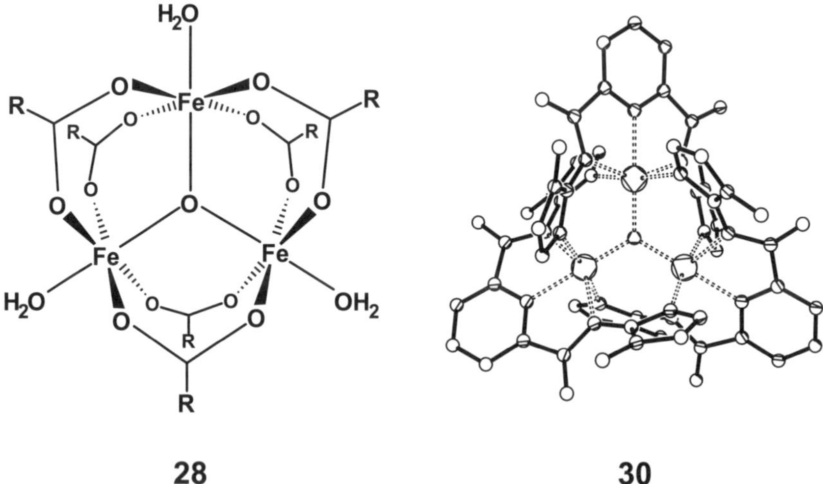

Figure 19 Left: schematic representation of $[Fe_3O(O_2CR)_6(H_2O)_3]$ **28**. Right: X-ray crystallographic structure of $[Fe_3OL^A{}_3]$ **30**.

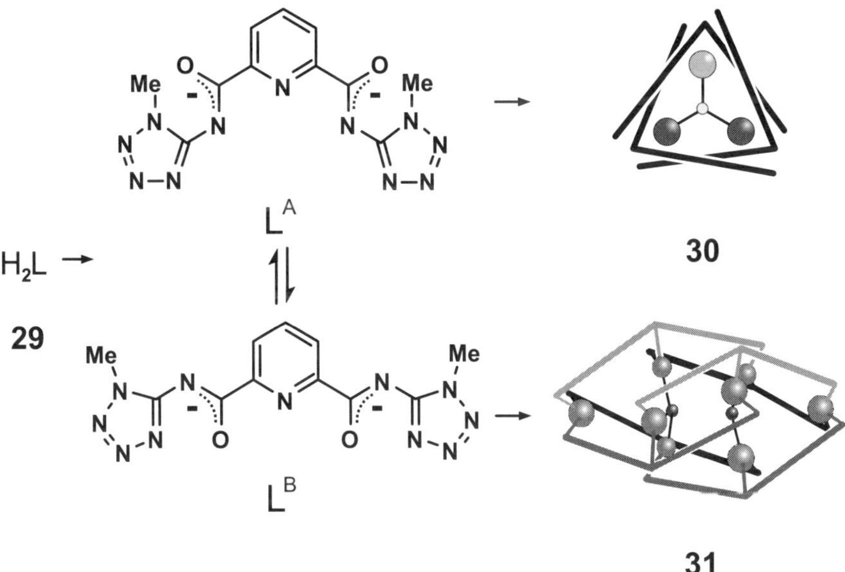

Scheme 3 Formation and schematic representation of μ_3-oxo-centered mixed-valent iron complex $[Fe_3OL^A{}_3]$ **30** and octanuclear bicapped bis[trinuclear-μ_3-oxo-centered) zinc complex $[Zn_8O_2L^B{}_6]$ **31**.

complex **30** is generated by an equilateral triangle with a μ_3-O^{2-} ion in the center and three iron ions in the vertices. Each of the doubly negatively charged pentadentate, tritopic ligands $(L^A)^{2-}$ links three metal centers. As a result, all iron ions are distorted octahedrons coordinated to five N donors and to a μ_3-O^{2-} ion shared by the three iron atoms. The iron centers of the racemic complex are homochiral since the tetrazolyl-N donors of the pentadentate tritopic ligands bind to neighboring iron ions at opposite sides of the (Fe,Fe,Fe)-triangular plane. When idealized symmetry is assumed, the triple-helical molecule belongs to the point group D_3. Alternatively, **30** may be considered as a toroidal helix. The lack of a counterion implies intramolecular charge compensation and therefore mixed-valent character for $[Fe_3OL^A{}_3]$ **30**. This is unequivocally confirmed by a Mössbauer spectrum.

According to an X-ray crystallographic analysis, $[Zn_8O_2L^B{}_6]$ **31** is present in the crystal as a neutral, octanuclear bis(triple-helical) chelate complex. The core of **31** consists of eight zinc(II) ions, forming a twofold capped, slightly twisted trigonal prism with a μ_3-O^{2-} ion in the center of each of the two inner faces (Figure 20; 3D-5). All of the six double negatively charged pentadentate ligands $(L^B)^{2-}$ link three zinc ions. The two antipodal Zn^{II} ions are coordinated by three μ_1- and three μ_2-oxygen chelate donors of three ligands. However, the six metal centers constituting the trigonal prism are coordinated by one pyridilene-nitrogen and two μ_2-oxygen

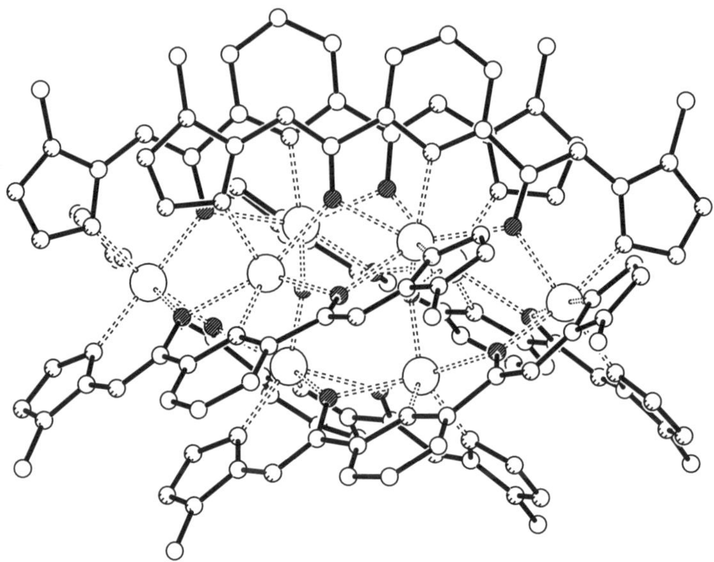

Figure 20 X-ray crystallographic structure of $[Zn_8O_2L^B{}_6]$ **31**.

atoms. Distorted octahedral coordination of these six Zn^{II} ions is achieved by two extra μ_3-O^{2-} ions. Alternatively, and in analogy to **30**, **31** may be considered as an octanuclear bicapped bis[trinuclear-μ_3-oxo-centered) zinc complex.

The self-assembly of the tris-2,2′-bipyridine ligands L^1 **32** and L^2 **35** (Scheme 4) with iron(II) salts yields polynuclear complexes displaying structures of a cyclic double-helix type. With L^1 in which the bipyridine units in the ligand are connected by ethane bridges, penta- or hexanuclear circular helicates $[5]^2$**cH 33** or $[6]^2$**cH 34** are obtained depending on the anion present during the self-assembly process. The tris-bipyridine ligand L^2 **35** with an oxypropylene spacer forms a tetranuclear circular helicate $[4]^2$**cH 36**. The effect of such a rather minor change (mere replacement of the CH_2CH_2 group of L^1 by a CH_2OCH_2 group in L^2) of the ligands on the product formation may be instructive for analyzing the structure/assembly relationship [34].

In the presence of the smaller chloride ion the self-assembly generates the pentanuclear circular helicate $[5]^2$**cH 33** (Figure 21). With the larger anions $SO_4{}^{2-}$, $BF_4{}^-$, and $SiF_6{}^{2-}$, the hexanuclear architecture is formed while the Br^- anion of intermediate size yields a mixture of $[5]^2$**cH** and $[6]^2$**cH**. The charge of the anion has apparently little influence on the structure formed as $[6]^2$**cH** is obtained with mono- and divalent anions. The structure depends rather on the size of the anion to be included in the circular helicates. These differences in the products formed, might be considered as resulting from a templating effect of the anion

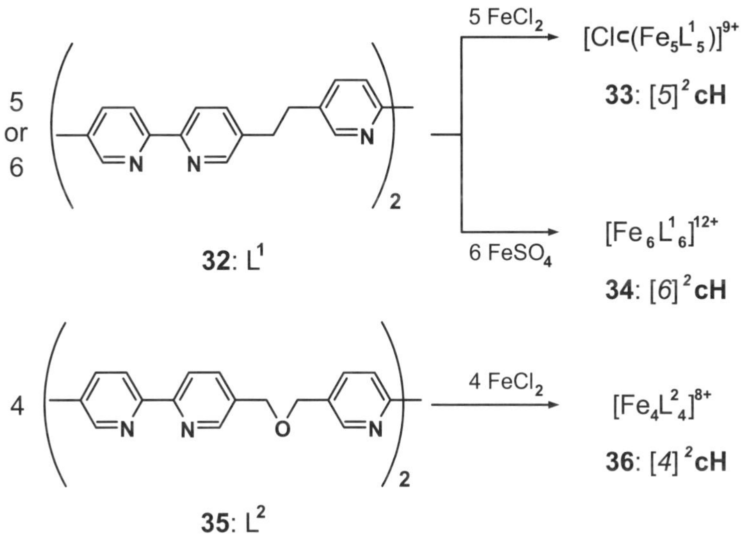

5 or 6 (ligand unit)$_2$ — 5 $FeCl_2$ → $[Cl\subset(Fe_5L^1_5)]^{9+}$ **33**: $[5]^2$**cH**

6 $FeSO_4$ → $[Fe_6L^1_6]^{12+}$ **34**: $[6]^2$**cH**

32: L^1

4 (ligand unit)$_2$ — 4 $FeCl_2$ → $[Fe_4L^2_4]^{8+}$ **36**: $[4]^2$**cH**

35: L^2

Scheme 4 Self-assembly of the circular helicates $[5]^2$**cH 33**, $[6]^2$**cH 34**, and $[4]^2$**cH 36**.

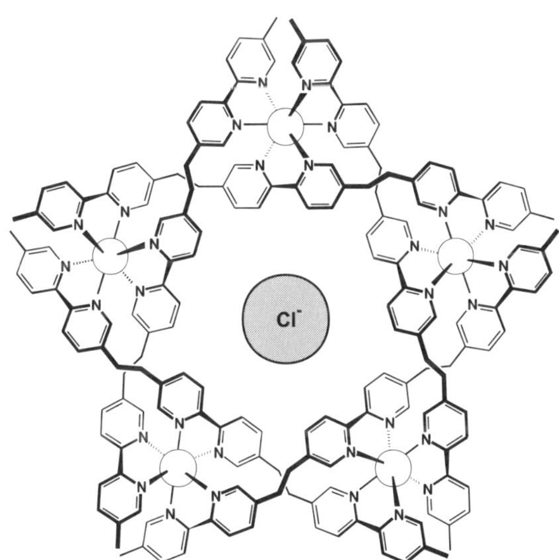

Figure 21 Cartoon representation of circular double-helicate $[5]^2$**cH 33** with included Cl^-.

during the formation of the circular helicate. Yet, the process can also be considered as the self-assembly of a receptor as a function of its substrate. Thus, the pentanuclear torus $[5]^2$**cH** **33** binds a Cl^- anion, and the larger hexanuclear torus $[6]^2$**cH** **34** accommodates, for example, a $SO_4{}^{2-}$ ion in its central cavity. The self-assembly process toward one or the other structure from the same components is determined by the substrate to be bound. It can thus be seen to present a procedure of selection from a virtual combinatorial library (VCL) consisting of the available components. From the point of view of the VCL concept, the chloride anion may act beyond templating and select $[5]^2$**cH** out of all the potential combinations. The formation of $[6]^2$**cH** in the case of $SO_4{}^{2-}$, $BF_4{}^-$, and $SiF_6{}^{2-}$ does not suggest that this structure is templated by the anions although this may be the case [34].

The principles of how the steric information contained in the number and orientation of ligand binding sites and the stereoelectronic preferences of the metal ion combine to give the final structure are becoming better understood [1–16]. In some systems there is only one optimal structure for the supramolecular species possible, in which all of the ligand binding sites are occupied, and all of the metal ions are coodinatively saturated. In contrast with other systems, a wide variety of stoichiometries and therefore many different structures are possible all of which fulfill the requirements necessary for the formation of a stable supramolecular system. For example, the ligand L^1 **32** (Scheme 4) has three bidentate chelating sites. Coordination to metal ions preferring sixfold coordination gives complexes with a 1 : 1 metal-to-ligand stoichiometry but with very different structures. Most notably are the triple helicate $[Ni_3L^1{}_3]^{6+}$ [72] and the pentanuclear circular helicate $[Cl \subset (Fe_5L^1{}_5)]^{9+}$ **33** [34].

Recently, reaction of the bis-bidentate ligand bis[3-(2-pyridyl)pyrazol-1-yl]dihydroborate $(L^3)^-$ with cobalt(II) salts resulted in the formation of the complex $[(ClO_4) \subset (Co_8L^3{}_{12})](ClO_4)_3$ **37** [65]. Despite the poor quality of the X-ray crystallographic data, the gross structure of the complex cation of **37** is quite clear (Figure 22) consisting of a $[Co_8L^3{}_{12}]$ ring with a perchlorate ion in the central cavity. Each ligand $(L^3)^-$ acts as a bridge between two adjacent cobalt ions, with an alternating pattern of one and two bridging ligands between each vicinal pair of metals. Each metal is six-coordinate with three bidentate chelating fragments, each from a different ligand, and all ligand binding sites are used. The complex cation is chiral with all eight metal centers having the same chirality (Δ in Figure 22).

A fascinating large inorganic architecture was generated by self-assembly from Cu^I ions and ligand **38** (Figure 23) [67]. The X-ray crystal structure data indicate that the product consists of a large complex cation $[Cu_{12}(\mathbf{38})_4]^{12+}$ **39** of toroidal shape. The central cavity is occupied by four $PF_6{}^-$ anions and solvent molecules. One can distinguish two types of the 12 Cu^I ions according to their environment. Four of them are complexed by two 2,2′-bipyridine (bipy) units, while the other eight are bound to a bipy and to a pyridine-pyrazine (pypz) unit. All Cu^I ions are in a distorted tetrahedral environment. The ligand strands are wrapped around each other forming four linked double helical sections with 12 crossing points.

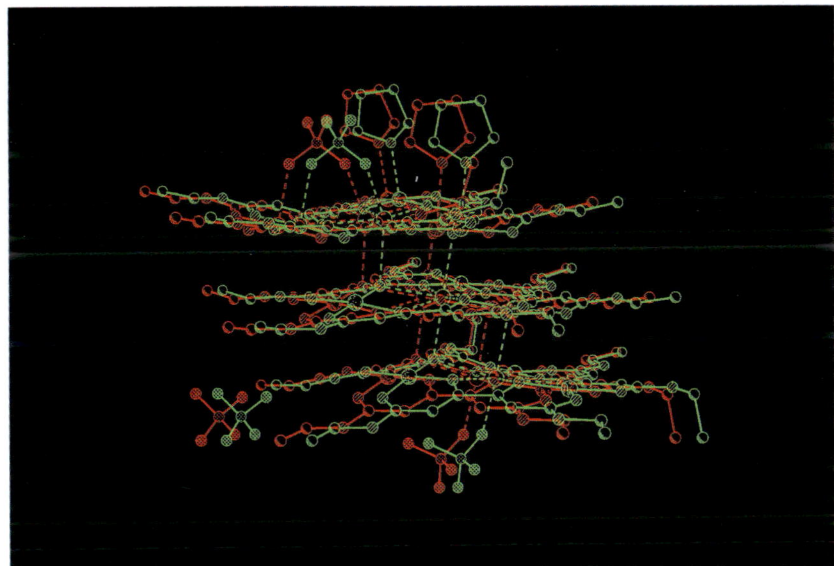

Figure 3D-1. Stereo representation of X-ray crystallographic structure of triple-decker **16.**

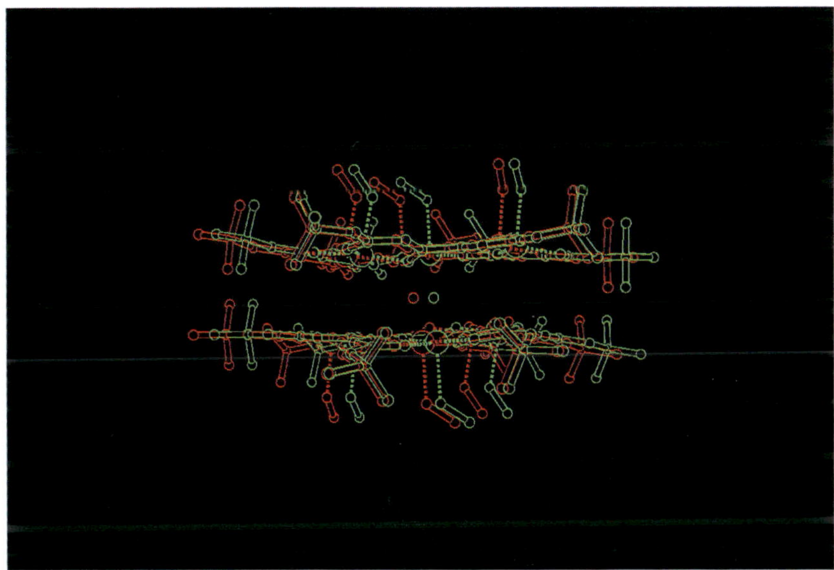

Figure 3D-2. Stereo representation of X-ray crystallographic structure of sandwich $[K\subset(Cu_3L_3)_2OMe]\bullet 6HOMe$ **17**, with K^+ and MeO^-disorderd.

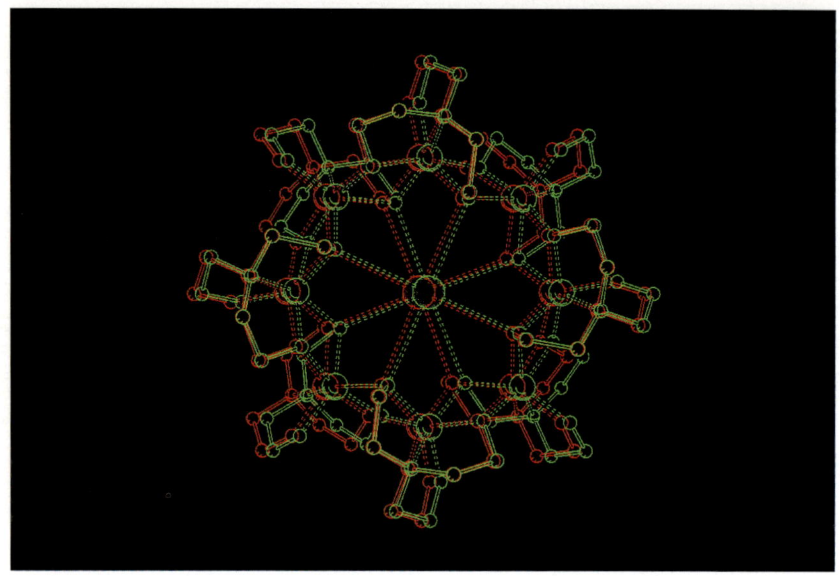

Figure 3D-3. Stereo representation of X-ray crystallographic structure of cation $\{Cs \subset Fe_8[N(CH_2CH_2O)_3]_8\}^+$ of **20.**

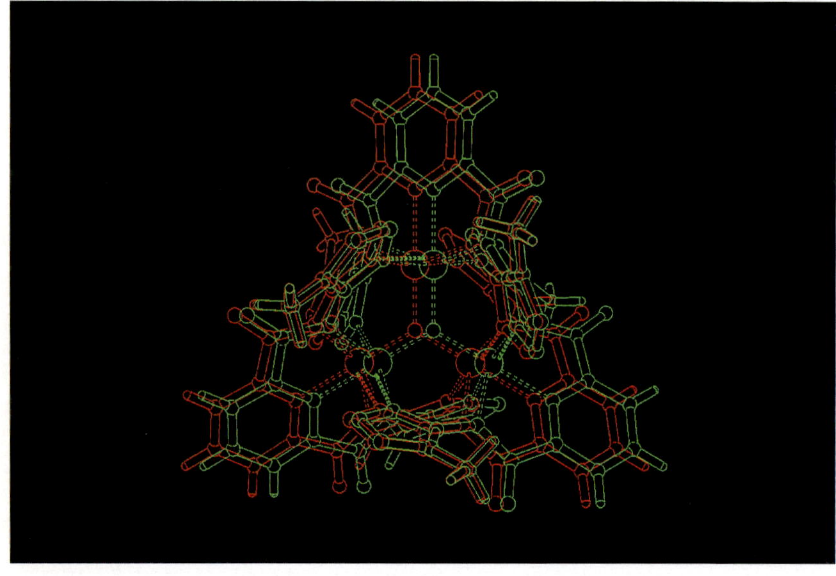

Figure 3D-4. Stereo representation of X-ray crystallographic structure of $[Fe_3OL^A{}_3]$ **30**, top view.

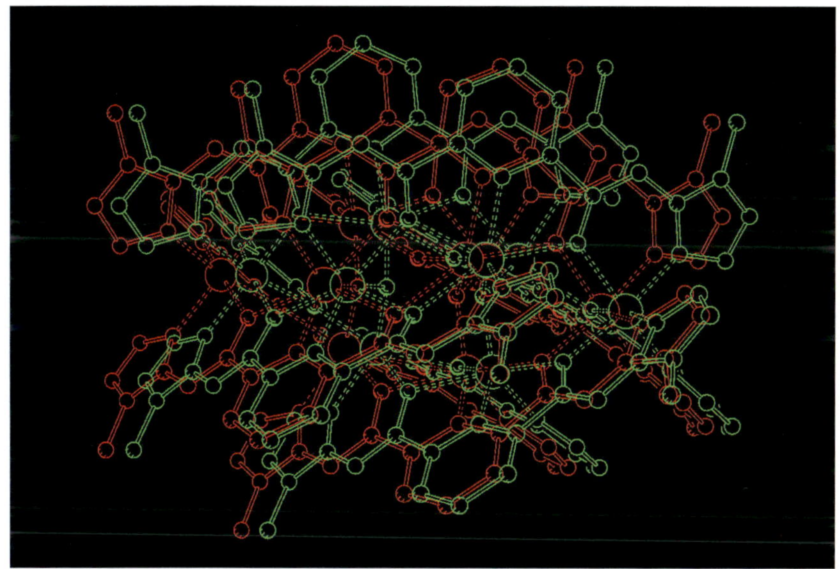

Figure 3D-5. Stereo representation of X-ray crystallographic structure of $[Zn_8O_2L^B{}_6]$ **31.**

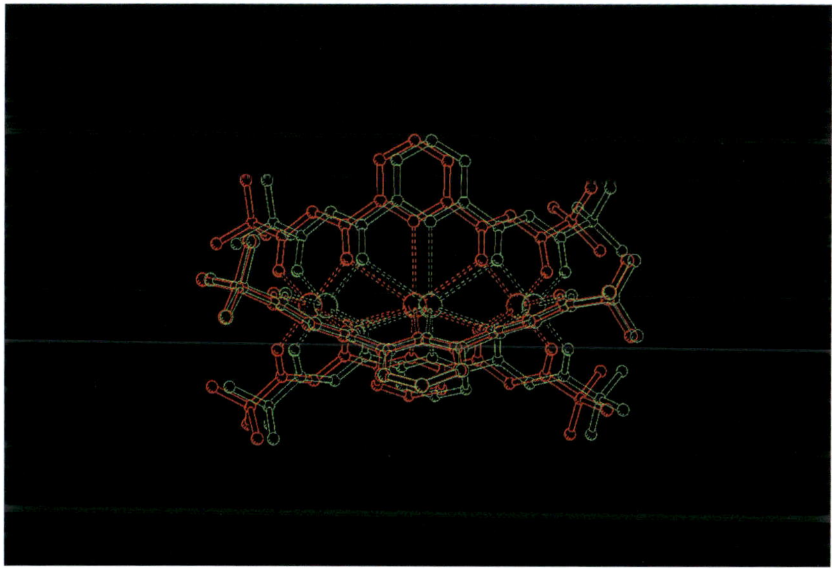

Figure 3D-6. Stereo representation of X-ray crystallographic structure of the monocation $[K\subset(Fe_2L^5{}_3)]^+$ of {2}-metallacryptate **65**, perpendicular to the Fe–Fe axis.

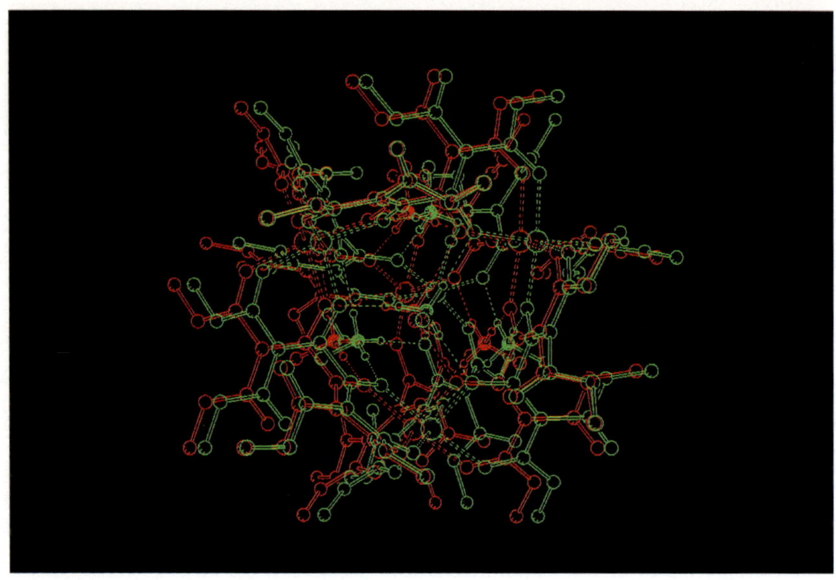

Figure 3D-7. Stereo representation of X-ray crystallographic structure of tetra-hemispheraplex $[(4NH_3\text{-}CH_3)\cap(Mg_4L^1{}_6)]$ **Mg-69'.**

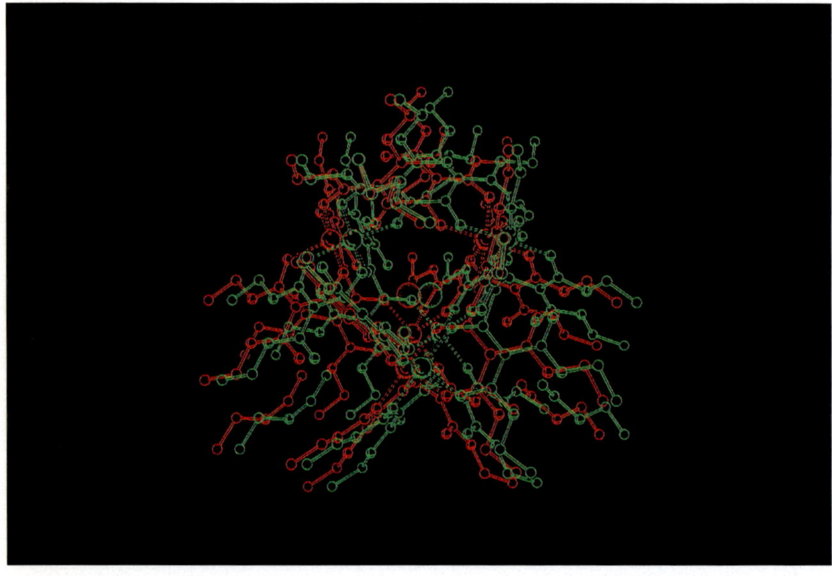

Figure 3D-8. Stereo representation of X-ray crystallographic structure of $[Cs\subset Fe^{II}Fe^{III}{}_3L_6]$ **Cs-81.**

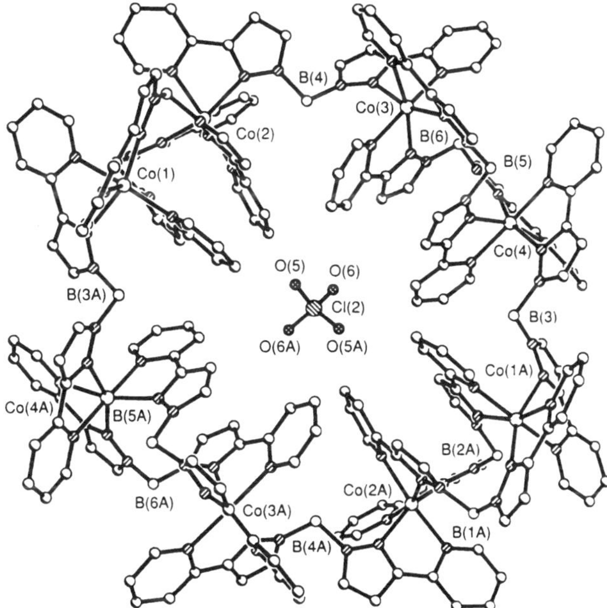

Figure 22 X-ray crystallographic structure of the trication $[(ClO_4) \subset (Co_8L^3{}_{12})]^{3+}$ of **37**.

2.2.2 Center-free circular helicates

In order to design species presenting specific structural and functional features, it is of great importance to establish the rules by which control of the self-assembly process can be achieved through chemical programming by means of suitable components and assembling algorithms. Copper(I) ions together with a certain quaterpyridine ligand self-assemble to generate in solution a mixture of a double helicate, a triangular circular helicate **40** (Figure 24) and a square grid complex. The double helicate could be isolated in crystalline form and its structure was elucidated by X-ray crystallographic analysis. On the basis of this structure as well as the spectroscopic data, the other two species present in solution were considered to be a square [2 × 2] grid and toroidal helicate **40** [68].

An X-ray structure determination of the colorless tetrahedral crystals obtained from the reaction of (*R*,*R*)-2,6-bis(4′-phenyloxazolin-2′-yl)pyridine **41** with $AgBF_4$ showed the formation of $[Ag_3(\mathbf{41})_3](BF_4)_3$ **42** (Figure 25). The complex consists of an equilateral triangle of silver ions with the ligands bridging the sides of the triangle. Each ligand coordinates through the oxazoline-N atoms with the pyridine-N donors lying much further away. Each ligand binds to one metal from below the plane of the silver atoms and to a second metal from above the plane. The structure may thus be considered as a triple helix in which the ligands are wrapped around the

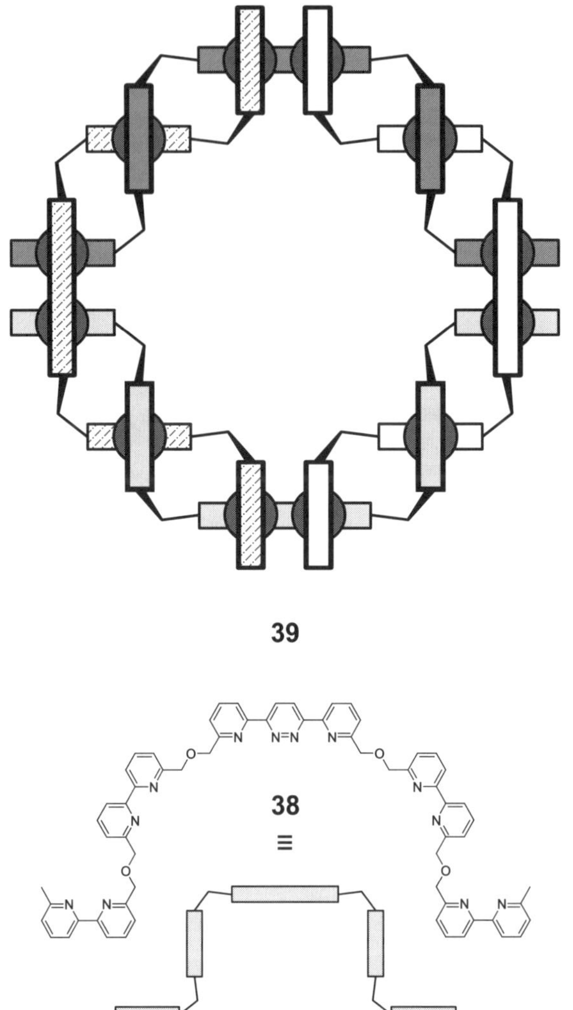

Figure 23 Ligand **38** and cartoon representation of cation $[Cu_{12}(\mathbf{38})_4]^{12+}$ **39**.

threefold axis. The toroidal helix **42** is generated enantioselectively and has *P* helicity induced by the stereogenic centers of enantiomerically pure **41** [69].

A further example of a stereospecific self-assembled circular helical structure was prepared from the chiral ligand **43** and silver ions. According to the X-ray crystal structure determination, the spontaneously formed sixfold circular single-stranded helicate $[Ag_6(\mathbf{43})_6](PF_6)_6$ **44** (Figure 26) is formed as a single diastereoisomer. The

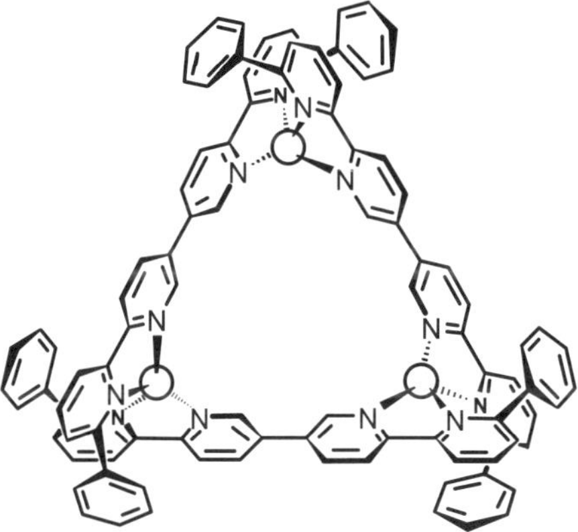

Figure 24 Schematic representation of circular helicate **40**.

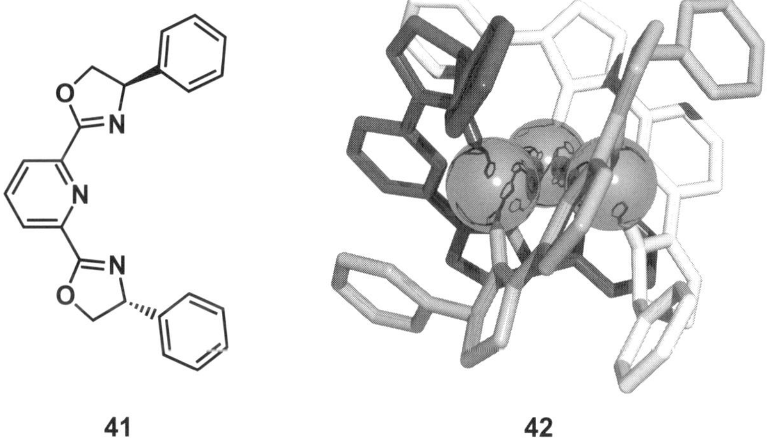

Figure 25 Schematic representation of ligand **41** and X-ray crystallographic structure of the trication $[Ag_3(\mathbf{41})_3]^{3+}$ of **42**.

bis-bidentate ligand forms a bridge between two adjacent silver ions. All stereogenic centers of the (−)-α-pinene groups of **43** are oriented inward, whereas, the terminal pyridines point outward. The silver ions are tetrahedrally coordinated by four nitrogen donor atoms of two different ligands. The handedness of the helix is *P* [70]. In contrast to most other circular helical systems, **44** assembles from the metal ions and the ligands alone without the support of an anionic template.

Medium-pressure liquid chromatography of the product isolated from the reaction of a slurry of $[FeCl_2(THF)_{1.5}]$ with $(TTDSi)Li_2$ $[(TTDSi)^{2-}$ = tetrahydro-4,4,8,8-tetramethyl-4,8-disila-*s*-indacenediyl] yields $[(TTDSi)Fe]_7$ **45** [73]. The ^{1}H and ^{13}C

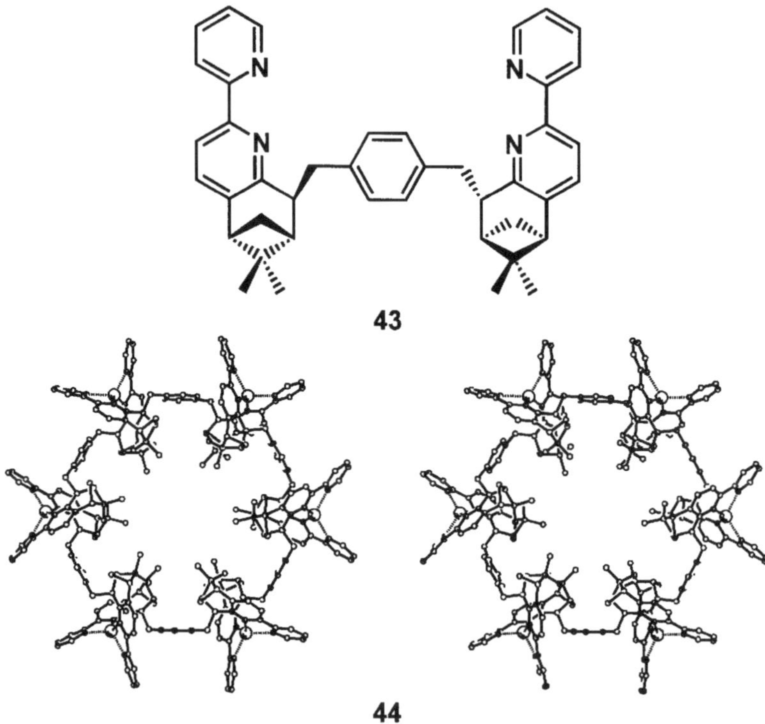

Figure 26 Schematic representation of ligand **43** and X-ray crystallographic structure of the hexacation $[Ag_6(\mathbf{43})_6]^{6+}$ of **44**, stereo representation.

NMR spectra showed only signals for bridging ligands. In particular, no signals of a chain-terminating ligand were recorded. Furthermore, the occurrence of two different methyl signals, in addition to two sets of signals indicating nonequivalent cyclopentadiene (Cp) rings, is consistent with a cyclic scaffold. The X-ray crystal structure of $[(TTDSi)Fe]_7$ **45** is represented in Figure 27. It consists of seven ferrocene units linked by seven pairs of $SiMe_2$ groups. Viewed ideally, the molecule belongs to the rare point group C_{7h}. Metallocenophane **45** may be regarded as a wheel with a hydrocarbon tire which provides good solubility.

2.3 Metallacryptands and {2}-Metallacryptates

To date, helicates represent the best-developed and most investigated supramolecular architectures [12,13,74]. Exploiting the experiences gained over the years, self-assembly phenomena through metal coordination have shown remarkable potential in the construction of molecular frameworks. Of these architectures, the following chapters deal with metallacryptands (helicates) and {2}-metallacryptates.

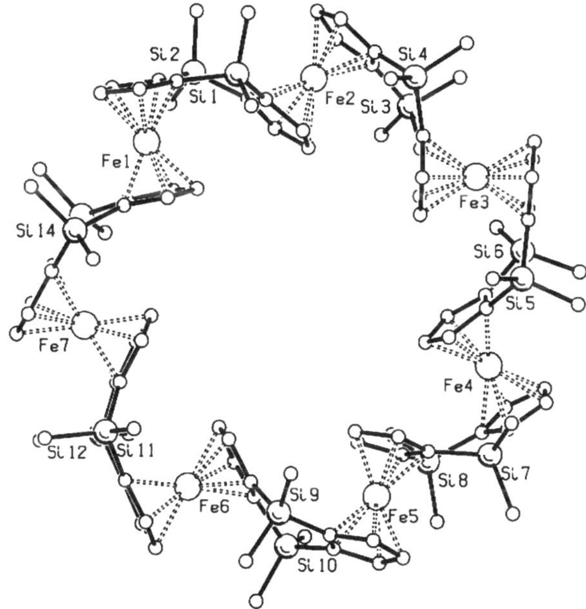

Figure 27 X-ray crystallographic structure of $[(TTDSi)Fe]_7$ **45**.

2.3.1 Helicates

Perhaps the earliest triple helicate **48** to be characterized, however, is that formed with rhodotorulic acid **46** (H_2L^1), the dihydroxamate siderophore produced by the yeast *Rhodotorula pilimanae* [75,76]. Subsequently, a related synthetic iron(III) triple helicate **49** based on diprotic tetradentate 1,2-hydroxypyridinone **47** (H_2L^2) was synthesized (Scheme 5) and characterized by an X-ray structure analysis.

The predictable nature of coordination chemistry for the generation of metallatopomers of {2}-cryptands has been successfully used for the preparation of numerous examples. However, supramolecular assemblies $[M_{2n}L_{3n}]$ are generated from octahedral metals and bis-bidentate ligands also for $n \neq 1$. The variety of stoichiometries observed with similar ligands or even identical ligands indicates that the factors discriminating such stoichiometries are quite subtle.

A key feature of stable self-assembled architectures is positive cooperativity, in which each step in the self-assembling process sets up and facilitates the subsequent step. In a rational design of triple-helical clusters, a series of bis(catecholamide) ligands (H_4L^n) **50–52** have been synthesized (Scheme 6). When three equivalents of any of these ligands are allowed to react with two equivalents of $M(acac)_3$ ($M = Fe^{3+}$, Al^{3+}, Ga^{3+}) and KOH in methanol, the $[K_6M_2L^n{}_3]$ triple helicates

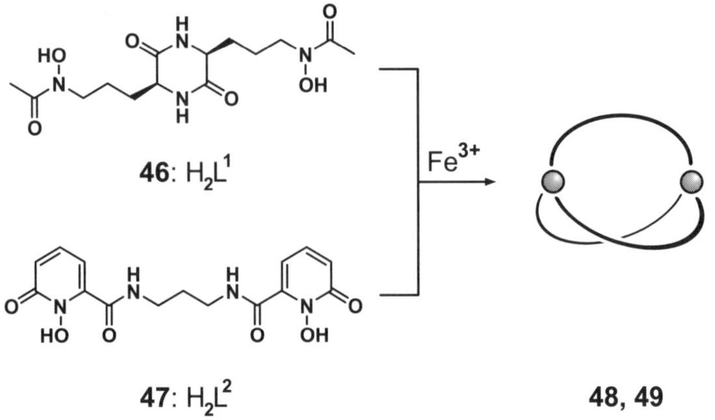

Scheme 5 Formation and schematic representation of homochiral triple helicate (Λ,Λ)-[$Fe_2L^1{}_3$] **48** and (Λ,Λ)-[$Fe_2L^2{}_3$] **49**.

can be isolated [76,77]. Remarkably, when mixtures of any two or all of the ligands are allowed to react at room temperature with $Ga(acac)_3$, only the individual complex hexaanions $[Ga_2L^n{}_3]^{6-}$ **53–55** are formed. No oligomeric mixed-ligand species are observed in solution. The structure of $[N(CH_3)_4]_6[Ga_2L^3{}_3]$ was established by single crystal X-ray crystallography.

The tris-catechol metal centers of these helicate complexes are chiral with either (Δ) or (Λ) absolute configuration. In the homochiral helicate structure, the chirality

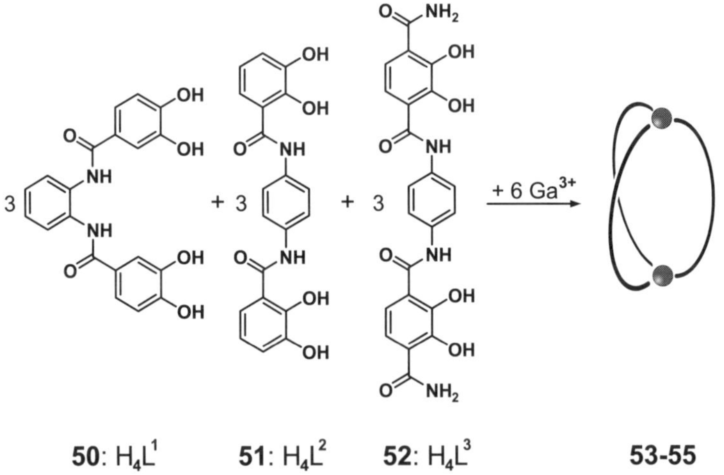

Scheme 6 Formation and schematic representation of homochiral triple-helical hexaanions $[Ga_2L^n{}_3]^{6-}$ **53–55** ($n = 1$–3).

of the first metal center induces the same chirality at the second metal center so that the only complexes present are (Δ,Δ) and (Λ,Λ) and constitute a racemic mixture. By kinetic investigations, it was shown that the proton-independent inversion of the (Δ,Δ)- and (Λ,Λ)-hexaanions $[Ga_2L^n{}_3]^{6-}$ **53–55** involve the heterochiral (Δ,Λ)-$[Ga_2L^n{}_3]^{6-}$ *meso*-anions as intermediates, which are produced by a single twist event along the reaction pathway. Thus the inversion proceeds by a stepwise mechanism without dissociation of the hexaanions. Most remarkably, it was also shown that in a proton-dependent reaction both metal centers are protonated and simultaneously invert to interchange the (Δ,Δ)- to (Λ,Λ)-helicate configuration.

2.3.2 {2}-Metallacryptates

Dinuclear triple-stranded metallacryptands are formed in spontaneous and cooperative self-assembly processes from three ligand strands, each bearing two chelating units, and two metal ions [12,13,74–81]. Recently alkyl-bridged bis(catecholate) ligands **56–58** were introduced, and it was shown that the absolute configuration around the two metal centers in the metallacryptand (triple helicate) **59** and metallacryptates **60** and **61** depends on the nature of the alkyl spacer (Scheme 7). Ligands with an even number of methylene units form a racemic mixture with either (Δ,Δ)- or (Λ,Λ)-configuration at the metal centers. Ligands with an odd number of methylene groups in the spacer, however, lead to achiral *meso*-metallacryptands with heterochiral (Δ,Λ)-configuration at the metal centers.

Reaction of bis(2,3-dihydroxyphenyl)methane H_4L^1 **56** with $[TiO(acac)_2]$ and Li_2CO_3 affords a well-defined product (Scheme 7). The diastereotopic behavior of

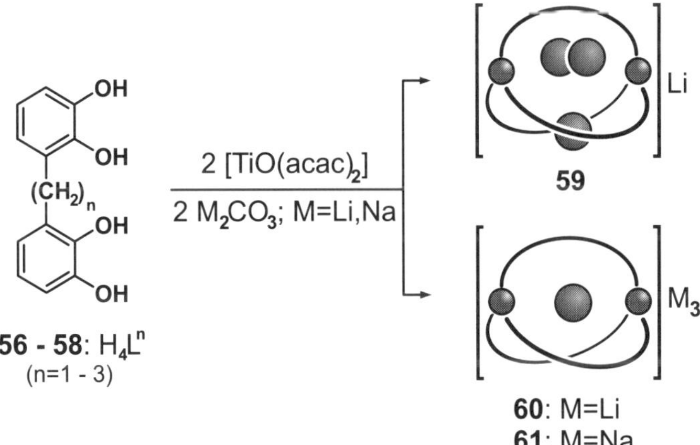

Scheme 7 Formation and schematic representation of (Δ,Λ)-$[Li][Li_3 \cap (Ti_2L^1{}_3)]$ **59**; (Δ,Δ)/(Λ,Λ)-$[Li]_3[Li \subset (Ti_2L^2{}_3)]$ **60**; and (Δ,Λ)-$[Na]_3[Na \subset (Ti_2L^3{}_3)]$ **61**.

the methylene protons in the ^1H NMR spectrum of **59** clearly reveals that the binuclear complex adopts the achiral *meso*-form and (Δ,Λ)-configuration at the metal centers. In the solid state, (Δ,Λ)-$[Li][Li_3 \cap (Ti_2L^1{}_3)]$ **59** shows the same absolute configuration as observed in solution (Figure 28). In the tetraanionic core $[Ti_2L^1{}_3]^{4-}$, two Ti^{4+} cations are linked by three bis(bidentate) catecholate ligands. Thus, each titanium ion is octahedrally coordinated by six oxygen atoms. Three of the Li^+ ions are *exohedrally* capping the squares of the trigonal prism formed by oxygen atoms and, in addition, are bound to DMF.

Reaction of three equivalents of 1,2-bis(2,3-dihydroxyphenyl)ethane **57** with two equivalents of $[TiO(acac)_2]$ and Li_2CO_3 affords only one diastereoisomer **60** (Scheme 7). The ^1H and ^{13}C NMR data of $[Li_3][Li \subset (Ti_2L^2{}_3)]$ **60** clearly reveal that the binuclear complex adopts the chiral $(\Delta,\Delta)/(\Lambda,\Lambda)$-configuration at the metal centers. In the solid state as well as in solution, the $[Ti_2L^2{}_3]^{4-}$ core encapsulates *endohedrally* one of the lithium counterions in its cavity. The additional three Li^+ ions connect the cores to give a polymeric overall structure in the solid state [79]. The corresponding (Δ,Λ)-$[Na_3][Na \subset (Ti_2L^3{}_3)]$ **61** (Scheme 7) was obtained from 1,3-bis(2,3-dihydroxyphenyl)propane **58**, and like **59** again is a *meso*-compound with two different configured pseudo-octahedral titanium(IV) tris(catecholate) moieties [80].

The use of suitable tailor-made ligands made available additional metallacryptands and monomeric {2}-metallacryptates [18,82]. A one-pot reaction of iron(III) chloride and triethylamine with 2,2′-dicyano-2,2′-isophthaloyldi(isopropyldicarboxylate) **62** (H_2L^4) after aqueous work-up yields dark red microcrystals $[Fe_2L^4{}_3]$ **64** in 92% yield (Scheme 8).

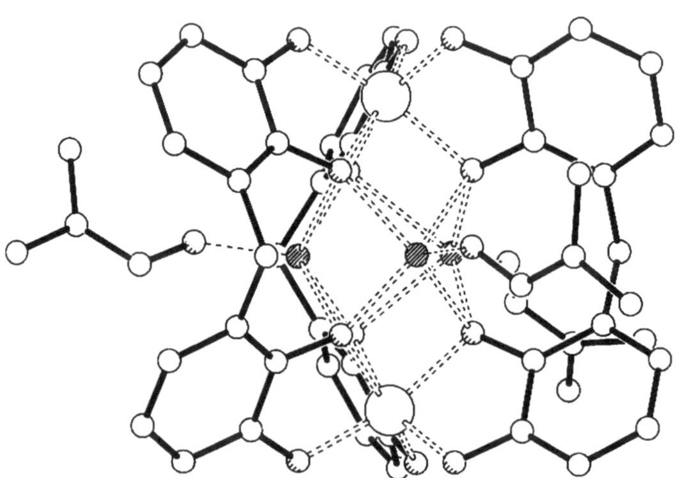

Figure 28 X-ray crystallographic structure of the anion $[Li_3 \cap (Ti_2L^1{}_3)]^-$ of **59**.

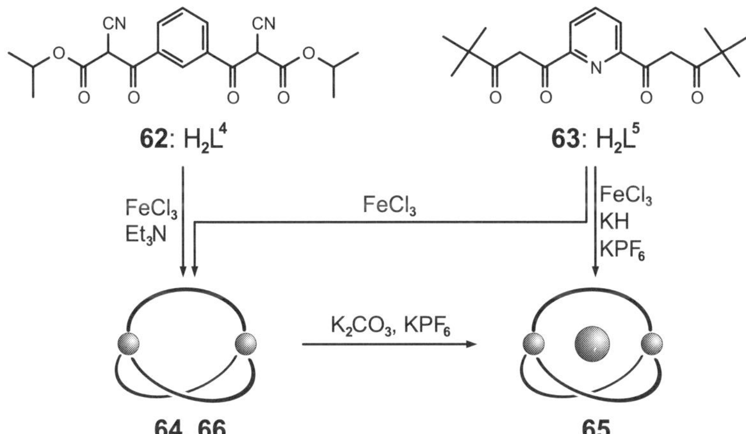

Scheme 8 Formation and schematic representation of {2}-metallacryptand $[Fe_2L^4{}_3]$ **64** and {2}-metallacrypate $[K \subset (Fe_2L^5{}_3)](PF_6)$ **65** and trispyridinium derivative $[Fe_2(HL^5)_3](FeCl_4)_3$ **66**.

An X-ray crystallographic analysis revealed that **64** is a {2}-metallacryptand (= triple helicate) (Figure 29). Each of the two iron centers is octahedrally surrounded by six oxygen atoms. In the chiral, racemic complex **64**, both iron centers are identically coordinated. Therefore, the {2}-metallacryptand is either a (Δ,Δ)-*fac* or (Λ,Λ)-*fac* triple helicate. The crystals obtained are composed of the homochiral {2}-metallacryptand **64**. The donating power within the interior of the complex is insufficient for the complexation of alkali cations. Above all, the three phenyl hydrogen atoms in **64** are directed inward towards the empty cavity in the center.

In order to synthesize the topologically equivalent metallacryptate **65** in an analogous manner to **64**, the *m*-phenylene spacer was substituted by an *m*-pyridylene spacer. Deprotonation of 1,1′-(2,6-pyridylene)bis-1,3-(4-dimethyl)pentanedione **63**

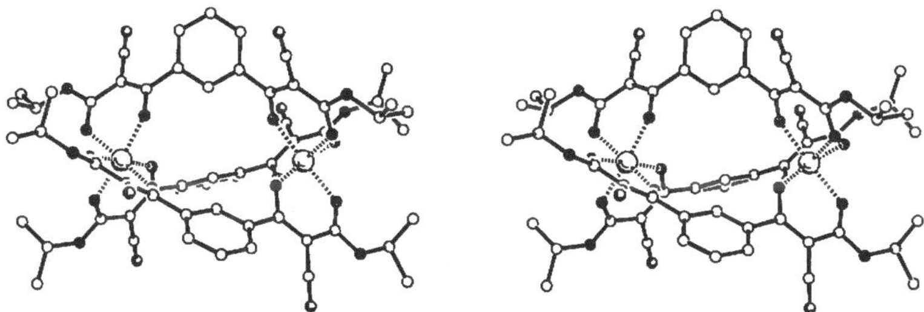

Figure 29 X-ray crystallographic structure of {2}-metallacryptand $[Fe_2L^4{}_3]$ **64**, stereo representation perpendicular to the Fe–Fe axis.

(H_2L^5) with potassium hydride followed by the addition of iron(III) chloride and subsequent work-up with aqueous potassium hexafluorophosphate solution produced the rust-colored microcrystals $[K \subset (Fe_2L^5{}_3)](PF_6)$ **65** (Scheme 8). According to an X-ray crystallographic analysis, **65** is a C_{3h}-symmetric {2}-metallacryptate (Figure 30; 3D-6). In the center of the cavity of **65** is a potassium ion with ninefold coordination to six ligand oxygens and to three pyridine nitrogen atoms. Three solvent molecules of dichloromethane are incorporated in the space between the individual molecules. A hexafluorophosphate ion acts as counterion. In contrast to *racemic* homochiral [(Δ,Δ)-/(Λ,Λ)-*fac*] **64**, the two iron centers in *meso* [(Δ,Λ)-*fac*] **65** have opposite chirality. It is noteworthy that treatment of H_2L^5 **63** with iron(III) chloride in the absence of a base yields the trispyridinium complex $[Fe_2(HL^5)_3](FeCl_4)_3$ **66**. On the other hand, potassium can be reencapsulated to reproduce **65** by simple reaction of **66** with a mixture of potassium carbonate and potassium hexafluorophosphate (Scheme 8) [83].

2.4 Self-assembled Adamantanoid Complexes

Biomimetic processes have gained significance over the past few years. One of the topics with relevance in this field is focused on polynuclear metal clusters [84]. The same holds true for tailor-made host compounds, which in solution selectively bind complementary guest molecules or incorporate them in concave cavities, similar to biological receptors or enzymes. In the process, host–guest complexes that depend on the formation of ion pairs or on hydrogen bonding require a smaller conformational adjustment to the receptor (induced fit) than systems that are based on van der Waals or charge-transfer interactions between π-electron donors and π-electron

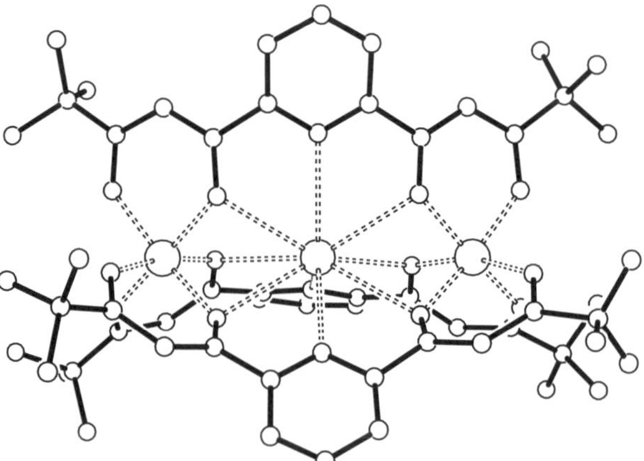

Figure 30 X-ray crystallographic structure of the monocation $[K \subset (Fe_2L^5{}_3)]^+$ of {2}-metallacryptate **65**.

acceptors. Metal-ligand interactions are highly directional and strong and can be used to direct cluster formation driven by the metal-ligand coordination bond.

One of the most interesting new aspects of synthetic chemistry is endohedral chemistry, that is the chemistry inside of cages [85]. Consequently, it is necessary to be able to build intramolecular cavities with variable diameters.

2.4.1 Tetrahemispheraplexes

Since a tetranuclear manganese cluster constitutes the active center in photosystem II, which carries out photooxidation of water in green plants and algae, tetranuclear manganese chelate complexes might serve as models for the elucidation of this fundamental biological process. The first tetranuclear metal chelate complex $[(4NH_4)\cap(Mg_4L^1{}_6)]$ **Mg-69** was prepared by serendipity [86], but this adventure had considerable impact on the development of the chemistry of metallaspherands and {3}-metallacryptates. In an improved, simple one-pot reaction, the tetrahemispheraplexes **M-69** of magnesium, manganese, cobalt and nickel were synthesized in high yields (80–90%) starting from dialkyl malonate **67**, $MeLi/MCl_2$ and oxalyl chloride **68** followed by work-up in aqueous ammonium chloride solution (Scheme 9) [86–89]. The doubly bidentate ligand **70** (L^1) is formally obtained by template coupling of two dialkyl malonate monoanions with oxalyl chloride. The cores of the tetranuclear complexes **M-69** are well suited to be tetrahemispherands. Therefore, the complexes $[(4NH_4)\cap(M_4L^1{}_6)]$ **M-69** are not salts in the usual sense, but tetrahemispheraplexes.

This is confirmed exemplarily by the X-ray structure of **Co-69** (Figure 31) [89]. The core of the complex is a tetrahedron composed of four cobalt(II) ions which are linked along each of the six edges of the tetrahedron by the doubly bidentate bridge

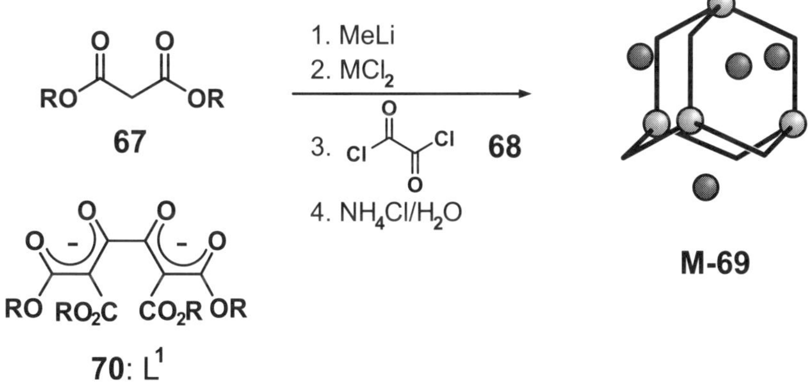

Scheme 9 Formation and schematic representation of tetrahemispheraplexes $[(4NH_4)\cap(M_4L^1{}_6)]$ **M-69** ($M = Mg^{2+}, Co^{2+}, Ni^{2+}, Mn^{2+}$).

70, so that each of the four cobalt(II) ions is octahedrally coordinated by six oxygen atoms. Thus, chiral, racemic **Co-69** has almost *T* symmetry, in which all four metal centers are identically coordinated [(Δ,Δ,Δ,Δ)-*fac* or (Λ,Λ,Λ,Λ)-*fac*] [90]. The four imaginary tetrahedral surfaces of the tetranuclear framework of **Co-69** are not planar, but each form a half shell. In the center of each shell are three keto oxygen atoms in ideal position to bind an ammonium ion through three hydrogen bonds. The exchange of ammonium ions in **M-69** by alkylammonium groups was achieved by addition of excess alkylamines to give the corresponding alkylammonium derivatives **M-69′** (Figure 3D-7). Double deprotonation of diethyl ketipinate (H_2L^2) with MeLi and addition of $MgCl_2$ followed by work-up with aqueous NH_4Cl furnishes the tetrahemispheraplex $[(4NH_4) \cap (Mg_4L^2{}_6)]$ [47].

2.4.2 Metallaspherands

In using metals with an oxidation state of three instead of two, the formation of neutral tetranuclear chelate complexes was expected. In introducing a spacer, it was possible to tailor the size of the cavity. The tetranuclear iron(III) chelate complex

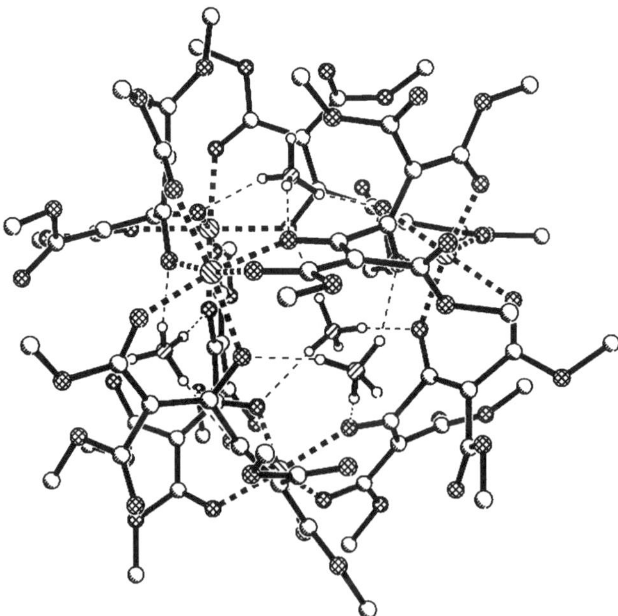

Figure 31 X-ray crystallographic structure of tetrahemispheraplex $[(4NH_4) \cap (Co_4L^1{}_6)]$ **Co-69**.

$[Fe_4L^1{}_6]$ **72** [90] was synthesized starting from tetramethyl 2,2′-terephthaloyldimalonate **71** (H_2L^1) (Scheme 10) and has exact S_4 symmetry in the crystal and is thus achiral (*meso*-form).

According to a single crystal X-ray diffraction, the ligands L^1 around the four iron centers in **72** are arranged facially and two of the four iron centers have the same configuration [(Δ,Δ)-/(Λ,Λ)-*fac*] (Figure 32). This is in contrast to the chiral, *T*-symmetric complexes **M-69**, in which all four metal centers are identically coordinated [(Δ,Δ,Δ,Δ)-*fac* or (Λ,Λ,Λ,Λ)-*fac*].

Molecular modeling was used to design the ligand isophthal-di-*N*-(4-methylphenyl)hydroxamic acid **73** (H_2L^2). The reaction of **73** with $[Ga(acac)_3]$ led to the white microcrystalline precipitate of $[Ga_4L^2{}_6]$ **74** (Scheme 10); whereas the dark red complex $[Fe_4L^2{}_6]$ was formed in a similar reaction with $[Fe(acac)_3]$ [91]. The

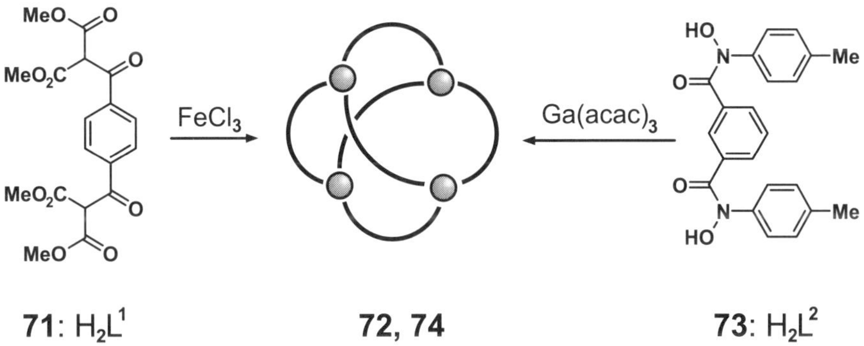

Scheme 10 Formation and schematic representation of complexes $[Fe_4L^1{}_6]$ **72** and $[Ga_4L^2{}_6]$ **74**.

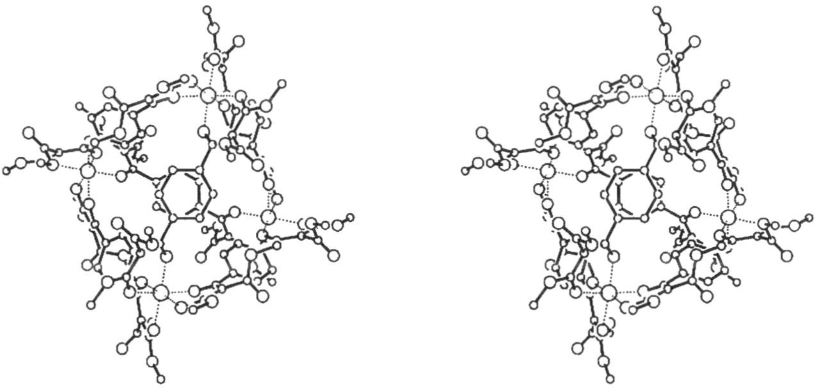

Figure 32 X-ray crystallographic structure of complex $[Fe_4L^1{}_6]$ **72**, stereo representation.

300 MHz ^{1}H NMR spectrum of **74** at room temperature established that all six ligands in the cluster are indistinguishable on the NMR time scale. The solid-state structure of **74** reveals S_4 symmetry. In each cluster four crystallographically identical DMF molecules fill the rigid cavities above the non-planar tetrahedral faces (Figure 33).

Recent results demonstrate that the specific stoichiometry of a supramolecular metal complex cannot definitively be predicted [92,93]. With a family of systematically varied bis-catecholamide ligands, gallium complexes were isolated, which were exclusively identified as metal complexes of two-to-three stoichiometry. Most interestingly, it was shown that the two chiral bis-catecholamide ligands **75** ($H_4{}^{(R,R)}L^3$) and **76** ($H_4{}^{(S,S)}L^4$), which differ only in the nature of the bridge, formed either the dinuclear triple-stranded helicate (Λ,Λ)-$(Et_4N)_6[Ga_2{}^{(R,R)}L^3{}_3]$ **77** or the adamantanoid tetranuclear cage compound $(\Lambda,\Lambda,\Lambda,\Lambda)$-$(Me_4N)_{12}[Ga_4{}^{(S,S)}L^4{}_6]$ **78** (Scheme 11). In the homochiral structures of **77** and **78**, the chirality of the metal centers is induced by the stereogenic centers of ligand H_4L^3 and H_4L^4. The chirality at the first metal center together with the stereogenic centers of the ligands induce the same chirality at the remaining three metal centers so that the only complexes present constitute an enantiomerically pure product as proven by X-ray crystallographic analyses [92,93]. An interesting difference between the dodeca anion $(\Lambda,\Lambda,\Lambda,\Lambda)$-$[Ga_4{}^{(S,S)}L^4{}_6]^{12-}$ of **78** and the hexaanion of (Λ,Λ)-$[Ga_2{}^{(R,R)}L^3{}_3]^{6-}$ of **77** is that the correlation of the stereocenters with the metal configuration is reversed [(*S*)-(Λ)] versus [(*R*)-(Λ)].

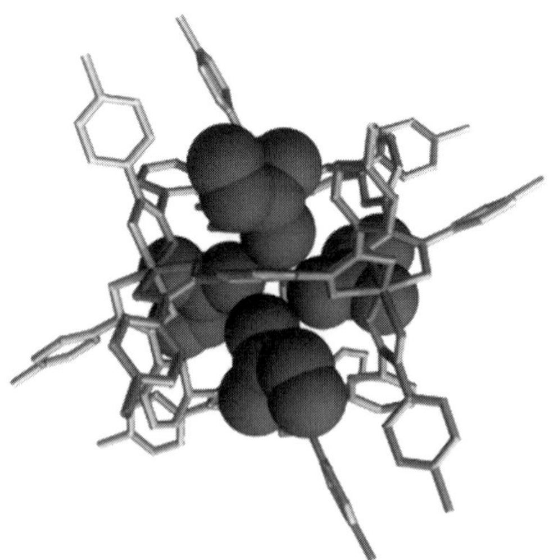

Figure 33 Stick figure of **74** including a space-filling representation of the absorbed DMF molecules.

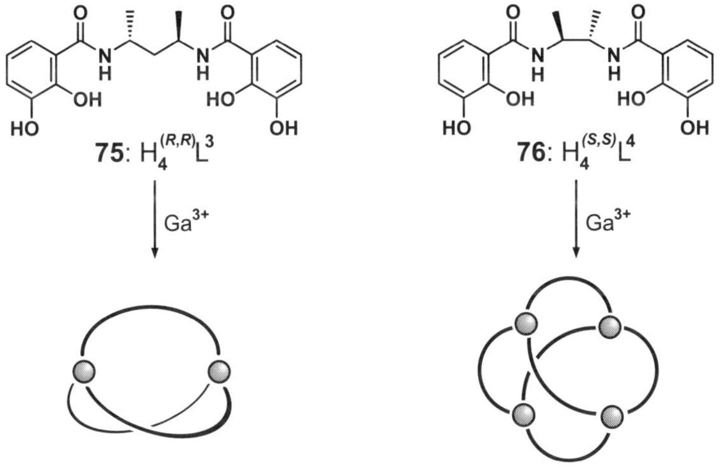

Scheme 11 Formation and schematic representation of complexes (Λ,Λ)-$(Et_4N)_6[Ga_2{}^{(R,R)}L^3{}_3]$ **77** and $(\Lambda,\Lambda,\Lambda,\Lambda)$-$(Me_4N)_{12}[Ga_4{}^{(S,S)}L^4{}_6]$ **78**.

2.4.3 {3}-Metallacryptates

Metal-ligand interactions are highly directional and cluster formation is driven by the coordination number and geometric preferences of the metal. Consequently, a strict geometric analysis was employed in the rational design of an M_4L_6 tetrahedral cluster, and computer modeling indicated that it would consist of a racemic mixture of homochiral $(\Delta,\Delta,\Delta,\Delta)$-*fac* or $(\Lambda,\Lambda,\Lambda,\Lambda)$-*fac* clusters of T symmetry [94]. The ligand H_4L **79** was synthesized from 2,3-dimethoxybenzoyl chloride and 1,5-diaminonaphthalene followed by BBr_3. Stirring of a methanol solution of stoichiometric amounts of $M(acac)_3$ ($M = Ga^{3+}$, Fe^{3+}), H_4L, KOH and Et_4NCl produced microcrystalline products, which analyzed as $K_5[Et_4N]_7[M_4L_6]$. The ultimate proof for the formation of the target cluster and the presence of a host–guest complex was provided by single crystal X-ray diffraction. The undeca cation $[(NEt_4) \subset Fe_4L_6]^{11-}$ **80** is a tightly closed box, with one Et_4N^+ ion located inside the cluster cavity (Figure 34).

2.4.4 Mixed-valent {3}-metallacryptates

The tetranuclear complexes mentioned above are endowed with a substantial cavity, which should, in principle, be suitable for the uptake of guest molecules. A notable feature of these adamantanoid complexes is their topological equivalence to

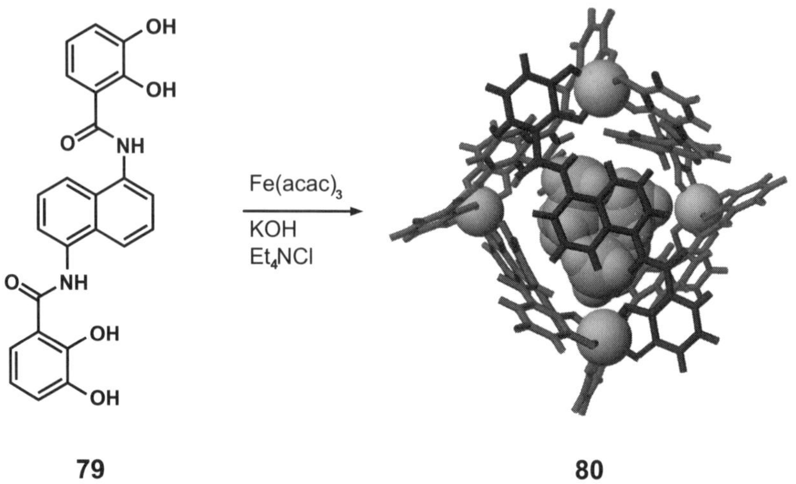

Figure 34 Formation and X-ray crystallographic structure of undeca cation $[(NEt_4) \subset Fe_4L_6]^{11-}$ **80**.

spherands. In order to test this hypothesis, dialkyl malonates **67** were reacted with methyl lithium, iron(II) chloride and oxalyl chloride **68**, followed by work-up with aqueous MCl. The doubly bidentate bridging ligand **70** (L) is formally obtained by template coupling of two dialkyl malonate monoanions with oxalyl chloride. This procedure gives access to mixed-valent {3}-metallacryptates $[M^I \subset Fe^{II}Fe^{III}{}_3L_6]$ **M-81** (Scheme 12) [95].

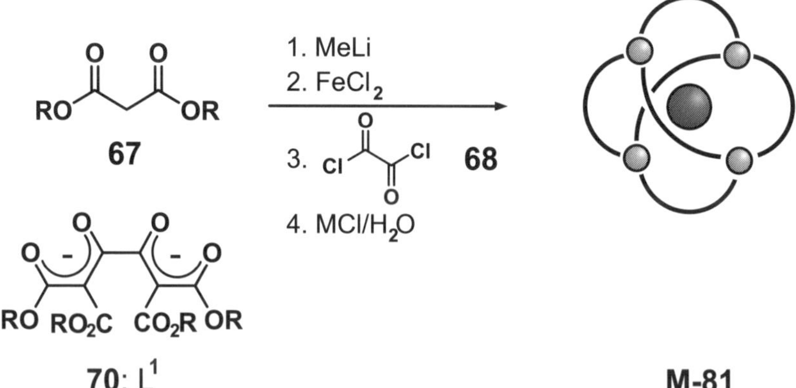

Scheme 12 Formation and schematic representation of {3}-metallacryptates $[M^I \subset Fe^{II}Fe^{III}{}_3L_6]$ **M-81** ($M^I = NH_4^+$, Na^+, K^+, Rb^+, Cs^+).

The results of Mössbauer spectroscopic investigations and X-ray crystallographic structure analyses showed that the complexes [$M^{I} \subset Fe^{II}Fe^{III}{}_{3}L_{6}$] **M-81** are present in the crystal as neutral, tetranuclear, mixed-valent chelate complexes with an endohedrally complexed cation for charge compensation. The receptor nucleus **Cs-81** (Figure 35; 3D-8) consists of a regular tetrahedron with vertices defined by one Fe^{II} and three Fe^{III} ions. In the center of the iron tetrahedron, there is a cesium ion which serves presumably as a template for the formation of **Cs-81** thereby becoming enclosed during the construction of the cage. All four iron centers are homochiral, hence **Cs-81** is present as a racemic mixture with (Δ,Δ,Δ,Δ)-*fac* or (Λ,Λ,Λ,Λ)-*fac* stereoisomers.

A decisive advantage of the synthetic strategy utilized here is that complex structures with defined architecture and specific properties are accessible by spontaneous self-assembly and without the inconvenience associated with multistep syntheses.

2.4.5 Inverted {3}-metallacryptates

Cations have been classically used to promote the assembly of ligands [5,18–21,38,45–47,53,79–81,86–89,95], but recently attention has been payed in using

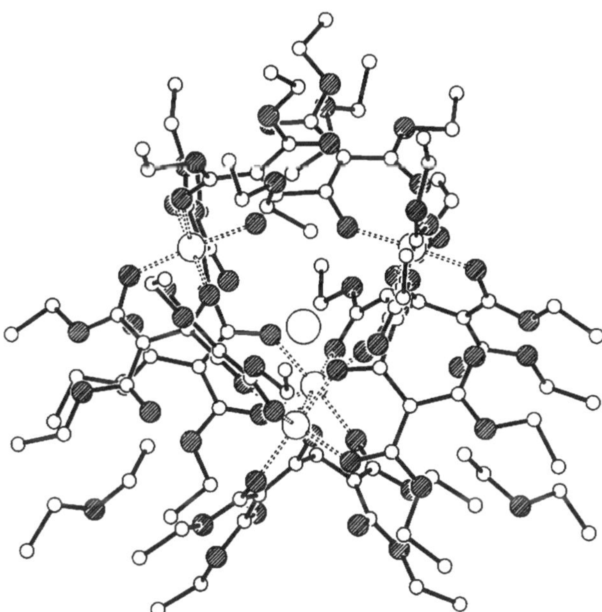

Figure 35 X-ray crystallographic structure of [$Cs \subset Fe^{II}Fe^{III}{}_{3}L_{6}$] **Cs-81**.

anions as templates for the formation of supramolecular entities [5,34,35,38,53,64–66,96].

The templating effect of the tetrafluoroborate anion leads to assembly of four Co(II) ions and six bridging ligands **82** around it to give the tetrahedral complex $[(BF_4) \subset Co_4L_6](BF_4)_7$ **83** with a bridging ligand along each edge and the anion trapped in the central cavity (Scheme 13) [97].

Tripod-metal templates $\{(tripod)M^{n+}[tripod = MeC(CH_2PPh_2)_3)]\}$ are stable units. Because of the fact that the tripod-iron(II) template binds relatively strong to three additional ligands these templates were used for the construction of three-dimensional cage compounds [98]. According to Scheme 13 six fumaronitrile ligands **84**, four iron(II) ions, four tripod ligands **85**, and eight tetrafluoroborate ions react to give the tetranuclear host–guest complex **86**.

The complex **86** was characterized by a complete elemental analysis as well as by NMR spectroscopy and X-ray structure analysis. The pictogram of the cationic core of the host–guest complex **86** is shown in Figure 36 together with the central encapsulated BF_4^- ion and four tetrahedron-face-capping BF_4^- ions. One of the reasons why 21 components, which exist isolated in solution, organize to give the trication $\{[(BF_4) \subset [MeC(CH_2PPh_2)_3Fe^{II}]_4(NCC_2H_2CN)_6] \cap (BF_4)_4\}^{3+}$ of the molecular host–guest complex **86**, appears to derive from a good size match between the guest BF_4^- anion and the cavity of the tetrahedral framework. The remaining three BF_4^- anions necessary for charge compensation occupy the tetrahedral and octahedral holes in an idealized densest packing.

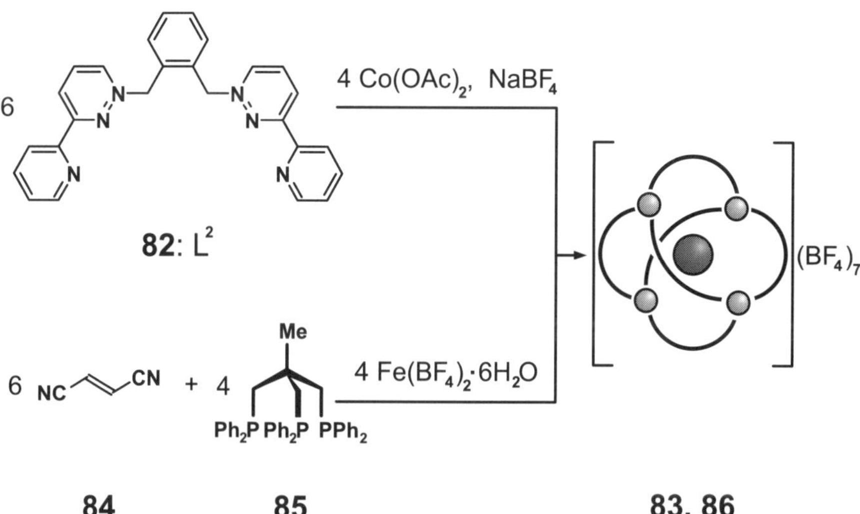

Scheme 13 Formation and schematic representation of $[BF_4 \subset Co_4L_6](BF_4)_7$ **83** and $\{[(BF_4) \subset [MeC(CH_2PPh_2)_3Fe^{II}]_4(NCC_2H_2CN)_6] \cap (BF_4)_4\}(BF_4)_3$ **86**.

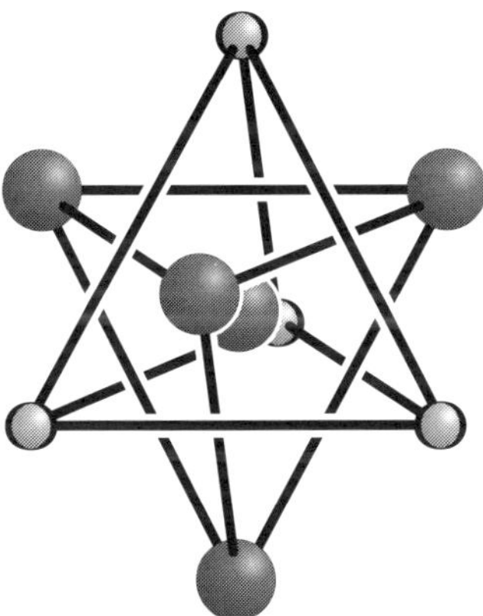

Figure 36 Pictorial representation of the iron tetrahedron with encapsulated BF_4^- ion and four tetrahedron-face-capping BF_4^- ions of the trication $\{[(BF_4) \subset [MeC(CH_2PPh_2)_3Fe^{II}]_4(NCC_2H_2CN)_6] \cap (BF_4)_4\}^{3+}$ of **86**.

2.4.6 {3}-metallacryptates of inverted stoichiometry

To date, three adamantanoid polyhedra of the type M_6L_4 have been reported [7,8,99]. They were obtained from Pd or Pt salts and the corresponding tridentate ligand by self-assembly. Reaction of 1,3,5-tris(pyrazol-1-ylmethyl)-2,4,6-triethylbenzene **87** (L^1) with palladium dichloride results in the self-assembly of a three-dimensional cage $[(PdCl_2)_6L^1{}_4]$ **88**. An X-ray crystal structure determination shows that the cage is composed of an octahedral arrangement of six palladium atoms bridged by a tetrahedral network of four molecules of the ligand (Figure 37).

In two other cases, ethylenediamine functions as ancillary ligand. For instance, reaction of (ethylenediamine)palladium(II) dinitrate with 2,4,6-tris(4′-pyridyl)-1,3,5-triazene **89** (L^2) yielded achiral $[(Pd\text{-}en)_6L^2{}_4](NO_2)_{12}$ **90** (Figure 38). This cage encapsulates four molecules of adamantane derivatives, as established by X-ray data. In a similar way, a chiral, discrete, nanoscale-sized supramolecular cage was prepared from the tridentate ligand 1,3,5-tris[(4′-pyridyl)ethynyl]benzene and [(*R*)-(+)-BINAP]Pd^{II} bis(triflate) (BINAP = 2,2′-bis(diphenylphosphino)-1,1′-binaphthyl). A unique feature of these chiral 3D metallacyclic polyhedra is that they belong to the *T* symmetry point group, which has so far only been observed in a very few covalent organic molecules [88,100].

2.4.7 {3}-metallacryptands of $[M_4L_4]$ stoichiometry

An alternative approach to the formation of symmetric tetrahedral clusters of $[M_4L_4]$ stoichiometry, namely one in which the pseudo-octahedral metals occupy the

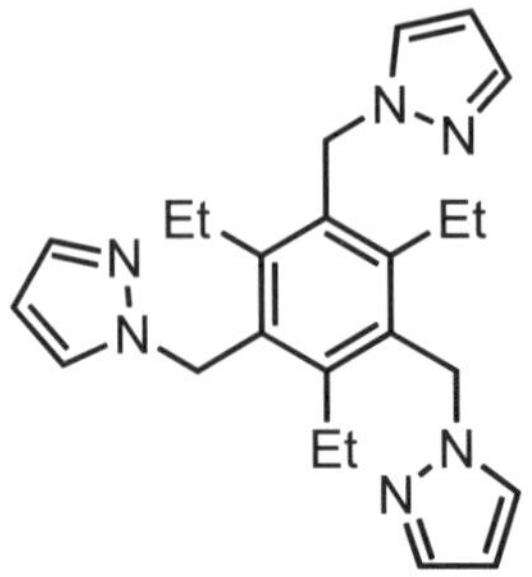

87: L^1

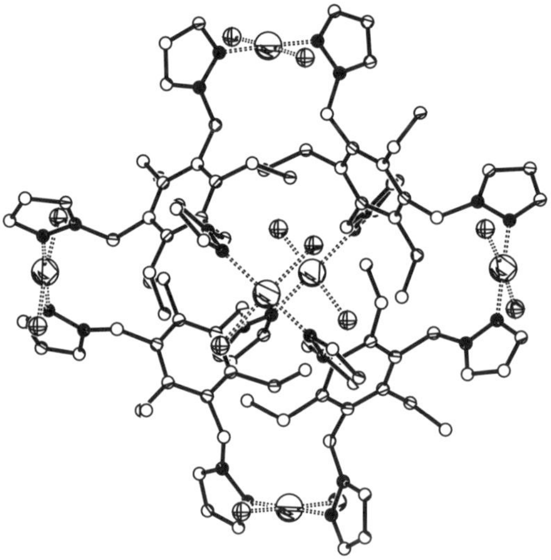

88

Figure 37 X-ray crystallographic analysis of $[(PdCl_2)_6L^1{}_4]$ **88**.

vertices of the tetrahedron and trigonally symmetric tris-bidentate ligands occupy the faces, was described [101].

Reaction of potassium tris[3-(2′-pyridyl)pyrazol-1-yl]hydroborate (KL^1) with one equivalent of $Mn(MeCO_2)_2{\cdot}4H_2O$ in methanol followed by treatment with KPF_6 afforded $[Mn_4L^1{}_4](PF_6)_4$ **91**. The complex contains four crystallographically independent Mn^{II} ions which are arrayed approximately in a tetrahedron. Each ligand $(L^1)^-$ sits above one triangular face of this tetrahedron and donates one bidentate arm to each of the three metal ions on that face. Each ligand therefore interacts with three different metal ions (Figure 39), and each metal ion is coordinated by one bidentate arm from each of three different ligands. The coordination geometries of the metal ions are highly distorted octahedra.

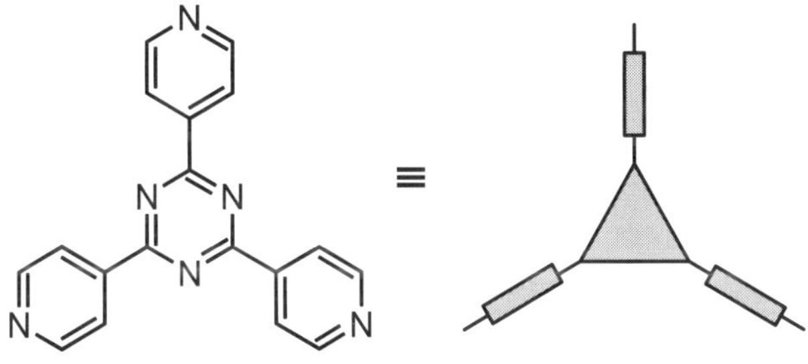

89: L^2

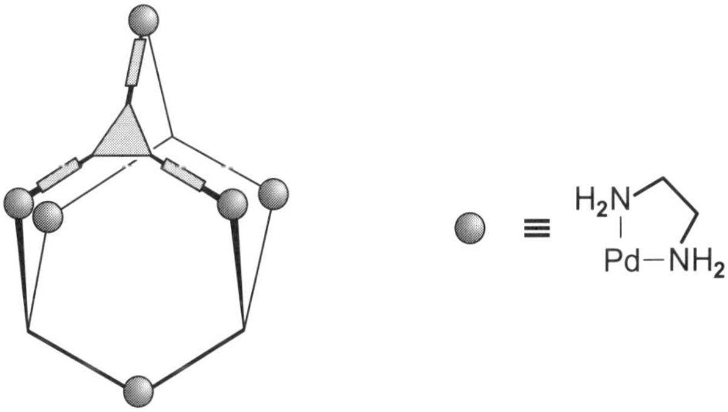

90

Figure 38 Pictorial representation of ligand **89**, and of the dodecacation $[(Pd\text{-}en)_6L^2{}_4]^{12+}$ of **90**.

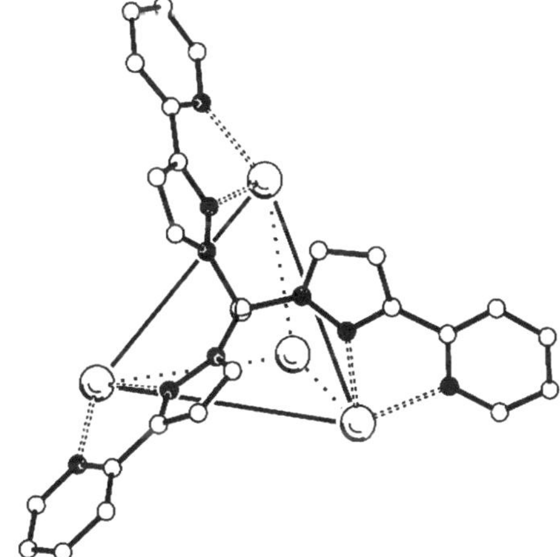

Figure 39 Pictorial representation of one triangular face of the tetracation $[Mn_4L^1{}_4]^{4+}$ of **91**.

The synthesis of ligand H_6L^2 **92** followed established routes. 1,3,5-Triaminobenzene was combined with 2,3-dimethoxybenzoic acid chloride, to produce the corresponding trisamide which was deprotected with BBr_3 to give triscatechol **92** in good yields. A solution of $Ti(OBu)_4$ in methanol was combined with **92** in methanol/triethylamine and after complex work-up racemic homochiral complex $[Ti_4L^2{}_4](HNEt_3)_8$ **93** was isolated. A single crystal structure provided the ultimate proof of the tetrahedral cluster. A stereo presentation is shown in Figure 40. The eight triethylammonium counterions are highly disordered, hydrogen bonded to both carboxyl and phenolic oxygens of the cluster. There is no evidence that the small cavity of the tetrahedron contains a guest, as observed in some larger tetrahedra [94,95,98].

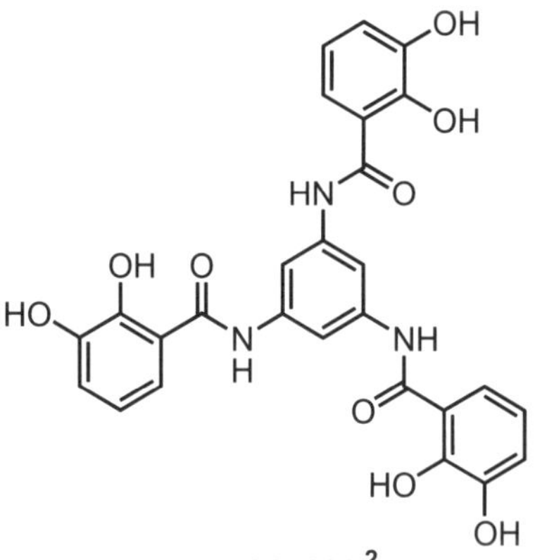

92: H_6L^2

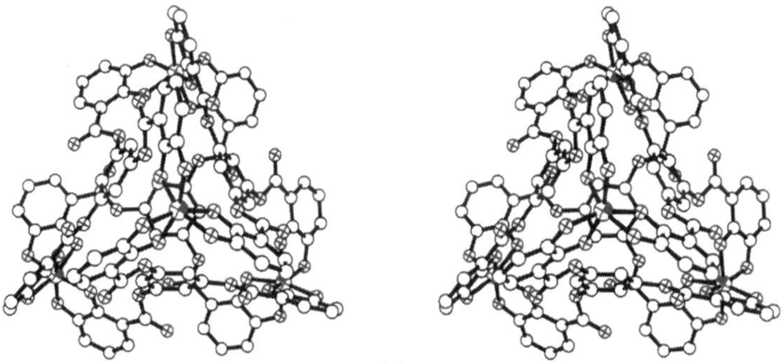

93

Figure 40 Ligand **92** and X-ray crystallographic structure of octaanion $[Ti_4L^2{}_4]^{8-}$ of **93**, stereo representation.

3 CONCLUSIONS AND OUTLOOK

Anion control in the self-assembly of cage coordination complexes is still in its infancy, but once it is applied efficiently, supramolecular chemistry will benefit enormously. Recently it has been shown [102], that coordination of ATU (ATU = amidinothiourea) ligands to nickel(II) cations occurs via both guanidino nitrogen atoms to form square-planar building blocks. The sulfur atoms of four $[Ni(ATU^-)_2]$ units **94** act as secondary donating sites to a further two nickel(II) ions to form in the presence of chloride ions the cage structure $[Cl \subset Ni_6(ATU)_8]Cl_3$ **95**, illustrated in Figure 41. In order to unequivocally establish the structure of the hexanuclear nickel complex **95**, X-ray diffraction was carried out. According to this analysis, the six nickel(II) cations are located in the vertices of a compressed tetragonal bipyramid. The two apical nickel(II) cations each are square-planar coordinated to the eight sulfur atoms of four $[Ni(ATU^-)_2]$ units.

A simple, general systematic strategy for the rational design of self-assembled supramolecular species has yet to be developed. However, a promising approach to the formation of novel supramolecular species, by spontaneous self-assembly of precursor building blocks under appropriate conditions, involves the coordination

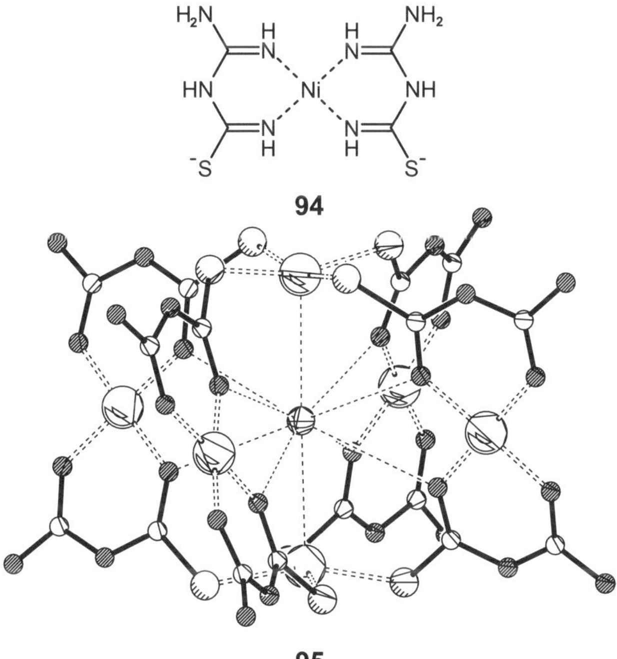

Figure 41 Schematic representation of building block **94** and X-ray crystallographic structure of trication $[Cl \subset Ni_6(ATU)_8]^{3+}$ of **95**.

motif, that is, the use of transition metals and multidentate ligands. It is evident that the concepts and principles known for the classical covalent supramolecular chemistry are valid and applicable to the coordination motif in the self-assembly of a nearly infinite number and variety of metal-containing supramolecular species. Additionally, coordination offers advantages and unique design features in the assembly of discrete supramolecular species. Just as in nature, function and use will follow and enrich such rapidly growing interdisciplinary fields as bioinorganic chemistry, and materials sciences.

ACKNOWLEDGEMENTS

Financial support of the Deutsche Forschungsgemeinschaft, the Fonds der Chemischen Industrie and the Volkswagen-Stiftung is greatly appreciated. Particular thanks are due to the enthusiastic co-workers mentioned, who have actively taken part in our own research. We also thank Sabine Kareth for her valuable assistance in preparing the manuscript.

REFERENCES

1. D. Philp, J. F. Stoddart, *Angew. Chem.* 1996, **108**, 1243; *Angew. Chem. Int. Ed. Engl.* 1996, **35**, 1154.
2. W. Hayes, J. F. Stoddart, *Large Ring Molecules* (Ed.: J. A. Semlyen), Wiley, New York, 1996, p. 433.
3. B. Dietrich, P. Viout, J.-M. Lehn, *Macrocyclic Chemistry*, VCH, Weinheim, 1993, p. 171 ff.
4. J.-M. Lehn, *Angew. Chem.* 1990, **102**, 1347; *Angew. Chem. Int. Ed. Engl.* 1990, **29**, 1304.
5. J.-M. Lehn, *Supramolecular Chemistry*, VCH, Weinheim, 1995, p. 17 ff.
6. F. Vögtle, *Supramolekulare Chemie*, Teubner, Stuttgart, 1992, p. 45 ff.
7. P. J. Stang, B. Olenyuk, *Acc. Chem. Res.* 1997, **30**, 502; P. J. Stang, *Chem. Eur. J.* 1998, **4**, 19.
8. C. M. Hartshorn, P. J. Steel, *Chem. Commun.* 1997, 541.
9. J. R. Fredericks, A. D. Hamilton, *Supramolecular Control of Structure and Reactivity* (Ed.: A. D. Hamilton), Wiley, New York, 1996, p. 1.
10. M. C. T. Fyfe, J. F. Stoddart, *Acc. Chem. Res.* 1997, **30**, 393.
11. P. N. W. Baxter, *Comprehensive Supramolecular Chemistry* (Eds.: J. L. Atwood, J. E. D. Davies, D. D. Macnicol, F. Vögtle, J.-M. Lehn), *Metal Ion Directed Assembly of Complex Molecular Architectures and Nanostructures*, Vol. 9 (Eds.: J.-P. Sauvage, M. W. Hosseini), 1st ed., Pergamon, Oxford, 1996, p. 165.
12. E. C. Constable, *Comprehensive Supramolecular Chemistry* (Eds.: J. L. Atwood, J. E. D. Davies, D. D. Macnicol, F. Vögtle, J.-M. Lehn), *Polynuclear Transition Metal Helicates*, Vol. 9 (Eds.: J.-P. Sauvage, M. W. Hosseini), 1st ed., Pergamon, Oxford, 1996, p. 213; E. C. Constable, *Tetrahedron* 1992, **48**, 10013.
13. C. Piguet, G. Bernardinelli, G. Hopfgartner, *Chem. Rev.* 1997, **97**, 2005.
14. M. Fujita, *Comprehensive Supramolecular Chemistry* (Eds.: J. L. Atwood, J. E. D. Davies, D. D. Macnicol, F. Vögtle, J.-M. Lehn), *Self-assembled Macrocycles, Cages, and*

Catenanes Containing Transition Metals in Their Backbones, Vol. 9 (Eds.: J.-P. Sauvage, M. W. Hosseini), 1st ed., Pergamon, Oxford, 1996, p. 253.
15. F. Vögtle, R. Hoss, M. Händel, *Applied Homogeneous Catalysis with Organometallic Compounds*, Vol. 2 (Eds.: B. Cornils, W. Hermann) VCH, Weinheim, 1996, p. 801.
16. J.-C. Chambron, C. Dietrich-Buchecker, J.-P. Sauvage, *Comprehensive Supramolecular Chemistry* (Eds.: J. L. Atwood, J. E. D. Davies, D. D. Macnicol, F. Vögtle, J.-M. Lehn), *Transition Metals as Assembling and Templating Species: Synthesis of Catenanes and Molecular Knots*, Vol. 9 (Eds.: J.-P. Sauvage, M. W. Hosseini), 1st ed., Pergamon, Oxford, 1996, p. 43.
17. A. Klug, *Angew. Chem.* 1983, **95**, 579; *Angew. Chem. Int. Ed. Engl.* 1983, **22**, 565.
18. R. W. Saalfrank, A. Dresel, V. Seitz, S. Trummer, F. Hampel, M. Teichert, D. Stalke, C. Stadler, J. Daub, V. Schünemann, A. X. Trautwein, *Chem. Eur. J.* 1997, **3**, 2058.
19. R. W. Saalfrank, N. Löw, S. Kareth, V. Seitz, F. Hampel, D. Stalke, M. Teichert, *Angew. Chem.* 1998, **110**, 182; *Angew. Chem. Int. Ed. Engl.* 1998, **37**, 172.
20. V. L. Pecoraro, A. J. Stemmler, B. R. Gibney, J. J. Bodwin, H. Wang, J. W. Kampf, A. Barwinski, *Prog. Inorg. Chem.* 1997, **45**, 83.
21. M. S. Lah, V. L. Pecoraro, *Comments Inorg. Chem.* 1990, **11**, 59.
22. H. Sleiman, P. Baxter, J.-M. Lehn, K. Rissanen, *J. Chem. Soc., Chem. Commun.* 1995, 715; G. S. Hanan, C. R. Arana, J.-M. Lehn, G. Baum, D. Fenske, *Chem. Eur. J.* 1996, **2**, 1292.
23. P. N. W. Baxter, G. S. Hanan, J.-M. Lehn, *Chem. Commun.* 1996, 2019.
24. M.-T. Youinou, N. Rahmouni, J. Fischer, J. A. Osborn, *Angew. Chem.* 1992, **104**, 771; *Angew. Chem. Int. Ed. Engl.* 1992, **31**, 733.
25. G. S. Hanan, D. Volkmer, U. S. Schubert, J.-M. Lehn, G. Baum, D. Fenske, *Angew. Chem.* 1997, **109**; *Angew. Chem. Int. Ed. Engl.* 1997, **36**, 1842; P. N. W. Baxter, J.-M. Lehn, J. Fischer, M.-T. Youinou, *Angew. Chem.* 1994, **106**, 2432; *Angew. Chem. Int. Ed. Engl.* 1994, **33**, 2284; P. N. W. Baxter, J.-M. Lehn, B. O. Kneisel, D. Fenske, *Angew. Chem.* 1997, **109**, 2067; *Angew. Chem. Int. Ed. Engl.* 1997, **36**, 1978; C. Duan, Z. Liu, X. You, F. Xue, T. C. W. Mak, *Chem. Commun.* 1997, 381.
26. R.-D. Schnebeck, L. Randaccio, E. Zangrando, B. Lippert, *Angew. Chem.* 1998, **110**, 128; *Angew. Chem. Int. Ed. Engl.* 1998, **37**, 119; J. R. Hall, S. J. Loeb, G. K. H. Shimizu, G. P. A. Yap, *Angew. Chem.* 1998, **110**, 130; *Angew. Chem. Int. Ed. Engl.* 1998, **37**, 121.
27. M. Fujita, O. Sasaki, T. Mitsuhashi, T. Fujita, J. Yazaki, K. Yamaguchi, K. Ogura, *Chem. Commun.* 1996, 1535; P. J. Stang, J. Fan, B. Olenyuk, *Chem. Commun.* 1997, 1453; H. Rauter, I. Mutikainen, M. Blomberg, C. J. L. Lock, P. Amo-Ochoa, E. Freisinger, L. Randaccio, E. Zangrando, E. Chiarparin, B. Lippert, *Angew. Chem.* 1997, **109**, 1353; *Angew. Chem. Int. Ed. Engl.* 1997, **36**, 1296.
28. P. Baxter, J.-M. Lehn, A. DeCian, J. Fischer, *Angew. Chem.* 1993, **105**, 92; *Angew. Chem. Int. Ed. Engl.* 1993, **32**, 69.
29. E. Leize, A. Van Dorsselaer, R. Krämer, J.-M. Lehn, *J. Chem. Soc., Chem. Commun.* 1993, 990; M. Fujita, F. Ibukuro, K. Yamaguchi, K. Ogura, *J. Am. Chem. Soc.* 1995, **117**, 4175.
30. B. Mohr, M. Weck, J.-P. Sauvage, R. H. Grubbs, *Angew. Chem.* 1997, **109**, 1365; *Angew. Chem. Int. Ed. Engl.* 1997, **36**, 1308; C. O. Dietrich-Buchecker, J.-P. Sauvage, J. P. Kintzinger, *Tetrahedron Lett.* 1983, **24**, 5095; C. O. Dietrich-Buchecker, J.-P. Sauvage, *Chem. Rev.* 1987, **87**, 795; J.-P. Sauvage, *Acc. Chem. Res.* 1990, **23**, 319; M. Fujita, F. Ibukuro, H. Seki, O. Kamo, M. Imanari, K. Ogura, *J. Am. Chem. Soc.* 1996, **118**, 899; M. Fujita, K. Ogura, *Coord. Chem. Rev.* 1996, **148**, 249.
31. P. N. W. Baxter, H. Sleiman, J.-M. Lehn, K. Rissanen, *Angew. Chem.* 1997, **109**, 1350; *Angew. Chem. Int. Ed. Engl.* 1997, **36**, 1294.
32. R. F. Carina, C. Dietrich-Buchecker, J.-P. Sauvage, *J. Am. Chem. Soc.* 1996, **118**, 9110.

33. G. R. Newkome, J. Groß, C. N. Moorefield, B. D. Woosley, *Chem. Commun.* 1997, 515; G. R. Newkome, F. Cardullo, E. C. Constable, C. N. Moorefield, A. M. W. Cargill Thompson, *J. Chem. Soc., Chem. Commun.* 1993, 925; E. C. Constable, P. Harverson, *Chem. Commun.* 1996, 33; H.-F. Chow, I. Y.-K. Chan, D. T. W. Chan, R. W. M. Kwok, *Chem. Eur. J.* 1996, **2**, 1085.
34. B. Hasenknopf, J.-M. Lehn, N. Boumediene, A. Dupont-Gervais, A. Van Dorsselaer, B. Kneisel, D. Fenske, *J. Am. Chem. Soc.* 1997, **119**, 10956; B. Hasenknopf, J.-M. Lehn, B. O. Kneisel, G. Baum, D. Fenske, *Angew. Chem.* 1996, **108**, 1987; *Angew. Chem. Int. Ed. Engl.* 1996, **35**, 1838.
35. R. W. Saalfrank, S. Trummer, H. Krautscheid, V. Schünemann, A. X. Trautwein, S. Hien, C. Stadler, J. Daub, *Angew. Chem.* 1996, **108**, 2350; *Angew. Chem. Int. Ed. Engl.* 1996, **35**, 2206.
36. R. D. Shannon, *Acta Crystallogr.* 1976, **A32**, 751.
37. V. L. Pecoraro, *Inorg. Chim. Acta* 1989, **155**, 171.
38. I. Haiduc, R. B. King, *Large Ring Molecules* (Ed.: J. A. Semlyen), Wiley, New York, 1996, p. 525.
39. B. R. Gibney, D. P. Kessissoglou, J. W. Kampf, V. L. Pecoraro, *Inorg. Chem.* 1994, **33**, 4840.
40. B. Kurzak, E. Farkas, T. Glowiak, H. Kozlowski, *J. Chem. Soc., Dalton Trans.* 1991, 163.
41. C. J. Pedersen, *Angew. Chem.* 1988, **100**, 1053; *Angew. Chem. Int. Ed. Engl.* 1988, **27**, 1021; D. J. Cram, *Angew. Chem.* 1988, **100**, 1041; *Angew. Chem. Int. Ed. Engl.* 1988, **27**, 1009; J.-M. Lehn, *Angew. Chem.* 1988, **100**, 91; *Angew. Chem. Int. Ed. Engl.* 1988, **27**, 89.
42. C. J. Pedersen, *J. Am. Chem. Soc.* 1967, **89**, 2495; G. Gokel, *Crown Ethers and Cryptands in Monographs in Supramolecular Chemistry* (Ed.: J. F. Stoddart), Black Bear Press, Cambridge, 1991, p. 64; E. Weber, F. Vögtle, *Crown-Type Compounds—An Introductory Overview*, *Top. Curr. Chem.* 1981, p. 1.
43. A. J. Stemmler, J. W. Kampf, V. L. Pecoraro, *Angew. Chem.* 1996, **108**, 3011; *Angew. Chem. Int. Ed. Engl.* 1996, **35**, 2841; A. J. Stemmler, A. Barwinski, M. J. Baldwin, V. Young, V. L. Pecoraro, *J. Am. Chem. Soc.* 1996, **118**, 11962; M. S. Lah, B. R. Gibney, D. L. Tierney, J. E. Penner-Hahn, V. L. Pecoraro, *J. Am. Chem. Soc.* 1993, **115**, 5857; A. J. Blake, R. O. Gould, C. M. Grant, P. E. Y. Milne, D. Reed, R. E. P. Winpenny, *Angew. Chem.* 1994, **106**, 208; *Angew. Chem. Int. Ed. Engl.* 1994, **33**, 195; J. A. Blake, R. O. Gould, P. E. Y. Milne, R. E. P. Winpenny, *J. Chem. Soc., Chem. Commun.* 1991, 1453; G. M. Gray, C. H. Duffey, *Organometallics* 1995, **14**, 245.
44. A. Caneschi, A. Cornia, S. J. Lippard, *Angew. Chem.* 1995, **107**, 511; *Angew. Chem. Int. Ed. Engl.* 1995, **34**, 467.
45. R. W. Saalfrank, N. Löw, F. Hampel, H.-D. Stachel, *Angew. Chem.* 1996, **108**, 2353; *Angew. Chem. Int. Ed. Engl.* 1996, **35**, 2209.
46. R. W. Saalfrank, I. Bernt, E. Uller, F. Hampel, *Angew. Chem.* 1997, **109**, 2596; *Angew. Chem. Int. Ed. Engl.* 1997, **36**, 2482.
47. R. W. Saalfrank, N. Löw, B. Demleitner, D. Stalke, M. Teichert, *Chem. Eur. J.* 1998, **4**, 1305.
48. K. L. Taft, C. D. Delfs, G. C. Papaefthymiou, S. Foner, D. Gatteschi, S. J. Lippard, *J. Am. Chem. Soc.* 1994, **116**, 823.
49. C. Benelli, S. Parsons, G. A. Solan, R. E. P. Winpenny, *Angew. Chem.* 1996, **108**, 1967; *Angew. Chem. Int. Ed. Engl.* 1996, **35**, 1825; A. Caneschi, A. Cornia, A. C. Fabretti, D. Gatteschi, *Angew. Chem.* 1995, **107**, 2862; *Angew. Chem. Int. Ed. Engl.* 1995, **34**, 2716.
50. R. Sessoli, D. Gatteschi, A. Caneschi, N. A. Novak, *Nature* 1993, **365**, 141; M. W. Wemple, D. M. Adams, K. S. Hagen, K. Folting, D. N. Hendrickson, G. Christou,

J. Chem. Soc., Chem. Commun. 1995, 1591; H. J. Eppley, H.-L. Tsai, N. de Vries, K. Folting, G. Christou, D. N. Hendrickson, *J. Am. Chem. Soc.* 1995, **117**, 301; D. P. Goldberg, A. Caneschi, S. J. Lippard, *J. Am. Chem. Soc.* 1993, **115**, 9299; R. Cortés, L. Lezama, J. L. Pizarro, M. I. Arriortua, T. Rojo, *Angew. Chem.* 1996, **108**, 1934; *Angew. Chem. Int. Ed. Engl.* 1996, **35**, 1810.

51. A. J. Blake, C. M. Grant, S. Parsons, J. M. Rawson, R. E. P. Winpenny, *J. Chem. Soc., Chem. Commun.* 1994, 2363.
52. A. Caneschi, A. Cornia, A. C. Fabretti, S. Foner, D. Gatteschi, R. Grandi, L. Schenetti, *Chem. Eur. J.* 1996, **2**, 1379.
53. D. Gatteschi, A. Caneschi, R. Sessoli, A. Cornia, *Chem. Soc. Rev.* 1996, 101.
54. A. Caneschi, A. Cornia, A. C. Fabretti, D. Gatteschi, W. Malavasi, *Inorg. Chem.* 1995, **34**, 4660.
55. C. A. Christmas, H.-L. Tsai, L. Pardi, J. M. Kesselman, P. K. Gantzel, R. K. Chadha, D. Gatteschi, D. F. Harvey, D. N. Hendrickson, *J. Am. Chem. Soc.* 1993, **115**, 12483.
56. M. Tesmer, B. Müller, H. Vahrenkamp, *Chem. Commun.* 1997, 721.
57. M. A. Bolcar, S. M. J. Aubin, K. Folting, D. N. Hendrickson, G. Christou, *Chem. Commun.* 1997, 1485.
58. A. K. Powell, S. L. Heath, D. Gatteschi, L. Pardi, R. Sessoli, G. Spina, F. Del Giallo, F. Pieralli, *J. Am. Chem. Soc.* 1995, **117**, 2491.
59. J. K. Beattie, T. W. Hambley, J. A. Klepetko, A. F. Masters, P. Turner, *Chem. Commun.* 1998, 45.
60. P. Klüfers, J. Schumacher, *Angew. Chem.* 1995, **107**, 2290; *Angew. Chem. Int. Ed. Engl.* 1995, **34**, 2119.
61. S. P. Watton, P. Fuhrmann, L. E. Pence, A. Caneschi, A. Cornia, G. L. Abbati, S. J. Lippard, *Angew. Chem.* 1997, **109**, 2917; *Angew. Chem. Int. Ed. Engl.* 1997, **36**, 2774.
62. S. M. J. Aubin, M. W. Wemple, D. M. Adams, H.-L. Tsai, G. Christou, D. N. Hendrickson, *J. Am. Chem. Soc.* 1996, **118**, 7746.
63. S. Bénard, P. Yu, T. Coradin, E. Rivière, K. Nakatani, R. Clément, *Adv. Mater.* 1997, **9**, 981; S. Decurtins, H. W. Schmalle, H. R. Oswald, A. Linden, J. Ensling, P. Gütlich, A. Hauser, *Inorg. Chim. Acta.* 1994, **216**, 65.
64. R. W. Saalfrank, N. Löw, S. Trummer, G. M. Sheldrick, M. Teichert, D. Stalke, *Eur. J. Inorg. Chem.* 1998, 559.
65. P. L. Jones, K. J. Byrom, J. C. Jeffery, J. A. McCleverty, M. D. Ward, *Chem. Commun.* 1997, 1361.
66. M. A. Houghton, A. Bilyk, M. M. Harding, P. Turner, T. W. Hambley, *J. Chem. Soc., Dalton Trans.* 1997, 2725.
67. D. P. Funeriu, J.-M. Lehn, G. Baum, D. Fenske, *Chem. Eur. J.* 1997, **3**, 99.
68. P. N. W. Baxter, J.-M. Lehn, K. Rissanen, *Chem. Commun.* 1997, 1223.
69. C. Provent, S. Hewage, G. Brand, G. Bernardinelli, L. J. Charbonnière, A. F. Williams, *Angew. Chem.* 1997, **109**, 1346; *Angew. Chem. Int. Ed. Engl.* 1997, **36**, 1287; see also: G. Hopfgarten, C. Piguet, J. D. Henion, *J. Am. Soc. Mass Spectrom.* 1994, **5**, 748.
70. O. Mamula, A. von Zelewsky, G. Bernardinelli, *Angew. Chem.* 1998, **110**, 301; *Angew. Chem. Int. Ed. Engl.* 1998, **37**, 290.
71. S. J. Lippard, J. M. Berg, *Bioanorganische Chemie*, Spektrum, Heidelberg 1995, Chap. 5, p. 136 ff.; S. M. Gorun, G. C. Papaefthymiou, R. B. Frankel, S. J. Lippard, *J. Am. Chem. Soc.* 1987, **109**, 4244; S. J. Lippard, *Angew. Chem.* 1988, **100**, 353; *Angew. Chem. Int. Ed. Engl.* 1988, **27**, 344; K. Wieghardt, K. Pohl, I. Jibril, G. Huttner, *Angew. Chem.* 1984, **96**, 66; *Angew. Chem. Int. Ed. Engl.* 1984, **23**, 77; I. Shweky, L. E. Pence, G. C. Papaefthymiou, R. Sessoli, J. W. Yun, A. Bino, S. J. Lippard, *J. Am. Chem. Soc.* 1997, **119**, 1037.

72. R. Krämer, J.-M. Lehn, A. De Cian, J. Fischer, *Angew. Chem.* 1993, **105**, 764; *Angew. Chem. Int. Ed. Engl.* 1993, **32**, 703.
73. B. Grossmann, J. Heinze, E. Herdtweck, F. H. Köhler, H. Nöth, H. Schwenk, M. Spiegler, W. Wachter, B. Weber, *Angew. Chem.* 1997, **109**, 384; *Angew. Chem. Int. Ed. Engl.* 1997, **36**, 387.
74. J.-M. Lehn, A. Rigault, J. Siegel, J. Harrowfield, B. Chevrier, D. Moras, *Proc. Natl. Acad. Sci. USA* 1987, **84**, 2565; A. Pfeil, J.-M. Lehn, *J. Chem. Soc., Chem. Commun.* 1992, 838; R. Kramer, J.-M. Lehn, A. Marquis-Rigault, *Proc. Natl. Acad. Sci. USA* 1993, **90**, 5394; J.-M. Lehn, A. Rigault, *Angew. Chem.* 1988, **100**, 1121; *Angew. Chem. Int. Ed. Engl.* 1988, **27**, 1059; U. Koert, M. M. Harding, J.-M. Lehn, *Nature* 1990, **346**, 339; A. F. Williams, *Chem. Eur. J.* 1997, **3**, 15; A. F. Williams, *Pure Appl. Chem.* 1996, **68**, 1285; L. J. Charbonnière, A. F. Williams, U. Frey, A. E. Merbach, P. Kamalaprija, O. Schaad, *J. Am. Chem. Soc.* 1997, **119**, 2488; C. Piguet, E. Rivara-Minten, G. Bernardinelli, J.-C. G. Bünzli, G. Hopfgartner, *J. Chem. Soc., Dalton Trans.* 1997, 421; C. Piguet, G. Bernardinelli, A. F. Williams, B. Bocquet, *Angew. Chem.* 1995, **107**, 618; *Angew. Chem. Int. Ed. Engl.* 1995, **34**, 582; E. C. Constable, A. J. Edwards, P. R. Raithby, J. V. Walker, *Angew. Chem.* 1993, **105**, 1486; *Angew. Chem. Int. Ed. Engl.* 1993, **32**, 1465; E. C. Constable, M. G. B. Drew, M. D. Ward, *J. Chem. Soc., Chem. Commun.* 1987, 1600; E. C. Constable, M. D. Ward, D. A. Tocher, *J. Am. Chem. Soc.* 1990, **112**, 1256; E. C. Constable, R. Chotalia, D. A. Tocher, *J. Chem. Soc., Chem. Commun.* 1992, 771; E. C. Constable, J. V. Walker, D. A. Tocher, M. A. M. Daniels, *J. Chem. Soc., Chem. Commun.* 1992, 768; J. Libman, Y. Tor, A. Shanzer, *J. Am. Chem. Soc.* 1987, **109**, 5880; L. Zelikovich, J. Libman, A. Shanzer, *Nature* 1995, **374**, 790.
75. C. J. Carrano, K. N. Raymond, *J. Am. Chem. Soc.* 1978, **100**, 5371; C. J. Carrano, S. R. Cooper, K. N. Raymond, *J. Am. Chem. Soc.* 1979, **101**, 599; R. C. Scarrow, D. L. White, K. N. Raymond, *J. Am. Chem. Soc.* 1985, **107**, 6540.
76. D. L. Caulder, K. N. Raymond, *Angew. Chem.* 1997, **109**, 1508; *Angew. Chem. Int. Ed. Engl.* 1997, **36**, 1440.
77. B. Kersting, M. Meyer, R. E. Powers, K. N. Raymond, *J. Am. Chem. Soc.* 1996, **118**, 7221; K. N. Raymond, G. Müller, B. F. Matzanke, *Top. Curr. Chem.* 1984, **123**, 49.
78. M. Albrecht, M. Schneider, H. Röttele, *Chem. Ber./Recueil* 1997, **130**, 615.
79. M. Albrecht, S. Kotila, *Angew. Chem.* 1996, **108**, 1299; *Angew. Chem. Int. Ed. Engl.* 1996, **35**, 1208.
80. M. Albrecht, S. Kotila, *Angew. Chem.* 1995, **107**, 2285; *Angew. Chem. Int. Ed. Engl.* 1995, **34**, 2134; M. Albrecht, H. Rötele, P. Burger, *Chem. Eur. J.* 1996, **2**, 1264.
81. M. Albrecht, *Chem. Eur. J.* 1997, **3**, 1466; M. Albrecht, S. Kotila, *Chem. Commun.* 1996, 2309.
82. V. A. Grillo, E. J. Seddon, C. M. Grant, G. Aromí, J. C. Bollinger, K. Folting, G. Christou, *Chem. Commun.* 1997, 1561; M. J. Hannon, C. L. Painting, A. Jackson, J. Hamblin, W. Errington, *Chem. Commun.* 1997, 1807; M. Albrecht, O. Blau, *Chem. Commun.* 1997, 345; M. Albrecht, R. Fröhlich, *J. Am. Chem. Soc.* 1997, **119**, 1656; M. Albrecht, C. Riether, *Chem. Ber.* 1996, **129**, 829.
83. R. W. Saalfrank, V. Seitz, D. L. Caulder, K. N. Raymond, M. Teichert, D. Stalke, *Eur. J. Inorg. Chem.* **1998**, 1313.
84. K. Wieghardt, *Angew. Chem.* 1989, **101**, 1179; *Angew. Chem. Int. Ed. Engl.* 1989, **28**, 1153; W. F. Beck, J. Sears, G. W. Brudvig, R. J. Kulawiec, R. H. Crabtree, *Tetrahedron* 1989, **45**, 4903.
85. Highlights on this topic: T. Weiske, T. Wong, W. Krätschmer, J. K. Terlouw, H. Schwarz, *Angew. Chem.* 1992, **104**, 242; *Angew. Chem. Int. Ed. Engl.* 1992, **31**, 183; H. Hopf, *Angew. Chem.* 1991, **103**, 1137; *Angew. Chem. Int. Ed. Engl.* 1991, **30**, 1117; C. Seel, F. Vögtle, *Angew. Chem.* 1992, **104**, 542; *Angew. Chem. Int. Ed. Engl.*

1992, **31**, 528; *Frontiers in Supramolecular Organic Chemistry and Photochemistry* (Eds.: H.-J. Schneider, H. Dürr), VCH, Weinheim, 1991; M. T. Pope, A. Müller, *Angew. Chem.* 1991, **103**, 56; *Angew. Chem. Int. Ed. Engl.* 1991, **30**, 34; A. Müller, R. Rohlfing, E. Krickemeyer, H. Bögge, *Angew. Chem.* 1993, **105**, 916; *Angew. Chem. Int. Ed. Engl.* 1993, **32**, 909; M. I. Khan, A. Müller, S. Dillinger, H. Bögge, Q. Chen, J. Zubieta, *Angew. Chem.* 1993, **105**, 1811; *Angew. Chem. Int. Ed. Engl.* 1993, **32**, 1780; K. Worm, F. P. Schmidtchen, A. Schier, A. Schäfer, M. Hesse, *Angew. Chem.* 1994, **106**, 360; *Angew. Chem. Int. Ed. Engl.* 1994, **33**, 327; Y.-D. Chang, J. Salta, J. Zubieta, *Angew. Chem.* 1994, **106**, 347; *Angew. Chem. Int. Ed. Engl.* 1994, **33**, 325.

86. R. W. Saalfrank, A. Stark, K. Peters, H. G. von Schnering, *Angew. Chem.* 1988, **100**, 878; *Angew. Chem. Int. Ed. Engl.* 1988, **27**, 851.
87. R. W. Saalfrank, A. Stark, R. Burak, *Inorg. Synth.* 1992, **29**, 275.
88. R. W. Saalfrank, A. Stark, M. Bremer, H.-U. Hummel, *Angew. Chem.* 1990, **102**, 292; *Angew. Chem. Int. Ed. Engl.* 1990, **29**, 311.
89. R. W. Saalfrank, R. Burak, S. Reihs, N. Löw, F. Hampel, H.-D. Stachel, J. Lentmaier, K. Peters, E.-M. Peters, H. G. von Schnering, *Angew. Chem.* 1995, **107**, 1085; *Angew. Chem. Int. Ed. Engl.* 1995, **34**, 993.
90. R. W. Saalfrank, B. Hörner, D. Stalke, J. Salbeck, *Angew. Chem.* 1993, **105**, 1223; *Angew. Chem. Int. Ed. Engl.* 1993, **32**, 1179; R. W. Saalfrank, R. Harbig, J. Nachtrab, W. Bauer, K.-P. Zeller, D. Stalke, M. Teichert, *Chem. Eur. J.* 1996, **2**, 1363.
91. T. Beissel, R. E. Powers, K. N. Raymond, *Angew. Chem.* 1996, **108**, 1166; *Angew. Chem. Int. Ed. Engl.* 1996, **35**, 1084.
92. E. J. Enemark, T. D. P. Stack, *Angew. Chem.* 1995, **107**, 1082; *Angew. Chem. Int. Ed. Engl.* 1995, **34**, 996.
93. E. J. Enemark, T. D. P. Stack, *Angew. Chem.* 1998, **110**, 977; *Angew. Chem. Int. Ed. Engl.* 1998, **37**, 932.
94. D. L. Caulder, R. E. Powers, T. N. Parac, K. N. Raymond, *Angew. Chem.* 1998, **110**, 1940; *Angew. Chem. Int. Ed. Engl.* 1998, **37**, 000, 1840.
95. R. W. Saalfrank, R. Burak, A. Breit, D. Stalke, R. Herbst-Irmer, J. Daub, M. Porsch, E. Bill, M. Müther, A. X. Trautwein, *Angew. Chem.* 1994, **106**, 1697; *Angew. Chem. Int. Ed. Engl.* 1994, **33**, 1621.
96. M. C. T. Fyfe, P. T. Glink, S. Menzer, J. F. Stoddart, A. J. P. White, D. J. Williams, *Angew. Chem.* 1997, **109**, 2158; *Angew. Chem. Int. Ed. Engl.* 1997, **36**, 2068.
97. J. S. Fleming, K. L. V. Mann, C.-A. Carraz, J. C. Jeffery, E. Psillakis, J. A. McCleverty, M. D. Ward, *Angew. Chem.* 1998, **110**, 1315; *Angew. Chem. Int. Ed. Engl.* 1998, **37**, 1279.
98. S. Mann, G. Huttner, L. Zsolnai, K. Heinze, *Angew. Chem.* 1996, **108**, 2983; *Angew. Chem. Int. Ed. Engl.* 1996, **35**, 2808.
99. M. Fujita, D. Oguro, M. Miyazawa, H. Oka, K. Yamaguchi, K. Ogura, *Nature* 1995, **378**, 469; P. J. Stang, B. Olenyuk, D. C. Muddiman, R. D. Smith, *Organometallics* 1997, **16**, 3094.
100. M. Nakazaki, *Top. Stereochem.* 1984, **15**, 199.
101. A. J. Amoroso, J. C. Jeffery, P. L. Jones, J. A. McCleverty, P. Thornton, M. D. Ward, *Angew. Chem.* 1995, **107**, 1577; *Angew. Chem. Int. Ed. Engl.* 1995, **34**, 1443; C. Brückner, R. E. Powers, K. N. Raymond, *Angew. Chem.* 1998, **110**, 1937; *Angew. Chem. Int. Ed. Engl.* 1998, **37**, 1837.
102. R. Vilar, D. M. P. Mingos, A. J. P. White, D. J. Williams, *Angew. Chem.* 1998, **110**, 1323; *Angew. Chem. Int. Ed. Engl.* 1998, **37**, 1258.

Chapter 2

Bistability in Iron(II) Spin-Crossover Systems: A Supramolecular Function

JOSÉ ANTONIO REAL

Universitat de València, Spain

INTRODUCTION

The idea that a single molecule or an assembly of molecules might function as an electronic device has been of interest since long ago [1–10]. In this regard, the molecular switch is one of the current approaches in molecular electronics. Bistability for the binary change of state in a molecular system is the prerequisite for an operational switch, according to Haddon and Lamola [11]. Furthermore, three additional essential conditions have to be fulfilled in order to get an operational molecular switch: (i) the switching must be controllable, (ii) the state of the switch must be readable; and (iii) the first two conditions must be executable at the molecular level. Kahn and Launay [12] defined molecular bistability as the property of a molecular system to evolve from one stable state to another stable state in a reversible and detectable fashion in response to an appropriate perturbation.

Spin-crossover phenomenon represents probably one of the best examples of molecular and supramolecular bistability. This phenomenon is observed both in solution as well as in the solid state. In the first case, the process is essentially molecular, owing to the isolation of molecules. In the solid state, the situation is in general quite different and involves cooperative character for the phenomenon.

Transition Metals in Supramolecular Chemistry, edited by J. P. Sauvage.

Cooperativity is one of the most appealing and elusive facets of the spin-crossover phenomenon. It is a main aspect because discontinuity in the magnetic and optical properties along with thermal hysteresis confer to these systems potential memory effect. Nevertheless, because most of the spin-crossover systems are discrete in nature, cooperativity stems from assemblies of molecules held together by noncovalent interactions and, consequently, difficult to control.

The cooperative mechanism of the spin-crossover phenomenon is well understood in some aspects. For instance, theoretical models account for the thermal evolution of the system. However, due to their phenomenologic character, the key parameters accounting for intermolecular interactions do not reflect the relevance of the microscopic details which could orientate us in designing molecules that can recognize each other and combine to produce systems with prescribed characteristics. Consequently, very general principles guide the synthetic chemistry in the search for suitable spin-crossover compounds; often trial and error or serendipity are the sole strategies.

The last three decades have witnessed the conceptual progression of supramolecular chemistry from simple recognition processes to the synthesis of complex interlocked and intertwined structures [13–18]. The methods discovered by supramolecular chemists to assemble units in a perfectly controlled manner may be of utmost importance for better understanding of the cooperative mechanism of the spin-crossover phenomenon. On the one hand, the understanding and utilization of intermolecular interactions in the context of crystal packing should allow us to synthesize solids with desired spin-crossover regimes. On the other hand, coordination chemistry of metal ions has provided a powerful tool to build a great variety of many novel molecular and supramolecular architectures. That is the control of self-assembly based on the use of the coordinating preferences of the metal ions (coordination number and geometry) as well as in the choice of polydentate ligands (type, number and distribution of donor atoms). That should allow us to design polynuclear spin-crossover systems aiming at exploring cooperativity.

We are concerned with the synthetic approaches which could clearly identify the parameters controlling the spin-crossover characteristics (the sharpness of the thermal spin conversion, the temperature range width of the hysteresis as well as the value of the critical temperature $T_{1/2}$). The ultimate goal is to tune the synthesis in order to provide compounds with the proper value of those parameters for both theoretical understanding and exploration of applications.

The present chapter is organized in three sections: the first section is devoted to basic theoretical background concerning the spin-crossover phenomenon, viz ligand field theory, thermodynamics and cooperativity; the second section reports on some examples that we feel are particularly relevant to illustrate the main directions in which research on synthetic aspects of cooperativity is being directed; finally, the third section describes three approaches, up to now reported, concerning spin bistability in supramolecular and molecular systems and memory effect.

1 THEORETICAL BACKGROUND

1.1 Spin-Crossover and Ligand Field Theory

Octahedral first-row transition metal complexes having d^4–d^7 configurations can exist in two different ground states with different spin multiplicities depending on the magnitude of the ligand field strength Δ. If Δ is greater than the mean spin-pairing energy, P, the d electrons tend to occupy the orbitals of lower energy: the metal ion is in its low-spin state (LS). For $\Delta < P$ the d electrons obey the first Hund rule and the metal ion is in its high-spin state (HS). Figure 1 show both types of configurations and a simplified Tanabe–Sugano diagram for an octahedral $3d^6$ complex. When Δ and P have the same order of magnitude, which happens for Δ values in the vicinity of the crossing point Δ_c, the two spin states may interconvert by the action of external constraints such as temperature, pressure and hence, light radiation. This phenomenon is called spin transition or spin equilibrium, the former expression being more appropriate for abrupt interconvertions and the latter for gradual ones. In general the phenomenon is termed spin-crossover.

Fe^{2+} ($3d^6$), Fe^{3+} ($3d^5$) and Co^{2+} ($3d^7$) are the most common transition-metal ions for which spin-state interconversion have been observed, a minor role being played by Co^{3+} ($3d^6$), Mn^{3+} ($3d^4$) and Cr^{2+} ($3d^4$). We shall concentrate on six-coordinate iron(II) spin-crossover compounds because they are by far the more investigated ones [19–22].

For these systems ${}^1A_{1g} \leftrightarrow {}^5T_{2g}$ spin conversion leads to a drastic change of the relevant physical properties. An abrupt change of color accompanies the spin transition in absence of allowed metal-to-ligand charge-transfer bonds hiding the much less intense parity-forbidden d–d bands. This pronounced thermochromism

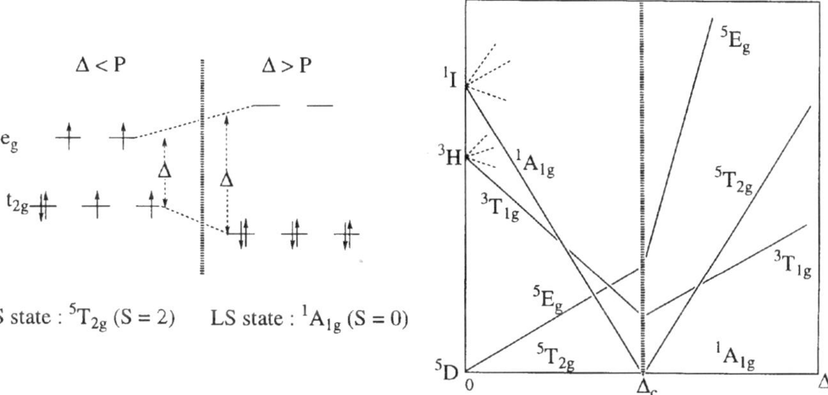

Figure 1 HS and LS configurations (left) and simplified Tanabe–Sugano diagram (right) of a $3d^6$ system.

between purple–red and white colors arise from the $^1A_{1g} \rightarrow {}^1T_{1g}$ and $^5T_{2g} \rightarrow {}^5E_g$ d–d absorptions taking place in the regions near 19 000 cm^{-1} and 12 500 cm^{-1}, respectively. This is illustrated in Figure 2 where the absorption spectra of $[Fe(ptz)_6](BF_4)_2$ at 290 K (HS) and 45 K (LS), are depicted [23].

On the other hand, the magnetic properties of iron(II) spin-crossover compounds change reversibly from diamagnetic, $^1A_{1g}$ (LS), to paramagnetic, $^5T_{2g}$ (HS). Variable temperature molar magnetic susceptibility measurements, χ_M, is the simplest way to follow the spin change. The $(\chi_M T)_{LS}$ product (T is the absolute temperature) for the LS state is zero while $(\chi_M T)_{HS}$ is found in the 3–3.6 $cm^3\ mol^{-1}\ K$ range for the HS state depending on the orbital contribution to the magnetic susceptibility. As an example, Figure 3 shows the temperature dependence of the $\chi_M T$ product for $[Fe(phen)_2(NCS)_2]$ (phen = 1,10-phenanthroline) [24]. In many instances, the molar fraction of HS molecules, n_{HS}, may be straightforwardly deduced from the thermal variation of $\chi_M T$ according to:

$$n_{HS} = [\chi_M T - (\chi_M T)_{LS}]/[(\chi_M T)_{HS} - (\chi_M T)_{LS}]$$

Alternatively, ^{57}Fe-Mössbauer spectroscopy is an extremely powerful technique for spin-transition studies on iron(II) compounds. The different iron(II) spin states are unambiguously identified on the basis of the hyperfine interaction parameters allowing us to evaluate directly n_{HS}.

At molecular level the spin-crossover phenomenon can be considered like an intraionic electron transfer between the e_g and the t_{2g} orbitals. The HS → LS spin-state change thus invariably implies a change in the population of the σ-antibonding e_g orbitals, which are directed toward the ligands. Along with this, an opposite change in the occupancy of the t_{2g} orbitals occurs affecting the electron back-donation between the metal ion and the vacant π* orbitals of the ligands. Both factors contribute to a change of the metal-ligand bond length. Thus, the spin-

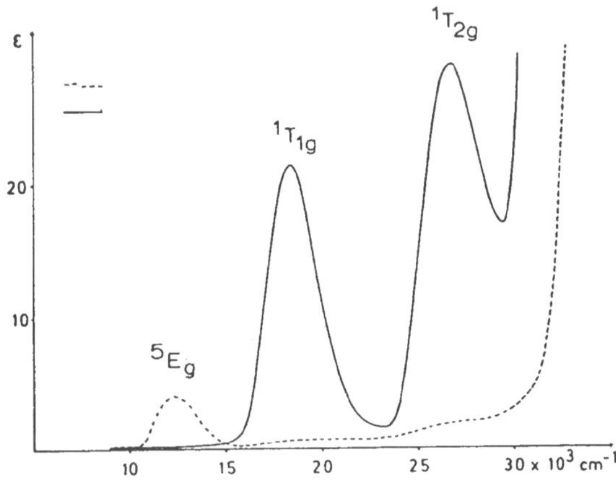

Figure 2 Absorption spectra in the region of d–d absorption of the HS(···, T = 290 K) and LS (——, T = 45 K) $[Fe(ptz)_6](BF_4)_2$ forms. Reproduced with permission from ref. 23.

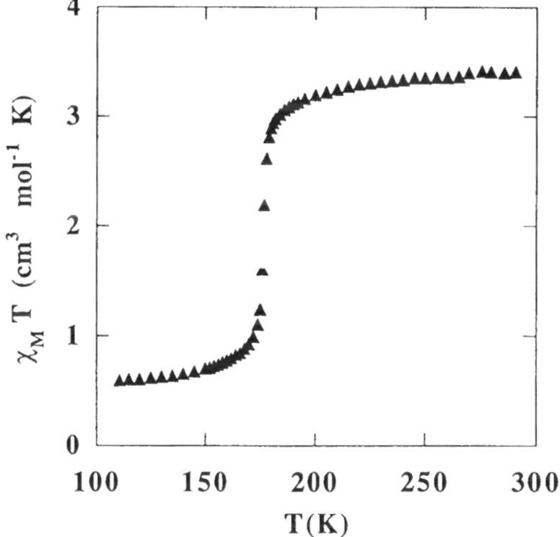

Figure 3 Temperature dependence of $\chi_M T$ for a polycrystalline sample of $[Fe(phen)_2(NCS)_2]$. Reproduced with permission from ref. 24.

crossover phenomenon is always accompanied by structural modifications of the molecule. Figure 4 shows two projections of the asymmetric unit of $[Fe(phen)_2(NCS)_2]$ at 293 K and 130 K. The figure depicts the modifications in the intramolecular geometry when passing from the HS to the LS form, i.e. the

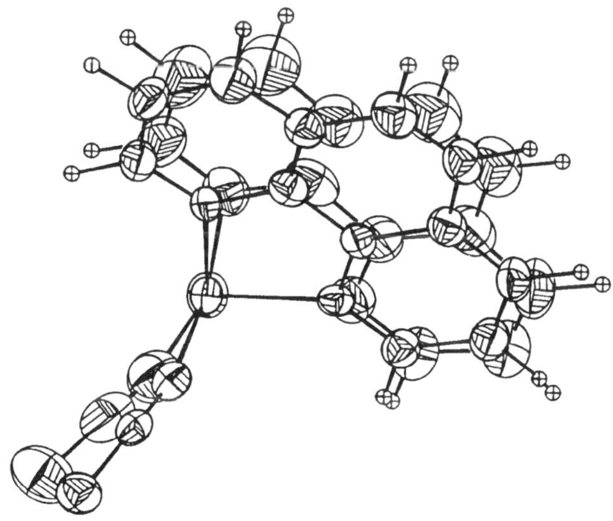

Figure 4 Schematic drawing of both HS (T = 293 K, lower) and LS (T = 130 K, upper) forms of the $[Fe(phen)_2(NCS)_2]$ asymmetric unit. Reproduced with permission from ref. 24.

shortening of Fe–N distances and the remarkable variation of N–Fe–N angles (*vide infra*). Typical values for the mean metal-ligand distances, R, in iron(II) spin-crossover complexes are close to 2 Å and 2.2 Å for R_{LS} and R_{HS}, respectively. In ligand field theory, Δ depends inversely on the fifth power of the metal-ligand distance R [25]. Consequently, each molecule experiences a drastic change of Δ upon spin conversion which is estimated to be $\Delta_{LS} \approx 1.6\Delta_{HS}$.

It is instructive to consider the energy of each complex, in each spin state, to be characterized by a simplified one-dimensional potential-energy diagram (Figure 5). The potential energy of the complex is plotted as a function of an internal coordinate. The symmetric metal-ligand stretching mode is the most relevant one involved in the spin-state interconversion. It is worth noting that the crossing of the two parabols corresponds to the crossing point Δ_c in Figure 1. This singular point represent an instable region where the transient species change geometry.

1.2 Thermodynamics for an Assembly of Non-Interacting Spin-Crossover Molecules

If we consider the Avogadro constant (N_A) of molecules, each individual molecule may exist in the HS state or in the LS state according to the transformation (1)

$$\text{LS} \rightleftharpoons \text{HS} \tag{1}$$

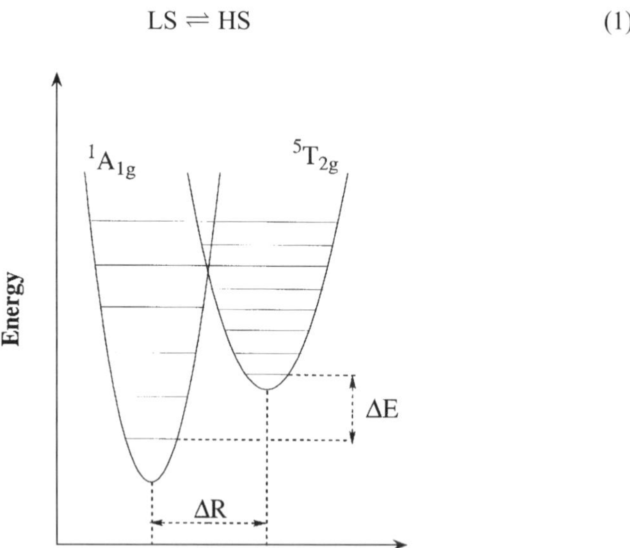

Figure 5 Schematic representation of the potential wells for the $^1A_{1g}$ and the $^5T_{2g}$ states of an iron(II) spin-crossover complex.

The relevant energy variation is the Gibbs free energy G. At constant pressure the free Gibbs energy per mole of compound may be expressed as

$$G = n_{HS}G_{HS} + (1 - n_{HS})G_{LS} + \Gamma(n_{HS}) - TS_{mix} \tag{2}$$

In equation (2) G_{HS} and G_{LS} are the standard Gibbs free energies in the absence of any interaction for N_A molecules in the HS and LS states, respectively. $\Gamma(n_{HS})$ is an interaction term which reflects the departure of the system from an ideal solution. S_{mix} is the mixing entropy. For a regular solution of molecules, S_{mix}, is determined by

$$S_{mix} = k[N_A \ln N_A - n_{HS}N_A \ln n_{HS}N_A - (1 - n_{HS})N_A \ln(1 - n_{HS})N_A] \tag{3}$$

Introducing the thermodynamical equilibrium condition in equation (2)

$$(\partial G/\partial n_{HS})_{T,P} = 0 \tag{4}$$

produces the relation

$$\ln[n_{HS}/1 - n_{HS}] = -[\Delta G + \partial\Gamma(n_{HS})/\partial n_{HS}]/RT \tag{5}$$

where $\Delta G = G_{HS} - G_{LS}$. According to whether ΔG is negative or positive the most stable phase is HS or LS, respectively. The singular temperature $T_{1/2}$ for which there is the same amount of LS and HS molecules, ($n_{HS} = n_{LS}$), is defined by $\Delta G = 0$, that is

$$T_{1/2} = \Delta H/\Delta S \tag{6}$$

In the limiting case of zero interactions between molecules the temperature dependence of n_{HS} is expresed as follows

$$n_{HS} = 1/(1 + \exp[\Delta G/RT]) = 1/(1 + \exp[\Delta H/RT - \Delta S/R]) \tag{7}$$

Figure 6 shows an example of the temperature dependence of n_{HS}, where $\Gamma(n_{HS}) = 0$ and ΔH and ΔS have been taken equal to 8.3 kJ mol^{-1} and 55 J K^{-1} mol^{-1} ($T_{1/2} = 160$ K), respectively. All the molecules are LS at low temperatures. On the other hand at high temperature, the transformation is incomplete (as expected for a Gibbs–Boltzmann law), the high-spin molar fraction tending to $[1 + \exp(-\Delta H/RT_c)]^{-1}$ instead of to unity. Furthermore, the transformation is very smooth taking place on a large temperature range. Such smooth transitions are usually observed in solution state but are uncommon in the solid state.

The entropy variation ΔS may be written as the sum of electronic ΔS_{el} and vibrational, ΔS_{vib}, contributions,

$$\Delta S = \Delta S_{el} + \Delta S_{vib} \tag{8}$$

ΔS_{el} is related to the orbital and spin degeneracies of the two electronic levels. Generally, the orbital contribution to ΔS_{el} is expected to be small due to the commonly low symmetry of the coordination core, consequently

$$\Delta S_{el} \approx \Delta S_{el,spin} = N_A k \ln[(2S + 1)_{HS}/(2S + 1)_{LS}] \tag{9}$$

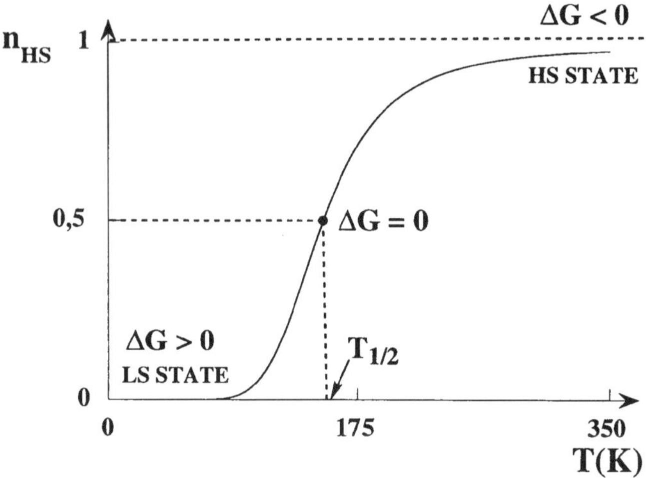

Figure 6 Thermal dependence of the HS molar fraction, n_{HS}, for an assembly of non-interacting spin-crossover molecules.

where $(2S+1)_{HS}$ and $(2S+1)_{LS}$ are the spin multiplicities of the HS and LS spin states, respectively. ΔS_{vib} includes the contributions arising from the changes in molecular and lattice vibrational modes associated with the spin conversion, $\Delta S_{vib}{}^{mol}$ and $\Delta S_{vib}{}^{latt}$, respectively. The main contibutions to the total entropy variation, $\Delta S_{el,spin}$ and $\Delta S_{vib}{}^{mol}$, are always positive. For instance, $\Delta S_{el,spin} = N_A k \ln(5/1) = 13.45\ J\ K^{-1}\ mol^{-1}$ for an $S = 0 \leftrightarrow S = 2$ spin conversion. $\Delta S_{vib}{}^{mol}$ is also positive because the mean metal-ligand distance is larger for the HS molecules and, consequently the disorder is more pronounced in the HS state than in the LS state. Hence, it follows that $\Delta H = N_A(\Delta E)$ is positive and consequently the minimun of the LS potential energy curve must be slightly lower than the minimum of the HS potential energy curve, as shown in Figure 5. So, the entropy gain $T\Delta S$ compensates this difference in energy. In summary, thermal spin-crossover phenomenon is an entropy-driven process.

Pressure-induced spin conversion at constant temperature can be understood by considering:

$$\Delta H = \Delta H_{HS} - \Delta H_{LS} = \Delta E + P\Delta V \qquad (10)$$

The internal energy change, $\Delta E = E_{HS} - E_{LS}$, should vary little since the molecules are expected not to be significantly affected by the moderate pressures required to induce the spin conversion. The term $P\Delta V$ is always positive since $V_{HS} - V_{LS} > 0$. At low applied pressures, $P\Delta V$ is comparatively small with regard ΔH and $\Delta H \approx \Delta E$. However, as pressure increases, higher values of ΔH and hence a more pronounced stabilization of the LS state is achieved.

1.3 Cooperative Effect: Interacting Molecules

The origin of the spin-crossover phenomenon is molecular but its manifestation depends also on intermolecular interactions. The change of metal-ligand bond lengths which accompanies a spin-state transition invariably produces a change of the molecular volume of the complex which spreads in the whole crystal by means of intermolecular interactions. Therefore, the molecular process described above only applies actually to systems in which molecules do not significantly interact, i.e. diluted systems. The existence of more or less strong interactions between molecules results in a more or less pronounced cooperative effect. The greater the cooperative effect, the more abrupt is the n_{HS} versus T curve. Eventually, thermal hysteresis is observed for a critical set of parameters.

A simple thermodynamical model derived from the theory of regular solutions [26], can account for the main features of both continuous and discontinuous transitions. In this theoretical framework, so-called Slitcher–Drickamer model, the interaction term in equation (2), $\Gamma(n_{HS})$, is defined as $\gamma n_{HS}(1 - n_{HS})$ where γ is an interaction parameter which reflects the tendency for molecules of one type to be surrounded by like molecules ($\gamma > 0$). So, equation (5) becomes:

$$\ln[n_{HS}/(1 - n_{HS})] = [\Delta H + \gamma(1 - 2n_{HS})]/RT - \Delta S/R \tag{11}$$

Figure 7 displays the $G = \mathrm{f}(n_{HS})$ curves, for several temperatures close to $T_{1/2}$, for $\Delta H = 9$ kJ mol^{-1}, $\Delta S = 60$ J K^{-1} mol^{-1} and $\gamma = 3.6$ kJ mol^{-1}. Within the temperature range considered, G exhibits two minima, one for n_{HS} close to zero and the other one for n_{HS} close to one. These two minima correspond to the LS and the HS phases, respectively. For $T < T_c$, the former represents a stable state and the latter a metastable state. The reverse situation holds for $T > T_c$. When $T = T_c$, the two minima are symmetrically situated with respect to $n_{HS} = 0.5$. Hysteresis may be observed (see Figure 8) when the energy barrier represented by the interaction term surpasses a critical value, namely when $\gamma > 2RT_c$. In Figure 8 the calculated points between the two vertical tangents have no physical meaning. In order to get a reasonable estimation of the thermodynamical parameters the two vertical tangents must cross the experimental hysteresis branches at the inflexion point ($T_{1/2}\downarrow$, $T_{1/2}\uparrow$).

It is important to note that Slitcher and Drickamer's model does not provide any information on the mechanism leading to the cooperativity term $\gamma(n_{HS})$. In order to give microscopic physical meaning to the interaction term, several approaches have been proposed [27–32]. The so-called "lattice expansion and elastic interaction model" appears as the most successful to interpret the available data [27–29]. In this approach the difference in volume between HS and LS molecules is assumed to cause long-range elastic deformations, leading to the cooperative spin-transition mechanism.

An experimental support for the existence of long-range intermolecular interactions responsible for cooperativity consists of moving the spin-crossover molecules

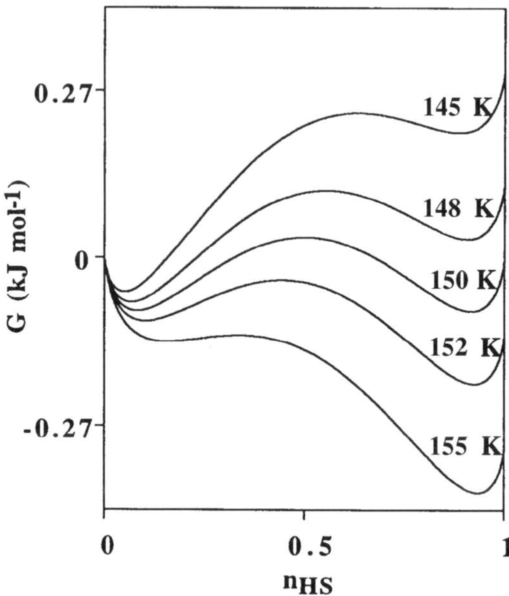

Figure 7 HS molar fraction dependence of the Gibbs free energy at various temperatures for an assembly of interacting molecules ($\Delta H = 9$ kJ mol^{-1}, $\Delta S = 60$ J K^{-1} mol^{-1} and $\gamma = 3.6$ kJ mol^{-1}).

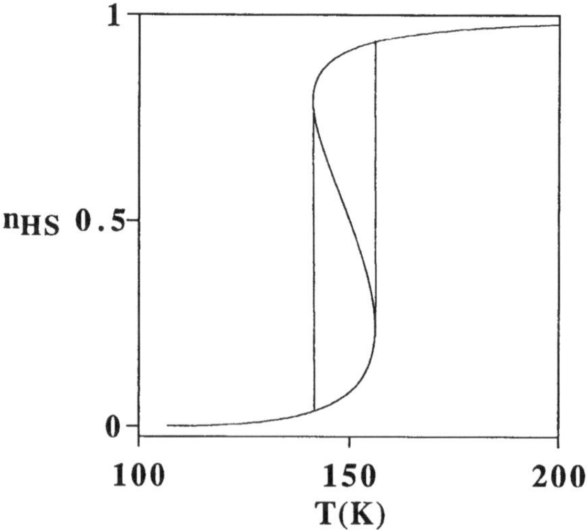

Figure 8 Thermal dependence of the HS molar fraction, n_{HS}, for an assembly of spin-crossover molecules ($\Delta H = 9$ kJ mol^{-1}, $\Delta S = 60$ J K^{-1} mol^{-1} and $\gamma = 3.6$ kJ mol^{-1}).

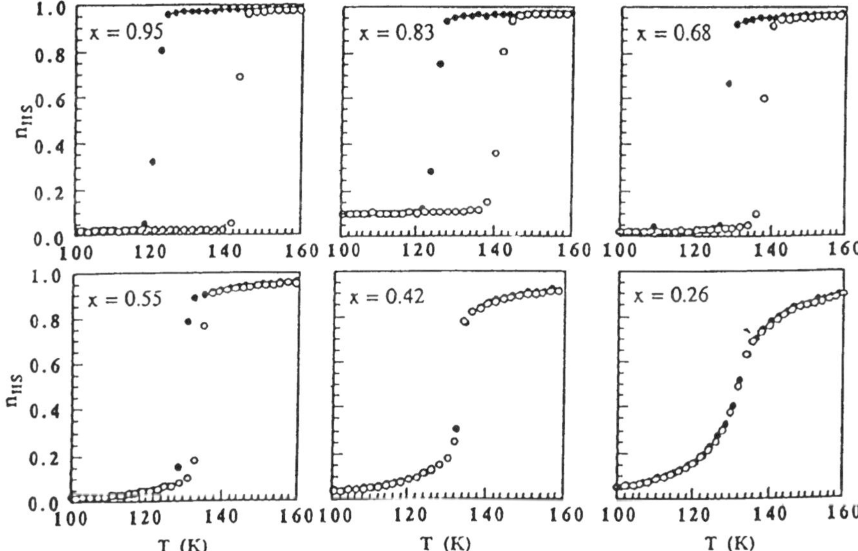

Figure 9 Thermal dependence of the HS molar fraction, n_{HS}, for $[Fe_xNi_{(1-x)}(btrz)_2(NCS)_2]\cdot H_2O$ obtained from magnetic susceptibility measurements in the cooling (●) and heating (○) modes. Reproduced with permission from ref. 45.

away each other by decreasing iron(II) concentration in $[Fe_xM_{(1-x)}]$ solid state solutions [33–44]. That is nicely illustrated by Figure 9 where the temperature dependence of n_{HS} both in the cooling and in the heating modes is shown for $[Fe_xNi_{(1-x)}(btr)_2(NCS)_2]\cdot H_2O$ mixed compounds (btr = 4,4′-bis(1,2,4-triazole)) (*vide infra*). When x decreases the n_{HS} versus T curves become comparatively more gradual and the hysteresis width gets narrower and cancels out for $x \approx 0.45$ [45].

2 SELECTED EXAMPLES OF IRON(II) SPIN-CROSSOVER COMPOUNDS

2.1 Cooperative Mechanism in $[Fe(Phen)_2(NCS)_2]$ and $[Fe(btz)_2(NCS)_2]$

Most of the spin-crossover compounds are mononuclear. In these systems the cooperative mechanism depends on the coupling between individual molecules held together by intermolecular forces. The rational control of these forces is not an obvious task and becomes more complicated when non-coordinate counterions and/or solvent molecules are included in the lattice. Much effort has been done in order to establish reliable connections between molecular and supramolecular structures on the basis of intermolecular interactions [18]. Nevertheless, in spite of this, to predict whether a specific crystal packing will lead to a specific cooperativity is still beyond our possibilities.

A way to get insights into the cooperative mechanism is to closely follow the molecular and crystalline changes which could play a relevant role in the cooperative mechanism. Multi-temperature X-ray structure results clearly establish the volume expansion within the course of the LS $\rightarrow$ HS spin conversion. The variation of metal-ligand bond length ΔR appears to be the most significant modification taking place within individual molecules. Thus, the volume difference between LS and HS species depends to a large extent on ΔR. The observed crystal volume is in fact composed of fractional contributions from the unit cell volumes of the HS and LS species and a linear volume change with temperature. The variation of the unit cell volume $V(T)$ may be reproduced by the following expression [46]

$$V(T) = V_{\mathrm{LS}}[1 + \alpha_{\mathrm{V}} T + \varepsilon n_{\mathrm{HS}}] \tag{12}$$

Here, α_V is the thermal expansion coefficient which is assumed to be equal in both lattices and ε is the dilatation coefficient $(V_{\mathrm{HS}} - V_{\mathrm{LS}})/V_{\mathrm{LS}}$, where V_{LS} and V_{HS} are the unit cell volumes of the pure LS and HS species at 0 K, respectively. In order to account for the anisotropy of the lattice, the thermal expansion α_V and dilatation ε coefficients must be introduced as tensors instead of scalars. Similarly, an equivalent expression could be defined for pressure-induced spin conversions.

Multi-temperature X-ray diffraction data for a series of spin-crossover complexes differing in cooperativity indicates that the molecule and crystal volume variations upon spin conversion are similar in all the cases irrespective of the cooperative nature of the spin conversion [47]. So, a systematic structural analysis of specifically designed spin-crossover compounds should be of utmost importance to establish correlations between intermolecular interactions and cooperativity. The comparative structural study of $[Fe(phen)_2(NCS)_2]$ and $[Fe(btz)_2(NCS)_2]$ where btz = 2,2′-bi-4,5-dihydrothiazine (Figure 10) represents the sole example so far reported oriented in this direction [48,49]. It illustrates the dependence of the nature of the phenomenon on the efficiency of the intermolecular contacts in transmiting the intramolecular reorganization upon spin conversion.

$[Fe(phen)_2(NCS)_2]$ has been the most investigated since it is considered as a model compound among iron(II) spin-transition complexes. It undergoes an abrupt LS $\leftrightarrow$ HS transition at a temperature $T_{1/2} \approx 176$ K [50–65]. The transition has also

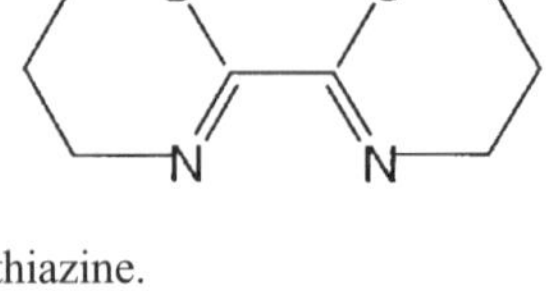

Figure 10 2,2′-Bi-4,5-dihydrothiazine.

been induced by pressure [66–68] and by light radiation [70]. Less attention, on the other hand, has been devoted to [Fe(btz)$_2$(NCS)$_2$] probably because it undergoes a very smooth LS ↔ HS conversion. This is centered around $T_{1/2} \approx 225$ K [71].

The opportunity of such comparative investigation appeared at the first stages of their structure determination: [Fe(phen)$_2$(NCS)$_2$] and [Fe(btz)$_2$(NCS)$_2$] have the same molecular arrangement and are isostructural, with similar lattice parameters and molecular packing. This similitude in their basic structural properties clearly contrast with respect to the difference in their magnetic behaviors (Figure 11) and more precisely in their cooperativities.

Multi-temperature and multi-pressure single-crystal X-ray diffraction studies have been carried out for [Fe(phen)$_2$(NCS)$_2$] and [Fe(btz)$_2$(NCS)$_2$] [48,49]. Let us describe their crystal structures and structural modifications under constraint. For both compounds, no change of symmetry was evidenced over the temperature range 300–130 K or the pressure range 0–1.30 GPa, the space group remaining orthorhombic Pbcn and the molecular packings being very similar whatever the temperature or pressure may be. Figure 12(a) and (b) show the molecular units [Fe(phen)$_2$(NCS)$_2$] and [Fe(btz)$_2$(NCS)$_2$] and the projections of crystal structures along their *a* axis, respectively. Each metal ion is surrounded by six nitrogen atoms belonging to two NCS$^-$ groups in *cis*-position and two phen (btz) ligands. In both cases the metal atom is located on a two-fold axis: two ligands of the same nature belonging to the same molecular unit deduce from one another by this axis.

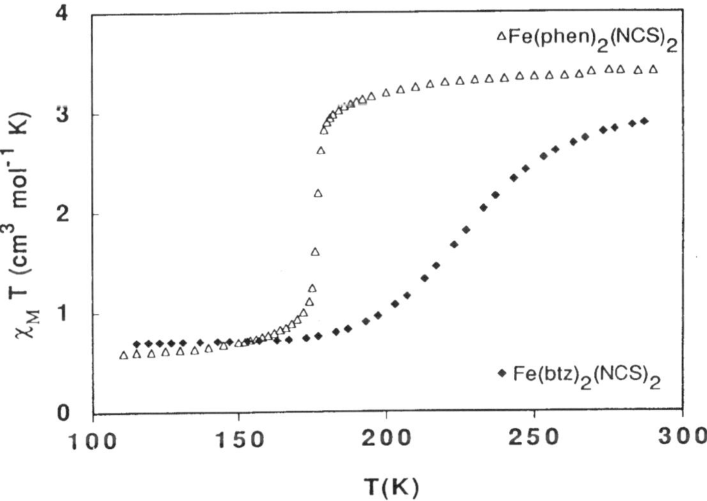

Figure 11 Comparison of the temperature dependence of χ_MT for polycrystalline samples of [Fe(phen)$_2$(NCS)$_2$] (△) and [Fe(btz)$_2$(NCS)$_2$] (◆). Reproduced with permission from ref. 48.

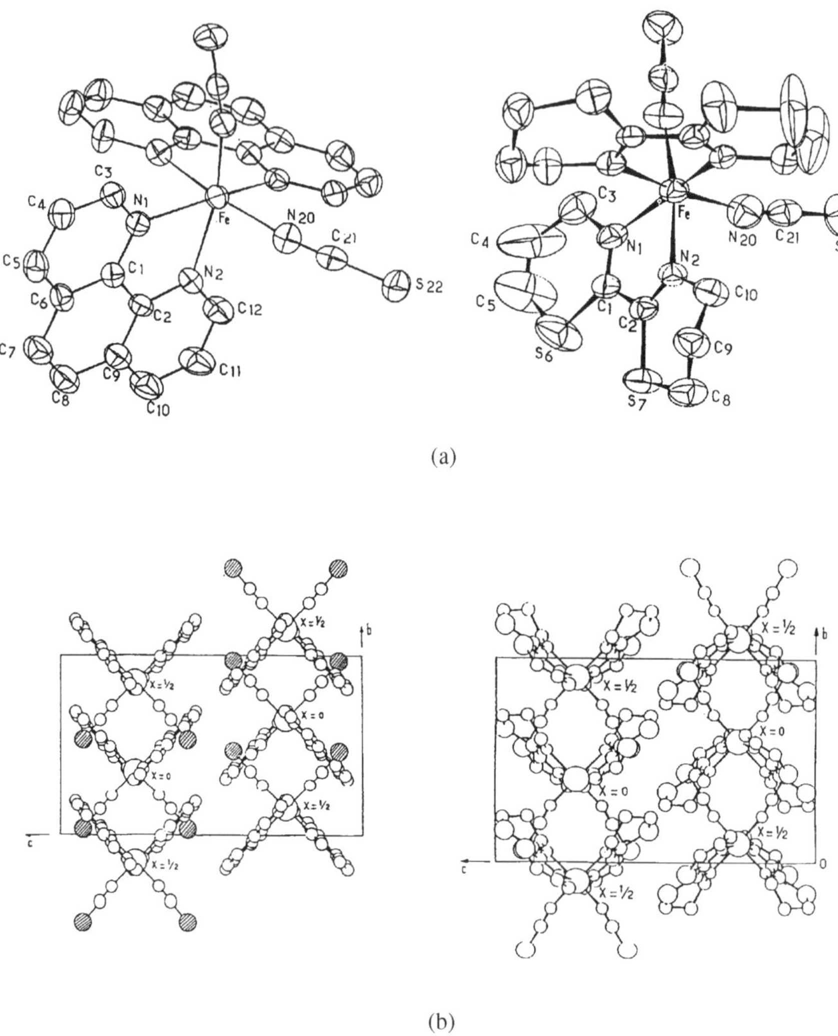

Figure 12 (a) Drawings of the $[Fe(phen)_2(NCS)_2]$ (left) and $[Fe(btz)_2(NCS)_2]$ (right) units showing the 50% probability ellipsoids. Hydrogen atoms have been omitted for clarity. (b) Projection along the *a* axis of the crystal structures of $[Fe(phen)_2(NCS)_2]$ (left) and $[Fe(btz)_2(NCS)_2]$ (right). Reproduced with permission from ref. 48.

In the HS state (300 K, ambient pressure) both complexes exhibit a strong distortion of the $[FeN_6]$ octahedron. The change of spin state induces noticeable shortening of the Fe–N bonds, which results in close values of the average Fe–L bond distance variations: $\Delta R = 0.170$ Å and 0.164 Å for $[Fe(phen)_2(NCS)_2]$ and $[Fe(btz)_2(NCS)_2]$, respectively. Important N–Fe–N bond angles variations are also observed. These modifications lead to a much more regular shape of the $[FeN_6]$ core of each compound. The structure determinations performed at a pressure of 1.0 GPa yield very similar $[FeN_6]$ octahedron characteristics for both compounds, characteristics which are very close to that observed at low temperature.

In both compounds the molecular packing can be similarly described as sheets of molecules parallel to the (a, b) planes, which stack along the c axis. However, the intermolecular interactions and their modification upon the spin change are found to widely differ. The intermolecular contacts are classified into two kinds: those that occur within (a, b) sheets (intrasheets contacts) and those that concern molecular units which belong to different (a, b) sheets (intersheets contacts). This classification allows us to point out that $[Fe(phen)_2(NCS)_2]$ has predominant contacts inside the (a, b) sheets, while $[Fe(btz)_2(NCS)_2]$ exhibits many more intersheet contacts. This difference is reinforced at low temperature or high pressure.

The variation of the lattice parameters a, b, c, and $V(T)$ as a function of temperature is shown in Figure 13. The cell parameters for $[Fe(btz)_2(NCS)_2]$ evolve almost continuously in the whole temperature range. The values of the linear thermal expansion coefficients along the cell axes are listed in Table 1. The relative variations of these parameters lead to the following anisotropy ratios (defined as $3\alpha_n/\Sigma\alpha_n$): 1.06: 0.98: 0.96 at 293 K and 0.83: 1.15: 1.03 at 130 K. These ratios indicate there is no preferential direction along which the structural arrangement is much more affected by temperature. The evolution of the a and b parameters for $[Fe(phen)_2(NCS)_2]$ as a function of temperature strongly differs from that observed for $[Fe(btz)_2(NCS)_2]$. It clearly shows a discontinuity in the close vicinity of the spin transition. The variation of the two parameters are of different amplitude and opposite direction. On the other hand, the evolution of the c unit cell parameter follows that observed in $[Fe(btz)_2(NCS)_2]$. In this direction the temperature dependence is continuous over the 130–293 K temperature range. This anisotropic behavior is reflected in the anisotropy ratio deduced from the corresponding α_n values (see Table 1): 0.30: 1.14: 1.57 and 1.16: 0.72: 1.12 at 293 and 130 K, respectively.

The results obtained from the pressure dependence of the lattice parameters agree with those obtained from thermal studies (see Table 1). A significant anisotropy is deduced for the linear compressibility coeficients of $[Fe(phen)_2(NCS)_2]$ whereas $[Fe(btz)_2(NCS)_2]$ exhibits an almost isotropic compressibility. The anisotropy of $[Fe(phen)_2(NCS)_2]$ is slightly reduced at high pressure, however, a remains the stiffest direction while c is the more compressible one.

Qualitatively, there is a correlation between such a lattice anisotropy and the spatial distribution of intermolecular contacts. In the case of $[Fe(phen)_2(NCS)_2]$, the lowest values of linear compressibility, which are associated with the stiffest lattice directions, concern axes a and b: these correspond to intrasheet directions in which many intermolecular contacts are observed; k_c, on the other hand, is larger than k_a and k_b, and corresponds to the intersheet direction for which many fewer contacts are observed. In the case of $[Fe(btz)_2(NCS)_2]$, the proportion of inter- and intrasheet

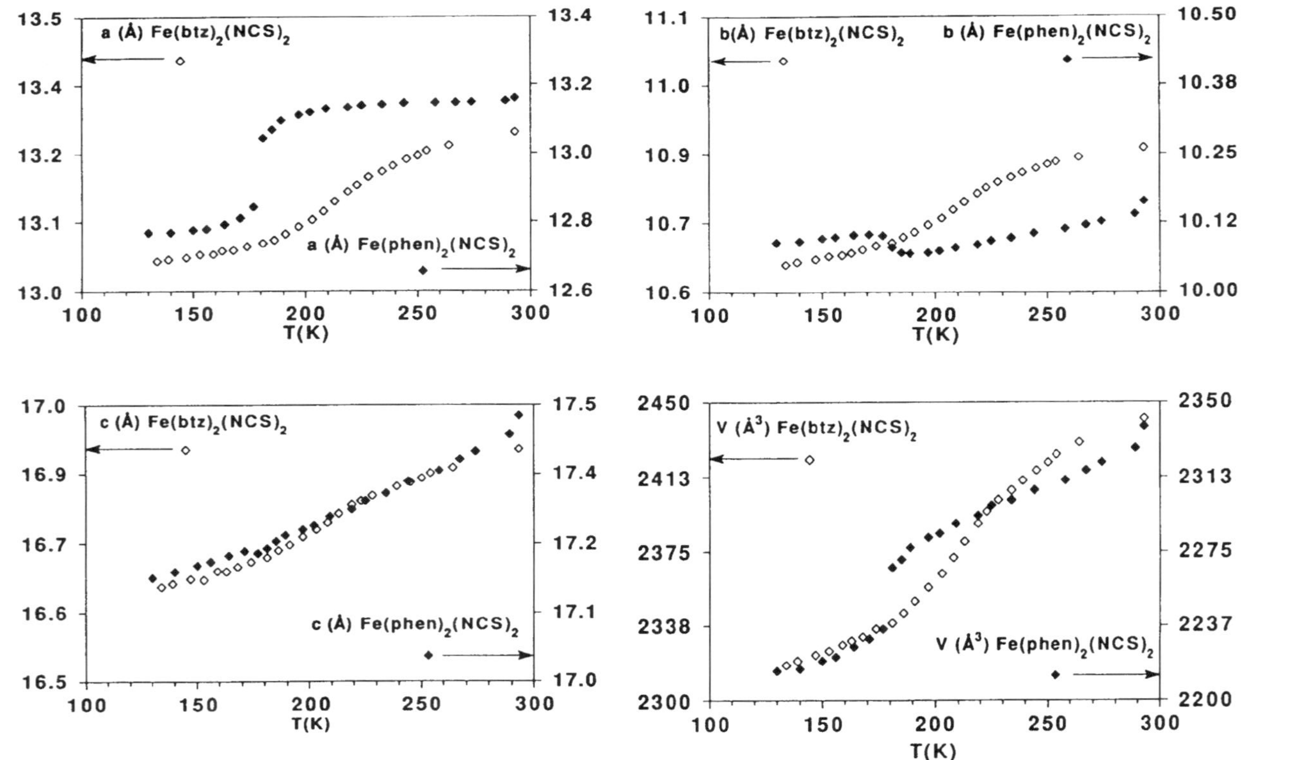

Figure 13 Temperature dependence of the lattice parameters of [Fe(phen)$_2$(NCS)$_2$] (◆) and [Fe(btz)$_2$(NCS)$_2$] (◇). Reproduced with permission from ref. 48.

Table 1 Linear and volumic thermal expansion α_i ($\times 10^4$, K^{-1}) and compressibility k_i ($\times 10^{-1}$, GPa^{-1}) coefficients.[a]

Compound	T(K)	α_a	α_b	α_c	α_V	Pressure	k_a	k_b	k_c	k_V
$[Fe(phen)_2(NCS)_2]$	293	0.22	0.83	1.15	2.20	10^3 HPa	0.21	0.33	0.53	1.07
	130	0.72	0.45	0.70	1.87	1 GPa	0.16	0.28	0.38	0.82
$[Fe(btz)_2(NCS)_2]$	293	0.80	0.74	0.72	2.26	10^3 HPa	0.41	0.43	0.37	1.21
	130	0.47	0.65	0.58	1.70	1 GPa	0.28	0.33	0.28	0.89

[a] $\alpha_i = (1/I)(\partial I/\partial T)_P$; $k_i = (1/I)(\partial I/\partial P)_T$; $(I = a, b, c, V)$.

contacts is almost reversed, and quasi-isotropic lattice with respect to pressure effects is observed. Despite these differences, both compounds exhibit very close values of volumic compressibility at low and high pressure $[Fe(phen)_2(NCS)_2]$ being a little stiffer than $[Fe(btz)_2(NCS)_2]$.

We conclude underlining that $[Fe(phen)_2(NCS)_2]$ and $[Fe(btz)_2(NCS)_2]$ have very similar structural features in both HS and LS states. The variations of the mean metal-ligand bond lengths as well as those of the unit cell volumes associated with the LS $\leftrightarrow$ HS conversion are found to be comparable. So, they are not relevant parameters to account for the difference in cooperativity. Moreover, the transition is neither associated with a crystallographic phase change nor triggered by structural order–disorder transition in the case of $[Fe(phen)_2(NCS)_2]$. Evidence is only provided for a large rearrangement of the iron atom environment without any orientation of the $[FeN_6]$ core. Lattice anisotropy illustrated by the relative values of the linear thermal expansion and compressibility coefficients is well correlated with the number and spatial distribution of the shortest intramolecular distances. It follows that the key factor governing the cooperativity of the process might be, in the present case, the number and strength of the intermolecular contacts. That strongly depends on the nature of the ligands surrounding iron(II). In the present case, π-overlapping between phen ligand fragments, taking place more efficiently in the *a* direction, are the most important intermolecular contacts in $[Fe(phen)_2(NCS)_2]$. On the other hand, the lack of aromaticity in the btz ligand prevents the occurrence of an equivalent π-stacking which leads to a more "relaxed" isotropic distribution of the intermolecular contacts in $[Fe(btz)_2(NCS)_2]$.

2.2 2,2′-Bipyrimidine-Bridged Iron(II) Spin-Crossover Complexes

A different strategy used to explore cooperativity is the study of polynuclear spin-crossover compounds. This can be achieved by using appropriate bridging ligands able to induce spin change in the subsequent polynuclear (dinuclear, trinuclear, etc.) compound. So, cooperativity can be enhanced by controlling both intra- and intermolecular interactions.The synthesis of 2,2′-bipyrimidine-bridged iron(II) spin-crossover complexes illustrates the first results along this line.

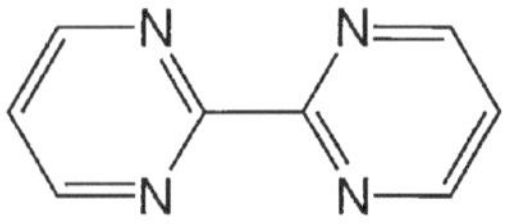

Figure 14 2,2′-Bipyrimidine.

2,2′-bipyrimidine (Figure 14) acts as a strong field ligand in its iron(II) complexes as evidenced by the structural and magnetic data of the tris-chelate $[Fe(bpym)_3](ClO_4)_2 \cdot 1/4H_2O$ (Figure 15) [72]. This compound is diamagnetic in full agreement with the average Fe–N bond-distance [1.970 Å]. The replacement of two bpym ligands by suitably less strong field ligands such as NCS^- and pyridine (py) affords the compound $[Fe(bpym)(py)_2(NCS)_2] \cdot 1/4py$ [73]. A perspective view of this molecule is shown in Figure 16. Iron(II) is surrounded by six nitrogen atoms belonging to a chelating bpym ligand, two thiocyanate groups in *cis* position and two pyridine molecules in *trans* position. The mean Fe–N bond distance is 2.186 Å (at room temperature) a value which is consistent with the high-spin state. The magnetic measurements show the occurrence of a very abrupt spin transition near 114 K. About 75% of the spin transition occurs within 1 K. As the temperature is increased, the transition is again very abrupt but occurs near 116 K. A hysteresis of *c*. 3 K is therefore observed. This compound can be viewed, at least formally, as a spin-crossover precursor of bipyrimide-bridged polynuclear compounds.

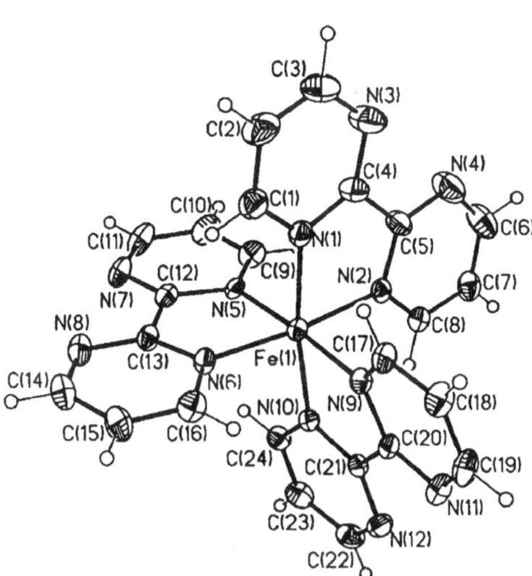

Figure 15 Perspective drawing of the $[Fe(bpym)_3]^{2+}$ cation. Thermal ellipsoids are drawn at the 30% probability. Reproduced with permission from ref. 72.

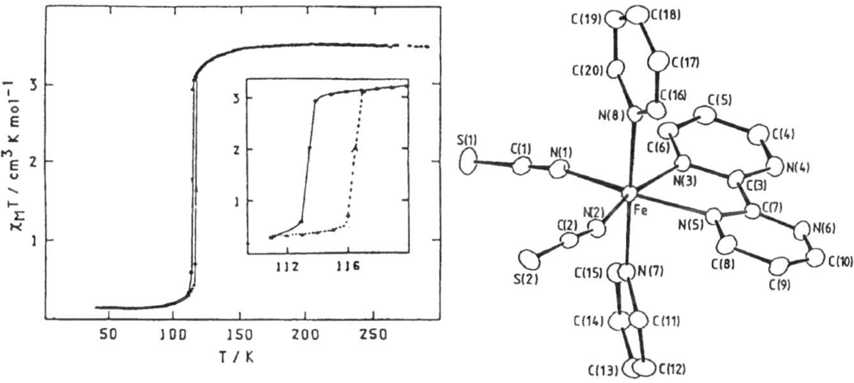

Figure 16 Temperature dependence of $\chi_M T$ and perspective view of the $[Fe(py)_2$-(bpym)$(NCS)_2]$ molecule. Falling and rising arrows indicate decreasing and increasing temperatures, respectively ($T_{1/2}\downarrow = 113.5$ K; $T_{1/2}\uparrow = 116.5$ K). Reproduced with permission from ref. 73.

It deserves to be noted that the ability of bpym to act as bis-chelating ligand and to mediate electronic effects between centers in the resulting polymetallic species is well documented [74–79]. In this respect, a series of bpym-bridged iron(II) complexes of formula $\{[Fe(L)(NCX)_2]_2$bpym$\}$ where L = bpym, bt (2,2′-bi-2-thiazoline) (Figure 17) and X = S, Se have been synthesized and characterized [76,80,81]. The molecular unit of $\{[Fe(bpym)(NCS)_2]_2$bpym$\}$ is sketched in Figure 18 and consists of centrosymmetric bpym-bridged iron(II) dinuclear entities. Two NCS^- ligands in the *cis* position and a peripheral bpym ligand acting as a bidentate ligand complete the octahedron around each metal. No spin transition is observed (see Figure 19) [76]. The iron(II) ions are HS in the whole temperature range and they are antiferromagnetically coupled. The exchange coupling constant being $J = -4.1$ cm^{-1}. At first sight this is a rather unexpected result because the metal environment in the dinuclear compound is close to that found in the complex $[Fe(bpy)_2(NCS)_2]$ (bpy = 2,2′-bipyridine), which is isostructural with $[Fe(phen)_2(NCS)_2]$, and also exhibits an abrupt spin transition (at 215 K) [82]. A comparison between the structural data of $\{[Fe(bpym)(NCS)_2]_2$bpym$\}$ and $[Fe(bipy)_2(NCS)_2]$ suggests that the bpym ligand and particularly the bridging

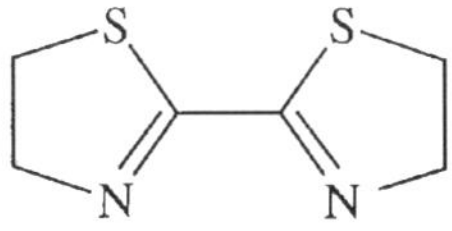

Figure 17 2,2′-Bi-2-thiazoline.

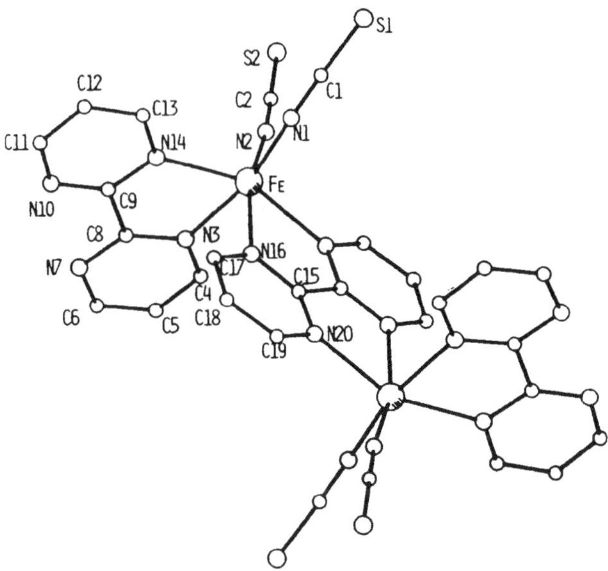

Figure 18 Perspective view of $[Fe(bpym)(NCS)_2]_2bpym$. Reproduced with permission from ref. 76.

one exerts a weaker field than the related chelating bpy. Most likely, this feature accounts for the absence of any spin transition in $\{[Fe(bpym)(NCS)_2]_2bpym\}$. However, the possibility that the intramolecular magnetic exchange interaction could stabilize the high-spin state should not be excluded.

In contrast to what is observed for the complex $\{[Fe(bpym)(NCS)_2]_2bpym\}$, the other three members of this series $\{[Fe(bpym)(NCSe)_2]_2bpym\}$, $\{[Fe(bt)(NCS)_2]_2$-bpym$\}$, and $\{[Fe(bt)(NCSe)_2]_2bpym\}$ show spin-crossover behavior [80,81]. Let us briefly comment on the physical characterization of these compounds. Growing single crystals has not yet been possible and structural information has been obtained by X-ray absorption techniques. The EXAFS data led to a rather accurate description of the basic structure and its modification upon spin conversion. They entirely match the structural features of $\{[Fe(bpym)(NCS)_2]_2bpym\}$ and show that the average Fe–N bond length, close to 0.2 Å, is greater for the HS state than for the LS state, as expected.

The magnetic properties of the series are depicted in Figure 19. The magnetic behavior of $\{[Fe(bpym)(NCSe)_2]_2bpym\}$ shows an abrupt spin transition in the 125–115 K temperature region with a 2.5 K width hysteresis loop. Only 50% of iron(II) ions undergoes the spin conversion. The further decreasing of magnetic properties at low temperature does not correspond either to a second step spin transition or to an intramolecular coupling exchange between bipyrimidine-bridged HS iron(II) ions. In fact, the Mössbauer spectrum at 4.2 K (see Figure 20) is consistent with the

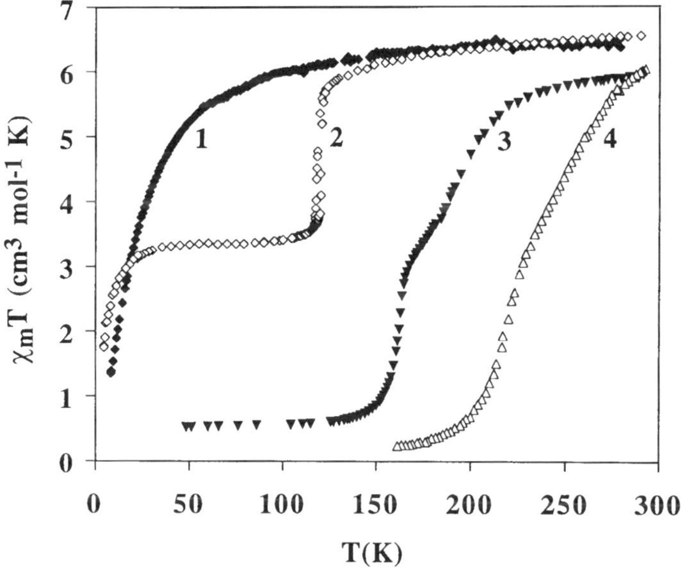

Figure 19 Thermal variation of χ_MT of $[Fe(bpym)(NCS)_2]_2bpym$ (**1**), $[Fe(bpym)(NCSe)_2]_2bpym$ (**2**), $[Fe(bt)(NCS)_2]_2bpym$ (**3**) and $[Fe(bt)(NCSe)_2]_2bpym$ (**4**). Reproduced with permission from ref. 81.

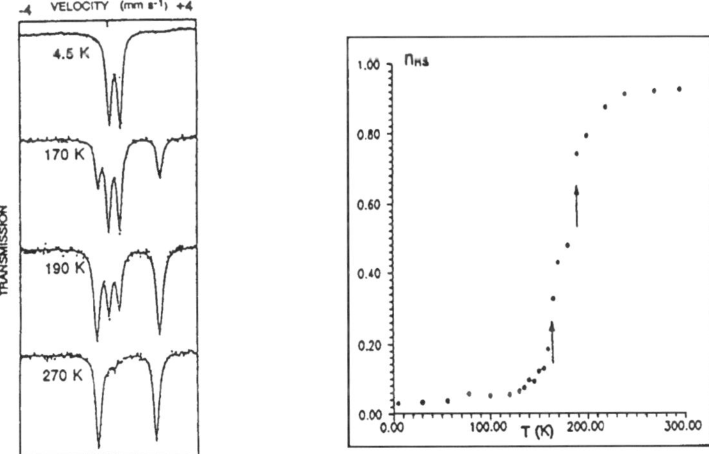

Figure 20 Selected Mössbauer spectra and thermal variation of n_{HS} deduced from the ratio of the Mössbauer absorption areas, corrected for Lamb–Mössbauer factors, of $[Fe(bt)(NCS)_2]_2bpym$. Reproduced with permission from ref. 80.

occurrence of 50% of HS ions. On the other hand, the simulation of apparent antiferromagnetic coupling exchange for the remaining 50% HS iron(II) ions in the 4.2–100 K region gives unreasonable physical parameters. This contrasts with the excellent fit obtained when zero-field splitting for uncoupled HS iron(II) ions is considered. These features strongly support the idea that only one iron(II) ion per dimer undergoes spin conversion, the remaining HS iron(II) ion experiences zero-field splitting.

The magnetic properties of {$[Fe(bt)(NCS)_2]_2$bpym}, and {$[Fe(bt)(NCSe)_2]_2$-bpym} are very similar showing the remarkable singularity of taking place in two steps. Steps 1 and 2 are centered around the temperatures $[T_{1/2}]_1 = 163$ and $[T_{1/2}]_2 = 197$ K, respectively, for {$[Fe(bt)(NCS)_2]_2$bpym}. The intermediate plateau between the two spin conversion (*c*. 169–181 K) corresponds to an HS iron(II) fraction of 50%. For {$[Fe(bt)(NCSe)_2]_2$bpym}, the spin conversion starts at higher temperatures, the two steps being separated by a less marked plateau: $[T_{1/2}]_1 = 223$ K and $[T_{1/2}]_2 = 265$ K.

Mössbauer spectroscopy confirms the magnetic data. As an example, typical spectra for {$[Fe(bt)(NCS)_2]_2$bpym} (left) are represented in Figure 20 along with the thermal variation of n_{HS} deduced from the ratio of Mössbauer absorption areas (right). The low-temperature and high-temperature main doublets are typical for the LS and HS states of iron(II), respectively. At intermediate temperatures, both contributions are clearly resolved, which indicates the coexistence of LS and HS iron(II) ions.

As stated above, thermally induced spin transition is an entropy-driven process. The entropy variation allows us to get insights into the long-range correlations which are characteristic of the cooperativity of the phenomenon. Differential scanning calorimetry (DSC) provides the total entropy and enthalpy variations upon spin conversion. The DSC curves for {$[Fe(bpym)(NCSe)_2]_2$bpym}, {$[Fe(bt)(NCS)_2]_2$-bpym}, and {$[Fe(bt)(NCSe)_2]_2$bpym} are shown in Figure 21. The relevant thermodynamic data deduced from these curves are given in Table 2. These data reveal the role played by the peripheral ligands. Thus, S $\leftrightarrow$ Se replacement in the NCX^- counterion has a small effect on the enthalpy but strongly influences the entropy. This effect is probably due to the vibrational changes derived from the significant atomic mass difference between S and Se. In contrast, bpym $\leftrightarrow$ bt replacement markedly affects the enthalpy. This result suggests that the bt ligand induces a stronger ligand field than the bpym one.

Two-step spin conversions are scarce. They have been first observed in mononuclear compounds. $[Fe(2\text{-pic})_3Cl_2]\cdot C_2H_5OH$ (2-pic = 2-picolyl-amine, (left) Figure 22) is one of the examples most investigated [19,22,43]. The molecular $[Fe(2\text{-pic})_3]^{2+}$ unit consist of a $[Fe–N_6]$ distorted octahedron with the 2-pic ligands in *mer* arrangement. Cooperativity is assured by a two-dimensional hydrogen-bonding network which links the amine hydrogen atoms of the complexes, the non-coordinating chloride anions and ethanol solvent molecules. The layers are linked by van der Waals interactions. From detailed experimental and theoretical studies it

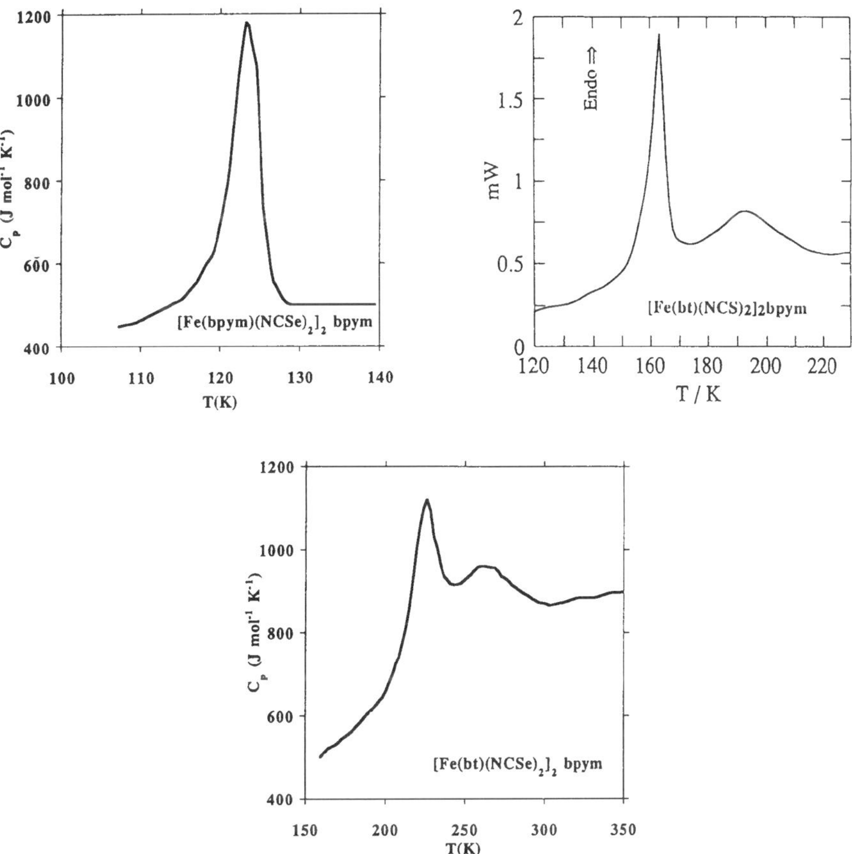

Figure 21 DSC curves obtained for $[Fe(bpym)(NCSe)_2]_2$bpym$[Fe(bt)(NCS)_2]_2$bpym and $[Fe(bt)(NCSe)_2]_2$bpym. Reproduced with permission from ref. 81.

Table 2 Thermodynamic parameters obtained from DSC curves for the dinuclear spin-crossover compounds.

Compound	STEP 1			STEP 2		
	$(\Delta H)_1$ kJ/mol	$(\Delta S)_1$ kJ/K mol	$(T_{1/2})_1$ K	$(\Delta H)_2$ kJ/mol	$(\Delta S)_2$ kJ/K mol	$(T_{1/2})_2$ K
$[Fe(bpym)(NCSe)_2]_2$bpym	—	—	—	3.0 ± 0.1	$25 + 1$	120
$[Fe(bt)(NCS)_2]_2$bpym	5.4 ± 0.5	41 ± 3	163	7.9 ± 0.5	41 ± 3	197
$[Fe(bt)(NCSe)_2]_2$bpym	5.7 ± 0.3	25.5 ± 1	223	6.6[a]	25[a]	265

[a] ΔH_2 has been estimated from$(T_{1/2})_2 = 265$ K and assuming $\Delta S_2 = 25$ J K^{-1} mol^{-1}.

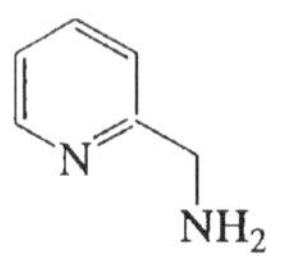

2-pic 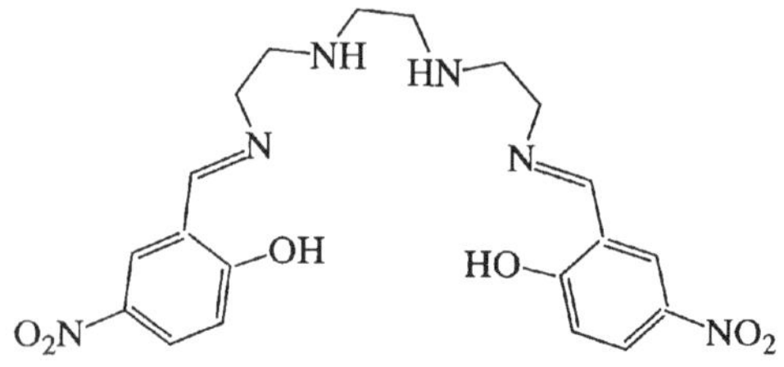 5-NO_2-sal-N(1,4,7,10)

Figure 22 (left) 2-Picolyl-amine, (right) 5-NO_2-sal-N(1,4,7,10).

can be concluded that the appearance of two steps in mononuclear systems requires a cooperative spin-state conversion mechanism with long-range elastic interactions related to lattice effects. Assumption of a sublattice structure for the HS and LS complexes leads to two-step behavior in the thermal variation of n_{HS} [83,84,85]. The sublattices originate from an anticooperative interaction between complexes, where LS–HS pairs are preferred. This is particularly true in the case of the mononuclear compound [Fe(5-NO_2-sal-N(1,4,7,10))] (see Figure 22 right). As in [Fe(2-pic)$_3$Cl$_2$]·C_2H_5OH, the molecular units are efficiently connected by strong hydrogen linkages NH···O–NO affording infinite chains of doubly hydrogen-bonded complex molecules. For this compound it has been clearly established that the two-step spin conversion, involving each 50% of molecules, occurs in the same temperature range as the two structural phase transitions, which clearly allows us to distinguish two equally distributed sets of molecules in this compound. Furthermore, the X-ray molecular structure determinations illustrate the role of intermolecular interactions in the cooperative character of the conversion mechanism through modifications of the hydrogen-bond network at each step of the spin-state conversion [86].

On the basis of subtle variations of the Mössbauer parameters, the "macroscopic" steps detected by magnetic, Mössbauer and calorimetric measurements for bpym-bridged iron(II) dinuclear compounds were found to essentially reflect the "microscopic" two steps of the following intramolecular spin conversions

$$[\mathrm{LS{-}LS}] \underset{\text{step 1}}{\longleftrightarrow} [\mathrm{LS{-}HS}] \underset{\text{step 2}}{\longleftrightarrow} [\mathrm{HS{-}HS}]$$

Hence, the two-step character of the spin conversion is intimately related to the binuclear nature of the compound. The half transition exhibited by {[Fe(bpym)-(NCSe)$_2$]$_2$bpym} has a clear meaning in the frame of the microscopic description: at low temperatures the system is in the mixed LS–HS state. The mixed state may be trapped as a result of the slow kinetics of the spin conversion at low temperature, which may prevent the first step from occurring.

Two equivalent models [80,87] have been developed to account for this peculiarity. The main idea concerning these models is that the enthalpy of the LS–HS species may not be exactly halfway between the enthalpies of the LS–LS and HS–HS like spin species. A two-step transition may be expected when enthalpy of LS–HS species, H(LS–HS), is lower than [H(LS–LS)–H(HS–HS)]/2. This condition, however, may not be sufficient; in addition, a significant cooperativity within the crystal lattice is required. In other words, the two-step character arises from the coupling between intramolecular anticooperative interactions favoring the LS–HS species which appear in the plateau and intermolecular cooperative interactions favoring phases of dominating LS–LS or HS–HS like spin species. Figure 23 shows computed conversion curves using the Ising-like model in the mean-field approach fitted to the experimental Mössbauer data; parameter values are listed in Table 3. J_{AB}, and J are the intramolecular and intermolecular coupling parameters, respectively, $\Delta E = E_{HS} - E_{LS}$ is the energy gap of the isolated spin-crossover atoms and g_{HS}/g_{LS} is the ratio of "effective" degeneracies accounting for both electronic degeneracies and densities of vibrational states [87,88].

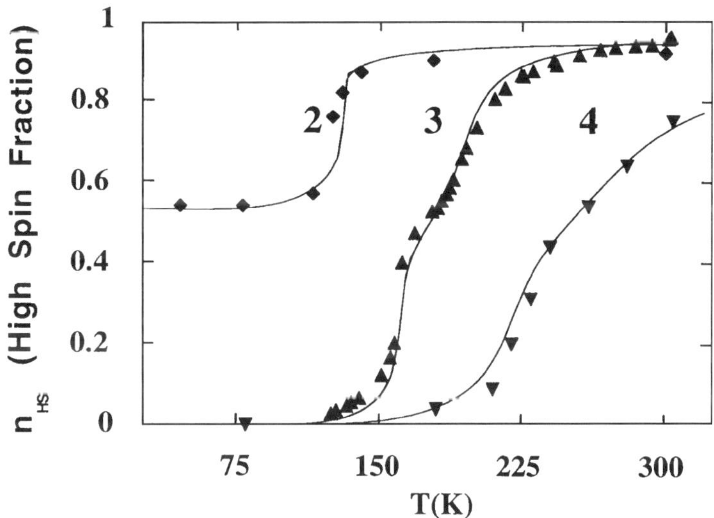

Figure 23 Computed conversion curves of $[Fe(bpym)(NCSe)_2]_2bpym$ (**2**), $[Fe(bt)(NCS)_2]_2bpym$ (**3**) and $[Fe(bt)(NCSe)_2]_2bpym$ (**4**), fitted to the experimental Mössbauer data; parameter values are listed in Table 3. Reproduced with permission from ref. 81.

Table 3 Parameters of the Ising-like binuclear model, treated in the mean-field approximation.

Compound	$\Delta E = E_{HS} - E_{LS}$ (cm^{-1})	J (cm^{-1})	J_{AB} (cm^{-1})	g_{HS}/g_{HS}
$[Fe(bpym)(NCSe)_2]_2bpym$	186	90	− 100	20
$[Fe(bt)(NCS)_2]_2bpym$	540	99	− 85	138
$[Fe(bt)(NCSe)_2]_2bpym$	520	111	− 122	20

2.3 Polymeric Spin-Crossover Systems

A further step consists of extending the connectivity between iron(II) metal ions by polymerizing of a molecular fragment, that could undergo spin conversion, to achieve nD ($n = 1$–3) spin-crossover systems.

Up to now, three examples of spin-crossover polymeric systems have been reported. The iron(II)-(4-R-1,2,4-triazole) (see Figure 24) interaction affords a series of linear chain compounds [89–99] in which iron(II) is triple bridged by triazole ligands through the nitrogen atoms occuping the 1-and 2-positions [99] (Figure 25). The nature of the spin transition depends on the substituent in position 4, the counterion and the noncoordinating solvent molecules. Some compounds of this series show very abrupt spin transitions with thermal hysteresis width up to 35 K (Figure 26).

The above-mentioned $[Fe(btr)_2(NCS)_2]{\cdot}H_2O$ compound where btr is (4,4-bis-1,2,4-triazole, Figure 27) represents the first example of a 2D polymeric spin-crossover compound [100]. Its structure and magnetic properties are depicted in Figure 28. Each iron(II) atom is surrounded by six nitrogen atoms belonging to two NCS^- groups in *trans* position and four btr ligands. The btr ligands link each metal ion to the other four defining a two-dimensional grid. The different stacked layers are connected by van der Waals forces and by hydrogen bonding through the water molecules. This compound shows an extremely abrupt spin transition with an hysteresis 21 K wide.

Finally, the compound $[Fe(tvp)_2(NCS)_2]{\cdot}CH_3OH$ (tvp = 1,2-di-(4-pyridyl)-ethylene, Figure 29) is unusual in that it is made up of interlocking 2D networks constituted by parallel layers [101]. Let us describe with some detail this structure. The iron(II) ion lies in a compressed octahedron with two *trans*-thiocyanato ligands filling the axial positions and four pyridine nitrogen atoms building the basal plane. Each tvp ligand connects two iron(II) ions defining the edges of a $[Fe(II)]_4$ rhombus. The edge shared rhombuses define the grid-layered structures mentioned above with all the iron(II) ions in a coplanar sheet. Parallel sheets are displaced so that the iron centers of the first sheet are vertically above the third, fifth and further odd-numbered sheets, while vertically above the mid-points of $[Fe(II)]_4$ rhombuses of the even-

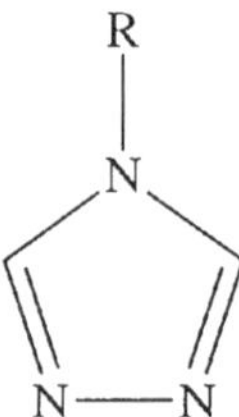

Figure 24 4-R-1,2,4-Triazole.

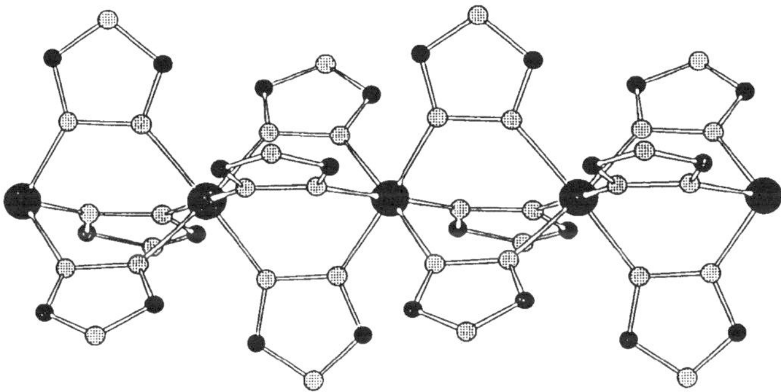

Figure 25 Structure of the polymeric Fe(II)-1,2,4-triazole spin-crossover compound as deduced from EXAFS data. The large and small black balls stand for iron and carbon atoms, respectively. The hatched balls stand for nitrogen atoms. Reproduced with permission from ref. 99.

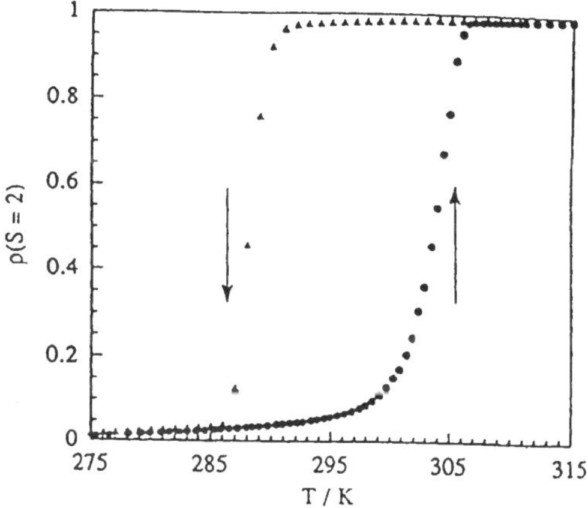

Figure 26 Temperature dependence of the HS molar fraction for $[Fe(4\text{-}Htrz)_{3-3x}\text{-}(4\text{-}NH_2trz)_{3x}](ClO_4)_2$ whose composition is adjusted in such a way that room temperature falls exactly in the middle of the thermal hysteresis loop. Reproduced with permission from ref. 98.

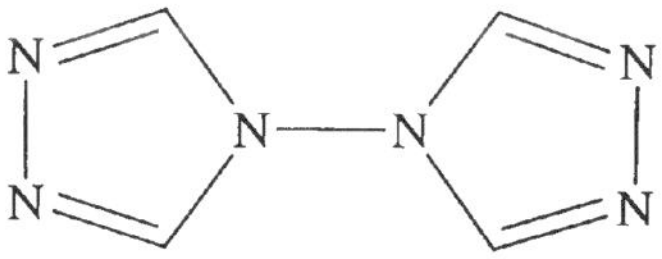

Figure 27 4,4-Bis-1,2,4-triazole.

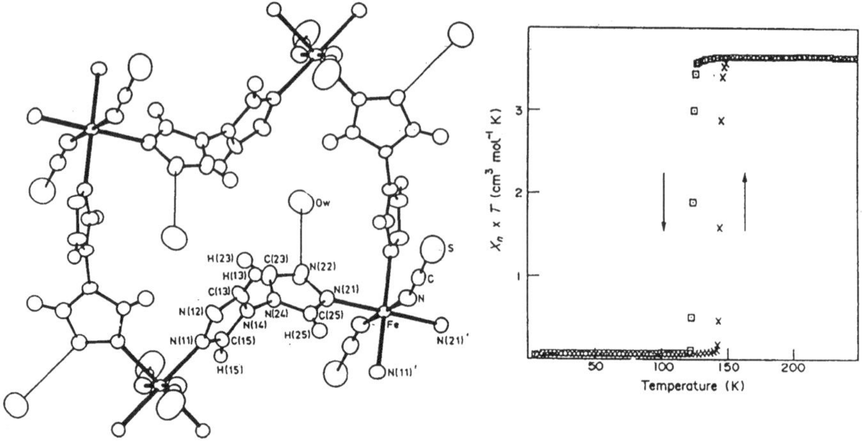

Figure 28 Structure of $[Fe(btr)_2(NCS)_2]\cdot H_2O$ and $\chi_M T$ versus T curves in the warming and cooling modes for this compound. Reproduced with permission from ref. 100.

numbered sheets (see Figure 30). An equivalent stack of sheets is found in planes perpendicular to the first set defining large square channels depicted in Figure 31.

The magnetic measurements show that the spin conversion is very sensitive to the sample preparation showing different high- and low-spin residual fractions at low and high temperature as well as different cooperativity. No detectable hysteresis is observed. Figure 32 (left) illustrates the magnetic behavior for two samples of different texture. Sample B corresponds to large single crystals which crack as a result of partial desolvatation. Sample A corresponds to a polycrystalline powder. The presence of residual paramagnetism is not uncommon in spin-crossover systems and is generally attributed to subtle effects induced by the presence of crystalline defects and molecular inclusions. The temperature dependence of the Mössbauer spectra of sample A is shown in Figure 32 (right). The dominant doublet observed at room temperature decreases progressively and a typical $S=0$ low-spin ground state doublet appears between the two former peaks. The intensity of which increases at the expense of the former with decreasing temperature.

Concerning the understanding of the spin-crossover mechanism in polymeric systems, the results are still limited in number but give support to the idea that

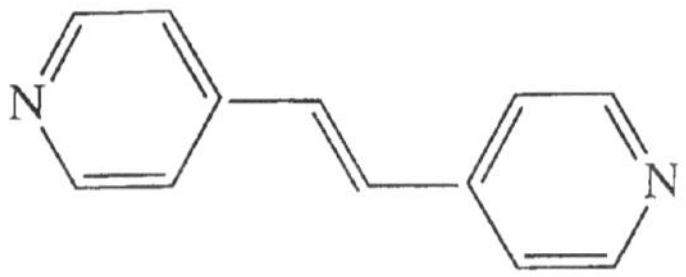

N N

Figure 29 1,2-Di-(4-pyridyl)-ethylene.

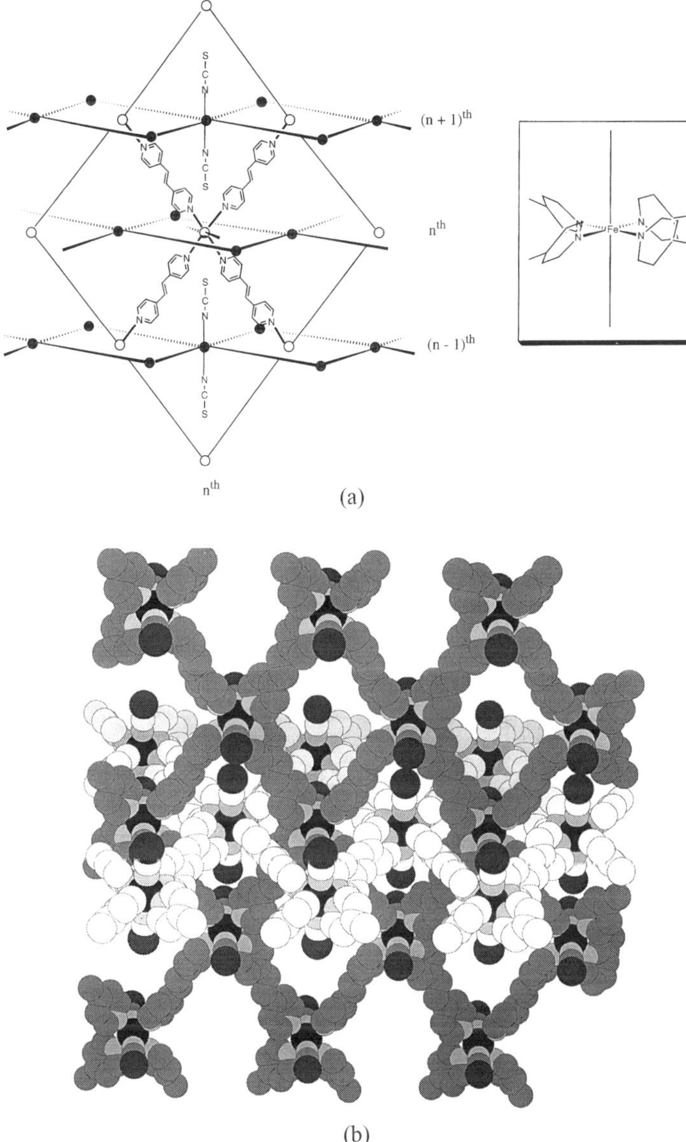

Figure 30 (a) Scheme of the interpenetrating sheets in the 110 direction: (left) view showing the particular disposition of the tvp ligands leading to the rhombus grid, the alternatively displaced sheets (along 001 direction), and the orientation of the trans NCS anions along the large diagonal of the $[Fe]_4$ rhombuses. Black and white spheres represent iron(II) ions which belong to different perpendicular networks, the straight lines denotes the tvp ligand (right) view of the propeller arrangement of the four basal pyridines. (b) Space-filling representation of the interlocking of two layers. Reproduced with permission from ref. 101.

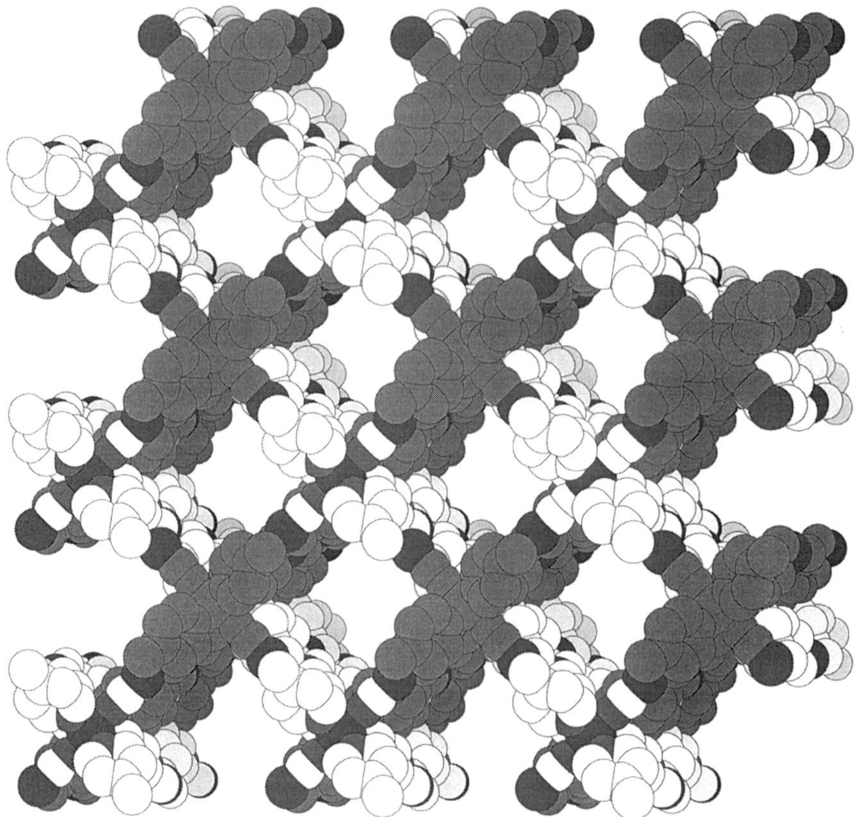

Figure 31 Space-filling representation of the structure in the 110 direction emphasizing the large channels formed by the interlocking of the two bidimensional nets. Reproduced with permission from ref. 101.

cooperativity (anticooperativity) may be magnified when spin-crossover metal ions are covalently linked by multi-atom (extended) bridges.

3 SPIN BISTABILITY AND MOLECULAR DEVICES

One of the most exciting challenges the chemist faces today is to design and synthesize new molecular materials suitable for handling information at a molecular level. Spin-crossover compounds are particularly appealing for this end. Bistability confers on them the ability to change drastically their magnetic and optical properties by the action of temperature, pressure and light radiation. Up to now, three

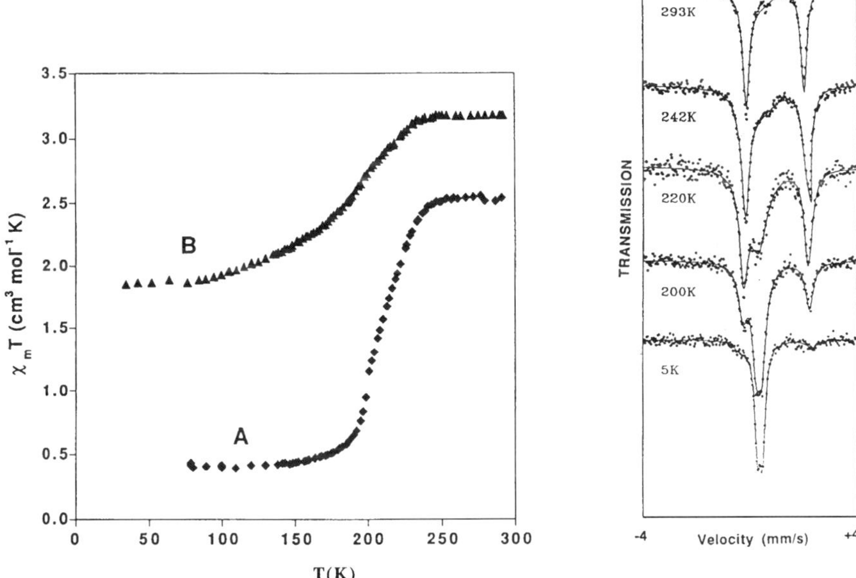

Figure 32 Thermal variation of χ_MT (samples A and B) and selected Mössbauer spectra (sample A), of $[Fe(tvp)_2(NCS)_2]{\cdot}CH_3OH$. Reproduced with permission from ref. 101.

approaches based on spin-crossover phenomenon have been explored: thermal addressing, light-induced excited spin-state trapping (LIESST) and ligand-driven light-induced spin change (LD-LISC).

According to Kahn, thermal addressing is possible when the spin transition fulfills the following requirements [102]: (i) the spin transition must be abrupt; (ii) it must present a thermal hysteresis (a 50 K width hysteresis loop is required to be operative); (iii) room temperature should be located as close as possible to the center of the hysteresis loop; and (iv) the transition must be accompanied by pronounced thermochromism which allows the reading of the information. The working principle of thermal addressing of a spin-crossover system, first described in ref 103, is schematized in Figure 33. An increase $+\Delta T$ followed by a decrease $-\Delta T$ of temperature, with $\Delta T > |T_{1/2}\downarrow - T_{1/2}\downarrow|$, leads to the HS state, whatever the starting state may be. Conversely, a decrease $-\Delta T$ followed by an increase $+\Delta T$ of temperature leads to the LS state. A prototype of a room temperature display based on thermal hysteresis has already been developed using $[Fe(4\text{-}R\text{-}1,2,4\text{-}triazole)_3A_2]{\cdot}nH_2O$ system. For some of these compounds, room temperature falls exactly at the middle of the hysteresis loop. These materials are violet or white at room temperature depending on their history. So, bistability can be detected optically.

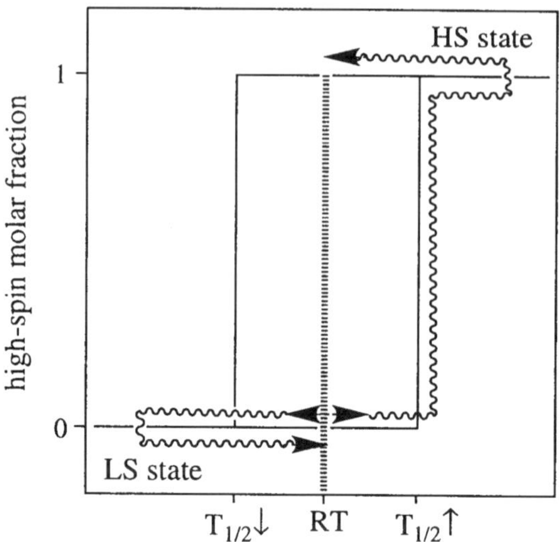

Figure 33 Scheme of the addressing principle of a spin-crossover compound where the starting state is LS. The ΔT-ΔT operation addresses the HS state, and the $-\Delta T + \Delta T$ operation addreses the LS state (adapted from ref. 102).

Optical excitation of the LS($^1A_{1g}$) ground state at very low temperature in some thermally driven spin-crossover compounds may result in the formation of a trapped HS($^5T_{2g}$) metastable state. This effect, termed "light-induced excited spin-state trapping" (LIESST), has been discovered and described by Gütlich and co-workers [20,22,23]. The mechanism can be explained on the basis of Figure 34. The process is initiated by irradiation into the spin-allowed $^1A_{1g} \rightarrow {}^1T_{1g}$ absorption band at temperatures well below the thermal transition temperature $T_{1/2}$. The excited singlet state is short-lived and quickly decays to the $^1A_{1g}$ ground state. Alternatively, the system attains the triplet $^3T_{1g}$ state via intersystem crossing. This in turn decays via a second intersystem crossing either to the $^1A_{1g}$ ground state or to the metastable $^5T_{2g}$ state where a nonequilibrium population is accumulated. The system remains trapped in the HS state due to the lack of thermal energy to overcome the structure reorganization energy barrier which separates HS and LS potential surfaces. The process is reversible: the $^5T_{2g}$ state may be converted back into $^1A_{1g}$ state by irradiating into the $^5T_{2g} \rightarrow {}^5E_g$ absorption band. The LS $\leftrightarrow$ HS optical switching process is very fast, of the order of a nanosecond, and the quantum yield is high. The photoinduced spin change can be monitored in several ways. For instance, Figure 35 illustrates the temperature-dependent magnetic moment of $[Fe(ptz)_6](BF_4)_2$ (ptz is 1-propyltetrazole, Figure 36). The spin transition occurs near 120 K. At temperatures lower than 50 K the LS state is fully converted to the HS state by irradiation with green light [20,22]. These kind of compounds, in principle, can be used as optical devices. However, some fundamental problems have to be solved, particularly the very low working temperature required to preclude fast HS $\rightarrow$ LS thermal relaxation.

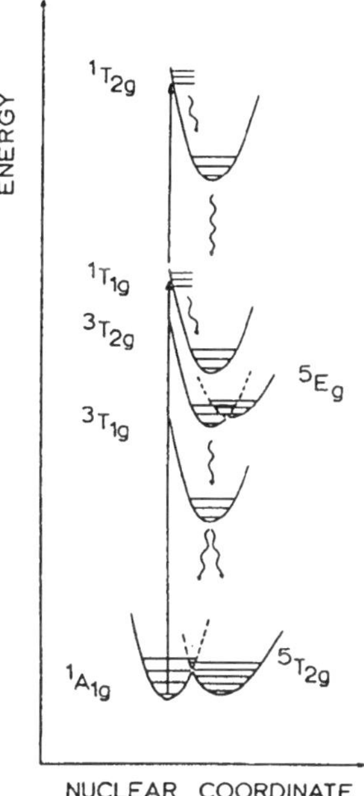

Figure 34 Schematic representation of the potential wells of the excited ligand field states in a d^6 spin-crossover system. Arrows indicate the mechanism for LIESST and reverse-LIESST. Reproduced with permission from ref. 23.

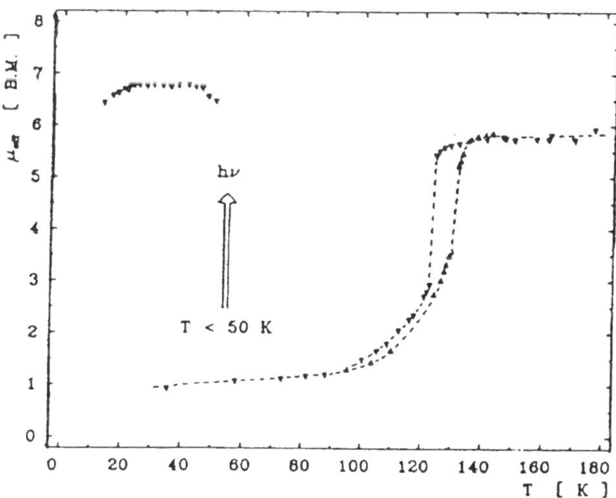

Figure 35 Effective magnetic moment of a polycrystalline sample of $[Fe(ptz)_6](BF_4)_2$ as a function of temperature and excitation with laser light. Reproduced with permission from ref. 23.

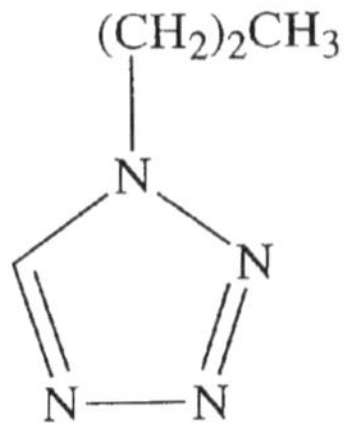

Figure 36 1-Propyltetrazole.

Recently, an alternative approach for optical addressing, which may work at room temperature, has been developed by Zarembowitch and Boillot and co-workers. It is called "ligand-driven light-induced spin change" (LD-LISC). This strategy consists in designing molecular spin-crossover compounds for which the reversible spin change is triggered by ligand-field strength modulation resulting from a photochemical alteration of the ligands. The first system found to be adapted to the observation of the LD-LISC effect is based on the *trans-cis* isomerization of 4-styrylpyridine (stpy) (see Figure 37). The two isomeric compounds have the formula $[Fe(stpy)_4(NCS)_2]$ [104]. Iron(II) ion is surrounded by four stpy ligands lying in the equatorial plane and two NCS^- groups filling the axial positions. Figure 38 depicts the molecular structures for the *cis*-and *trans*-isomers. The *trans*-isomer exhibits a thermally induced spin conversion centered around $T_{1/2} = 108$ K, while the *cis*-stpy derivative is in the HS state at any temperature. Figure 39 shows the temperature dependence of the $\chi_M T$ product for the two compounds as well as a scheme of the photoconversion working principle. The highest temperature at which LD-LISC might be observed is near 90 K where the *trans*-stpy derivative has completely attained the LS state. At this temperature photoisomerization quantum yields are expected to be very low. Keeping this in mind, these authors have significantly shifted $T_{1/2}$ of the *trans*-stpy derivative toward higher temperature, by replacing NCS^- by $NCB(Ph)_3{}^-$, up to 190 K. For this compound photoisomerization of the stpy ligand at 140 K, in the complex embedded within a cellulose acetate matrix, is effectively shown to trigger the spin-state change of the ion(II) ions [105].

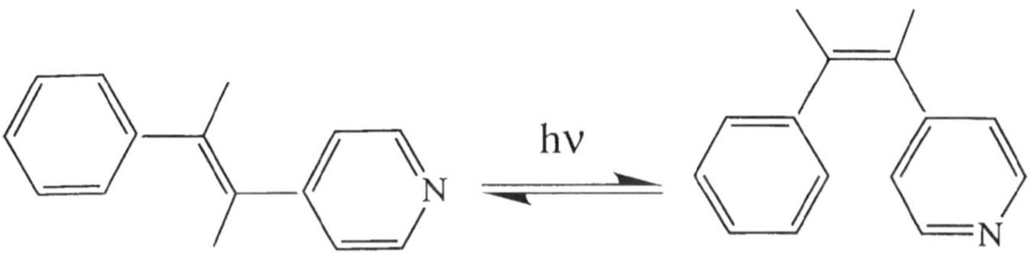

Figure 37 *Trans-cis*-isomers of 4-styrylpyridine.

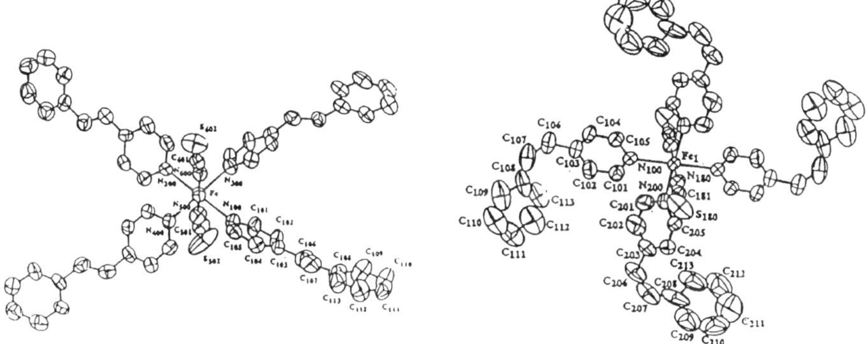

Figure 38 Perspective drawing of the [Fe(*trans*-stpy)$_4$(NCS)$_2$] (left) and [Fe(*cis*-stpy)$_4$(NCS)$_2$] (right) molecules. Reproduced with permission from ref. 104.

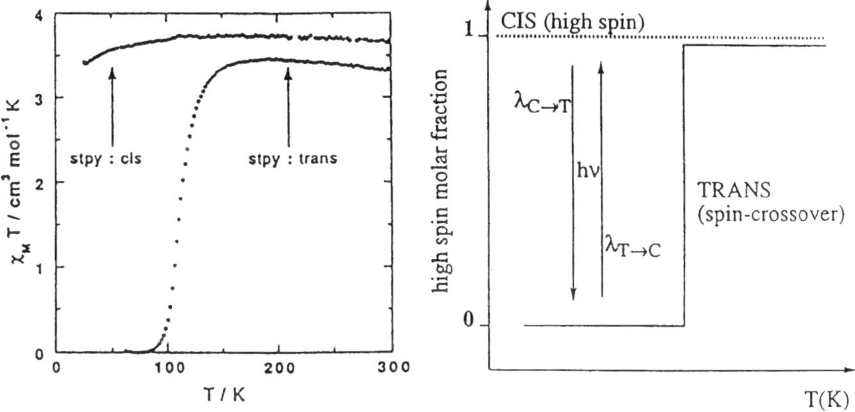

Figure 39 (left) Temperature dependence of $\chi_M T$ for the [Fe(stpy)$_4$(NCS)$_2$] compounds formed with *trans*-stpy and *cis*-stpy (right). Scheme of the LD-LISC addressing principle. Reproduced with permission from ref. 104.

CONCLUSIONS

The aim that has motivated the present contribution is to place the spin-crossover phenomenon in the context of supramolecular chemistry. There are two reasons for doing this. Firstly, cooperativity depends on both the intermolecular and intramolecular interactions occurring in discrete mononuclear and polynuclear as well as in extended *n*D ($n = 1, 2, 3$) polymeric spin-crossover systems. Thus, understanding cooperativity requires knowing the factors that govern the aggregation of molecules into assemblies. That should provide us with rational strategies to design molecular solids with potentially interesting properties and, particularly, cooperative spin transitions. Incorporation of the synthetic approaches stemming from supramolecular chemistry should allow us to build chemical networks through a careful selection of the building blocks that will assemble into structures with specific and desired topologies. Secondly, one of the main challenges in supramolecular chemistry is understanding non-covalent interactions to build superstructures resembling that of biological systems and imparting to them some kind of function. In this respect, it is well established that iron(II) spin-crossover compounds are switchable systems that express their properties at molecular level and it is possible to gain access to these properties. Hence, spin-crossover compounds offer real possibilities to supramolecular chemists in order to design functionalized supramolecular systems.

ACKNOWLEDGEMENTS

This work was financially supported by the Dirección General de Investigación Científica y Técnica (DGICYT) (Spain) through Project PB94-1002 and the Human Capital and Mobility Program (Network on Magnetic Molecular Materials from ECC) through grant ERBCHRXCT920080. We also wish to thank to Dr J. Zarembowitch (Orsay), Dr M. L. Boillot (Orsay), Dr R. Ruiz (Orsay-Valencia) and Professor M. Julve (Valencia) for their helpful discussions.

REFERENCES

1. R. Feynman, *Miniaturization*, ed. A. Gilbert (Reinhold, New York) 1961, 282–296.
2. W. A. Little, *Phys. Rev.* 1964, **A134**, 1416–1424.
3. W. A. Little, *J. Polym. Sci., Polym. Lett. Ed.* 1970, 29.
4. K. E. Drexler, *Proc. Natl. Acad. Sci. USA* 1981, **78**, 5275–5278.
5. F. L. Carter, Ed., *Molecular Electronic Devices* (Dekker, New York) 1982.
6. F. L. Carter, Ed., *Molecular Electronic Devices II* (Dekker, New York) 1987.
7. J. S. Miller, *Adv. Mater.* 1990, **2**, 378.
8. J. S. Miller, *Adv. Mater.* 1990, **2**, 495.
9. J. S. Miller, *Adv. Mater.* 1990, **2**, 601.
10. D. Goldhaber-Gordon, M. S. Montemerlo, J. C. Love, G. J. Opiteck, J. C. Ellenbogen, *Proceeding of the IEEE* 1997, **85**, 521–540.
11. R. C. Haddon, A. A. Lamola, *Proc. Natl. Acad. Sci. USA* 1985, **82**, 1874.
12. O. Kahn, J. P. Launay, *Chemtronics 3* 1988, 151.
13. J. M. Lehn, *Supramolecular Chemistry. Concepts and Perspectives*, VCH, Weinheim, 1995.
14. J. P. Sauvage, *Acc. Chem. Res.* 1990, **23**, 319–327.
15. G. M. Whitesides, E. E. Simanek, J. P. Mathias, C. T. Seto, D. N. Chin, M. Mammaen, D. M. Gordon, *Acc. Chem. Res.* 1995, **28**, 37–44.
16. E. C. Constable, *Prog. Inorg. Chem.* 1994, **42**, 67–138.

17. D. B. Amabilino, J. F. Stoddart, *Chem. Rev.* 1995, **95**, 2725–2828.
18. G. R. Desiraju, *Angew. Chem. Int. Ed. Engl.* 1995, **34**, 2311.
19. P. Gütlich, *Struct. Bonding* (Berlin) 1981, **44**, 83.
20. P. Gütlich, A. Hauser, H. Spiering, *Angew. Chem. Int. Ed. Engl.* 1994, **33**, 2024.
21. E. König, *Struct. Bonding* (Berlin) 1991, **76**, 51.
22. P. Gütlich, A. Hauser, *Coord. Chem. Rev.* 1990, **97**, 1.
23. S. Decurtins, P. Gütlich, K. M. Hasselbach, A. Hauser, H. Spiering, *Inorg. Chem.* 1985, **24**, 2174.
24. B. Gallois, J. A. Real, C. Hauw, J. Zarembowitch, *Inorg. Chem.* 1990, **29**, 1152.
25. H. L. Schläfer, G. Gliemann, *Basic Principles of Ligand Field Theory* (Wiley, London) 1972.
26. C. P. Slichter, H. G. Drickamer, *J. Chem. Phys.* 1972, **56**, 2142.
27. H. Spiering, E. Meissner, H. Köppen, E. W. Müller, P. Gütlich, *Chem. Phys.* 1982, **68**, 65.
28. N. Willenbacher, H. Spiering, *J. Phys. C.* 1988, **21**, 1423.
29. H. Spiering, N. Willenbacher, *J. Phys.: Condens. Matter* 1989, **1**, 10080.
30. T. Kambara, *J. Phys. Soc. Japan* 1980, **74**, 4199.
31. T. Kambara, *J. Chem. Phys.* 1981, **74**, 4199.
32. S. Ohnishi, S. Sugano, *J. Phys. C: Solid State Phys.* 1981, **14**, 39.
33. M. S. Haddad, W. D. Federer, M. W. Lynch, D. N. Hendrickson, *Inorg. Chem.* 1981, **20**, 131.
34. P. Ganguli, P. Gütlich, E. W. Müller, *Inorg. Chem.* 1982, **21**, 3429.
35. H. Spiering, E. Meissner, H. Köppen, E. W. Müller, P. Gütlich, *Chem. Phys.* 1982, **68**, 65.
36. I. Sanner, E. Meissner, H. Köppen, H. Spiering, P. Gütlich, *Chem. Phys.* 1984, **86**, 227.
37. P. Adler, L. Wiehl, E. Meissner, C. P. Köhler, H. Spiering, P. Gütlich, *J. Phys. Chem. Solids* 1987, **48**, 517.
38. P. S. Rao, A. Reuveni, B. R. McGarvey, P. Ganguli, P. Gütlich, *Inorg. Chem.* 1981, **20**, 204.
39. M. Sorai, J. Ensling, P. Gütlich, *Chem. Phys.* 1976, **18**, 199.
40. P. Gütlich, R. Link, H. G. Steinhauser, *Inorg. Chem.* 1978, **17**, 2509.
41. P. Gütlich, H. Köppen, R. Link, H. G. Steinhauser, *J. Chem. Phys.* 1979, **70**, 3977.
42. R. Jakobi, H. Spiering, L. Wiehl, E. Gmelin, P. Gütlich, *Inorg. Chem.* 1988, **27**, 1823.
43. C. P. Köhler, R. Jakobi, E. Meissner, L. Wiehl, H. Spiering, P. Gütlich, *J. Phys. Chem. Solids* 1990, **51**, 239.
44. R. Jakobi, H. Spiering, P. Gütlich, *J. Phys. Chem. Solids* 1992, **53**, 267.
45. J. P. Martin, J. Zarembowitch, A. Dworkin, J. G. Haasnoot, E. Codjovi, *Inorg. Chem.* 1994, **33**, 2617–2623.
46. L. Wiehl, G. Kiel, C. P. Köhler, H. Spiering, P. Gütlich, *Inorg. Chem.* 1986, **25**, 1565.
47. E. Köning, *Prog. Inorg. Chem.* 1987, **35**, 527.
48. J. A. Real, B. Gallois, T. Granier, F. Suez-Panama, J. Zarembowitch, *Inorg. Chem.* 1992, **31**, 4972.
49. T. Granier, B. Gallois, J. Gaultier, J. A. Real, J. Zarembowitch, *Inorg. Chem.* 1993, **32**, 5305–5312.
50. W. A. Baker, H. M. Bobonich, *Inorg. Chem.* 1964, **3**, 1184.
51. E. König, K. Madeja, *Chem. Commun.* 1966, *3*, 61.
52. E. König, K. Madeja, *Inorg. Chem.* 1967, **6**, 48.
53. A. T. Casey, F. Isaac, *Aust. J. Chem.* 1967, **20**, 2765.
54. P. Ganguli, P. Gütlich, *J. Phys.* 1980, **41**, C1–313.
55. P. Ganguli, P. Gütlich, E. W. Müller, W. Irler, *J. Chem. Soc., Dalton Trans.* 1981, 441.
56. E. W. Müller, H. Spiering, P. Gütlich, *Chem. Phys, Lett.* 1982, **93**, 567.

57. I. Dézsi, B. Molnar, T. Tarnoczi, K. Tompa, *J. Inorg. Nucl. Chem.* 1967, **29**, 2486.
58. S. Savage, Z. Jia-long, A. G. Maddock, *J. Chem. Soc., Dalton Trans.* 1985, 991.
59. E. König, K. Madeja, *Spectrochim. Acta* 1967, **23A**, 45.
60. J. H. Takemoto, B. Hutchinson, *Inorg. Nucl. Chem. Lett.* 1972, **8**, 769.
61. J. H. Takemoto, B. Hutchinson, *Inorg. Chem.* 1973, **12**, 705.
62. M. Sorai, S. Seki, *J. Phys. Chem. Solids* 1974, **35**, 555.
63. R. Herber, L. M. Casson, *Inorg. Chem.* 1986, **25**, 847.
64. M. Sorai, S. Seki, *J. Phys. Soc. Jpn.* 1972, **33**, 575.
65. C. Cartier, P. Thuéry, M. Verdaguer, J. Zarembowitch, A. Michalowicz, *J. Phys.* 1986, **46**, 563.
66. J. R. Fisher, H. G. Drickamer, *J. Chem. Phys.* 1971, **54**, 4825.
67. D. M. Adams, G. J. Long, A. D. Williams, *Inorg. Chem.* 1982, **21**, 1049.
68. J. Pebler, *Inorg. Chem.* 1983, **22**, 4125.
69. S. Usha, R. Srinivasan, C. N. R. Rao, *Chem. Phys.* 1985, **100**, 447.
70. S. Decurtins, P. Gütlich, C. P. Köhler, H. Spiering, *J. Chem. Soc., Chem. Commun.* 1985, 430.
71. G. Bradley, V. McKee, S. M. Nelson, J. Nelson, *J. Chem. Soc.* 1978, 522.
72. G. De Munno, M. Julve, J. A. Real, *Inorg. Chim. Acta* 1997, **255**, 185–188.
73. R. Claude, J. A. Real, J. Zarembowitch, O. Kahn, L. Ouahab, D. Grandjean, K. Boukheddaden, F. Varret, A. Dworkin, *Inorg. Chem.* 1990, **29**, 4442–4448.
74. G. De Munno, G. Bruno, *Acta Crystallogr. Sec. C* 1984, **40**, 2030.
75. G. Brewer, E. Sinn, *Inorg. Chem.* 1985, **24**, 4580.
76. J. A. Real, J. Zarembowitch, O. Kahn, X. Solans, *Inorg. Chem.* 1987, **26**, 2939.
77. E. Andrés, G. De Munno, M. Julve, J. A. Real, F. Lloret, *J. Chem. Soc., Dalton Trans.* 1993, 2169.
78. G. De Munno, M. Julve in *Metal-Ligand Interactions. Structure and Reactivity*, N. Russo, D. R. Salahub (Eds.). NATO ASI Ser. C Vol. 474 (Kluwer, Dordrecht) 1996, 193.
79. G. De Munno, F. Lloret, M. Julve in *Magnetism: a Supramolecular Function*, O. Kahn (Ed.). NATO ASI Ser. C Vol. 484 (Kluwer, Dordrecht) 1996, 555.
80. J. A. Real, H. Bolvin, A. Boussekson, A. Dworkin, O. Kahn, F. Varret, J. Zarembowitch, *J. Am. Chem. Soc.* 1992, **114**, 4650.
81. J. A. Real, I. Castro, A. Boussekson, M. Verdaguer, R. Buriel, M. Castro, J. Linares, F. Varret, *Inorg. Chem.* 1997, **36**, 455–464.
82. M. Kono, M. M. Kido, *Bull. Chem. Soc. Jpn.* 1991, **64**, 339.
83. V. V. Zelentsov, G. I. Lapouchkin, S. S. Sobolev, V. I. Shipilov, *Dokl. Akad. Nauk.* 1986, **289**, 393.
84. N. Sasaki, T. Kambara, *Phys Rev. B* 1989, **40**, 2442.
85. A. Boussekson, J. Nasser, J. Linares, K. Boukheddaden, F. Varret, *J. Phys. 1* 1992, **2**, 1381.
86. D. Boinnard, A. Boussekssou, A. Dworkin, J. M. Savariault, F. Varret, J. P. Tuchagues, *Inorg. Chem.* 1994, **33**, 271.
87. A. Boussekson, F. Varret, J. Nasser, *J. Phys. 1 (Paris)* 1993, **3**, 1463.
88. A. Boussekson, H. Constant-Machado, F. Varret, *J. Phys. 1* 1995, **5**, 747.
89. L. G. Lavrenova, V. N. Ikorskii, V. A. Varnek, I. M. Oglezneva, S. V. Larionov, *Koord. Khim.* 1986, **12**, 207.
90. L. G. Lavrenova, V. N. Ikorskii, V. A. Varnek, I. M. Oglezneva, S. V. Larionov, *J. Struct. Chem.* 1993, **34**, 960.
91. K. H. Sugiyarto, H. A. Googwin, *Aust. J. Chem.* 1994, **47**, 263.
92. J. Kröber, J. Audiere, E. Codjovi, O. Kahn, J. Haasnot, F. Groliere, C. Jay, A. Bousseksou, J. Linares, F. Varret, A. Gonthier-Vassal, *Chem. Mater.* 1994, **6**, 1404.

93. F. G. Lavrenova, V. N. Ikorskii, V. A. Varnek, I. M. Oglezneva, S. V. Larionov, *Koord. Khim.* 1990, **16**, 654.
94. L. G. Lavrenova N. G. Yudina, V. N. Ikorskii, V. A. Varnek, I. M. Oglezneva, S. V. Larionov, *Polyhedron* 1995, **14**, 1333.
95. V. A. Varnek, L. G. Lavrenova, *J. Struct. Chem.* 1995, **36**, 104.
96. E. Codjovi, L. Sommier, O. Kahn, C. Jay, *New J. Chem.* 1996, **20**, 503.
97. O. Kahn *et al.* in *Molecule-Based Magnetic Materials* (eds, M. M. Turnbull, T. Sugimoto, L. K. Thompson), Symposium Series No. 644, American Chemical Society, Washington DC, 1996, 298.
98. J. Kröber, E. Codjovi, O. Kahn, F. Grolière, C. Jay, *J. Am. Chem. Soc.* 1993, **115**, 9810.
99. A. Michalowicz, J. Moscovici, B. Ducourant, D. Cracco, O. Kahn, *Chem. Mater.* 1995, **7**, 1833.
100. W. Vreugdenhil, J. H. Van Diemen, R. A. G. De Graaff, J. G. Haasnoot, J. Reedijk, A. M. Van DerKraan, O. Kahn, J. Zarembowitch, *Polyhedron* 1990, **9**, 2971–2979.
101. J. A. Real, E. Andres, M. C. Muñoz, M. Julve, T. Granier, A. Boussekson, F. Varret, *Science* 1995, **268**, 265.
102. O. Kahn, J. Kröber, C. Jay, *Adv. Mater.* 1992, **4**, 718.
103. J. Zarembowitch, O. Kahn, *New. J. Chem.* 1991, **15**, 181.
104. C. Roux, J. Zarembowitch, B. Gallois, T. Granier, R. Claude, *Inorg. Chem.* 1994, **33**, 2273.
105. M. L. Boillot, C. Roux, J. P. Audière, A. Dausse, J. Zarembowitch, *Inorg. Chem.* 1996, **35**, 3975.

Chapter 3

Fluorescent Sensors for and with Transition Metals

LUIGI FABBRIZZI, MAURIZIO LICCHELLI, PIERSANDRO PALLAVICINI, LUISA PARODI, AND ANGELO TAGLIETTI

Università di Pavia, Italy

1 THE DESIGN OF A FLUORESCENT CHEMICAL SENSOR

In a chemical context, sensing of a molecular substrate results from the combination of two different and well-defined functions: (1) recognition of the substrate; (2) signalling to the outside of the recognition event. Hence, the most simple and logical approach to the design of a molecular sensor would involve the coupling of two distinct components, one devoted to perform function (1), the other function (2). Nature and behaviour of such a two-component device is pictorially illustrated in Figure 1.

In the figure, the recognition is provided by a concave receptor (the host), which should interact selectively with the chosen substrate (the guest, indicated by a sphere). The sketch in the figure emphasizes the role of size-and-shape complementarity in the recognition process. Shape is important in the case of polyatomic substrates (e.g. organic anions, amino acids) and includes the proper spatial setting of the binding sites on the concave receptor surface. Spherical substrates (e.g. monoatomic ions, both positively and negatively charged) are discriminated only on the basis of their size: the oldest and best-known examples refer to the interaction of s block metal ions with cyclic and polycyclic polyethers (crowns [1] and cryptands [2]). Needless to say, the energy of the receptor-substrate interaction is an important factor in determining selectivity, too. This is especially true in the case of the guests considered in this chapter: transition metals. Transition metal ions of a given n*d*

Transition Metals in Supramolecular Chemistry, edited by J. P. Sauvage.

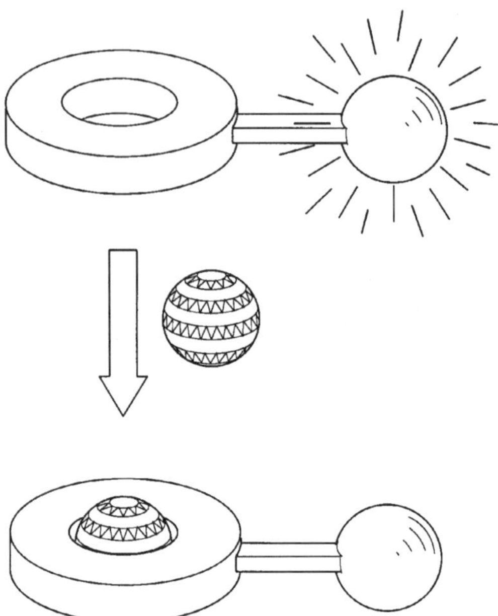

Figure 1 Recognition and sensing. The selective interaction of the substrate with the receptor subunit (ball in the hole), is communicated to the outside by an adjacent, covalently linked signalling subunit.

series and possessing the same electrical charge do not differ too much in size, but they can establish coordinative interactions of very different energy, depending upon their electronic configuration. These relatively large energy differences in the metal-ligand interaction (i.e. the substrate–receptor interaction) can be used for discriminative purposes.

The second subunit of the sensing device is expected to put the molecular life in contact with the macroscopic world, by communicating to the operator the occurrence of the receptor-substrate interaction. Communication takes place through the variation of a well-defined property of the signalling subunit. Any instrumentally detectable property can be appropriate (the intensity of either an absorption band or an emission band in the UV-visible region; the shift of either a voltammetric wave or an NMR line), provided its variation is tangible. Fluorescence emission is a very convenient property to be monitored, for a number of reasons, which include simplicity of instrumentation, high sensitivity which allows substrate sensing at trace level and, most intriguing, direct visual perception even in very diluted solutions.

Hopefully, the recognition process should induce a change of the magnitude of light emission intensity of two orders of magnitude or more. This generates a situation in which fluorescence is at first fully quenched (e.g. before substrate binding; OFF state), and then (after substrate binding) revived (ON state). Essentially for this reason, fluorescent sensors are also denoted as *switches* [3, 4]. Noticeably, ON/OFF fluorescence switching (which may be activated not only by ionic substrates, including H^+, but also by electrons, *vide infra*) has a potential for information processing, along the arduous route to molecular computers. Basically, a fluorescent chemical sensor for a metal ion could have a structural formula of the type reported below (**1**).

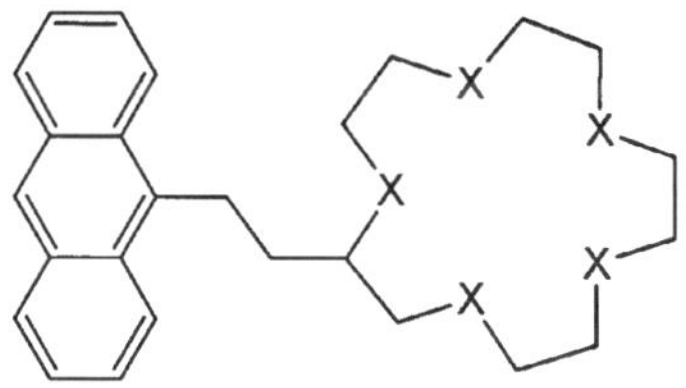

1 X = NH, O, S

The fluorescent fragment of system **1** is one of the most classical and strongly emitting organic fluorophores: anthracene; the receptor is a multi-dentate ligand whose donor atoms will be chosen depending upon the nature of the metal ion to be recognized (X = NH, O, S). The receptor can contain negatively charged groups, in order to favour the chelation of highly charged cations; it can be cyclic or polycyclic, a feature that increases the stability of the receptor-metal complex. Fragments of varying nature can be used to connect the fluorescent fragment and the metal-binding subunit. In the representative system **1**, the two subunits have been linked by an ethylenic chain.

2 ELECTRON AND ENERGY TRANSFER MECHANISMS

Simple linking of the receptor subunit to a signalling fragment does not necessarily make a molecular sensor. The two-component system should also provide an efficient mechanism by which things occurring within the recognizing portion (uptake/release of the substrate) properly modify the emission of the proximate subunit. Quite happily, the substrates considered in this chapter, transition metals, have two efficient mechanisms available for influencing the light emission of a fluorophore: the electron transfer (eT) and the energy transfer (ET) processes (Figure 2) [5].

The nature of a photo-induced eT process involving a fluorophore Fl and a metal centre M is illustrated in the orbital diagram in Figure 2a. If, for instance, the fluorophore is an aromatic fragment, as in the representative example **1**, fluorescence results from a $\pi^* \rightarrow \pi$ transition. Let us consider first the case of a metal centre displaying reducing tendencies: it will transfer an electron from a filled (or half-

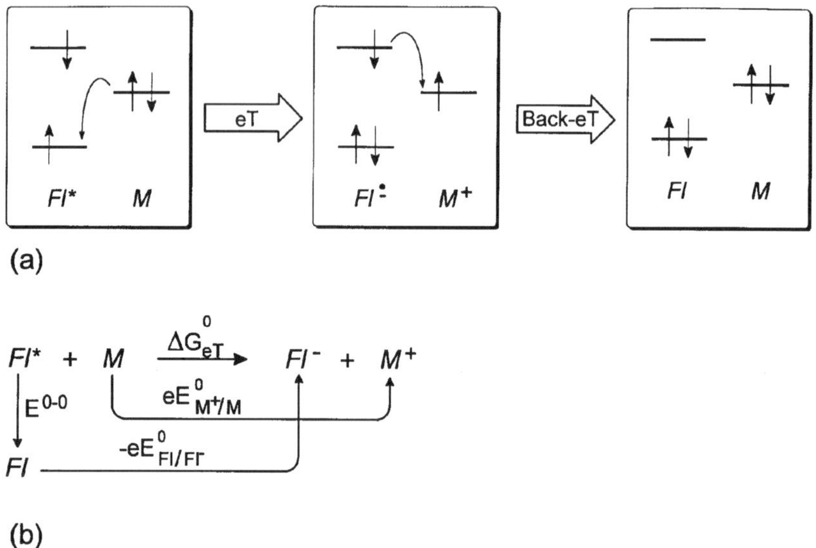

Figure 2 (a) Orbital scheme illustrating the quencing of a photo-excited fluorophore (Fl^*) by a metal centre (M), via an M–to–Fl^* electron transfer process (eT). The transient ion pair that forms, $\{Fl^-, M^+\}$ undergoes a non-radiative back-electon transfer process. Thus, the natural emission of Fl is quenched. (b) Thermodynamic bases of the quenching process via an eT mechanism: the free energy change associated to the eT process, $\Delta G°_{eT}$, is calculated through the combination of pertinent photo-physical and electrochemical quantities.

filled) d level to the half-filled π level of the photo-excited fluorophore, leading to the formation of the transient species $\{Fl^{\cdot-}, M^+\}$. Then, a back-electron transfer will take place from the π^* of $Fl^{\cdot-}$ to the half-filled orbital of M^+, a process that regenerates the original partners in their ground state. Thus, in presence of the reducing metal M, the photonic energy assumed by the fluorophore is no longer restored through the π^*–to–π radiative transition, but is released via a non-radiative pathway which involves the metal. As a consequence, the natural fluorescence of Fl is quenched.

The occurrence of the quenching process can be predicted on a thermodynamic basis. In fact, the free energy change associated to the key step, i.e. the M–to–Fl^* eT process, $\Delta G°_{eT}$, can be estimated through the combination of three quantities which can be obtained from simple photophysical and electrochemical experiments, as illustrated in the thermodynamic cycle of Figure 2b. In particular: $\Delta G°_{eT} = -E^{0\text{-}0} + eE°_{M^+/M} - eE°_{Fl/Fl^-}$, where $E^{0\text{-}0}$ is the photonic energy (also called spectroscopic energy), which is usually obtained by the energy of the emission band measured at the spectrofluorimeter; e is the electron charge; $E°_{M^+/M}$ and $E°_{Fl/Fl^-}$ are the electrode potential values associated to the $M^+ + e^- \rightarrow M$ and $Fl + e^- \rightarrow Fl^-$ redox changes, respectively. These latter quantities can be obtained from the $E_{1/2}$

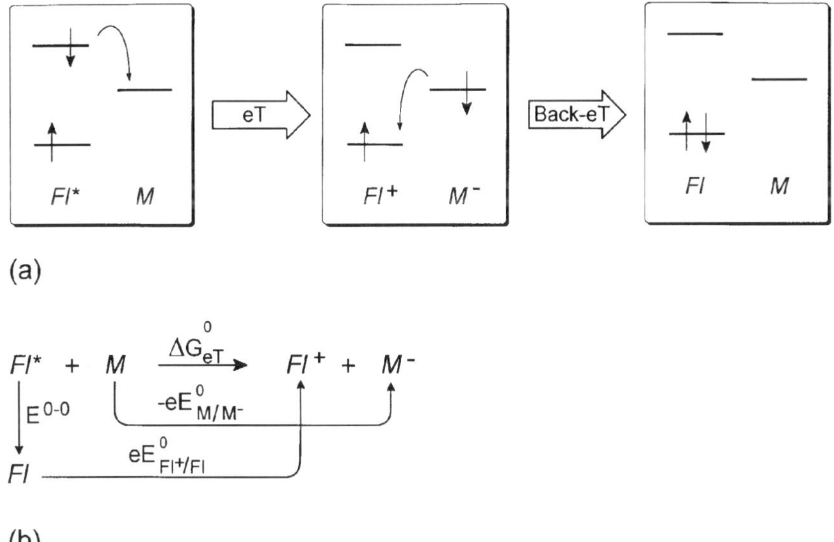

Figure 3 (a) Orbital scheme illustrating the quenching of a photo-excited fluorophore (Fl^*) by a metal centre (M): the metal has now oxidizing tendencies and a Fl^*–to–M eT process takes place. In the ion pair that forms, $\{Fl^+, M^-\}$, a back-eT process, from M^- to Fl^+ takes place: fluorescence is quenched. (b) Thermodynamic cycle for the calculation of ΔG°_{eT}.

values determined through cyclic voltammetry experiments (provided that the redox couple displays an electrochemically reversible behaviour; this is regularly observed with transition metals and their coordination complexes, less frequently observed with organic fluorophores). A distinctly negative value of ΔG°_{eT} accounts for the occurrence of the photo-induced eT process.

Figure 3a illustrates the other possible case of fluorescence quenching via an eT process. In this case, the metal centre M displays oxidizing tendencies. In particular, M offers an empty orbital of adequate energy to which an electron is transferred from the π^* level of the proximate photo-excited fluorophore Fl^*. Again, thanks to the neighbouring metal, the photonic energy is deactivated through a non-radiative pathway, which involves the formation of the transient state $\{Fl^{\cdot +}, M^-\}$ and the subsequent back-electron transfer process. At the end, the partners are restored to their ground state and fluorescence has been quenched. Combination of the appropriate photophysical and electrochemical quantities gives an estimation of the ΔG°_{eT} value for the Fl^*–to–M eT process (Figure 3b, $\Delta G^{\circ}_{eT} = -E^{0\text{-}0} - eE^{\circ}_{M^+/M} + eE^{\circ}_{Fl^+/Fl}$): a negative value accounts for the occurrence of the fluorescence quenching.

However, the fact the ΔG°_{eT} value is distinctly negative does not necessarily mean that the eT process takes place (and the fluorescence is quenched). Let us imagine that the metal ion is incorporated within the coordinating receptor subunit in a two-component system as that illustrated in Figure 1, and that metal-ligand interactions address the redox tendencies of the metal towards a one-electron release. To reach the π level of Fl^*, the electron can move from M along the fragment linking the two subunits. This process is feasible if the linker is not too long, is rigid and electron-permeable. The two latter features should be observed in presence of multiple bonds, which stiffen the bridge and provide vacant π^* orbitals. If the linker is an aliphatic chain, 'through-bond' electron transfer is much slower. However, the flexible chain may allow the two-component system to fold, occasionally bringing the fluorophore and the metal centre to a van der Waals distance and favouring the occurrence of a direct and fast electron transfer. Curiously, this eT process is said to occur 'through space', even if the electron does not fly through the solution, but is brought to its destination by a randomly moving piece of the molecule.

The other mechanism which may be responsible for the fluorescence quenching by a transition metal centre is energy transfer (ET): in particular, the type defined as *electronic* energy transfer [5]. This mechanism is illustrated by the orbital scheme in Figure 4. The metal centre M possesses some empty or half-filled energy levels, whose energies are intermediate between π^* and π. This situation is easily achieved in presence of classical aromatic fluorophores, as the energy of the π bond, to which the $\pi^*-\pi$ separation is related, is much greater than the energy of the metal-ligand interactions, which determines the separation between d orbitals. In these circumstances, a simultaneous exchange of two electrons takes place (from the π^* orbital to the empty level of the metal and from the filled level of the metal to π, see Figure 4). The circular double-electron exchange restores Fl in its ground state and produces the metal-centred excited state M^*. Metal-centred (d–d) excited states in most cases give rise to a non-radiative decay, thus fluorescence is quenched. Any metal centre can undergo this electronic ET process, provided it contains at least a half-filled orbital of not too high an energy. This is the situation of cations having a d^{1-9} electronic configuration (transition metals). The condition for the occurrence of such

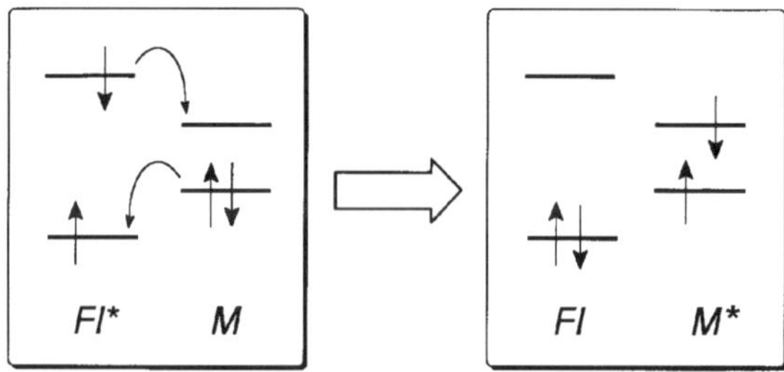

Figure 4 Orbital scheme illustrating the quenching of a photo-excited fluorophore Fl^* by a nearby metal centre M via an electronic energy transfer (ET) mechanism. A simultaneous exchange of two electrons takes place, one from Fl^* to M, one from M to Fl. Following this circular electron motion, Fl^* is deactivated. The excited M^* centre which is obtained can emit and relax to its ground state, but in most cases undergoes a non-radiative decay.

a process, as outlined by the orbital scheme in Figure 4, is that the spectroscopic energy of the fluorophore, $E^{0\text{-}0}$, is greater than the energy associated to the d–d transition on the metal, $h\nu_{d-d}$. This means, in practice, that the d–d absorption band of M must be situated at a distinctly higher wavelength than the emission band of Fl. In particular, the free energy change of the electronic ET process is given by: $\Delta G^{\circ}_{ET} = -E^{0\text{-}0} + h\nu_{d-d}$. The occurrence of this double-electron transfer requires the spatial overlap of the orbitals (π^* and d, in this particular case). Thus, Fl* and M must be in close contact. In a two-component system such as **1**, such contact can take place only if the system occasionally folds. This is a first indication that flexible linkers may favour the occurrence of intramolecular ET processes of the double-electron exchange type.

3 DISCRIMINATING eT AND ET MECHANISMS

In the previous section it has been shown that the interaction of a transition metal ion with the recognizing portion of a two-component system such as **1** may induce the quenching of the light emission of the proximate fluorophore. Determining whether fluorescence quenching is due to an eT or an ET mechanism is not a trivial problem from an experimental point of view.

The occurrence of an ET process can be unambiguously assessed if, after irradiation of the solution at the wavelength of the absorption band of Fl, one observes the emission of the metal-centred excited state M*. It has been already pointed out that the highly distorted d–d excited states are rarely emissive and undergo non-radiative decay processes. One of the few exceptions is Cr^{III}.

An intramolecular electronic energy transfer leading to Cr^{III} emission takes place in the multi-component system **2** [6]. After the irradiation on the $[Ru^{II}(bpy)_2(CN)_2]$ luminescent fragment, the excitation is transferred to one of the $-[CNCr^{III}$(cyclam)CN] subunits, whose emission spectrum is observed. However, it should be noted that system **2** does not really belong to the class of two-component sensors as

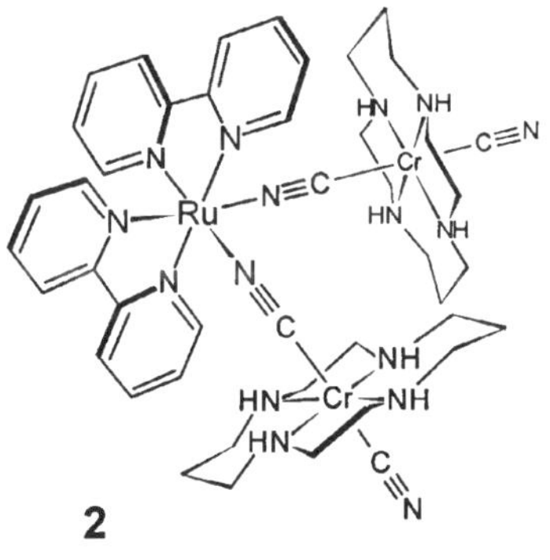

2

it does not present an empty cavity for the reversible incoming of the metal guest (Cr^{III}), but it results from the assembling of two preformed fragments ($[Ru^{II}(bpy)_2(CN)]-$ and $-[Cr^{III}(cyclam)CN]$), which share the ambidentate ligand CN^-.

The direct characterization of an eT mechanism requires a much more complicated technique: time-resolved spectroscopy. The solution containing the system under investigation is irradiated by a laser pulse, and the absorption spectra of the solution are consecutively recorded at chosen and very short time intervals (e.g. every 10 ns). If, in the envisaged two-component system $Fl \sim M$, an M–to–Fl eT process takes place upon illumination, one should be able to measure the absorption spectra of Fl^- and M^+, as well as their decay, which allows the determination of the lifetime of the transient species $Fl^- \sim M^+$. It goes without saying that very sophisticated and expensive instrumentation is required to carry out this type of experiment. Moreover, the smaller the fluorophore lifetime and the faster the back-electron transfer process, the more rapid and expensive the data acquisition equipment required. In particular, narrow laser pulses and especially fast data collections are needed for systems such as **1**, where a short-living polyaromatic fluorophore (anthracene, $\tau = 5$ ns) is linked to the electron donor (or acceptor) group by a rather short carbon chain.

Happily, an empirical procedure exists, that allows one to discriminate eT and ET mechanisms simply using a conventional spectrofluorimeter equipped with an accessory for measurements at low temperatures (a liquid nitrogen cryostat). In particular, the fluorescence spectrum of the investigated system dissolved in a polar solvent is measured both at room temperature and at liquid nitrogen temperature. Obviously, a solvent capable of vitrifying on cooling must be chosen: not water, but ethanol; not acetonitrile, but butyronitrile. The solution does not fluoresce at room temperature as either an eT or an ET mechanism operates within the $Fl \sim M$ system: if freezing makes the fluorescence revive, room temperature quenching is due to an eT mechanism; if not, an ET mechanism is active.

A genuine eT mechanism generates separation of electrical charges, to give $Fl^- \sim M^+$ (or $Fl^+ \sim M^-$). This process is accompanied by the simultaneous rearrangement of the solvent molecules around the system. At liquid nitrogen temperature, solvent molecules are immobilized and their rearrangement, as well as the eT process, are prevented. Hence, the excited fluorophore can restore the photonic energy in the more natural and preferred way, by emitting light. This situation is pictorially illustrated in Figure 5.

The case illustrated in Figure 5 (the $Fl \sim M$ system hosts a reducing metal centre with zero charge) is especially sensitive to the state of the solvent, as one moves from a poorly solvated neutral species to the strongly solvated ion pair $Fl^- \sim M^+$. However, when the metal centre is already positively charged, e.g. a divalent cation, the eT process generates a redistribution of the electrical charge and a consequent drastic rearrangement of the solvation sphere.

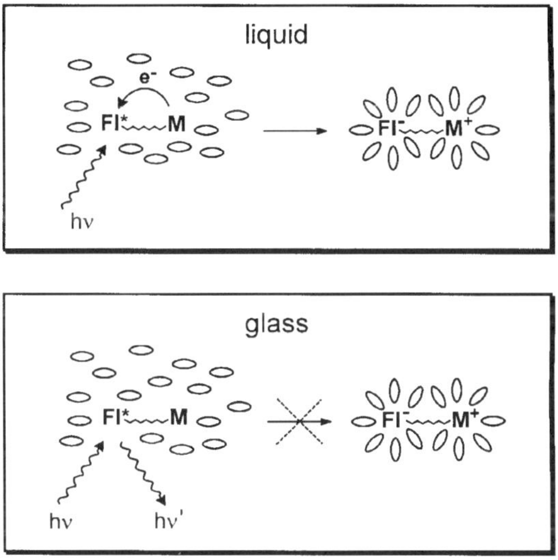

Figure 5 In a covalently linked two-component system Fl ~ M, an eT process from the donor subunit M to the nearby photo-excited fragment Fl* induces a drastic rearrangement of solvent molecules around the $Fl^- \sim M^+$ ion pair (upper part of the figure). Freezing of the solution (e. g. at liquid nitrogen temperature) immobilizes solvent molecules, thus preventing the occurrence of the eT process and allowing the Fl* subunit to fluoresce (lower part of the figure).

The solvent freezing effect can be interpreted also on the basis of the Jablonski diagram sketched in Figure 6. In a fluid solution, the ion pair $Fl^- \sim M^+$ is strongly stabilized by the solute–solvent interactions and has a lower energy than the photo-excited system $Fl^* \sim M$, which makes the eT process thermodynamically feasible. In a frozen medium, the solvational stabilization effect is lacking, so that the energy of the ion pair is higher than that of the $Fl^* \sim M$ form and the eT process is prevented.

It should also be considered that a change in the oxidation state of the transition metal ion M typically induces a drastic change of its stereochemical preferences and a substantial steric rearrangement of the coordinating subunit framework. Again,

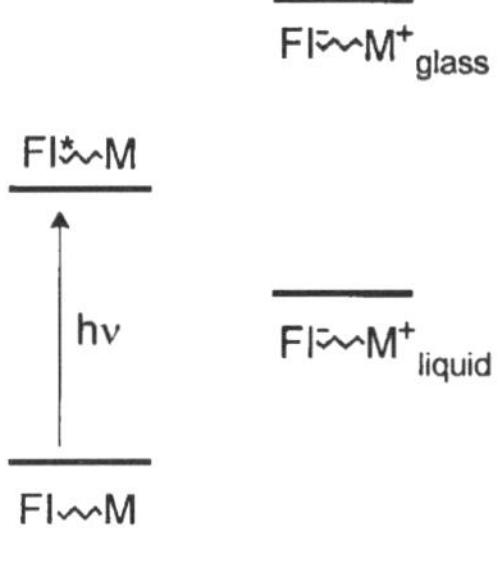

Figure 6 Jablonski diagram illustrating the eT process from the M subunit to the photo-excited Fl* subunit in a covalently linked two-component system Fl ~ M. In a fluid polar medium, the $Fl^- \sim M^+$ ion pair which form is stabilized by the interactions with solvent molecules and the eT process is thermodynamically favoured. If solvent molecules are immobilized, the $Fl^- \sim M^+$ ion pair is not stabilized by solute–solvent interactions and its energy is higher than that of the photo-excited system $Fl^* \sim M$: this prevents the occurrence of the eT process.

freezing prevents the ligand's reorganization, further disfavouring the occurrence of an eT process from/to the metal centre. Thus, for both solvational and coordinative factors, even eT processes characterized by remarkably negative ΔG°_{eT} values do not take place in a frozen polar medium, where we observe fluorescence rebirth.

4 TRANSITION METAL RECOGNITION AND SENSING: Ni^{II} AND Cu^{II}

Divalent cations of the 3d series, which are considered *borderline* in the hard and soft classification of metal centres [7], have the greatest affinity towards the *borderline* donor atom nitrogen, having either an sp^3 (amine) or an sp^2 (e.g. pyridine) hybridization. As most of the 3d metal ions profit from a tetragonal stereochemistry in terms of ligand field stabilization energy, a tetramine chelating agent seems a suitable receptor. In particular, among linear tetramines, **3** has the most favourable size to place its donor atoms in the coordination sites required by the metal centre, i.e. the corners of a square whose edge is about 3 Å (this corresponds to a M^{II}–N bond-length of 2.1 Å, a typically observed distance for amine complexes of divalent 3d metal ions).

Thus, appending a light-emitting subunit at the tetramine framework of **3** generates a potential fluorescent sensor for 3d metal ions. In particular, in system **4**, an anthracene fragment has been linked through a $-CH_2-$ group to one of the terminal amine groups of **3** [8]. Since in protic solvents, hydrogen ions compete with metal ions for the amine group, the acid-base behavior of **4** with respects to its emitting properties has to be preliminarily investigated. In this connection, it should

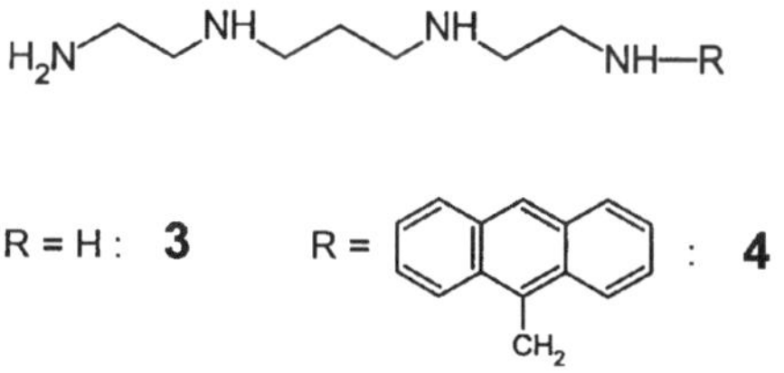

H_2N–NH–NH–NH–R

R = H: **3**   R = (9-anthryl)CH_2 : **4**

be noted that the amine group itself displays electron donor properties (which decrease along the series: tertiary > secondary > primary) and may be therefore involved in a photo-induced electron transfer process to a nearby fluorophore. Protonation cancels the reducing tendencies of the amine group and prevents the occurrence of the photo-induced eT process. Hence, the fluorescence of a given fluorophore can be switched ON/OFF at will through the protonation-deprotonation of an appended amine group. This is the basis of the design of simple and efficient fluorosensors of pH.

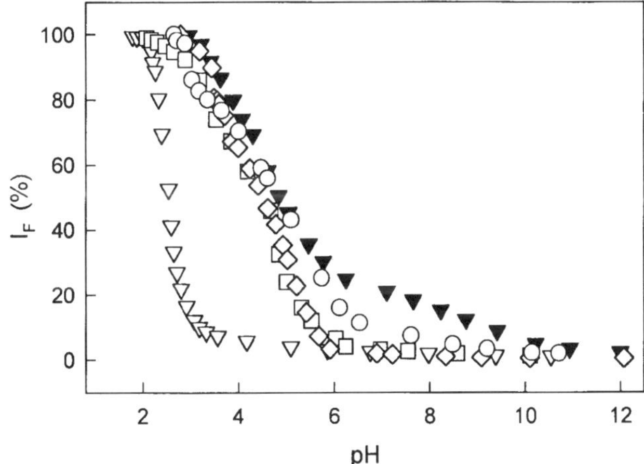

Figure 7 Spectrofluorimetric titration by standard base of the two-component system **4**, in aqueous MeCN: full triangles: **4** plus excess acid; open triangles: **4** plus 1 equiv. of Cu^{II} and excess acid; squares: **4** plus 1 equiv. of Ni^{II} and excess acid; diamonds: **4**, plus 1 equiv. of Co^{II} and excess acid; circles: **4**, plus 1 equiv. of Fe^{II} and excess acid.

Figure 7 displays the pH dependence of the fluorescence intensity of the anthracene subunit of **4**, as obtained by titrating with standard NaOH a solution containing **4** and excess acid, within a spectrofluorimetric cuvette (medium: $MeCN/H_2O$, 4 : 1 v/v). At $pH \leq 2$ all the amine groups are protonated and **4** displays its full fluorescence. On increasing pH, the fluorescence intensity, I_F decreases: complete quenching is achieved at $pH \geq 10$. The plot of I_F vs. number of added equivalents of OH^- (not reported here) shows that fluorescence quenching takes place with the neutralization of second and third ammonium groups of **4**. It is probable that these groups are the secondary ones closest to the anthracene fragment. As soon as these nitrogen atoms deprotonate, they can release an electron to the proximate photo-excited fluorophore. When the titration experiment is carried out on a solution containing 1 equiv. of Cu^{2+}, the fluorescence decrease is anticipated and takes place according to a sharp sigmoidal profile in a narrow pH range (2–3.5, see Figure 7). Quenching has to be associated to the complexation of the Cu^{II} centre by the tetramine subunit. This is confirmed by the appearance in the absorption spectrum of a band centred at 520 nm: such a d–d band is typical of a square-planar $Cu^{II}N_4$ complex.

Titration of a Ni^{II} containing solution produces a different I_F vs. pH profile (see Figure 7): in its first part the profile almost superimposes with that observed with **4** alone, then it decreases more steeply, according to a sigmoidal pattern. This behaviour is consistent with the formation of a complex of Ni^{II} with the tetramine subunit which is less stable than the Cu^{II} analogue. In aqueous solution log K values for the formation of the Cu^{II} and Ni^{II} complexes with the tetramine precursors are

23.2 and 16.1 [9]. Thus, first I_F decreasing has to be ascribed to the fact that ammonium groups begin to deprotonate, but Ni^{II} ions are not yet coordinated. Hence, quenching takes place via the amine–to–An* eT process. At pH = 4.4, full tetramine coordination to the metal centre takes place: quenching of the anthracene fragment is now induced by the close Ni^{II} ion and the sigmoidal I_F decay reflects the sigmoidal increase of the concentration of the metal-receptor coordination complex. Fluorescence quenching is complete at a pH value 2 units higher than observed with Cu^{II}. In particular, when the Cu^{II} ion has just completed An* quenching (pH = 2.9), the Ni^{II} ion has yet to begin its job. This situation offers the opportunity to discriminate Cu^{II} and Ni^{II}, using **4** as a fluorosensor.

In fact, when a solution of **4**, which has been adjusted to pH = 2.9, is titrated with Ni^{II}, emission of the anthracene fragment is not modified even after the addition of several equiv. of the metal: at this pH the tetramine subunit is unable to coordinate Ni^{II} (see Figure 8). Then, if Cu^{II} is added to the same solution, I_F decreases lineraly and complete quenching is observed after the addition of 1 equiv.

The nature of the quenching mechanism can be ascertained by carrying out spectrofluorimetric investigations on butyronitrile solutions containing **4** and 1 equiv. of the metal (either Cu^{II} or Ni^{II}), frozen at 77 K. In both cases, no fluorescence revival was observed, which indicated the occurrence of an ET process.

Among other 3d divalent metal ions Co^{II} and Fe^{II} display a pH-dependent quenching profile similar to that observed for Ni^{II}, but displaced to slightly higher

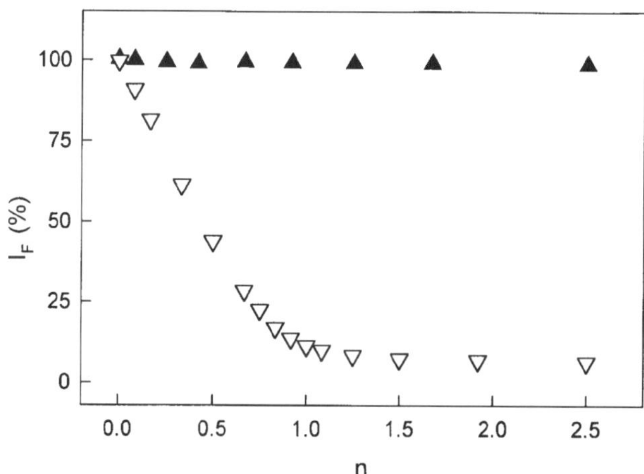

Figure 8 Spectrofluorimetric discrimination of Cu^{II} and Ni^{II} using fluorosensor **4**. The cuvette contains an aqueous MeCN solution of **4**, buffered to pH = 2. 9. The solution is first titrated with Ni^{II} and the fluorescent emission is not altered at all (at this pH, the Ni^{II} ion is not *recognized* by the receptor subunit of **4**). Then, on titration with Cu^{II}, a linear decrease of I_F is observed, with complete quenching after 1 equiv. addition.

pH values: this reflects the lower stability of the corresponding tetramine complexes, as accounted for on the basis of the Irving-Williams series.

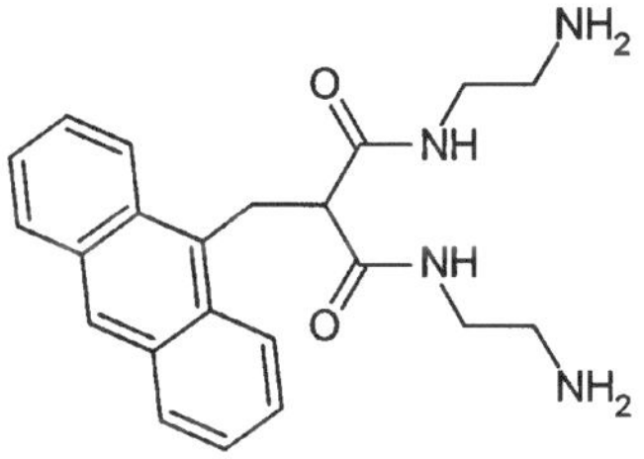

5

5 is another two-component system structurally very similar to **4**. The chelating subunit is still quadridentate, but the two middle amine nitrogen atoms have been replaced by amide groups. The amide group itself displays very poor coordinating tendencies, but if deprotonated, is a strong donor for transition metals.

The process of amide deprotonation and simultaneous metal coordination of the receptor subunit of system **5** is sketched in Figure 9. A square-planar complex species forms: the negative charge of the deprotonated amide group does not reside on the nitrogen atom, but is delocalized over the entire N–C–O fragment. Due to π-delocalization the portion of the receptor framework containing the two amide groups, –NCOCHCON–, is planar and rigid. Ni^{II} and Cu^{II} ions are able to induce amide deprotonation and coordination in slightly acidic or neutral solution, according to the mechanism of Figure 9, but other divalent cations (e.g. Mn^{II}, Fe^{II}, Co^{II}) are not, even in basic solution. Deprotonation of the amide group itself is a very endoergonic process and can take place only if this negative contribution is more than compensated by a very exothermic metal–ligand interaction. This is the case of metal ions late in the transition series, Ni^{II} and Cu^{II}, that profit very much from ligand field stabilization effects, but not of cations earlier in the transition series. The combination of diamine–diamide subunit with a fluorescent fragment is expected to

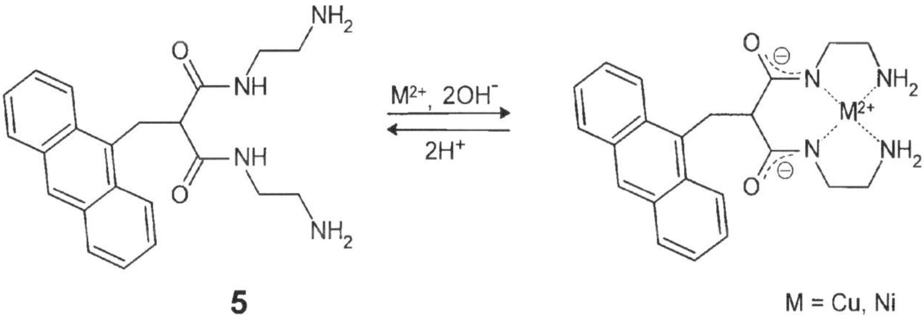

M^{2+}, $2OH^-$ ⇌ $2H^+$

5 M = Cu, Ni

Figure 9 The coordinating behaviour of diamine–diamide chelating agents. The coordination of a divalent transition metal M^{II} (M = Cu, Ni) involves the simultaneous deprotonation of the two amide group. A neutral species forms.

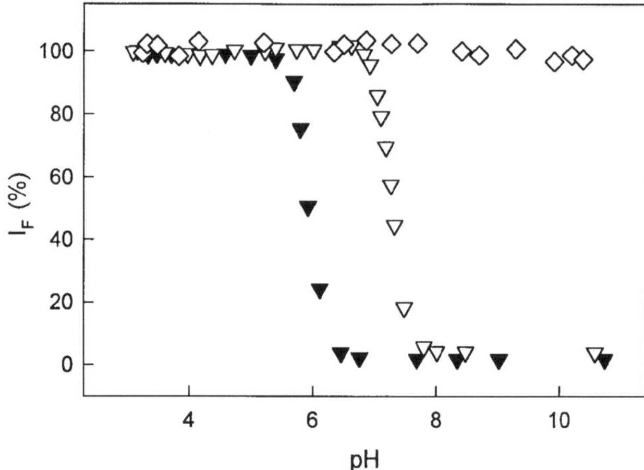

Figure 10 Spectrofluorimetric titration by standard base of the two-component system **5**, in aqueous MeCN: diamond: **5** plus excess acid; full triangles: **5**, plus 1 equiv. of Cu^{II} and excess acid; open triangles: **5**, plus 1 equiv. of Ni^{II} and excess acid.

produce a fluorescent sensor specific for the Cu^{II} and Ni^{II} ions. In **5**, an anthracene fragment has been linked through a $-CH_2-$ group to the carbon backbone of the quadridentate receptor (Figure 10) [10].

In contrast to what was observed with **4**, titration of **5** in the spectrofluorimetric cuvette does not induce any fluorescence quenching, even in basic solution. The nitrogen atoms closest to the fluorophore, which belong to amide groups, do not exhibit reducing tendencies and cannot be involved in any N–to–An* eT process. Primary amine nitrogen atoms have some reducing tendencies, but they are too far away from the fluorophore. Thus, full emission of the anthracene fragment is observed along the entire 2–12 pH range. On the other hand, titration in presence of Cu^{II} and Ni^{II} induces fluorescence quenching according to symmetrical sigmoidal profiles. In the case of Cu^{II} quenching takes place about 1.5 pH units before that observed for Ni^{II}, due to the higher thermodynamic stability of the metal–receptor complex. As observed with fluorosensor **4**, the separation of the sigmoidal I_F vs. pH profiles allows **5** to discriminate Cu^{II} and Ni^{II} cations. In particular, in a solution of **5** buffered to a pH equidistant between the two profiles, 7.1, addition of Ni^{II} does not modify fluorescence intensity, whereas addition of Cu^{II} induces a linear decrease of I_F.

Unlike system **4**, sensor **5** is totally insensitive to Fe^{II} and Co^{II}. The higher selectivity is related to the fact that amide deprotonation is promoted only by metals forming very strong coordinative interactions, a privilege reserved to metals late in the 3d series. In this way, system **5** does not recognize or sense size or shape, but merely the position in the Periodic Table.

Moreover, in contrast to what was observed with system **4**, freezing at 77 K of a solution containing equimolecular amounts of **5** and of the metal (either Ni^{II} or Cu^{II})

induces a fluorescence revival: this unequivocally indicates that quenching in a fluid solution, e.g. at room temperature, is due to an eT process. The occurrence of an intramolecular M–to–An* eT process is accounted for on a thermodynamic basis: corresponding ΔG° values are distinctly negative: −0.50 eV for Cu^{II} and −0.35 eV for Ni^{II}. Undoubtly, the well-known tendency of the deprotonated amino group to stabilize unusually high oxidation states such as Cu^{III} and Ni^{III} favours the M^{II}–to–M^{III} redox change and the occurrence of the photo-induced electron transfer process [11]. The possibility that the quenching process is due to an An*–to–M^{II} intramolecular eT process is to be ruled out as reduction to the M^{I} state is especially disfavoured in systems such as these and is not observed even at strongly negative values of the potential of the working electrode in voltammetric experiments (down to the solvent cathodic discharge in MeCN, −2.1V vs. SCE).

Interestingly, systems **4** and **5**, which show completely analogous sensing behaviour towards Ni^{II} and Cu^{II} ions, use different signal transduction mechanisms: ET and eT, respectively.

According to the modular approach, components of the fluorosensor can be changed at will. For instance, it could be of some interest to replace the quadridentate receptors of systems **4** and **5** by their cyclic counterparts, to obtain **6** and **7** [8]. The reason of the interest is that, *ceteris paribus*, cyclic ligands form more stable metal complexes than their open-chain analogues (the thermodynamic macrocyclic effect [12]). The tetramine receptor in **6** has the skeleton of the classical 14-membered macrocycle *cyclam*, whereas the receptor subunit of **7** refers to the other well-known object of macrocyclic chemistry *dioxocyclam*.

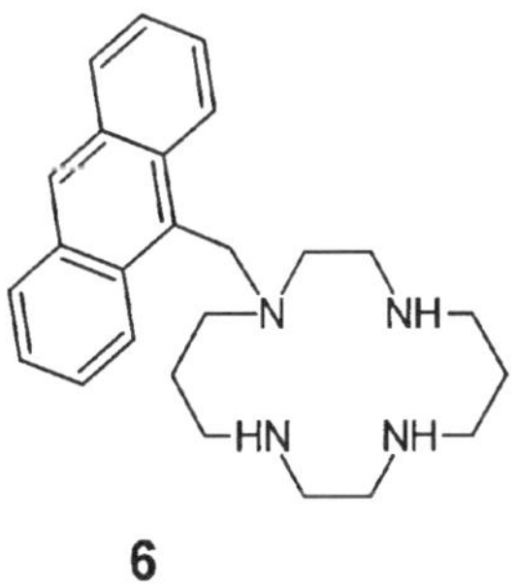

6 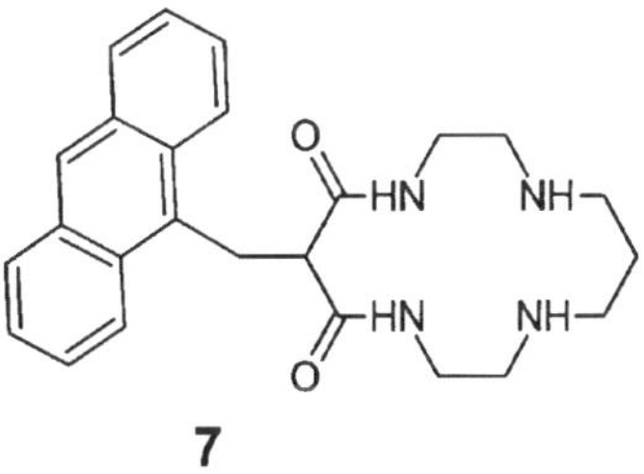 **7**

The fluorescent behaviour of **6** is quite similar to that observed with the parent non-cyclic system **4**. In particular, in the presence of equimolar amounts of either Cu^{II} or Ni^{II}, sigmoidal I_F vs. pH profiles were observed on titration with standard base and complete fluorescence quenching was observed at pH ≥ 4 and pH ≥ 7, respectively. However, **6** presents some unfavourable features that preclude its use as a sensor.

The efficiency of a receptor to be used as a subunit of a sensor is strictly related to the facility and simplicity of the binding process. In particular, substrate binding has to be fast and reversible. For instance, the same sigmoidal I_F vs. pH profile, as observed in the spectrofluorimetric titration experiments involving sensors **4** and **5** and illustrated in Figures 7 and 9, is obtained both on titration with standard base (pH increase) and with standard acid (pH decrease), demonstrating fast reversibility. This is not the case of system **6**. In fact, if the basic non-fluorescent solution (e.g. pH = 12) is back-titrated with standard acid, fluorescence is not restored, even in strongly acidic solution and after waiting several hours. This behaviour reflects the extreme inertness of $[M^{II}(cyclam)]^{2+}$ complexes towards demetallation: $[Ni^{II}(cyclam)]^{2+}$ lasts in 1M $HClO_4$ with a half-time of approximately 30 years [13]! This kinetic macrocyclic effect has to be associated to the close fit of the 14-membered ring for Cu^{II} and Ni^{II} ions and results from the ligand difficulty to rearrange and expose nitrogen atoms to the incoming H^+ ions. It has been previously pointed out that quick binding reversibility is an essential prerequistite in the design of sensory materials. The high resistance of the macrocyclic tetramine receptor to release the imprisoned metal prevents system **6** from being used as a sensor.

On the contrary, the other system bearing a cyclic receptor subunit, **7**, displays fully reversible behaviour: superimposable I_F vs. pH profiles are obtained in the spectrofluorimetric titrations in presence of either Cu^{II} and Ni^{II}, both when increasing and decreasing pH. The rapid demetallation of the macrocyclic complexes has to be related to the fact that the H^+ ions go to bind the very accessible and partially negatively charged carbonyl oxygen atoms of the two deprotonated amide groups of the dioxocyclam subunit: this increases the double bond character of the C–N bonds, drastically reduces the coordinating tendencies of the nitrogen atoms and promotes metal extrusion from the ring (see the sketch in Figure 11) [14]. Due to its unique electronic features, the dioxocyclam fragment is the only tetra-aza macrocycle that can act as a receptor for transition metal ions and can be used to build a sensor.

The nature of the metal-induced quenching mechanisms in systems **6** and **7** is the same as that observed with open-chain analogues: ET for the tetramine derivative, eT for the deprotonated diamide–diamine derivative. We can try to associate the different behaviour to different structural features present in the two classes of receptors.

First, it has to be noted that in the case of the cyclam-containing systems $[M^{II}(\mathbf{6})]^{2+}$, which are actually quenched via an ET mechanism, the occurrence of an An^*–to–M^{II} electron transfer would be favoured from a thermodynamic point of view (M = Cu, $\Delta G^\circ = -1.1$ eV; M = Ni, $\Delta G^\circ = -0.3$ eV). Predominance of the ET mechanism over the feasible eT processes in systems $[M^{II}(\mathbf{4})]^{2+}$ and $[M^{II}(\mathbf{6})]^{2+}$ can be ascribed to the fact that the fluorophore and the metal centre are quite close and occasional rotation of the bridging $-CH_2-$ group may bring the anthracene fragment at a van de Waals contact with the metal, the required condition for the ET process to take place [15]. On the other hand, in systems $[M^{II}(\mathbf{5})]$ and $[M^{II}(\mathbf{7})]$ an ET mechanism is still possible, due to the availability of metal-centred empty or half-filled orbitals of low energy, but actually it is an eT process that quenches the

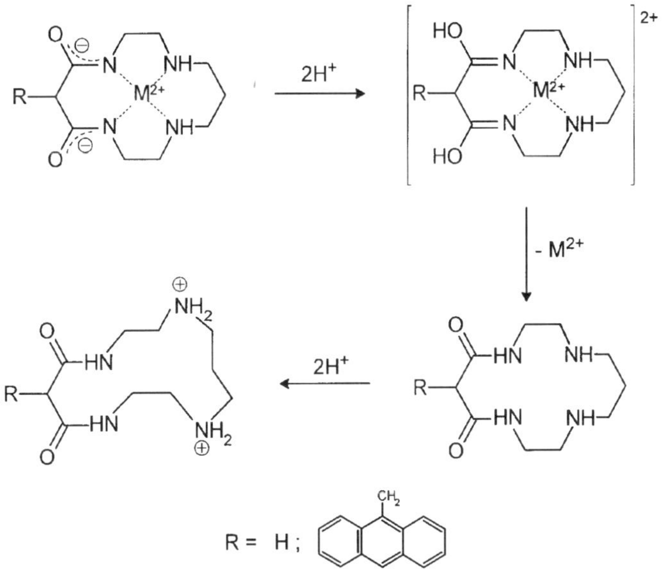

Figure 11 Fast demetallation in acidic solution of [M^{II}(dioxocyclamato(2^-))] complexes (M = Cu, Ni). Two protons attack the very accessible and partially negative oxygen atoms of the carbonyl groups, which induces a drastic reduction of the coordinating tendencies of the adjacent nitrogen atoms and promotes metal extrusion.

anthracene fluorescence. In this case, the fluorophore is much more distant from the metal centre than observed with tetramine systems **4** and **6**. Moreover, the portion of the ligand framework, to which the anthracene fragment is appended, stiffened through π-delocalization, prevents metal and An^* from coming to van der Waal's contact, precluding the occurrence of an ET process. On the other hand, the availability of the ligand's π^* orbitals makes the thermodynamically favoured electron travel from M to An^* especially fast and comfortable.

In conclusion, the flexible $-CH_2-$ linker allows a direct contact between metal and fluorophore favouring the occurrence of an ET process in systems **4** and **6**. Sensors **5** and **7** take profit from the rigidity and electron permeability of their spacers and operate via the eT mechanism.

5 TRANSITION METAL RECOGNITION AND SENSING: Fe^{III}

Following the general principles outlined in the previous section, one could design efficient molecular fluorosensors for any transition metal: simply append one (or more) luminescent fragment to a ligating framework displaying selective tendencies towards the envisaged metal centre. A successful example refers to iron(III).

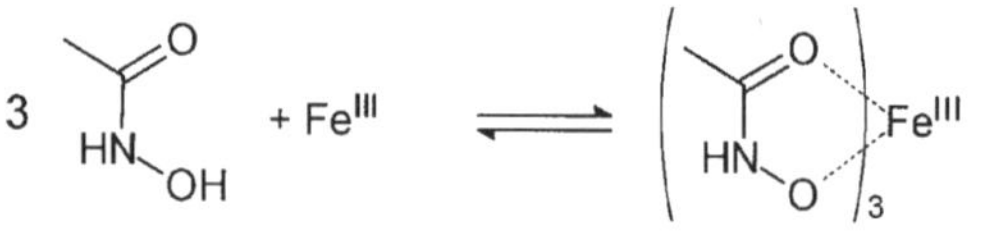

The hydroxamate chelating agent shows a very high affinity towards Fe^{III}: in particular, three hydroxamate ligands bind Fe^{III}, to give a very stable octahedral complex, whose global formation constant, β_3, is $10^{-28.3}$ in water [16]. The hydroxamate subunit is present in siderophores, low molecular weight multidentate systems responsible for iron(III) storage and transport in plants and microorganisms [17]. Hydroxamate-containing systems form coordination complexes so stable that are able to solubilize $Fe(OH)_3$ (**8**).

Appending three hydroxamate subunits to the tripodal platform $CH_3C(CH_2OCH_2CH_2-)_3$ offers a convenient stereochemical arrangement for full coordination of the Fe^{III} ion [18]. To provide signalling functionality, it appeared synthetically convenient to append to the receptor framework not one but three fluorescent fragments. In particular, a pyrene fragment was linked, via a $-CH_2-$ spacer, to each hydroxamate subunit [19].

Noticeably, the presence of three linked and proximate fluorophores led to the formation of excimers upon irradiation of a metal free solution of **8** in MeOH/water (4 : 1, v/v). In particular, the strong excimer band at 485 nm (excimer-to-monomer intensity ratio = 17) does not vary over the 2–8 pH interval. At pH $\geq$ 8 the emission is quenched according to a sigmoidal profile (see Figure 12).

Quenching is associated to the stepwise deprotonation of the three hydroxamic acid fragments of **8**, whose pK_A values range between 8.2 and 9.5. In particular, the

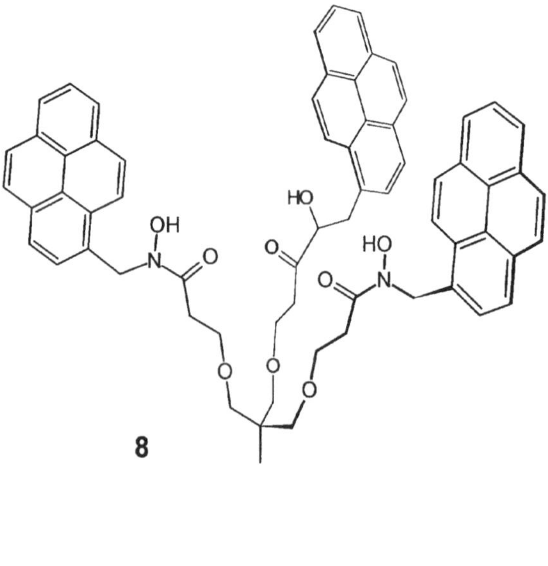

8

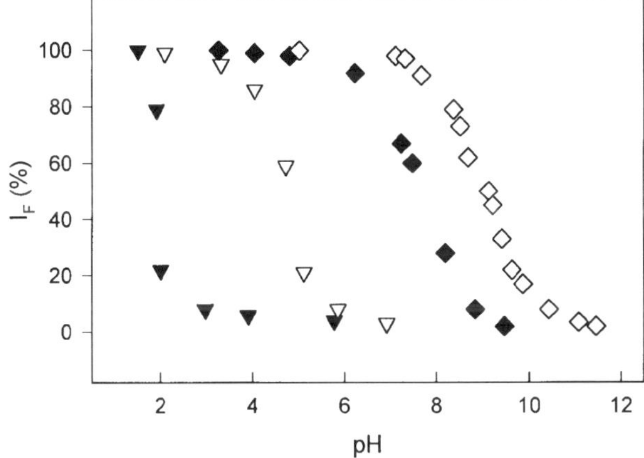

Figure 12 Fluorosensing of transition metals by the tripodal tri-hydroxamate system **8**. Spectrofluorimetric titration by standard base of **8**, in aqueous MeOH: diamonds: **8** plus excess acid; full triangles: **8**;, plus 1 equiv. of Fe^{III} and excess acid; open triangles: **8**, plus 1 equiv. of Cu^{II} and excess acid; full diamonds: **8** plus 1 equiv. of Ni^{II} and excess acid. Co^{II} and Fe^{II} do not modify the titration profile obtained with **8** alone (open diamonds).

electron-rich hydroxamate groups (in particular, their deprotonated oxygen atoms) can act as electron donors and quench, through an eT mechanism, the nearby photo-excited fluorophores. In presence of 1 equiv. of Fe^{III}, the sigmoidal fluorescence quenching profile is anticipated to pH = 2, due to the formation of a very stable metal complex. The nature of the quenching mechanism has changed, as it takes place via an ET process (double electron exchange type) involving the orbitals of the $3d^5$ cation. The hydroxamate ligand exhibits a fairly large binding affinity towards Cu^{II} and pH titration in presence of 1 equiv. of Cu^{II} gives rise to a sigmoidal quenching profile centred at pH = 4.5. A further shift of the quenching profile toward higher pH values is observed with Ni^{II}. Apparently, Co^{II} and Fe^{II} do not modify the fluorescent emission, as the I_F vs. pH profile is coincident with that observed for the titration of **8** alone.

The large separation of the quenching profiles of Fe^{III} and Cu^{II} allows an easy discrimination of the two metal ions. For instance, on titration with Cu^{II} of a solution of **8** buffered at pH = 3, the strong excimer emission remains unchanged (see Figure 13). On the other hand, titration with Fe^{III} induces linear I_F decrease and complete quenching with 1 equiv. addition.

6 FLUORESCENCE QUENCHING OR ENHANCEMENT FOR METAL ION SENSING?

Fluorosensors **5** and **8** can make a useful concerted job, as **5** senses Cu^{II}, but neither recognizes nor senses Fe^{III}; on the other hand, **8** senses Fe^{III} in presence of Cu^{II},

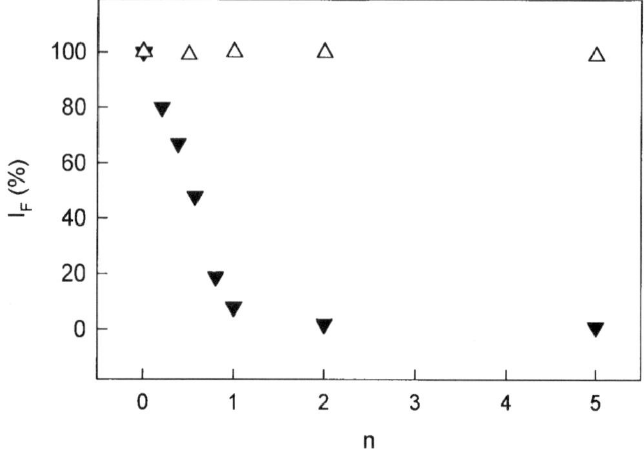

Figure 13 Spectrofluorimetric discrimination of Fe^{III} and Cu^{II} using fluorosensor **8**. The cuvette contains an aqueous MeOH solution of **8**, buffered to pH = 3. The solution is first titrated with Cu^{II} and the fluorescent emission is not altered at all (at this pH, the Cu^{II} ion is not *recognized* by the receptor subunit of **8**). Then, on titration with Fe^{III}, a linear decrease of I_F is observed, with complete quenching after 1 equiv. addition.

when operating at the appropriate pH. The two systems nicely demonstrate how efficiently the sensor selectivity can be designed by taking into account the coordinative preferences of the envisaged metal ion. One hundred years of coordination chemistry have taught chemists how to fit the electronic and stereochemical demands of the metal centre by modulating the structural features of the receptor: type and number of donor atoms, whether the multidentate ligand backbone has to be linear, branched or cyclic, whether it has to be flexible or rigid. All this experience is very useful for designing the selectivity of the recognition process of most transition metal ions. On the other hand, signalling is not a problem in view of the spontaneous tendency of d block metals to quench the emission of a nearby fluorophore, via either an eT or an ET mechanism (or both simultaneously).

However, quenching does not appear as the most desirable choice of fluorescent signalling of a recognition event in solution. As a matter of fact, other species in solution, which may be present in a concentration much larger than that of the investigated analyte, can interfere, by quenching themselves fluorescence (for being redox active, containing an heavy atom, possessing a radical nature, like dioxygen). Signalling through fluorescence enhancement (OFF–ON) seems a much more beneficial option, as it suffers the competition of a much lower number of exotic agents [21].

Metal ions such as Na^+, K^+, Ca^{2+}, etc., which have a close shell electronic configuration (ET precluded) and lack of any redox activity (eT prevented), are photophysically inactive. This makes possible the design of efficient OFF–ON fluorosensors. A classical example is given by de Silva's fluorophore-spacer-receptor system **9**, which senses K^+ through a 47 times enhancement of the fluorescence of the anthracene fragment [22]. In absence of metal, the fluorophore is quenched by an eT process from the proximate oxidizable tertiary amine nitrogen atom. When the K^+ ion is encircled by the macrocycle, the nitrogen lone pair becomes involved in the metal-ligand interaction and cannot release electrons, making fluorescence revive. Recognition selectivity is thus provided by the favourable matching of cation and NO_5 18-membered ring diameters. The nitrogen atom of the ring acts as a *relay* in the signal transduction process.

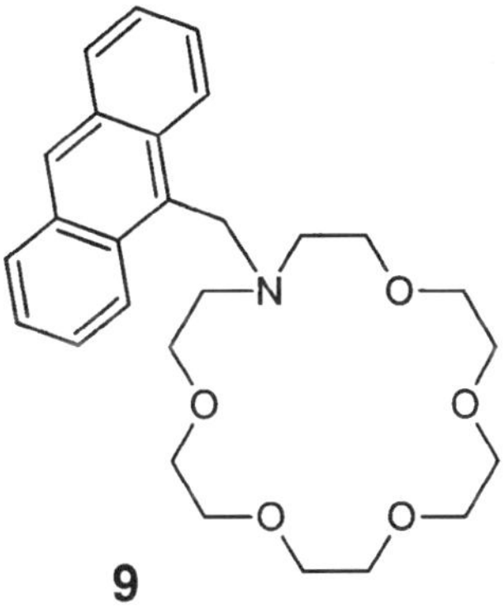

Transition metal sensing via enhancement (OFF–ON) rather than quenching of fluorescence (ON–OFF) may seem an elusive target, due to the photophysical activity of metal itself. An example in this sense has been observed with the cage system **10**, which in a metal-free THF solution is poorly fluorescent (due to the occurrence of an eT process from the tertiary amine group to the proximate anthracene fragment) [23]. Inclusion of either Ni^{II} or Cu^{II} into the cage, still in a THF solution, induces a sharp enhancement of the fluorescence of the appended anthracene fragments. However, the apparent paradox of fluorescence enhancement by a proximate transition metal may be explained by considering that, before coordination, a very efficient quenching mechanism (eT from a close tertiary amine nitrogen atom) operates and quenches most of the light emission. After coordination, this mechanism is replaced by another one, either ET or eT in nature, which involves a more distant metal centre and which is definitely less efficient. Thus, following metal binding, we do not observe a full revival of fluorescence, but simply a less efficient quenching of the fluorophore (which results in a definite enhancement of the light emission). Comparison of the OFF–ON system **10** with the ON–OFF systems **4** and **8** is misleading as in the latter cases the receptor, prior to metal binding, has its donor atoms protonated ($-NH^+$ and $-OH$ groups), so that the nearby fluorophore can display its full luminescence. On coordination, either an eT or ET mechanism involving the metal centre becomes operative and causes a drastic reduction of the emission intensity. Thus, signalling of transition metal recognition through fluorescence enhancement is not a chimaera, provided the donor atoms of

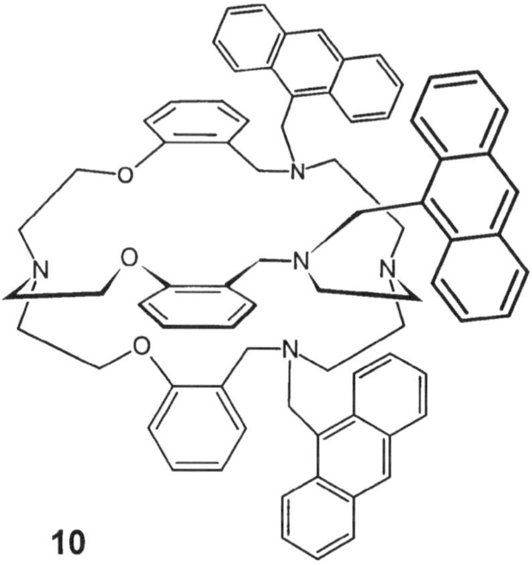

10

the receptor moiety, prior to coordination, are allowed to quench the proximate fluorophore. In the case of anthryl-polyamine derivatives this situation can be achieved by avoiding the protonation of the amine groups, for instance carrying out titration experiments in an aprotic medium. It is disappointing that sensing methodologies are in most cases required in aqueous or water-containing solutions.

7 RECOGNITION AND SENSING OF Zn^{II}

The Zn^{II} ion, which has a $3d^{10}$ electronic configuration, cannot be strictly considered a transition metal ion. However, even if it does not profit at all from ligand field stabilization effects, it forms complexes with amine ligands whose solution stability is comparable to that of 3d metal ions. For instance, the $\log K_1$ associated to the binding of the first molecule of ammonia (equilibrium: $M^{II}+NH_3=[M^{II}(NH_3)]^{2+}$), 2.32, is lower than that observed for Cu^{II} (4.14) and Ni^{II} (2.72), but larger than for Co^{II} (2.11), Fe^{II} (1.53) and other divalent 3d cations [24]. Moreover, Zn^{II} possesses a unique prerogative of interest in the world of fluorescent sensing: due to its closed shell electronic configuration and unconditional resistance to any redox activity, it is absolutely inoffensive from a photophysical point of view. As a consequence, Zn^{II} addresses the design of the most valuable fluorosensors of the OFF–ON type, similar to those commonly used for s block cations. A clear example is given by the previously discussed system **4**, whose tetramine receptor moiety displays fairly strong coordinating tendencies towards Zn^{II}.

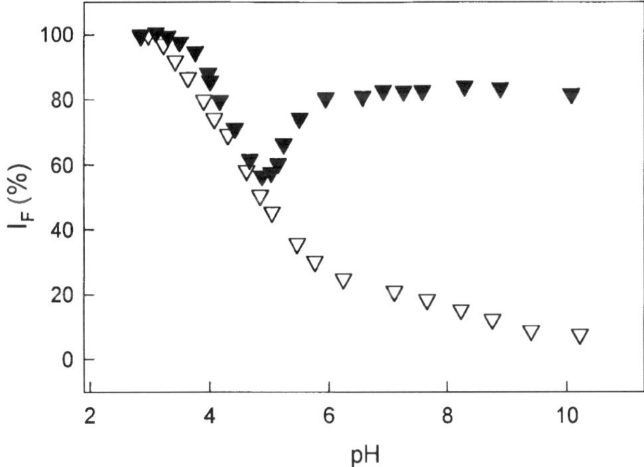

Figure 14 Fluorescence enhancement associated to Zn^{II} coordination to **4**. Full triangles refer to the spectrofluorimetric titration by standard base in aqueous MeCN of the two-component system **4**, plus 1 equiv. of Zn^{II} and excess acid; the titration profile obtained for a metal-free solution (open-triangles) is reported for comparative purposes. Coordination to Zn^{II}, a pH $\geq$ 4.5, interrupts the eT process from the amine group adjacent to the fluorophore and awakens the fluorescence.

Figure 14 shows the I_F vs. pH profile obtained by titrating with standard base an $MeCN/H_2O$ solution containing equimolar amounts of **4** and Zn^{II}, plus excess acid. Until pH = 5, I_F decreases, according to a profile which superimposes well on that obtained in absence of metal. At pH = 5, I_F stops decreasing, then increases again to reach, at pH = 6, a limiting value corresponding to about 80% of the original intensity. The left edge of the I_F well describes the situation in which the ammonium groups close to the luminescent fragment deprotonate, but do not get bound to Zn^{II}, yet. At pH = 5, the Zn^{II} tetramine complex begins to form and, accordingly, the amine–to–An* eT process is suspended, which, in the absence of any other quenching mechanism, makes fluorescence revive. The complex formation and emission revival is complete at pH = 6.

The profile in Figure 14 shows that at pH $\geq$ 6, **4** can act as an OFF–ON sensor of Zn^{II}. As a matter of fact, titration with Zn^{II} of a solution of **4** buffered to pH = 8.1 (see Figure 15) induces a linear increase of I_F, to reach a limiting value after the addition of 1 equiv. Cu^{II} and Ni^{II}, if present in solution, would preclude Zn^{II} sensing by **4**, as they form more stable complexes than Zn^{II} with the tetramine receptor and quench the fluorosensor.

System **4** can be modified in such a way than Zn^{II} recognition can induce fluorescence quenching. In particular, the classical donor group *N,N*-dimethylaniline (DMA) can be appended to the other side of the tetramine receptor to give **11** [25].

$R = N(CH_3)_2$: **11** $R = NO_2$: **12**

In the typical pH-titration experiment, I_F does not stop decreasing at pH = 5, but keeps going down until complete quenching (Figure 16).

This behaviour is ascribed to the fact that, when the Zn^{II} tetramine complex forms, the DMA donor fragment is brought close enough to the anthracene subunit to allow the occurrence of a through-space eT process. Figure 17 sketches very roughly the mechanism of the Zn^{II} driven quenching mechanism. In the sketch, the tetramine subunit linking the donor group and the fluorophore coordinates the Zn^{II} ion according to a square planar stereochemistry.

However, the non-transitional Zn^{II} ion would rather prefer either a tetrahedral or a five-coordinate stereochemistry (in the latter case, a solvent molecule would complete the coordination polyhedron). Figure 18 shows the CPK model for the tetrahedral arrangement, as calculated by a semi-empirical molecular model. In this situation, the nitrogen atom of the donor group is only 5.1 Å distant from the closest

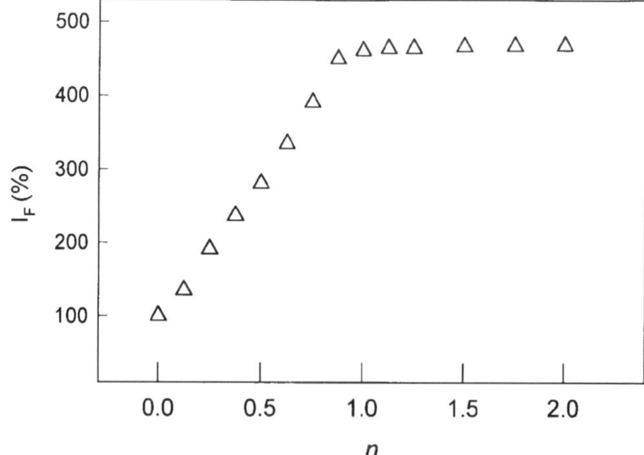

Figure 15 Fluorosensing of Zn^{II} through fluorescence enhancement. On titration with Zn^{II} of a solution of **4** in aqueous MeCN, a linear increase of I_F is observed, with complete fluorescence restoring after 1 equiv. addition.

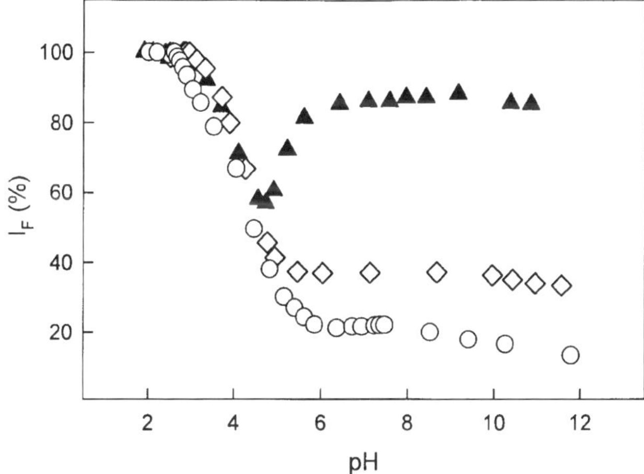

Figure 16 Zn^{II} addressed intramolecular electron transfer involving ω,ω′-substituents of a linear tetramine. Spectrofluorimetric titration by standard base of disubstituted tetramines **11** (diamonds) and **12** (circles), in aqueous MeCN, in presence of 1 equiv. of Zn^{II}. Triangles refer to the titration experiment with reference system **4**. For systems **11** and **12**, I_F keeps decreasing at pH $\geq$ 4.5, since the eT process from the adjacent amine group of An* is replaced by an intracomplex eT process involving either the $-NMe_2$ (**11**) or the $-NO_2$ (**12**) substituent.

carbon atom of the anthracene fragment, C(9). This distance should be compared with that calculated for the fully stretched system **11**: 20.6 Å. Such a situation is achieved in a strongly acidic solution, when all the amine groups are protonated and the anthracene subunit displays full emission.

Thus, when a solution of **11**, buffered to pH = 8.1, it titrated with Zn^{II}, only a minuscule increase of I_F is observed. This discourages the use of **11** as a fluorosensor of Zn^{II}. However, the poorly discernible equivalent point in the plot in Figure 19 is interesting from a mechanistic point of view as it indicates that the nitrogen–to–An* eT process has been completely replaced by the intracomplex DMA–to–An* eT mechanism. A similar behaviour is observed with system **12**, in which the DMA donor group of **11** has been replaced by the acceptor group nitrobenzene (NB) (see Figures 16 and 19). The Zn^{II} addressed fluorescence quenching is now ascribed to an intracomplex eT process from the excited fluorophore. It should be noted that both eT processes (DMA–to–An* and An*–to–NB) are distinctly favoured from a thermodynamic point of view: $\Delta G^{\circ}_{eT} = -0.4$ and -1.0 eV, respectively.

Appending two anthracene subunits to an open-chain tetramine receptor brings some new interesting effects related to Zn^{II} complexation [26]. The disubstituted system **13** in an aqueous basic solution shows (i) a weak structured emission band with $\lambda_{max} = 414$ nm, which is due to the normal anthracene emission and is quenched in most part via the amine–to–An* eT mechanism, and (ii) a weak

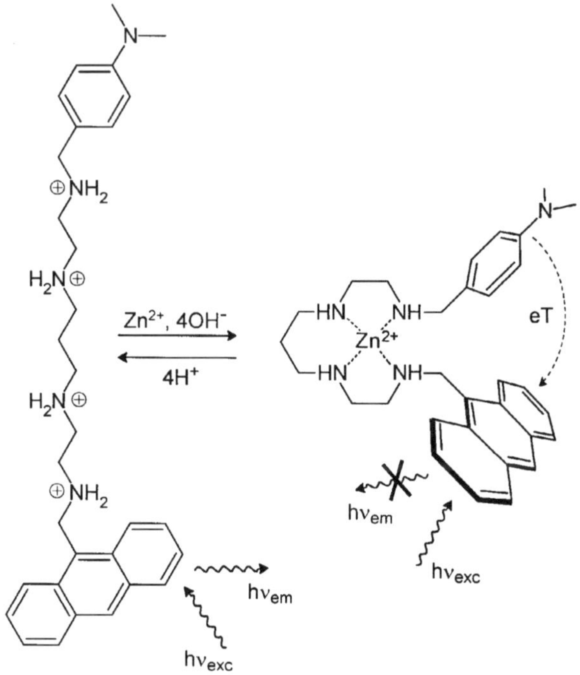

Figure 17 A pictorial view of the mechanism by which Zn^{II} induces quenching of anthracene fluorescence of three-component system **11**. On coordination to the metal centre, the *N,N*-dimethylaniline subunit and anthracene fluorophore are brought close enough to allow a 'through-space' eT process to take place.

non-structured emission centred at 500 nm. This latter emission is ascribed to the excimer which occasionally forms with the folding of the tetramine linker. On addition of Zn^{II} to the basic solution of **13**, the excimer band intensity increases, to reach a four-fold intensity after the addition of 1 equiv. of the metal. The enhanced

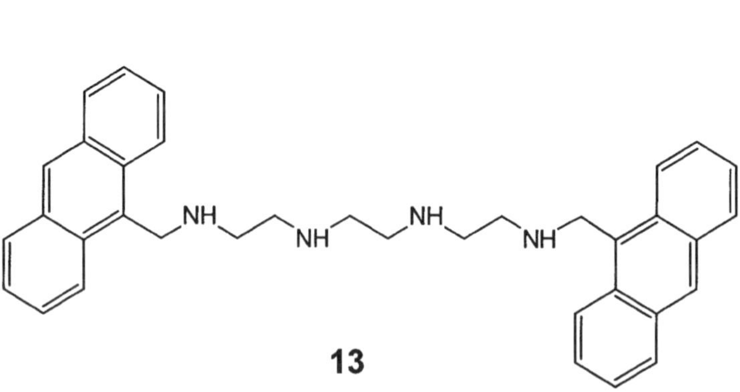

13

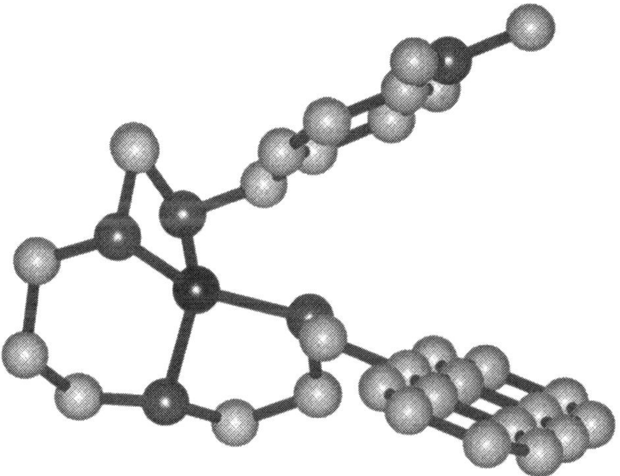

Figure 18 The ball-and-stick model of the Zn^{II} complex of the disubstituted tetramine **11**, as calculated through a semi-empirical molecular orbital method. Hydrogen atoms have been omitted for clarity. The Zn^{II} centre (black ball) chose a tetrahedral stereochemistry. Five-coordination (including a solvent molecule) is possible, too. Both stereochemical arrangements put the *N,N*-dimethylaniline and anthracene subunits at a distance favourable to the occurrence of a donor-to-fluorophore eT process.

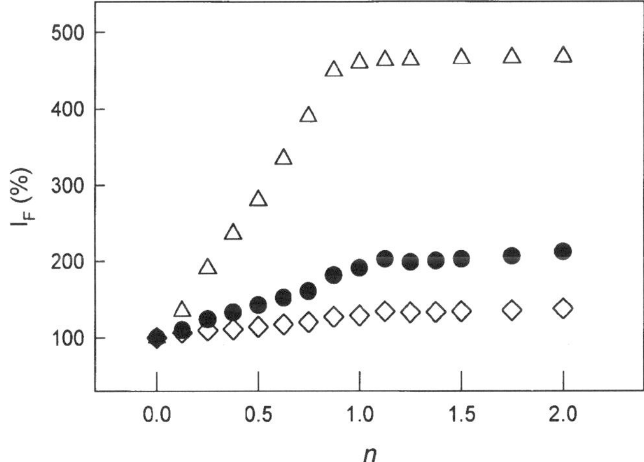

Figure 19 Spectrofluorimetric titration with Zn^{II} of a solution of **4** (triangles), **11** (diamonds) and **12** (circles) in aqueous MeCN. In the case of the disubstituted tetramines **11** and **12**, metal complexation does not induce the fluorescence revival observed with **4**, due to the occurrence of an intracomplex eT process.

excimer emission has to be ascribed to the beneficial structural effect induced by Zn^{II} coordination. As outlined in the case of the donor-acceptor interactions discussed previously, the coordination to the metal of the tetramine linker places the two substituents at a close distance: this makes the otherwise seldom intramolecular interaction of an anthracene fragment with its excited counterpart much easier and more probable, leading to a substantial increase of the excimer emission. This situation allows the analyst to detect the Zn^{II} concentration from the ratio of the intensities on the bands at 500 and 414 nm. Ratiometric methodologies are especially desired for analytical purposes, since they grant a safe determination of the analyte, even when the concentration of the sensor is unknown [27].

8 ANION SENSING BASED ON THE METAL–LIGAND INTERACTION

Nothing hinders the utilization of the two-component (fluorophore–receptor) approach to the design of fluorosensors for anions. In this case, the light-emitting fragment has to be linked to a receptor subunit capable of interacting selectively (recognizing) with negatively charged substrates. Most examples of anion recognition involve receptor-substrate electrostatic interactions, including hydrogen bonding [28]. In particular, the receptor should offer a concave array of positively charged sites (e.g. ammonium groups), whose shape and size are complementary to those of the anion. A large number of receptors for any kind of inorganic and organic anions has been made available during the last two decades. However, very few of them have been functionalized with a light-emitting fragment: anion fluorosensors are rare.

An example is given by the two-component system **14**, in which the anthracene subunit has been linked to a tripodal tetramine (Figure 20) [29]. At pH = 6, all the amine groups are protonated, but the anthrylamine one. Due to the familiar amine–to–An* eT process, the fluorescence of the anthracene fragment is almost completely quenched. On addition of the HPO_4^{2-} anion, fluorescence is fully restored. This is explained by assuming that the three oxygen atoms of the phosphate anion, each one formally possessing one-third of negative charge, interact with the ammonium groups of the receptor subunit, through hydrogen bonding. On the other hand, the –OH group of the anion establishes a hydrogen bonding interaction with the anthrylamine group. The latter interaction withdraws the amine nitrogen electron pair from the eT process and brings back fluorescence. Thus, system **14** is an efficient OFF–ON fluorosensor of phosphate in neutral solution, resembling in the signal transduction mechanism the fluorosensors of the photophysically inactive metal ions (alkaline, alkaline-earth and Zn^{II}),

Electrostatic interactions are poorly energetic and in protic solvent they often do not outbalance the endothermic effects associated to anion dehydration. Thus, recognition of anions has been carried out in most cases in aprotic and poorly polar solvents.

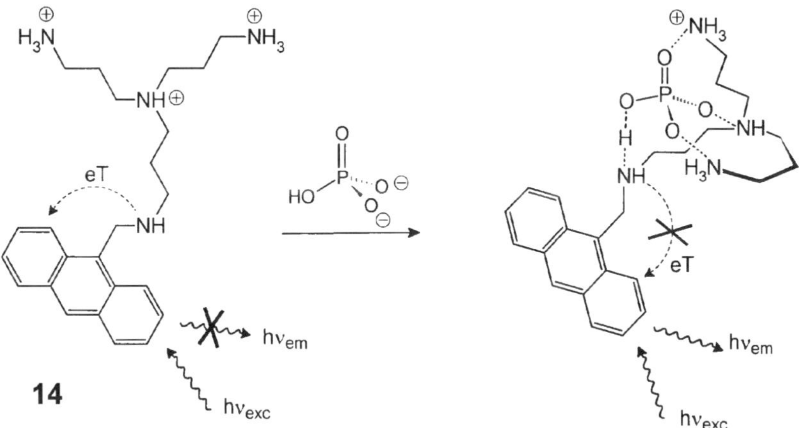

Figure 20 A two component fluorosensor of the HPO_4^{2-} ion. When involved in a hydrogen bonding interaction with the –OH group of the hydrogen-phosphate anion, the proximate amine group no longer transfers an electron to the nearby An* fragment. Thus, anion recognition is signalled through fluorescence enhancement.

Metal-ligand interactions are in general much more energetic than electrostatic interactions. Therefore, the use of metal centres as binding sites for anions would allow the recognition process to take place even in very polar and protic media, including water. Receptors of this kind should contain one or more metal centres; the metal should be bound to donor atoms present on the receptor framework, but it should possess also a vacant or temporarily occupied coordination site, reserved to a donor atom of the anion. In the world of fluorescent sensing, the choice of the metal is mandatory: Zn^{II}. Happily, Zn^{II}, besides its known photophysical innocence, has a high affinity towards negatively charged ligands.

In the two-component system **15**, a tren fragment is linked through a $-CH_2-$ to the anthracene fluorophore [30]. The tripodal tetramine receptor forms with Zn^{II} a stable complex of trigonal bipyramidal stereochemistry, whose all but one coordination sites are held by the ligand's nitrogen atoms. The remaining position, which in solution can be provisionally occupied by a solvent molecule, is available for anion binding. The $[Zn^{II}(tren)]^{2+}$ moiety of the fluorescent $[Zn^{II}(\mathbf{15})]^{2+}$ derivative typically displays a good affinity towards the carboxylate group, but when, say, a benzoate anion is bound, in an ethanolic solution, to give a fairly stable 1 : 1 adduct ($\log K = 4.7$) the appended reporter does not send any message. Indeed, there is no reason why the anthracene fluorescence has to be altered by benzoate coordination to the metal. The incoming substrate, in order to modify the emission, should provide an interfering mechanism with the photo-excited fragment, for instance either releasing or uptaking an electron; but this is not the case of plain benzoate. However, titration of a solution of $[Zn^{II}(\mathbf{15})]^{2+}$ with the 4-*N*,*N*-dimethylaminebenzoate anion induces a linear decrease of fluorescence, with a sharp equivalent point correspond-

15 $\xrightarrow{Zn^{2+}}$ $[Zn^{II}(\mathbf{15})]^{2+}$

ing to the formation of a 1 : 1 adduct (Figure 21). The same behaviour is observed when titrating with 4-nitrobenzoate.

Molecular modelling (as roughly sketched in Figure 22) shows that anion coordination to the metal brings the donor group $-N(CH_3)_2$ (as well as the acceptor group $-NO_2$) close enough to the anthracene fragment to allow the occurrence of a 'through-space' eT process to (or from) the photo-excited fluorophore.

Thus, the $[Zn^{II}(\mathbf{15})]^{2+}$ system could be considered as a rather rudimentary ON–OFF fluorosensor for aromatic carboxylates. Apparently, it recognizes any kind of benzoate derivatives, but it signals the recognition only of those substrates displaying

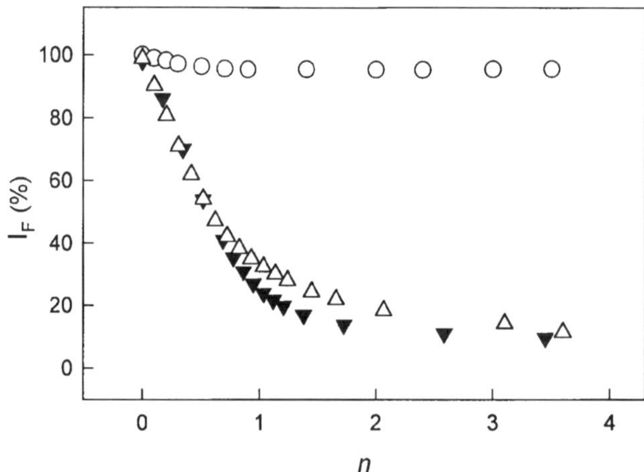

Figure 21 Spectrofluorimetric titration of a solution of $[Zn^{II}(\mathbf{15})]^{2+}$ with a benzoate anion in MeOH solution. Titration with plain benzoate (circles) does not alter the fluorescence emission of the anthracene subunit. Titration with a benzoate ion bearing either a donor (4-*N,N*-dimethylaminebenzoate, full triangles) or an acceptor substituent (4-nitribenzoate, open triangles) induces fluorescence quenching.

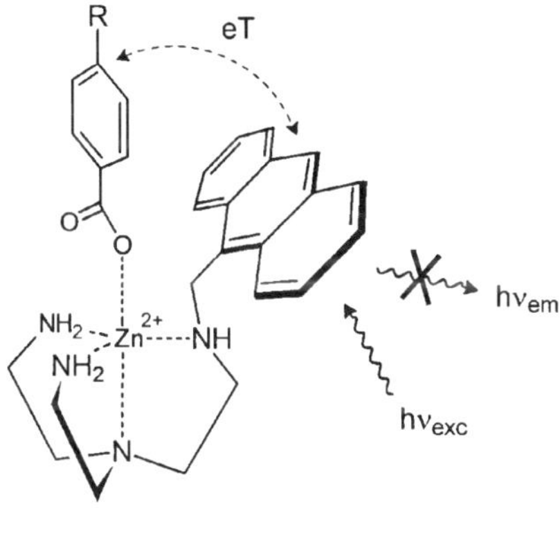

Figure 22 Pictorial view of the intracomplex photo-induced eT process from/to the donor/acceptor R substituent on benzoate, responsible for the quenching of anthracene fluorescence.

either electron donor or electron acceptor properties. Really, this is not exactly what one would expect from a sensor worth its name.

9 METAL-CONTAINING FLUOROSENSORS FOR AMINO ACIDS

Perhaps the most remarkable application of fluorescent molecular sensors concerns intracellular analysis. In particular, fluorescence microscopy allows the quantitative determination, as well as the spatial characterization, of the analyte inside the cell. One of the most investigated intracellular substrates is Ca^{II} and very efficient two-component fluorosensors have been designed in order to monitor and map its concentration inside the cell, say during muscle contraction [31]. Fluorescent molecular sensors for any kind of intracellular analytes are highly solicited by cell biology researchers: among which are sensors of amino acids. Since any amino acid bears a carboxylate donor group, one could suggest the use of a system such as $[Zn^{II}(\mathbf{15})]^{2+}$ as a sensor. Indeed, $[Zn^{II}(\mathbf{15})]^{2+}$ in an ethanol/water mixture (4 : 1, v/v), buffered to pH = 6.8, forms weakly stable 1 : 1 adducts with natural amino acids ($\log K > 2$) whatever the amino acid is. Substrate binding does not modify the emission of the reporter anthracene fragment ($\log K$ values had been determined through spectrophotometric titration experiments, monitoring absorption bands in the UV region). The scarce affinity may depend upon the electrostatic repulsions between the metal ion and the ammonium group on the amino acid. Moreover, selectivity seems precluded by the fact that the recognizing site, the Zn^{II} centre, looks at a part common to all the investigated substrates: the carboxylate group. In order to increase the affinity towards the substrate and to improve selectivity, one should insert in the receptor portion of the sensor further binding sites displaying specific affinity towards the envisioned amino acid.

A successful example refers to system **16**, in which the tren framework has been armed with two antracenyl and one benzyl substituents [32]. The corresponding Zn^{II} complex, $[Zn^{II}(\mathbf{16})]^{2+}$, displays a high affinity towards natural amino acids bearing aromatic substituents: phenylalanine (phe, **17**) and tryptophane (trp, **18**).

16

Log$K$ values associated with the formation of the $[Zn^{II}(\mathbf{16})(phe)]^{2+}$ and $[Zn^{II}(\mathbf{16})(trp)]^{2+}$ adducts, in 4 : 1 ethanol/water mixture, buffered to pH = 6.8, are 4.48 ± 0.09 and 4.21 ± 0.01, respectively (calculated through spectrophotometric titration experiments). These values are distinctly higher than those observed for all the other amino acids (e.g. glycine, log$K = 3.06 \pm 0.16$).

The rather high solution stability of $[Zn^{II}(\mathbf{16})(phe)]^{2+}$ and $[Zn^{II}(\mathbf{16})(trp)]^{2+}$ adduct can be ascribed to the establishing of π-stacking interactions between the aromatic

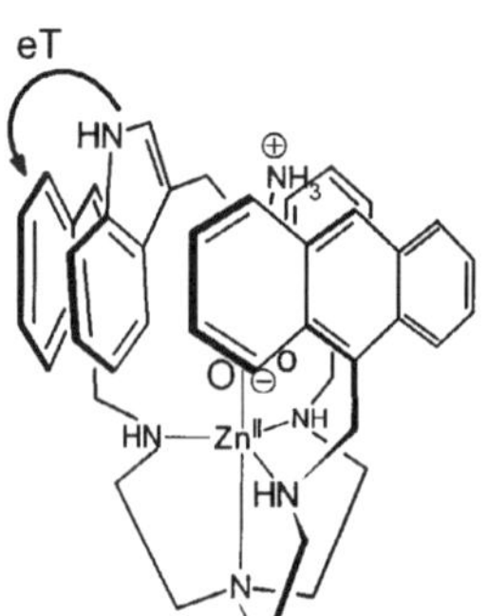

Figure 23 Hypothesized structure of the $[Zn^{II}(\mathbf{16})(trp)]^{2+}$ adduct. Tryptophane (trp) is recognized by the Zn^{II} tetramine receptor through: (i) the formation of a metal-carboxylate coordinative bond; (ii) the establishing of π-stacking interactions between the aromatic part of the amino acid and one of the facing polyaromatic substituents of the tripodal tetramine framework.

17 **18** **19**

part of the Zn^{II}-bound amino acid and one of the facing polyaromatic substituents of the tren framework. This binding situation is tentatively sketched in Figure 23.

Most interestingly, from the point of view of fluorosensing, titration of a solution of $[Zn^{II}(\mathbf{16})]^{2+}$ with trp induces fluorescence quenching (see Figure 24). The log$K$ value obtained from the spectrofluorimetric titration profile (4.28 ± 0.03) is coincident with that obtained from absorption spectra. On the contrary, titration with the other amino acid bearing an aromatic subsituent, phe, does not alter the anthracene emission. Fluorescence quenching in the $[Zn^{II}(\mathbf{16})(\mathrm{trp})]^{2+}$ adduct can be ascribed to a 'through-space' eT process from the secondary amine nitrogen atom of the trp subunit to the facing photo-excited An fragment.

The selective behaviour of the $[Zn^{II}(\mathbf{16})]^{2+}$ receptor relies on its capability to establish two distinct interactions with the $NH_3^+CH(\mathbf{R})COO^-$ substrate: (i) the $Zn^{II}-COO^-$ metal–ligand interaction and (ii) the π-stacking interaction between **R** and aromatic substituents on the receptor framework.

The example above fits well the most general rule of molecular recognition science: the higher number of the interaction points between receptor and substrate, the higher the selectivity. However, it is possible to achieve specific recognition of a

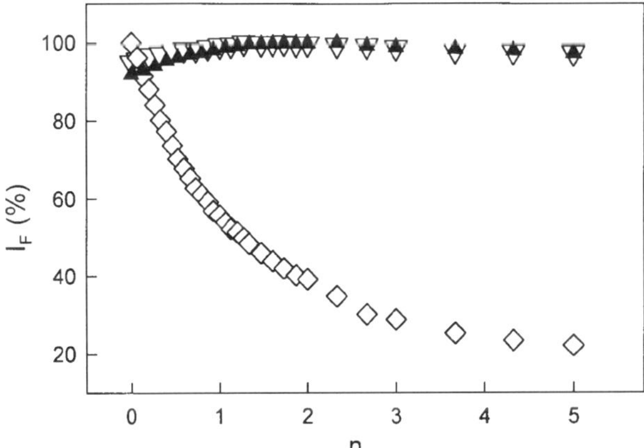

Figure 24 Titration of the $[Zn^{II}(\mathbf{16})]^{2+}$ system with amino acids in an MeOH solution: glycine (open triangles), phenylalanine (full triangles), tryptophane (diamonds). Only tryptophane, when bound to the metal centre, is able to transfer an electron to the facing An* subunit, quenching its fluorescence.

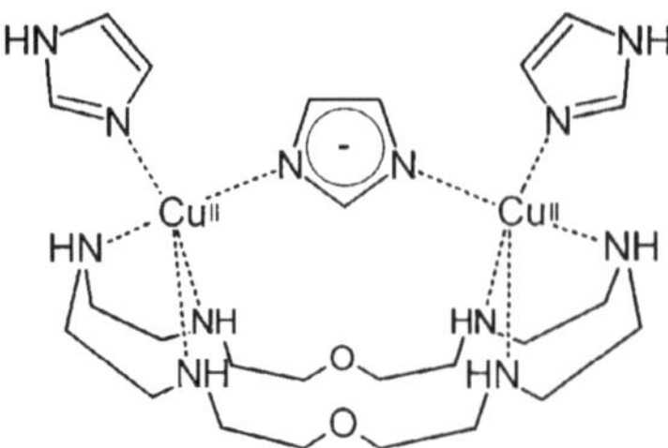

Figure 25 The imidazolate anion bridges two Cu^{II} centres prepositioned in an azamacrocycle. Each metal is five-coordinate, according to a trigonal bipyramidal stereochemistry (crystal structure in ref. [34]).

given amino acid based on a single point interaction. Specificity requires that the interaction involves the R portion of the amino acid.

A case story is represented by histidine (his, **19**), whose R substituent is an imidazolyl group. Imidazole itself (imH) is not an ambidentate ligand, but in presence of two Cu^{II} ions it releases a proton and, as the imidazolate anion (im^-), bridges the two metal centres [33]. Very importantly, the two Cu^{II} ions have to be prepositioned, i.e. they must occupy some fixed position in a given coordinative framework. Moreover, their distance must fit the 'bite' of the ambidentate im^- ligand. Finally, each copper centre must have a vacant coordination site.

One of the first reported examples of a dicopper(II) complex containing a bridging imidazolate anion is reported in Figure 25 [34]. The two Cu^{II} ions are hosted by a hexamine macrocycle and each nitrogen atom of the im^- anion occupies a vacant position of a distorted trigonal bipyramid. The very endothermic effect associated to the deprotonation of the imH acid is more than compensated for by the strong bonding interactions between the Cu^{II} ions and the partially negatively charged nitrogen atoms of im^-. A further stabilization effect must be associated to the electron delocalization over the entire Cu^{II}–NCN–Cu^{II} fragment. However, one cannot design a fluorosensor for imidazole and its derivatives based on the interaction of im^- with two prepositioned Cu^{II} centres, because of the already mentioned photophysical reasons. A useful tip comes from bioinorganic chemistry: an im^- fragment (from a histidine residue) bridges a Cu^{II} ion and a Zn^{II} ion in the copper-zinc superoxide dimutase metallo-enzyme. (Cu, Zn)SOD does not display any function related to imidazole recognition, but controls the Cu^{II}/Cu^{I} redox change in order to favour the quick disproportionation of the harmful O_2^- ion [35]. In any case, nature indicates that the imidazolate anion is a good ligand for Zn^{II}, too, and indirectly suggests the use of this metal in the design of a fluorosensor for histidine.

System **20** possesses some interesting features that recommend its use in the construction of a fluorosensor of imidazole [36]: (i) it has two distinct coordinating chambers and can host two metal ions; (ii) it leaves a coordination site to each metal centre available for further binding; (iii) the platform linking the two coordinating subunits is a fluorophore.

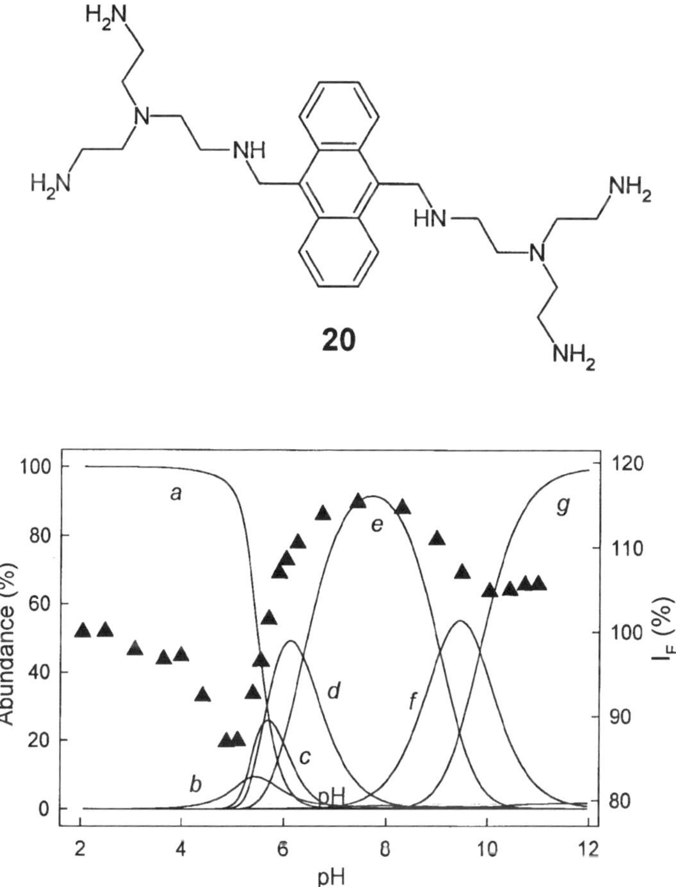

Figure 26 Formation of a dizinc(II) complex of the octamine **20** in aqueous solution: (a), left vertical axis: pH dependence of the concentration of the species present at equilibrium for the system **20** (L, 1 eqv.)/Zn^{II} (2 eqvs.) in an aqueous solution 0.1 M in $NaClO_4$, at 25°C: *a*: $[LH_6]^{6+}$; *b*: $[LH_5]^{5+}$; *c*: $[Zn^{II}_2(LH_2)]^{6+}$; *d*: $[Zn^{II}_2(LH)]^{5+}$; *e*: $[Zn^{II}_2L]^{4+}$; *f*: $[Zn^{II}_2L(OH)]^{3+}$; *g*: $[Zn^{II}_2L(OH)_2]^{2+}$; (b), right vertical axis, full triangles: pH dependence of the relative fluorescence intensity of the solution, I_F.

The formation of a stable dizinc(II) complex of **20** in aqueous solution is demonstrated by potentiometric titration experiments. Figure 26 shows the distribution diagram of the species present at the equilibrium in solution containing **20** and 2 equiv. of Zn^{II}, over the 2–12 pH range. The dimetallic species $[Zn^{II}_2(\mathbf{20})]^{4+}$ begins to form at pH = 5.5 and reaches its maximum concentration (85%) at pH = 7.5. This species should be more correctly written $[Zn^{II}_2(\mathbf{20})(H_2O)_2]^{4+}$, as the remaining axial

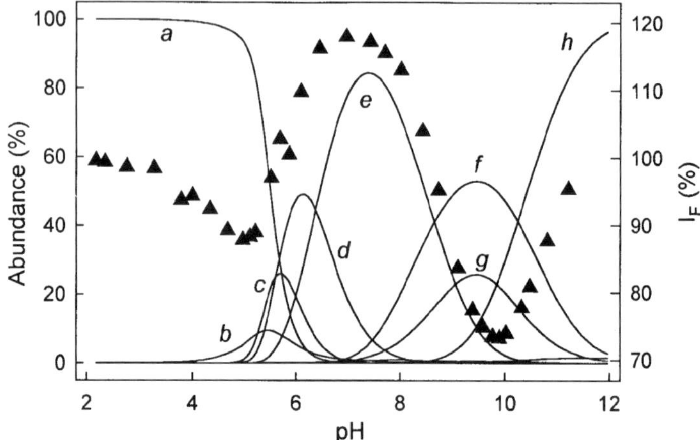

Figure 27 Recognition of the imidazolate anion in aqueous solution by the dizinc(II)-**20** system: (a) left vertical axis: pH dependence of the concentration of species present at equilibrium for the system **20** (L, eqv.)/Zn^{II} (2 eqvs.)/imidazole (imH) in an aqueous solution 0.1 M in $NaClO_4$, at 25°C: *a*: $[LH_6]^{6+}$; *b*: $[LH_5]^{5+}$; *c*: $[Zn^{II}{}_2(LH_2)]^{6+}$; *d*: $[Zn^{II}{}_2(LH)]^{5+}$; *e*: $[Zn^{II}{}_2L]^{4+}$; *f*: $[Zn^{II}{}_2L(im)]^{3+}$; *g*: $[Zn^{II}{}_2L(OH)]^{3+}$; *h*: $[Zn^{II}{}_2L(OH)_2]^{2+}$; (b), right vertical axis, full triangles: pH dependence of the fluorescence intensity of the solution, I_F.

position of each metal centre is occupied by a solvent molecule. On increasing pH, these water molecules deprotonate stepwise, to give the hydroxide-containing species: $[Zn^{II}_2(\mathbf{20})(H_2O)(OH)]^{3+}$ and $[Zn^{II}{}_2(\mathbf{20})(OH)_2]^{2+}$. The latter species is present at 100% after pH = 11. The I_F vs. pH profile for the Zn^{II}-**20** system, 2 : 1 molar ratio, also shown in Figure 26, is similar to that observed with the formation of the mono-nuclear Zn^{II} complex of the anthracene-tetramine system **4**, displayed in Figure 14. In particular, after decreasing and reaching a minimum, I_F increases again at pH ≥ 5. The distribution diagram in the same figure indicates that (i) the I_F increase corresponds to the incipient formation of the $[Zn^{II}{}_2(\mathbf{20})(H_2O)_2]^{4+}$ complex, and (ii) the progressive I_F decrease after pH = 8 is associated to the stepwise deprotonation of the axially bound water molecules. It is possible that an eT process from the electron-rich OH^- ion(s) to the nearby photo-excited platform is responsible for the latter effect.

The potentiometric titration experiment of an aqueous solution containing Zn^{II}, **20** and imH in a 2 : 1 : 1 molar ratio indicates that a stable $[Zn^{II}{}_2(\mathbf{20})(im)]^{3+}$ complex forms over the 7.5–11.5 pH range. In particular (see the distribution diagram in Figure 27) the imidazolate bridged species reaches its maximum concentration (55%) at pH = 9.5. Thus, not only copper(II), but also a pair of Zn^{II} ions are capable of promoting imH deprotonation and bridging, even if Zn^{II} does not profit from the ligand field effect and cannot be involved in any through-bridge electron delocalization and spin pairing.

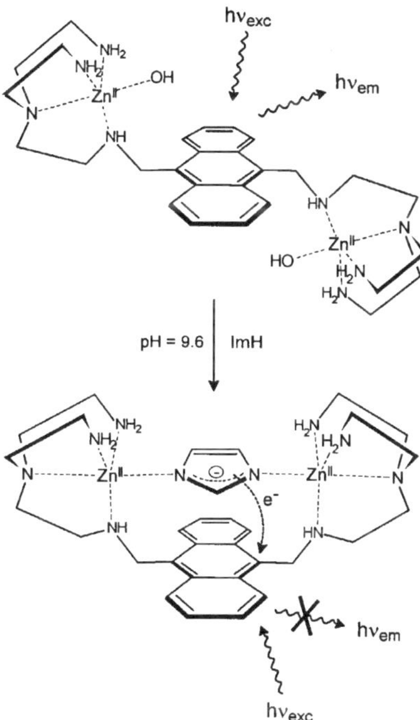

Figure 28 Sensing of the imidazolate ion by the $[Zn^{II}(\mathbf{20})(OH)_2]^{2+}$ complex. The metal-bound electron-rich imidazolate anion transfers an electron to the facing photo-excited anthracene subunit. Thus, recognition, is signalled through fluorescence quenching.

Figure 28 illustrates how system **20** rearranges in order to bring the two Zn^{II} centres at a convenient distance for im^- bridging. Very interestingly for sensing purposes, the formation of the $[Zn^{II}_2(\mathbf{20})(im)]^{3+}$ species induces a substantial decrease of the fluorescent emission, the greatest effect being observed at pH = 9.8.

Thus, if a solution containing **20** and 2 equiv. of Zn^{II}, buffered at pH = 9.8, is titrated with imidazole a decrease of fluorescence is observed (see Figure 29). Non-linear least-squares analysis of the I_F vs. equiv. profile indicates the formation of a 1 : 1 adduct with a log K value of 3.65 ± 0.04. An eT process from the electron-rich imidazolate moiety to the facing An* platform should be responsible for fluorescence quenching. Molecular modelling indicates that the im^- and An fragments are rather close (the shorter interatomic distance, i.e. that between the C(1) atom of im^- and the C(9) atom of anthracene, is 3.2 Å). Moreover, the planes containing the two subunits are nearly parallel, a situation favourable to the occurrence of a fast electron communication, through the π molecular orbitals of the two aromatic moieties.

The $[Zn^{II}_2(\mathbf{20})]^{4+}$ system should behave as an ON–OFF fluorosensor of any substrate bearing an imidazole residue. In fact, titration with histidine of a solution

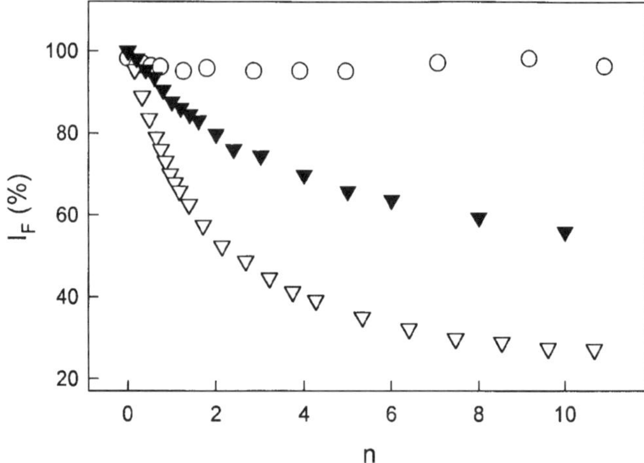

Figure 29 Spectrofluorimetric titration of the dizinc(II)-**20** system in an aqueous solution buffered to pH = 9.6, with imidazole (open triangles), histidine (full triangles), acetate (circles).

of $[Zn^{II}_2(\mathbf{20})]^{4+}$ buffered at pH = 9.6 induces fluorescence quenching, due to the formation of a 1 : 1 adduct. However, the I_F vs. equiv. profile is less steep than that observed with plain imidazole to which a lower value of log$K$ (2.92 ± 0.01) corresponds. This can be due to the steric repulsive effects exerted by the α-amino acid framework appended to the imidazole subunit. In any case, $[Zn^{II}_2(\mathbf{20})]^{4+}$ senses specifically histidine in presence of any other amino acid. In particular, (i) none of the other natural amino acids alters the fluorescence of the $[Zn^{II}_2(\mathbf{20})]^{4+}$ system; (ii) presence of a five-fold excess of any other amino acid in solution does not modify the I_F vs. equiv. profile of Figure 29, when titrating $[Zn^{II}_2(\mathbf{20})]^{4+}$ with histidine [36].

The origin of the specific recognition and sensing is straightforward: carboxylate, the donor group that other amino acids than histidine can offer is a much poorer ambidentate ligand than imidazolate, and is not able to force the endothermic rearrangement of the $[Zn^{II}_2(\mathbf{20})]^{4+}$ system which precedes the bridging of the two metal centres (outlined in Figure 28).

10 FLUORESCENT SENSORS OF THE REDOX POTENTIAL

Analytical chemists make use of indicators to visually detect the end point of redox titrations. These indicators are redox active substances, whose either oxidized or reduced form, or both, are intensely coloured, and their colours are different. Redox indicators are used when the titrating agent as well as the analyte are colourless or

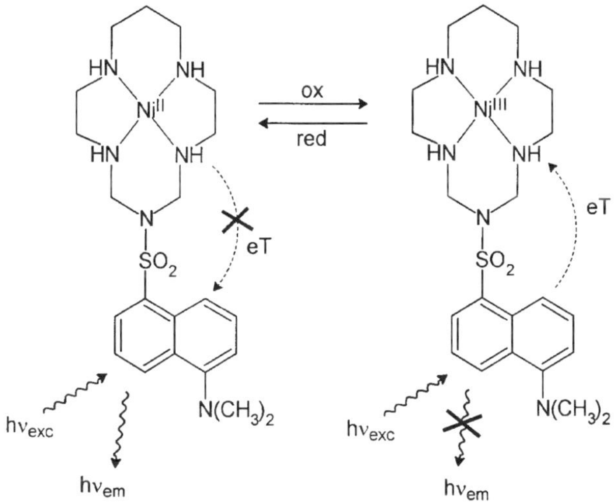

Figure 30 A two-component fluorescent sensor of the redox potential. The Ni^{III} derivative quenches the emission of the nearby dansyl subunit via a fluorophore-to-metal eT process, the Ni^{II} derivative does not. The Ni^{III}/Ni^{II} redox couple potential is 0.08 V vs. Fc^{+}/Fc. When containing a redox agent with a potential lower than 0.08 V, the solution will be fluorescent. If an oxidizing agent with a potential higher than 0.08 V is present, the solution will not emit any more.

poorly coloured in their oxidized and reduced forms. As an example, ferroins are 1 : 3 complexes of iron with substituted 1,10-phenanthrolines: the Fe^{II} form has a deep red colour (due to an MLCT transition), the Fe^{III} derivative is pale blue (much less intense d–d absorption). Depending upon the nature of substituents, ferroins have redox potentials ranging from 1.3 to 0.9 V and are used as indicators in titrations involving rather strong oxidizing agents (from Ce^{IV} to dichromate) [37]. Fluorescent redox indicators can be designed following the two-component approach outlined in the previous sections. In particular, a fluorescent subunit has to be covalently linked to a redox active fragment C, in which the reduced form, C_{red}, and the oxidized one, C_{ox}, have a comparable stability and are separated by a fast and reversible one-electron half-reaction. The ON–OFF fluorescent behaviour requires that one of the two forms quenches the emission of the nearby fluorophore, whereas the other does not. As transition metals are naturally inclined to act as one-electron redox active centres, one should link to the light-emitting subunit a receptor capable of hosting a transition metal centred redox couple. A cyclic arrangement of the receptor is recommended, in order to prevent metal extrusion by some exotic agent present in solution, which would cause indicator destruction.

An example of a two-component fluorescent redox indicator is illustrated in Figure 30 [38]. The redox active centre is represented by the Ni^{III}/Ni^{II} couple, which

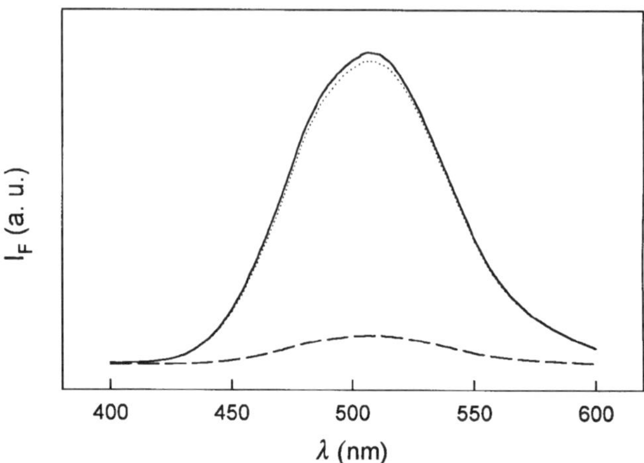

Figure 31 Redox switching of the dansyl fluorescence of the two-component system illustrated in Figure 30. The Ni^{II} derivative, in aqueous ethanol, displays the dansyl emission (solid line); on addition of $S_2O_8^{2-}$, quick one-electron oxidation to the Ni^{III} derivative takes place and fluorescence is quenched (dashed line); on subsequent addition of NO_2^-, Ni^{III} is reduced to Ni^{II} and fluorescence is revived (dotted line).

has been firmly encircled by a 14-membered tetra-aza macrocyclic subunit belonging to the family of cyclam. The fluorophore is the dansyl fragment, whose emission at 510 nm is due to a charge transfer excited state.

The reduced form (Ni^{II} derivative) is fluorescent, the oxidized one (Ni^{III} derivative) is not (see corresponding emission spectra in aqueous ethanol in Figure 31). One can move from one state to the other, thus switching ON–OFF fluorescence both electrochemically (carrying out a controlled potential electrolysis experiment) and chemically. In particular, on addition of $S_2O_8^-$ to the Ni^{II} derivative, a one-electron oxidation process takes place on the metal centre and fluorescence is quenched. One subsequent addition of NO_2^{2-}, Ni^{III} is reduced to Ni^{II} and fluorescence is revived. Quenching of the photo-excited dansyl fragment (dans) in the oxidized derivative is ascribed to a dans*–to–Ni^{III} eT process (in particular, it has been observed that when an ethanolic solution of the Ni^{III} form is frozen to 77 K fluorescence is fully restored).

The occurrence of an eT process in a fluid solution is accounted for by the strongly negative associated ΔG°_{eT} value, as calculated through the thermodynamic cycle reported in Figure 32. Rigidity of the linker (a sulphonamide group) as well as its permeability to electrons (due to the availability of π molecular orbitals) make the eT process compete successfully with the radiative decay. The occurrence of the other formally possible eT process, Ni^{II}–to–dans*, is prevented, due to the very unfavourable ΔG°_{eT} value (≥ 0.68 eV, see the cycle in Figure 32). Thus, if a solution containing the nickel containing two-component system emits light at 510 nm, it

$$Ni^{III} + {}^{*}Dns \xrightarrow{\Delta G^{0}_{eT}} Ni^{II} + Dns^{+}$$

E^{*}

$E^{0}(Ni^{III}/Ni^{II})$

Dns $-E^{0}(Dns^{+}/Dns)$

$$\Delta G^{0}_{eT} = -[E^{*} + eE^{0}(Ni^{III}/Ni^{II}) - eE^{0}(Dns^{+}/Dns)] = -1.93 \text{ eV}$$

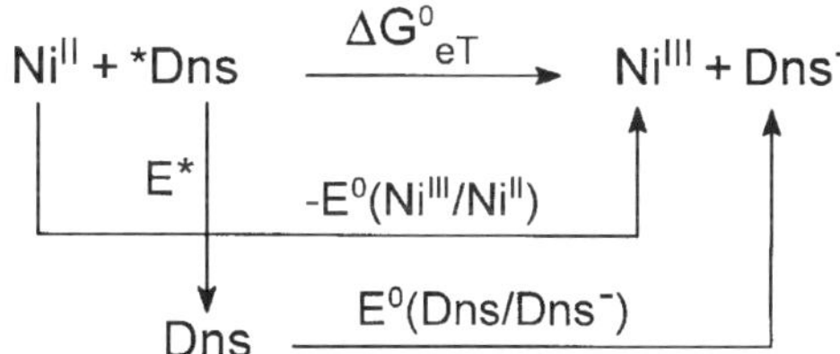

$$\Delta G^{0}_{eT} = -[E^{*} - eE^{0}(Ni^{III}/Ni^{II}) + eE^{0}(Dns/Dns^{-})] \geq 0.68 \text{ eV}$$

Figure 32 Thermodynamic bases of the fluorescence switching behaviour illustrated in Figure 30. The eT process from the Ni^{III} centre to the nearby photo-excited dansyl subunit (dns^{*}) is characterized by a strongly negative free energy change: $\Delta G^{\circ}_{eT} = -1.93$ eV, as calculated through the combination of pertinent photophysical and electro-chemical quantities. On the other hand, the dns^{*}-to-Ni^{II} eT process is very disfavoured ($\Delta G^{\circ}_{eT} \geq 0.68$ eV).

means that redox active substrates with a potential lower than 0.08 V (vs. Fc^{+}/Fc) are present. Quenching of the emission indicates the presence of oxidizing substates with a potential higher than 0.08 V (vs. Fc^{+}/Fc).

Systems such as this may be of no special interest in classical redox titrimetry. However, they have a great potential in molecular biology studies, for instance to evaluate (and to map) the redox potential in domains hardly assessed by an electrode, e.g. inside a cell. Life compatibility of both the active subunit and of the luminescent fragment should be preliminary considered in the design of a fluorosensor of the redox potential at a cell level.

REFERENCES

1. C. J. Pedersen, *J. Am. Chem. Soc.* 1976, **89,** 7017.
2. B. Dietrich, J.-M. Lehn and J.-P. Sauvage, *Tetrahedron Lett.* 1969, **34,** 2885.
3. L. Fabbrizzi and A. Poggi, *Chem. Soc. Rev.* 1995, 197.

4. A. P. de Silva, H. Q. N. Gunaratne, T. Gunnlaugsson, A. J. M. Huxley, C. P. McCoy, J. T. Rademacher and T. E. Rice, *Chem. Rev.* 1997, **97,** 1515.
5. V. Balzani and F. Scandola, *Supramolecular Photochemistry,* Ellis Horwood, London, 1991.
6. C. A. Bignozzi, M. T. Indelli and F. Scandola. *J. Am. Chem. Soc.* 1989, **111,** 5192.
7. R. G. Pearson, *J. Am. Chem. Soc.* 1963, **85,** 3533.
8. L. Fabbrizzi, M. Licchelli, P. Pallavicini, A. Perotti, A. Taglietti and D. Sacchi, *Chem. Eur. J.* 1996, **2,** 167.
9. D. C. Weatherburn, E. J. Billo, J. P. Jones and D. W. Margerum, *Inorg. Chem.* 1970, **9,** 1557.
10. L. Fabbrizzi, M. Licchelli, P. Pallavicini, A. Perotti and D. Sacchi, *Angew. Chem. Int. Ed. Engl.* 1994, **33,** 1975.
11. L. Fabbrizzi, A. Perotti and A. Poggi, *Inorg. Chem.* 1983, **22,** 1411.
12. D. K. Cabbiness and D. W. Margerum, *J. Am. Chem. Soc.* 1970, **92,** 2151.
13. E. J. Billo, *Inorg. Chem.* 1984, **23,** 236.
14. L. C. Siegfried and T. A. Kaden, *J. Phys. Org. Chem.* 1992, **5,** 549.
15. P. Suppan, *Chemistry and Light,* The Royal Society of Chemistry, Cambridge, 1994, pp. 65–68.
16. G. Schwarzenbach and K. Schwarzenbach, *Helv. Chim. Acta.* 1963, **46,** 1390.
17. E. C. Theil and K. N. Raymond, *Bioinorganic Chemistry,* I. Bertini, H. B. Gray, S. J. Lippard, J. S. Valentine, eds, University Science Books, Mill Valley, California, 1994.
18. R. J. Motekaitis, Y. Sun and A. E. Martell, *Inorg. Chem.* 1991, **30,** 1554.
19. F. Fages, B. Bodenant and T. Weil, *J. Org. Chem.* 1996, **61,** 3956.
20. F. Fages, *Chemosensors of Ion and Molecule Recognition,* J.-P. Desvergne and A. W. Czarnik, eds, Kluwer Academic Publishers, Dordrecht, 1997, pp. 221–240.
21. A. P. de Silva, H. Q. N. Gunaratne, T. Gunnlaugsson, A. J. M. Huxley, C. P. McCoy, J. T. Rademacher and T. E. Rice, *Chem. Rev.* 1997, **97,** 1515.
22. A. P. de Silva and S. A. de Silva, *J. Chem. Soc. Chem. Commun.* 1986, 1709.
23. P. Ghosh, P. K. Bharadway, S. Mandal and S. Ghosh, *J. Am. Chem. Soc.* 1996, **118,** 1553.
24. L. G. Sillen and A. E. Martell, *Stability Constants of Metal Ion Complexes,* The Chemical Society, Oxford, 1971.
25. L. Fabbrizzi, M. Licchelli, P. Pallavicini and A. Taglietti, *Inorg. Chem.* 1996, **35,** 1773.
26. J. A. Sclafani, M. T. Maranto, T. M. Sisk and S. A. Van Arman, *Tetrahedron Lett.* 1996, **37,** 2193.
27. *Fluorescent Chemosensors for Ion and Molecule Recognition* (ed, A. W. Czarnik), ACS Symposium Series 538, American Chemical Society, Washington, 1993, p. 7.
28. F. P. Schmidtchen and M. Berger, *Chem. Rev.* 1997, **97,** 1609–1646.
29. M. E. Huston, E. U. Akkaya and A. W. Czarnik, *J. Am. Chem. Soc.* 1989, **111,** 8735.
30. G. De Santis, L. Fabbrizzi, M. Licchelli, A Poggi and A. Taglietti, *Angew. Chem. Int. Ed. Engl.* 1996, **35,** 202.
31. G. Grynkiewicz, M. Poenie and R. Y. Tsien, *J. Biol. Chem.* 1985, **206,** 3440.
32. L. Fabbrizzi, M. Licchelli, G. Rabaioli and A. Taglietti, unpublished results.
33. P. K. Coughlin, S. J. Lippard, A. E. Martin and J. E. Bulkowski, *J. Am. Chem. Soc.* 1980, **102,** 7616.
34. P. K. Couglin, A. E. Martin, J. C. Dewan, E.-I. Watanabe, J. E. Bulkowski, J.-M. Lehn and S. J. Lippard, *Inorg. Chem.* 1984, **23,** 1004.
35. J. A. Cowan, *Inorganic Biochemistry,* VCH, New York, 1993, pp. 254–260.
36. L. Fabbrizzi, G. Francese, M. Licchelli, A. Perotti and A. Taglietti, *Chem. Commun.* 1997, 581.
37. *Vogel's Textbook of Quantitative Inorganic Analysis, 4th Edition,* Longman, New York, 1978, pp. 293–294.
38. G. De Santis, L. Fabbrizzi, M. Licchelli, N. Sardone and A. H. Velders, *Chem.-Eur. J.* 1996, **2,** 1243.

Chapter 4

The Chirality of Polynuclear Transition Metal Complexes

CHRISTOPHE PROVENT AND ALAN F. WILLIAMS

University of Geneva, Switzerland

1 INTRODUCTION

Coordination chemists have a somewhat ambivalent attitude towards chirality. Alfred Werner, in his seminal paper on the structure of coordination complexes [1,2], pointed out that an octahedral complex with two or three bidentate ligands will have two enantiomers, and indeed later used this to support his theory. He completed the remarkable synthesis and resolution of the enantiomers of the complex $[Co\{cis\text{-}Co(NH_3)_4(OH)_2\}_3]^{6+}$ [3], thereby establishing that the presence of organic carbon was not an essential prerequisite for chirality. In spite of this vital role in the early development of the subject, chirality is mentioned only irregularly in the literature of coordination chemistry. Simple complexes are generally achiral, and the most frequently studied properties such as stability, spectra and magnetism do not depend on chirality, while in mechanistic studies its presence would usually be a further complication in an already complicated system. The great majority of complexes studied are substitutionally labile, and consequently the separation and study of the properties of the enantiomers is impossible. It is only in the fields of bioinorganic chemistry, where the ligands are naturally chiral, and of enantioselective catalysis that chirality has occupied a central position.

If the neglect of chirality may be justified in mononuclear systems, this is not generally the case in polynuclear systems where several chiral centres may be present. This leads to a large number of possible diastereomers, with different symmetries, structures and properties. In general the complexes discussed in this

Transition Metals in Supramolecular Chemistry, edited by J. P. Sauvage.

chapter have been prepared by self-assembly routes [4], a euphemistic way of saying that a suitable mixture of ligands and metal cations will combine upon mixing to give a well-defined product. If this complex has several chiral centres, then clearly there must have been some communication between the different centres during the assembly to avoid the formation of an intractable mixture of diastereomers. A remarkable feature of this field of chemistry is how selective these reactions can be. Let us take as an example the beautiful hexameric complex $[NaFe_6(OCH_3)_{12}(dbm)_6]^+$ (dbm = dibenzoylmethane) reported by Lippard *et al.* [5] formed upon mixing of $FeCl_3$, dibenzoylmethane and sodium methoxide (Figure 1). Each iron atom is complexed by one dbm^- ligand, and two pairs of bridging methoxo ligands. It is thus in a *tris*(bidentate) environment, and has a defined chirality. Inspection shows that the chirality alternates on moving round the ring, leading to an overall symmetry D_{3d}. Once the chirality of one metal centre is defined, that of the five others is equally fixed.

Apart from the structural aspects and the mechanism of self-assembly, there is another reason to study the chirality of these systems. There is an increasing body of evidence that suggests that polynuclear complexes generated by self-assembly reactions may show much greater kinetic inertness than simple mononuclear

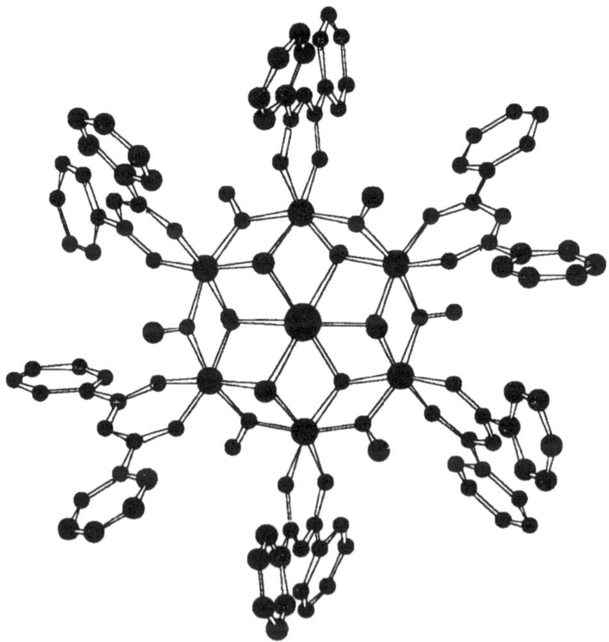

Figure 1 The structure of $[NaFe_6(OCH_3)_{12}(dbm)_6]^+$ (redrawn from reference 5). The sodium lies on a pseudo threefold axis at the centre of a hexagon of iron atoms, bridged by methoxy groups.

complexes. We discuss this point later in the chapter, but for the moment it suffices to note that the assumption that these complexes are rapidly exchanging racemates may not generally be justified.

Our intention in this chapter is to draw attention to the chiral aspects of self-assembled polynuclear complexes, and to show how the study of properties related to chirality can afford useful information. After a brief review of the nature of chiral centres in these complexes and the experimental methods used for studying them, we will discuss a number of structure types which are currently attracting interest, such as helicates, dendrimers, molecular boxes, and topologically complex molecules.

2 CENTRES OF CHIRALITY IN POLYNUCLEAR SYSTEMS

The complexes discussed in this chapter will have the general structure shown in Figure 2. Two (or more) metal ions are bound to the binding sites of a ligand molecule. Apart from the binding sites, the ligand possesses a bridging unit between the binding sites, and may carry a number of ancillary groups which play no direct role in the assembly, but which may be present to improve solubility, or indeed to introduce an element of chirality. All these components of the complex may contain chiral centres, and we shall discuss them in turn.

The metal centre. In organic chemistry, chiral centres are usually associated with an asymmetric carbon atom, but this notion is of limited use for metal ions. Most tetrahedral metal ions are extremely labile, although pseudo-tetrahedral complexes such as $(C_5H_5)MLL'L''$ may be resolved into enantiomers. Octahedral centres with

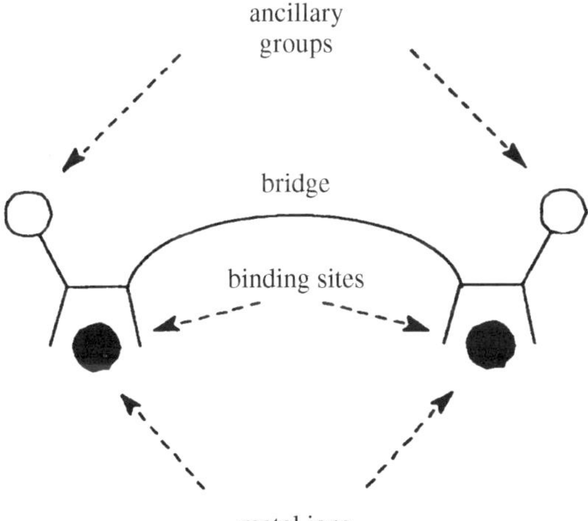

Figure 2 Schematic drawing of the components of a polynuclear complex.

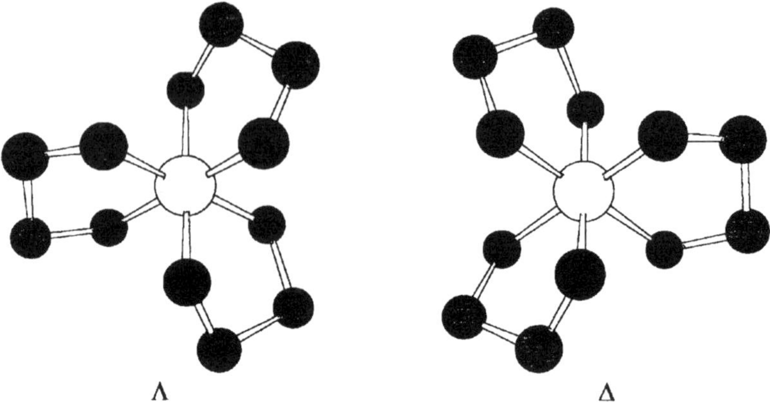

Figure 3 The two enantiomers of $[Co(en)_3]^{3+}$.

six different ligands will be chiral, but have little relevance to self-assembled complexes where the metal is normally bound to two or more identical binding sites, and consequently possesses a certain degree of symmetry. The classical example of chirality is the *tris*-bidentate octahedral unit such as $[Co(en)_3]^{3+}$ (en = 1,3-diaminoethane) for which the two enantiomers are shown in Figure 3.

Among other coordination spheres which may be chiral, there is the trivial case of the *cis*-*bis*-bidentate octahedral systems such as *cis*-$[Co(en)_2Cl_2]^+$, *bis*-bidentate tetrahedral [M(A-A)2] or *bis*-tridentate octahedral $[M(A\text{-}A\text{-}A)_2]$ complexes where the dihedral angle between the two chelate planes is different from 90°, as is often the case. An example, together with the *tris*-tridentate $[M(A\text{-}A\text{-}A)_3]$ tricapped trigonal prism geometry often found for lanthanide complexes, is shown in Figure 4. It will be seen that a wide variety of metal centres may be chiral, and it is consequently advisable to examine each centre carefully.

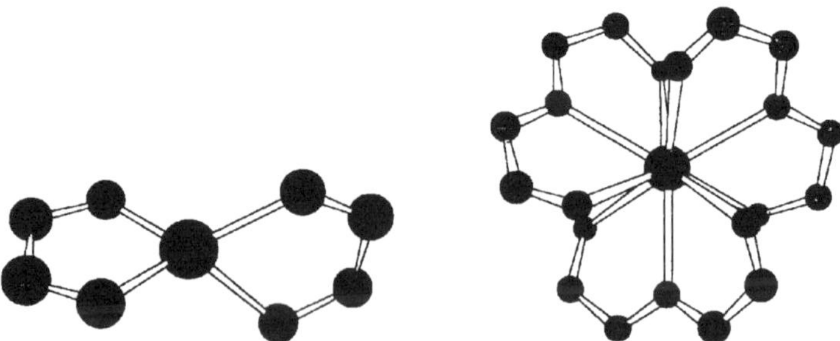

Figure 4 Chiral metal centres in tetrahedral geometry (left) and tricapped trigonal prismatic geometry (right). Both drawings have Δ configurations.

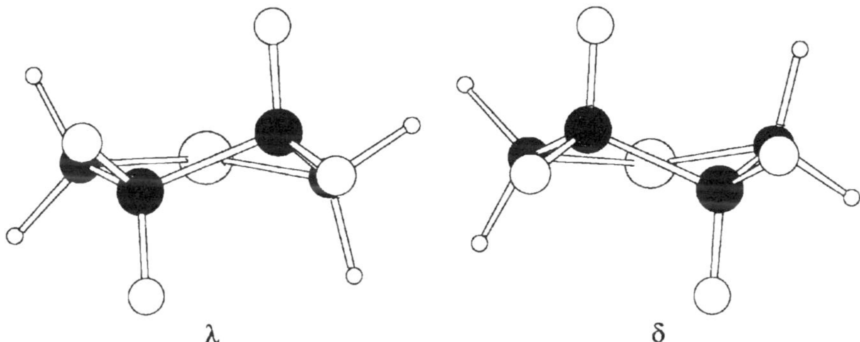

Figure 5 The $M(H_2NCH_2CH_2NH_2)$ chelate viewed along the twofold axis towards the M atom.

The binding site and the bridging unit. The formation of a chelate ring imposes a conformation upon the atoms of the ring, and this will in general be chiral. The ethylenediamine chelate ring is the simplest example, and the two enantiomers are shown in Figure 5.

The assignment of the chirality is based upon the 'skew lines' convention [2,6] in which two lines are chosen to characterize the structure (in this case the N–N axis and the C–C bond). The assignment is based on whether the line closest to the reader, in this case the C–C bond, follows a right-handed (δ) or left-handed (λ) helical direction about the second line. In a similar way, the formation of a polynuclear complex will impose a conformation on the atoms in the bridging unit, which may thereby become chiral. It should be noticed, however, that the significance of this chirality arising from conformation may be limited, especially if the conformation is flexible and exchange between the two enantiomers is rapid.

Ancillary groups. Finally it may be noted that the various substituents of the ligand may well carry chiral centres. This is indeed a traditional way of introducing a specific chirality into a ligand. Most frequently the source will be an asymmetric carbon atom, but there are other possibilities. The propeller-like arrangement of phenyl groups in triphenylphosphane results in two possible enantiomers, and this chirality may influence the rest of the complex.

This discussion shows that polynuclear systems can contain a very large number of chiral centres. Consider, for example, a triple helical dinuclear complex shown in Figure 6.

There are two metal centres, six chelate rings and three bridges, a total of 11 stereogenic centres, which leads to many possible isomers. Even if one limits the degrees of freedom of the chelate rings by using planar α,α'-diimine ligands, there are still 16 possible isomers. In practice only two enantiomers are usually observed,

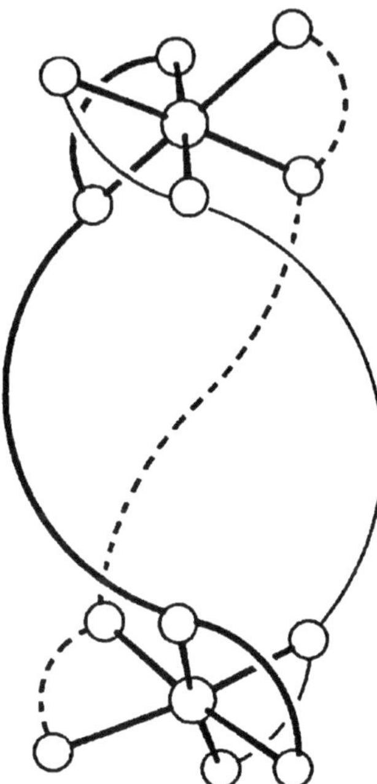

Figure 6 A triple helical complex in which three *bis*-bidentate ligands link two chiral octahedral centres.

indicating a remarkable degree of selectivity. This arises from the fact that the metal ions used are labile, and consequently the individual reaction steps may be regarded as reversible. Since the different chiral centres all interact, the chirality at one centre is strongly influenced by that at several others, and the thermodynamic driving force to adopt a particular chirality is magnified. Starting from achiral components, the creation of the first chiral centre will be random, but once this choice is made, the others will follow automatically.

We may, of course, introduce a predefined chirality in the components. In practice this is always done by introduction of one or more asymmetric carbons in the ligand. The introduction of an asymmetric carbon into the chelate ring in *S*-pdta^{4-}, **1**, or in (*S*,*S*)-cdta^{4-}, **2**, results in the selective formation ($>99\%$) of Λ-[Co(*S*-pdta)]$^{-}$ and Λ-[Co(*S*,*S*-cdta)]$^{-}$ [7] respectively. In both cases the substituents of the chelate ring seek to adopt an equatorial position, and thereby determine the chirality of the chelate ring as δ, which in turn determines the chirality at the metal centre.

An interesting example of the effect of introducing chirality into the bridging unit has been given by Suzuki *et al.* [8]. The ligand **3** contains two pyridine binding units separated by a chiral spacer. If (*R*,*R*)-**3** is reacted with Ag(I), crystals of an extended

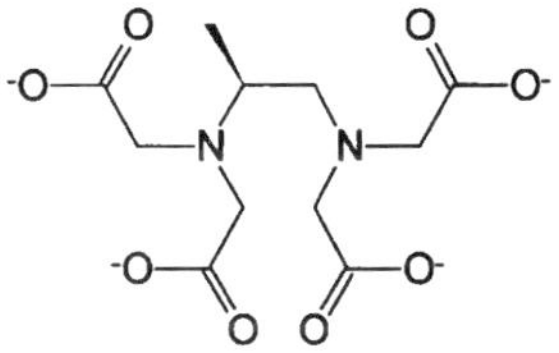

1

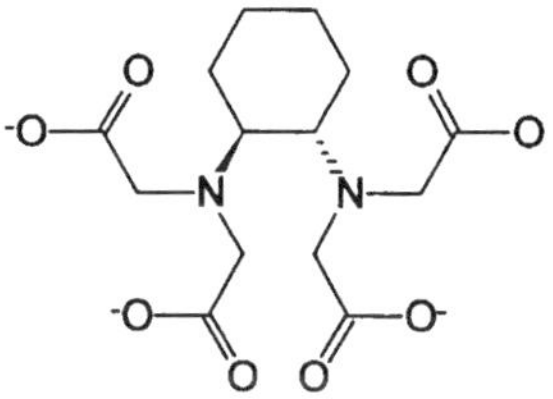

2

system are obtained, with Ag(I) ions bridging two ligands **3**, but if a racemic mixture of (*R*,*R*)- and (*S*,*S*)-**3** is used, a closed, centrosymmetric dinuclear complex $[Ag_2((R,R)\text{-}\mathbf{3})((S,S)\text{-}\mathbf{3})]^{2+}$ is formed.

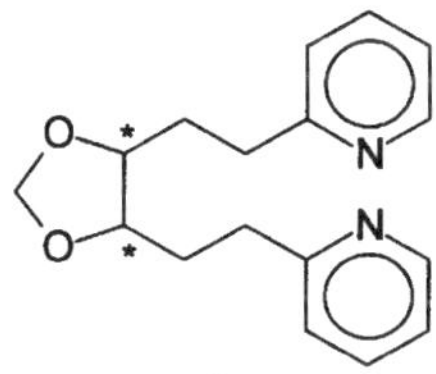

3

The use of enantiomerically pure ligands excludes the presence of symmetry elements such as centres of inversion or planes in the final assembled structure. With racemic ligands the assembly can follow either a homochiral path, in which ligands of the same chirality bind to the same metal, or a heterochiral path as in $[Ag_2((R,R)\text{-}\mathbf{3})((S,S)\text{-}\mathbf{3})]^{2+}$. Which way is chosen will depend on circumstances. Thus ligand **4** forms preferentially a heterochiral complex $[Co((R,R)\text{-}\mathbf{4})((S,S)\text{-}\mathbf{4})]^{2+}$ with Co(II) when R = Ph as a result of interligand repulsion, but if R = Me a mixture is observed. With Ag(I), however, homochiral assembly to $[Ag_2((R,R)\text{-}\mathbf{4})_2]^{2+}$ is observed for the ligand where R = Bz [9].

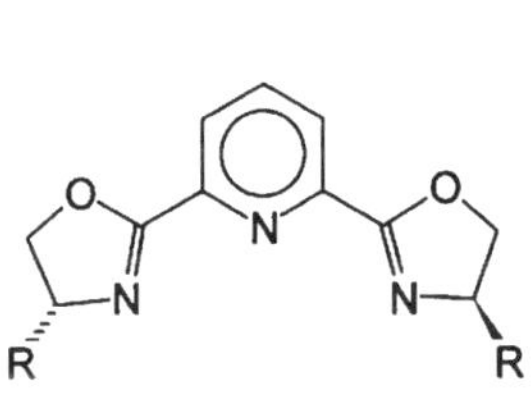

4

Introduction of chirality in ancillary groups often acts by steric repulsion between different ligands, but the podates form an important exception. These multidentate ligands have two or more ligating strands attached to a template, and if this is chiral, as for example in the natural iron chelating agent enterobactin, **5** [10], then the chirality at the metal centre is predetermined. This technique has been used to great effect in the helicates.

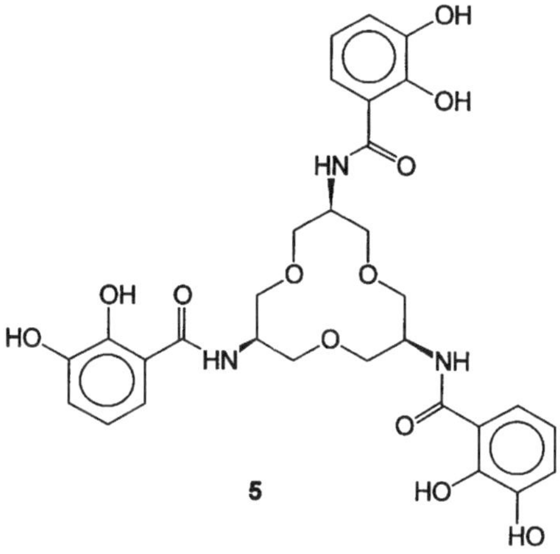

5

3 EXPERIMENTAL METHODS

Since coordination chemists tend to ignore the effects of chirality, it is appropriate to make a few remarks concerning its influence on the experimental methods generally used to study these systems.

X-ray crystallography. This is the dominant structural technique in this field. It should be noted that the crystallization of a compound from a solution may result in the preferential isolation of one of the different forms present in solution [11], and consequently it is dangerous to assume an exact parallel between solid state and solution structure. Assuming the complex to be chiral, we may ask what effect this will have on its crystallization and on the resolution of its crystal structure [12]. If the mother solution contained only one enantiomer, then the crystal will form in a space group with no improper rotation axis. If the mother solution is a racemate, containing equal numbers of both enantiomers, then it may crystallize either as a *racemic compound*, in which the crystals contain equal numbers of each enantiomer, generally related by a symmetry operation such as an inversion centre, mirror plane, or glide plane, or as a *racemic mixture* in which each crystal contains only molecules of one enantiomer. The most famous example is sodium ammonium tartrate which crystallizes as a racemic mixture if the temperature is not too high. Pasteur observed the formation of two slightly different crystal forms (corresponding to the two enantiomers) which he was able to separate with tweezers and show to possess equal and opposite optical rotations. This observation was in fact the starting point for studies of molecular chirality.

Space group assignments of racemic compounds give no special problems, but crystals containing only one enantiomer are not always straightforward. Some space groups are enantiomers of others, such as $P3_1$ and $P3_2$, and the choice of one space group rather than the other should be justified. However, if the crystal does contain only one enantiomer, then the absolute configuration of the structure may be determined as a consequence of anomalous dispersion of X-rays. The result of the anomalous dispersion is that the intensities of Friedel pairs of reflections *I*(*hkl*) and *I*(*-h-k-l*) are no longer equal. The magnitude of the anomalous dispersion increases with the atomic number of the scattering atom, and in general it is sufficiently important with transition metal compounds to be measured reliably. It is therefore strongly recommended to measure a number of Friedel pairs if the space group is non-centrosymmetric, choosing as far as possible relatively strong reflections whose intensity may be measured accurately. In the solution of the crystal structure, the simplest way of deciding which enantiomer is present is to refine an enantiomorph polarity parameter [13] whose value varies from 0 for one enantiomer to 1 for the other. Intermediate values indicate some problem with the crystal structure such as an incorrect space group assignment or the presence of twinning.

NMR spectrospcopy. This is the most powerful technique for studying structure in solution, and two methods have been used regularly to study chirality: the introduction of probe groups, and the use of chiral shift reagents. The most useful probe group is a methylene group attached to the ligand:

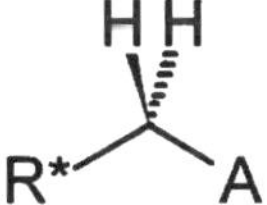

If the substituent R* is chiral, then the two protons are diastereotopic (that is, the substitution of one or the other will give two diastereomers), and show different resonances in the ^{1}H nmr spectrum. In the absence of coupling to R* or A, a well-resolved AB spectrum is observed. Judicious placing of CH_2 groups in the ligand gives a sufficient number of probes to characterize the chirality of the molecule as a whole.

6 can act as a *bis*-bidentate ligand, and with octahedral cations such as Fe(II) forms triple helical complexes such as $[Fe_2(\mathbf{6})_3]^{4+}$ (see Figure 6). The methylene protons **I** give an ABX_3 signal which indicates the chirality of the iron coordination sphere, while the bridging methylene protons **II** give a singlet, requiring them to be related by a symmetry element. This can only be a twofold axis, and therefore requires the chiralities at the two metal centres to be identical [14]. On the other

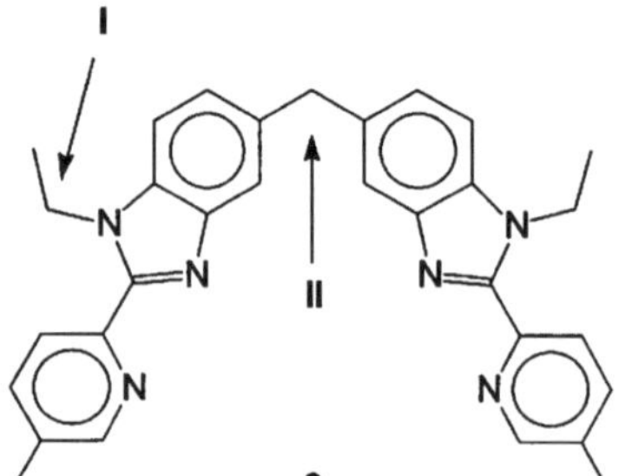

6

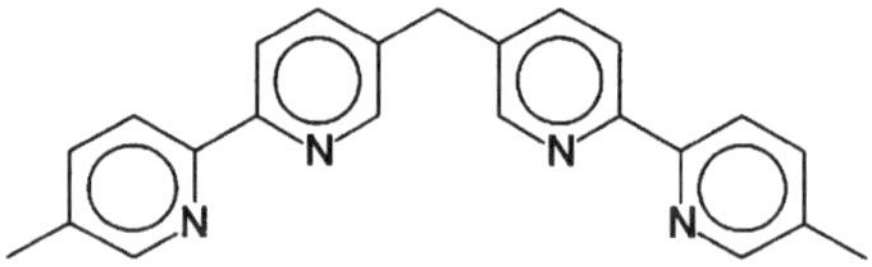

7

hand, the complex $[Fe_2(\mathbf{7})_3]^{4+}$ shows an AB signal for the bridging methylene unit; the twofold axis is therefore absent, and the chiralities of the two metal centres must be different [15].

Chiral reagents such as anthracenyl-1-trifluoromethyl ethanol [16] have also been used to establish the chirality of the self-assembled complexes [17], but their use has been less extensive than that of diastereotopic protons. Most chiral NMR reagents have been designed to complex organic substrates, and interact less strongly with transition metal complexes, but we have found that the chiral *tris*(tetrachlorobenzenediolato)phosphate anion developed by Lacour [18] interacts strongly with aromatic heterocyclic ligands used in this type of chemistry to give good splittings.

The observation of an AB signal, or a splitting upon addition of a chiral reagent, is positive evidence for the chirality of the complex, but care must be taken to distinguish local chirality from overall chirality: a *meso* complex will contain two chiral centres of opposite chirality, which may be indicated by an AB signal for diastereotopic protons, even though the complex as a whole is achiral. This is why the observation of the singlet for the bridging methylene protons in complexes of **6** is particularly important. Another point which is occasionally overlooked is that these systems may be undergoing a variety of dynamic processes which are rapid on a NMR timescale. In one pathological case encountered with a complex of ligand **4** (R = Ph), a simple spectrum at room temperature, albeit with rather broadened lines, gave a beautifully resolved spectrum at $-40°C$ showing the presence of three different complexes [9].

Resolution of enantiomers. There have been relatively few attempts to resolve the enantiomers of polynuclear complexes. Clearly the half-life for racemization should be several hours at least, and this would seem to exclude complexes of labile metal ions, although, as we discuss below for particular cases, there is evidence that many of these polynuclear complexes may have much greater kinetic inertness than the analogous mononuclear complexes. A second limitation is the fact that the traditional methods of resolution tend to be demanding in quantity of complex, which is a major barrier when the ligands require multi-step syntheses.

The standard method for resolution of coordination complexes involve crystallization of salts with chiral anions such as tartrate, ((+)-tartrato)antimonate(III) [19], or *tris*-(cysteinesulphinato)cobaltate [20], or *d*-$[Co(en)_2(NO_2)_2]^+$ as a chiral cation [7]. Alternatively, ion exchange chromatography may be used with a suitable chiral counter ion for elution.

Chiroptical properties. Two enantiomers may be distinguished by their different interactions with a third object which is itself chiral. If this third object is to be a light beam, then some sort of chirality must be imposed upon it. This may be achieved by using circularly polarized light, in which the electric dipole vector of the radiation describes a helical motion as it moves away from the observer (Figure 7).

When monochromatic circularly polarized light passes through an enantiomerically pure sample which absorbs at the corresponding frequency, rcp light will be absorbed to a greater or less extent than lcp light. The *circular dichroism* (CD) of the sample is defined as $\Delta\varepsilon = \varepsilon_L - \varepsilon_R$ where ε_L and ε_R are the molar absorption coefficients for lcp and rcp light, respectively.

In addition to a difference in absorption, rcp and lcp light will experience different retardations, and consequently there will be a phase shift between the two beams. Now a plane polarized light beam may be regarded as the sum of an lcp and an rcp beam of equal intensity. When a plane polarized light beam passes through an enantiomerically pure sample, the two beams will experience different phase shifts, and the result will be a rotation of the plane of polarization. This is the phenomenon of *optical rotation*, the rotation of the plane of polarization. Optical rotation varies strongly with the wavelength of the light, and the wavelength at which it was measured should always be stated. The plot of rotation as a function of wavelength is called an optical rotatory dispersion (ORD) curve.

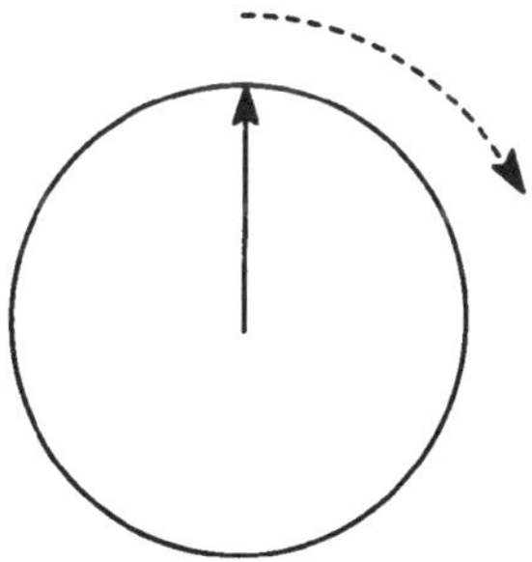

Figure 7 Right circularly polarized (rcp) light. The electric dipole vector of the radiation describes a clockwise movement as it moves away from the observer. Left circularly polarized (lcp) light will describe an anticlockwise spiral.

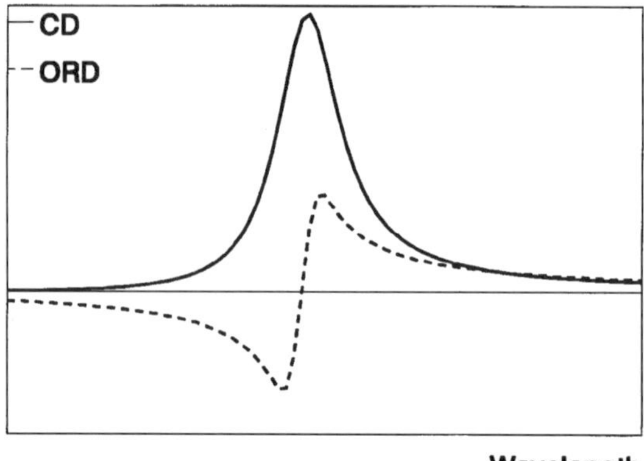

Figure 8 CD and ORD curves for an electronic transition. Note that several electronic transitions may be present in a real compound, and the CD and ORD curves will be the sum of the contributions from these transitions.

Circular dichroism and optical rotatory dispersion are related phenomena [20], and correspond respectively to the imaginary and real parts of the dielectric susceptibility. Their form for a positive CD effect is shown in Figure 8.

Both CD and ORD are associated with the electronic spectrum of the sample. While CD bands are usually localized close to the absorption maximum, optical rotations may be measured at some distance from the absorption wavelength. This explains why colourless sugars still give measurable optical rotations in the visible region. Although optical rotations are still measured to characterize compounds, circular dichroism is generally supplanting ORD. The magnitude of both effects is related to the rotational strength of the transition. This will be large when the translation is allowed both by an electric dipole and by a magnetic dipole mechanism. In practice CD is a rather weak effect, typically 1% or less of the absorption coefficient, but d–d bands are anomalous in that they are electric dipole forbidden, but magnetic dipole allowed. The CD of d–d bands may thus be as great as 10% of the absorption coefficient, and optical rotations of transition metal complexes measured near to d–d transitions may be very large.

A typical CD or ORD spectrum will show at longest wavelengths signals due to the d–d transitions. In the low symmetries associated with these chiral compounds there may be several d–d transitions within one band of the electronic spectrum. The CD bands associated with these transitions may not necessarily have the same sign and so several CD bands of differing sign may be seen within the envelope of one d–d transition. For (Δ)-$[Co(en)_3]^{3+}$ the first transition to the $^1T_1$ state in pure octahedral symmetry is split into two components (E and A_2) which have opposite and almost equal CD intensities. As one moves to shorter wavelengths, CD signals from metal-ligand charge transfer states will be observed, and in the UV signals due to ligand-ligand transitions will be seen.

4 CHIRALITY IN POLYNUCLEAR STRUCTURES

In the first part of this chapter we have tried to present the general features of chirality in transition metal complexes. In the remainder, we shall discuss a number of polynuclear complexes where chirality is important, either as a factor in analysing the structure or as a means of studying the properties.

4.1 The Helicates

The helicates have attracted considerable attention in supramolecular chemistry as good examples of products of self-assembly reactions in which a number of components are assembled to give a unique structure in high yield. The general structure of a helicate is given in Figure 9.

The strands of the helix are formed by one or more ligands carrying two or more metal binding sites. The ligand(s) twist around the helical axis along which are disposed the metal atoms. The basic structural principle is that the binding sites provided by the ligand strands should match the coordination requirements of the metal ions. The two examples in Figure 10 show two *bis*-bidentate ligands **8** providing a tetrahedral coordination of Cu(I) ions, and three *bis*-bidentate ligands **9** giving an octahedral coordination of two Co(II) ions.

The helical axis is in general a line, but a recent development is the synthesis of circular helicates in which the ligands twist about a circle, the metal ions being disposed at regular intervals, and these are discussed in section 4.2. Reviews of helicates have appeared recently, and the reader is referred to these [23–25] for a discussion of properties of these complexes.

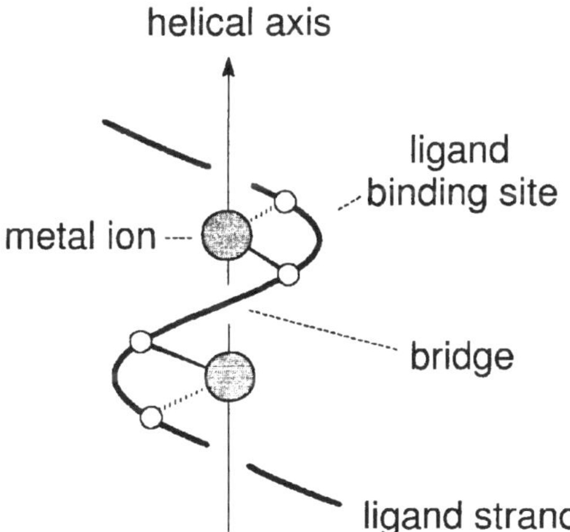

Figure 9 Schematic drawing of the features of a helicate. Only one strand is shown.

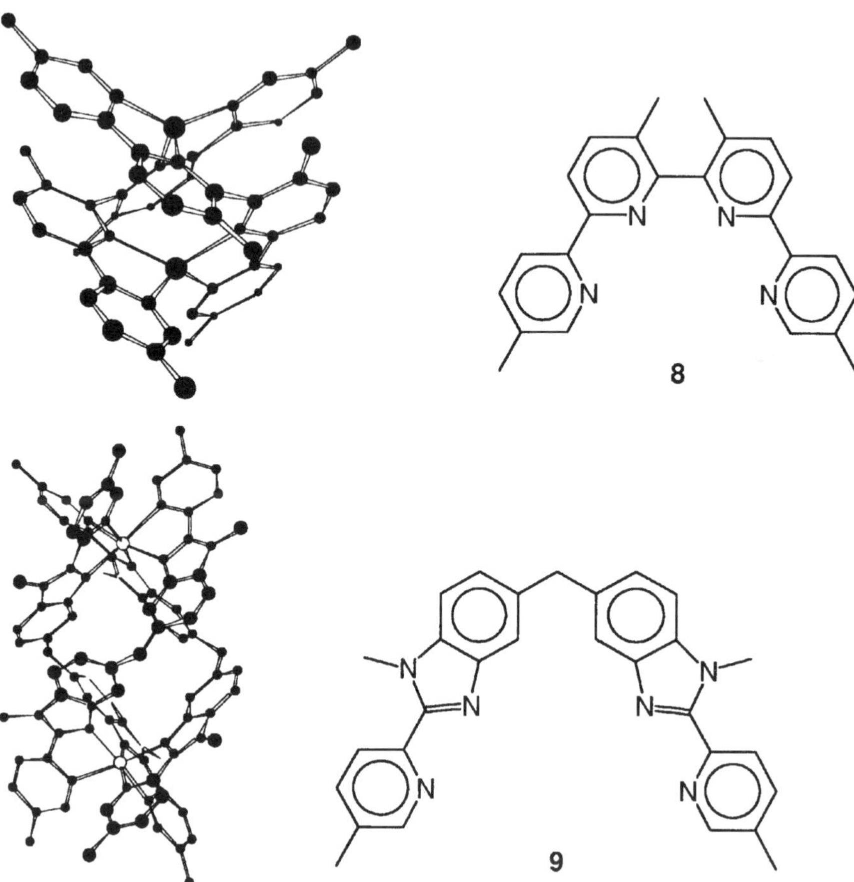

Figure 10 Two examples of helicates showing respectively tetrahedral coordination of Cu(I) in a double helicate *M*-$[Cu_2(\mathbf{8})_2]^{2+}$ (above) [21], and octahedral coordination of Co(II) in a triple helicate *P*-$[Co_2(\mathbf{9})_3]^{4+}$ (below) [22].

Following Cahn, Ingold and Prelog [26], the two enantiomers of the helix are denoted *P*- for a right-handed helix (as shown in Figure 9) or *M*- for a left-handed helix. A full description of the chirality is, however, more complicated since, as mentioned above, the metal ions, the chelate rings, and the bridging unit are all possible stereogenic centres. One of the remarkable features of the helicates is the high degree of diastereoselectivity often observed at the metal centres. Thus the helicates shown in Figure 10 are found only as ΔΔ or ΛΛ diastereomers, and the *meso* (ΔΛ) complexes are not observed. The degree and the nature of the diastereoselectivity depends strongly on the system. All systems so far reported with nuclearity higher than two seem to give exclusively homochiral complexes. With dinuclear triple helices, homochirality is normally observed but some exceptions are known: the heterochiral complex $[Fe_2(\mathbf{7})_3]^{4+}$ [15] has already been mentioned, and Albrecht has reported formation of *meso* ($\Delta\Lambda$) complexes for $[Ti_2(\mathbf{10})_3]^{4-}$ and $[Fe_2(\mathbf{11})_3]^{4+}$ [27,28].

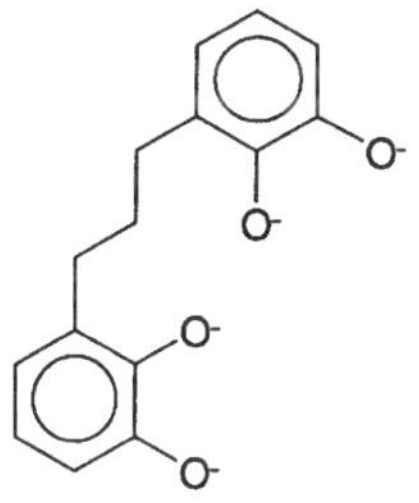

10

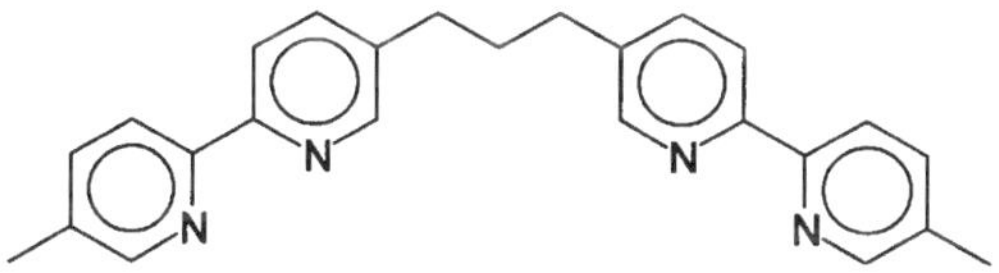

11

This inversion of the diastereoselectivity cannot be explained simply: the closely related complex $[Fe_2(\mathbf{12})_3]^{4+}$ shows $\Delta\Delta$ and $\Lambda\Lambda$ diastereoselectivity [29]. A slightly different variation is observed with $[Fe_2(\mathbf{13})_3]^{4+}$ [30], where the two metal centres are homochiral, but the ethylene bridge twists the opposite way about the axis. Saalfrank [31] has recently reported a remarkable system using the closely related ligands **14** and **15**; **14** gives a homochiral triple helicate $[Fe_2(\mathbf{14})_3]$ with Fe^{3+} whereas the pyridyl bridged ligand **15** can encapsulate a potassium ion inside the helicate and gives a *meso* complex ($\Delta\Lambda$)-$[KFe_2(\mathbf{15})_3]^+$.

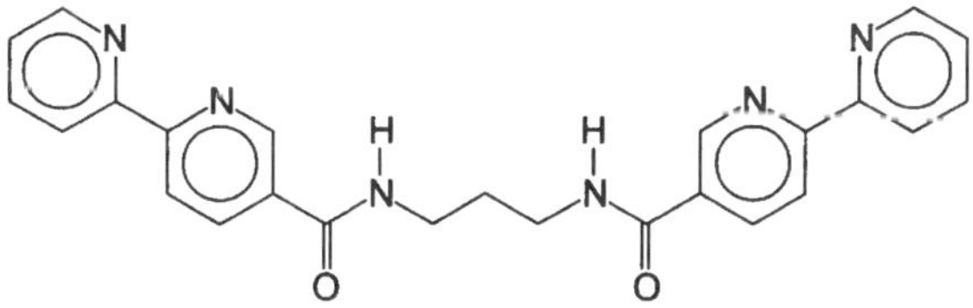

12

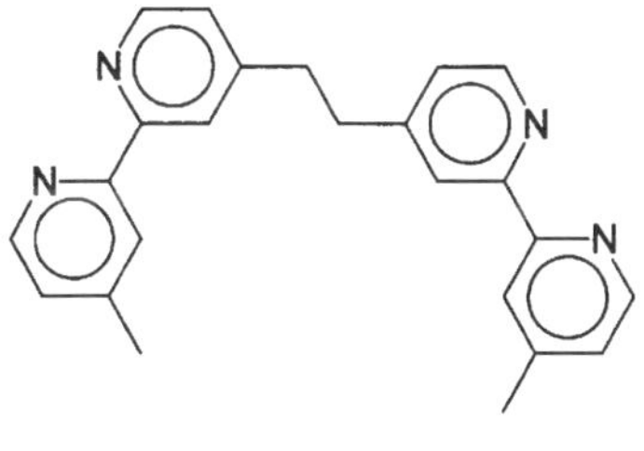

13

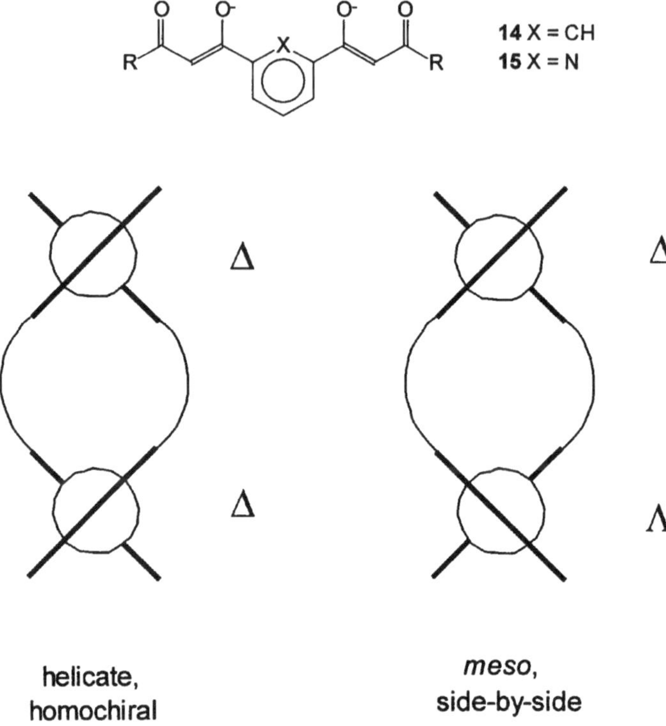

Figure 11 Schematic view of the two diastereomers for a dinuclear distranded complex.

While the triple-stranded systems tend to show high diastereoselectivity one way or another, tetrahedral double-stranded systems may give mixtures. The two possibilities are distinguished as helical or side-by-side complexes (Figure 11). Mixtures are particularly common when bridges and chelate rings are flexible. Thus the chiral ligand **16** apparently forms mainly helicates in solution with Ag(I) and Au(I), although crystallization gives the side-by-side structure with Au(I), but a mixture of side-by-side and helical complexes with Ag(I) [32]. In the same article, molecular mechanics was used to study the stability of the different conformation of the bridge and chelate rings, and the results agreed quite well with the observed crystal structures.

Modification of the ligand can switch the favored conformation. Thus ligand **17** forms equal amounts of helical and side-by-side complexes with Cu(I) [33], but

Ph Ph

Ph_2P P P PPh_2

16

introducing more rigidity as in ligand **18** favors the helical form which becomes the only form observable if *o*-dimethoxybenzene is added to the mixture, as a result of incorporation of the guest inside the bridge of the helix [34].

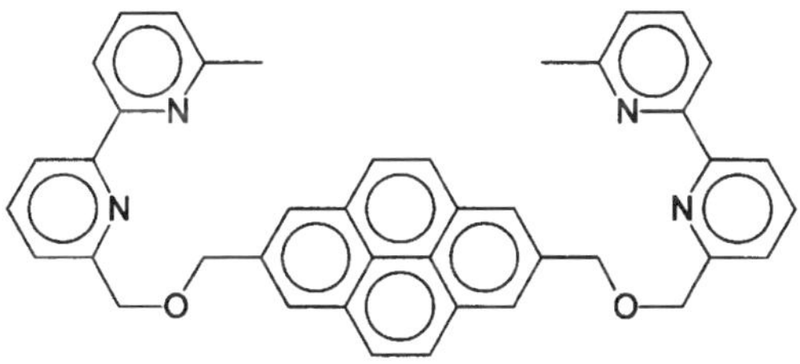

17

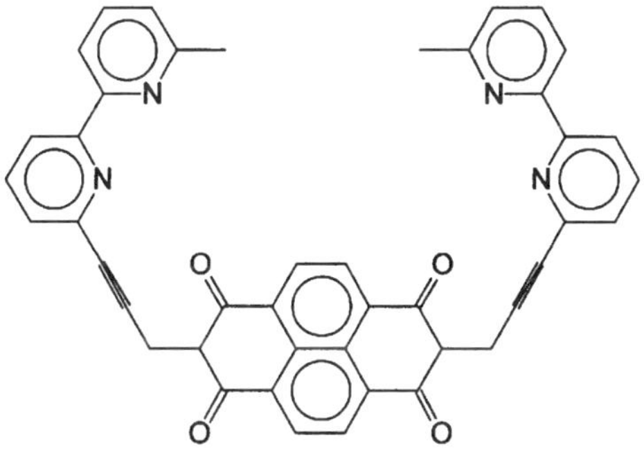

18

One of the most dramatic examples of switching from side-by-side to helical was found by Dietrich-Buchecker and Sauvage in their synthesis of a molecular knot, using a double helicate as a precursor. In the initial synthesis [35] a tetramethylene bridge formed helical and side-by-side complexes in a ratio of 1 : 8 (Figure 12); the desired precursor represented only *c*. 11% of the total, and the yield of the final knotted product was correspondingly low. Replacement of the flexible tetramethylene bridge by the more rigid 1,3-phenylene unit led to exclusive formation of the helical complex, and a final yield of the knotted product of around 30% [36].

Surprisingly little work has been carried out on the resolution of homochiral helicates into the two enantiomers. Self-resolution upon crystallization has been observed for two homonuclear triple helicates [37,38], but there seem to be only two well-authenticated cases of enantiomeric resolution, both using antimonyl tartrate: the complex $[Co_2(\mathbf{9})_3]^{6+}$, a dinuclear triple helix [39], and a trinuclear double helical complex of iron(II) $[Fe_3(\mathbf{19})_2]^{6+}$ with a *tris*-terpyridyl ligand [40]. The circular dichroism spectrum of $[Co_2(\mathbf{9})_3]^{6+}$ is shown in Figure 13.

The general features of the CD spectrum correspond to the description in section 3. The magnitude of the CD and ORD effects is roughly twice that observed for an

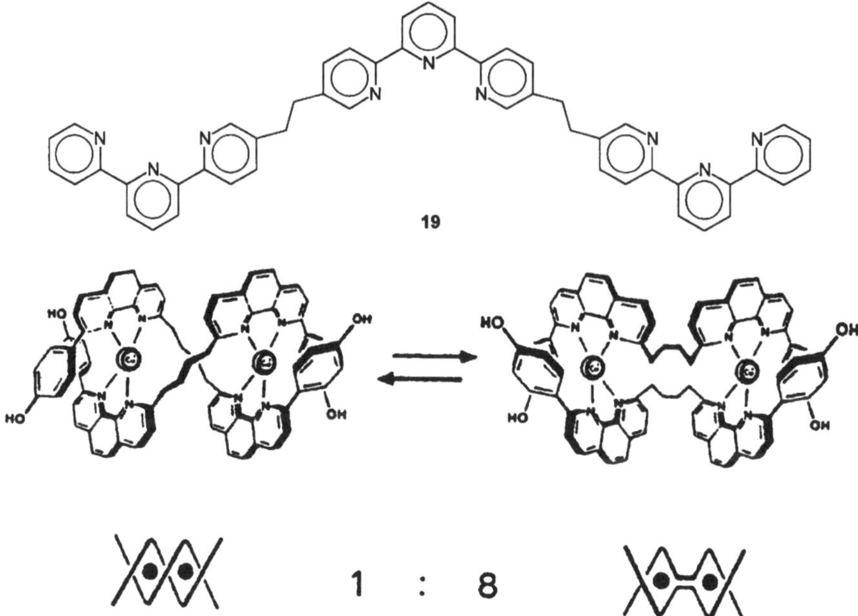

Figure 12 The initial synthesis of a molecular knot used the double helical complex (left) as a precursor, but this is only a minor product with the flexible $-(CH_2)_4-$ bridging unit. Reproduced with permission from reference 35.

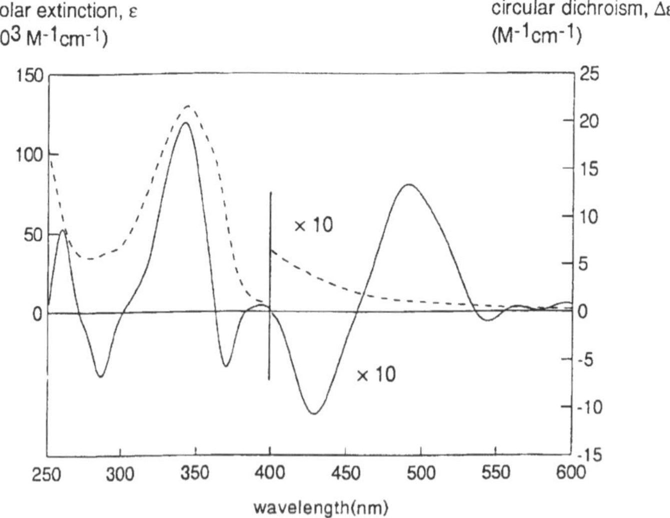

Figure 13 Circular dichroism (full line) and electronic absorption spectrum (dashed line) of $[Co_2(\mathbf{9})_3]^{6+}$. It will be noted that the d–d transitions in the region 400–500 nm present only a shoulder in the electronic spectrum but give rise to a strong CD feature.

isolated mononuclear complex such as $[Co(bipy)_3]^{3+}$, which implies that the transitions involved are not delocalized over the whole complex as in the helicenes [41] where very large chiroptical effects are observed. Upon reduction of the complex with dithionite, an electronic spectrum typical of the Co(II) complex $[Co_2(\mathbf{9})_3]^{4+}$ was observed, and the CD of the ligand-based bands below 350 nm showed that the configuration of the complex had been maintained [42].

The existence of enantiomers of helicates offers the possibility of studying their rate of racemization, and this can give information on the lability of the systems. The use of self-assembly reactions requires in principle the use of labile ions so that the potential energy hypersurface can be sufficiently explored to give the thermodynamic product, but it is often forgotten that this may result in considerable lability of the final product, and hence a high sensitivity to a change in external conditions [43]. The enantiomerically pure complexes $[Co_2(\mathbf{9})_3]^{6+}$ and $[Fe_3(\mathbf{19})_2]^{6+}$ mentioned above showed no change in optical rotation over several weeks. Both are low spin $3d^6$ systems where a degree of kinetic inertness might be expected. However, a study of the racemization of single crystals of a self-resolved trinuclear nickel triple helicate [38] had previously shown that optical activity was lost at a much slower rate than the analogous mononuclear complex. The most complete study to date has been made of the dinuclear triple helicate $[Co_2(\mathbf{9})_3]^{4+}$ containing the labile Co(II) ion. Although mononuclear complexes with a similar ligand sphere isomerize with rate constants of approximately $10\ s^{-1}$ at room temperature, the rate constant for racemization of $[Co_2(\mathbf{9})_3]^{4+}$ was $1.4 \times 10^{-5}\ s^{-1}$ at 25°C [42]. This remarkable difference was investigated in some detail by Charbonnière *et al.* [44], who established that the mechanism for racemization requires dissociation of a cobalt ion from the complex:

$$P\text{-}[Co_2(\mathbf{9})_3]^{4+} \rightarrow P\text{-}[Co(\mathbf{9})_3]^{2+} + Co^{2+} \rightarrow M\text{-}[Co(\mathbf{9})_3]^{2+} + Co^{2+} \rightarrow P\text{-}[Co_2(\mathbf{9})_3]^{4+}$$

Addition of free ligand **9**′ with a different substituent at the benzimidazole allowed the trapping of the free Co(II) and could be observed by electrospray mass spectrometry. The need to break up the dinuclear complex no doubt explains its relative inertness.

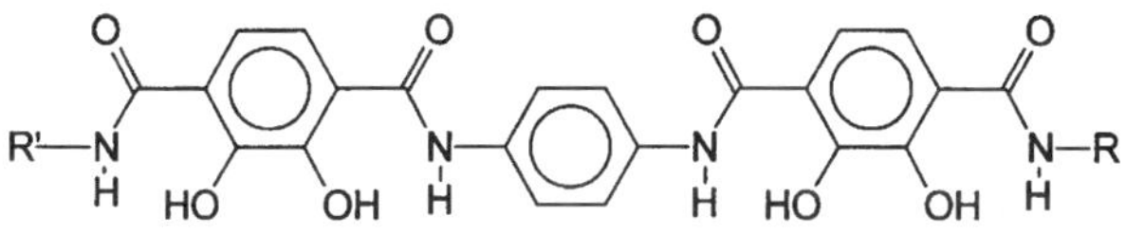

20

The catecholato ligands investigated by Raymond [45] show rather different behaviour. The triple helical complex $[Ga_2(\mathbf{20})_3]^{6-}$ undergoes racemization on the NMR timescale as studied by the coalescence of signals due to diastereotopic groups on the terminal amide ligands. The analogous mononuclear complex had been shown to racemize via an intramolecular Bailar twist [46], and the dinuclear helicate has a free energy barrier to inversion only 20% higher, suggesting that it passes through an achiral $\Delta\Lambda$ intermediate. Albrecht has also studied the racemization of dicatecholato triple-helical complexes of **21** [47] and **22** [48] with Ti(IV). Racemization is again observable on the NMR timescale, but in this case depends also on the nature of the alkali metal counter ion present, since this is partly encapsulated inside the helicate. The free energies of activation of trinuclear and dinuclear species are very similar, suggesting that this is another intramolecular racemization in which each stereogenic centre is inverted successively by a Bailar mechanism.

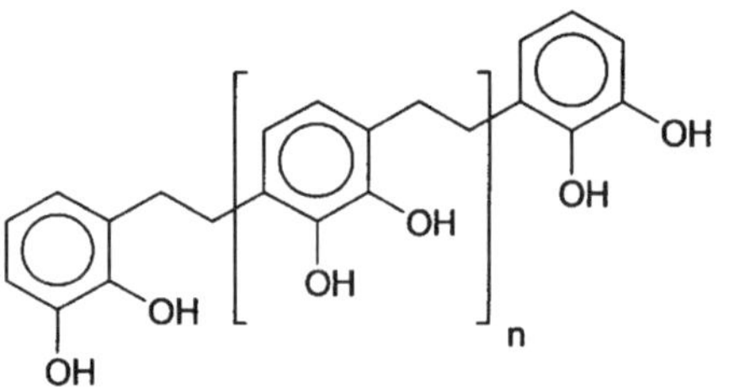

n = 0 **21**
n = 1 **22**

It will be noted that the difference in racemization rates are very considerable, and it has been suggested [44] that this arises from two factors: low barriers to the Bailar twist in d^0 and d^{10} systems, and flexibility in the ligand—models of **9** suggest that it is impossible to attain the trigonal prismatic transition state of the Bailar twist.

The final system in which racemization has been studied in detail concerns the ligands **23** and **24**. The complex $[Cu_2(\mathbf{23})_2]^{2+}$ has a double-helical structure in which the pyridine binds weakly to both copper ions [49], whereas the complex $[Cu_2(\mathbf{24})_2]^{2+}$ adopts a side-by-side *meso-* structure [50]. Figure 14 shows the two structures.

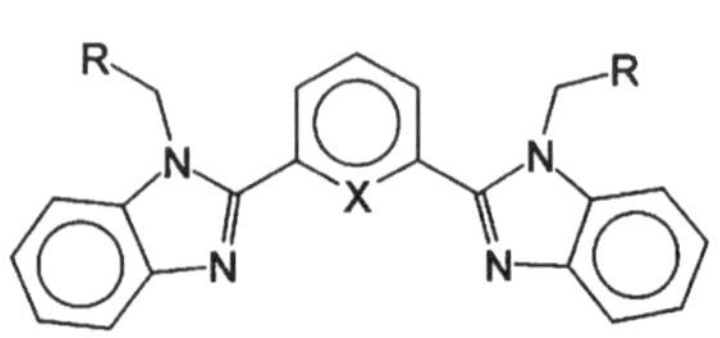

X = N **23**
X = CH **24**

In both structures the copper ions are stereogenic centres, and if the group R is not hydrogen, then the methylene protons are diastereotopic. Figure 15 shows the methylene region of the ^1H nmr spectrum of $[Cu_2(\mathbf{23})_2]^{2+}$ (R = 3,5-dimethoxyphenyl), showing the collapse of the AB spectrum as the temperature rises [51].

Even at low temperatures the methylene protons of $[Cu_2(\mathbf{24})_2]^{2+}$ (R = 3,5-dimethoxyphenyl) show only a singlet, implying greater lability than the helical complex $[Cu_2(\mathbf{23})_2]^{2+}$. A study of the effects of ligand derivatization and change of solvent on these complexes showed the helical structure to be both more stable and more inert, and suggested a dissociative mechanism for inversion [51]. The origin of

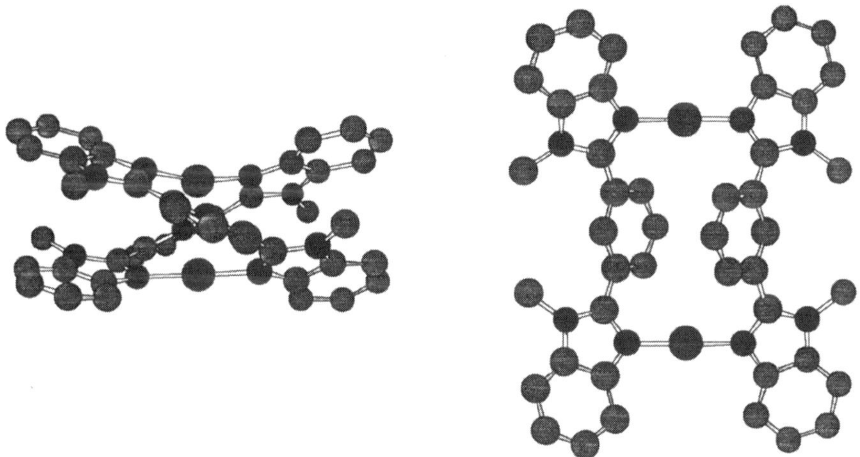

Figure 14 The structures of the two complexes $[Cu_2(\mathbf{23})_2]^{2+}$ and $[Cu_2(\mathbf{24})_2]^{2+}$.

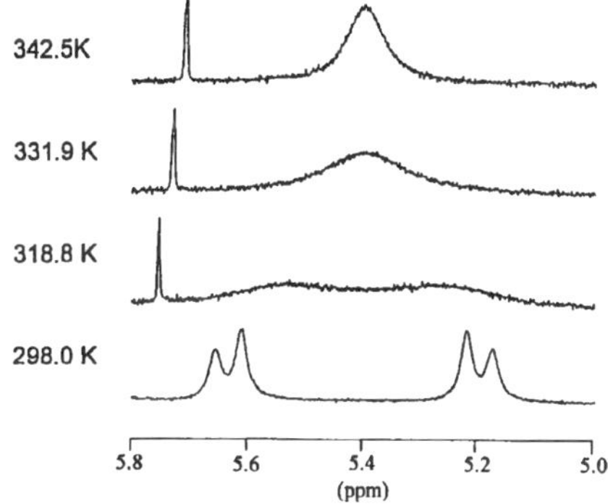

Figure 15 The methylene region of the ^1H nmr spectrum of $[Cu_2(\mathbf{23})_2]^{2+}$ (R = 3,5-dimethoxyphenyl), showing the collapse of the AB spectrum as the temperature rises [51].

the greater stabilty of the helical system lies in the weak bridging coordination of the central pyridine in $[Cu_2(\mathbf{23})_2]^{2+}$ and the existence of a stacking interaction between pyridine and benzimidazole moieties in the helical structure. Interestingly enough, the copper benzimidazole bond, as judged by Cu–N bond length, is actually slightly weaker in the helical structure than in $[Cu_2(\mathbf{24})_2]^{2+}$. These results lead to the general conclusion that polynuclear helicates are less labile than their mononuclear analogues, especially when the pitch is short. The study of racemization is easily the best way of establishing the labilty of these complexes, but their general intertness has also been shown by electrospray mass spectroscopy studies of the formation of helicates [52].

The selective synthesis of *P* or *M* helicates is an obvious challenge. To obtain such selectivity, a chiral element must be introduced into the components, and, since the metal centres used in self-assembly reactions are generally labile, the chirality must be introduced into the ligand. If we refer to the general scheme of Figure 2, this may be in the binding site, the bridge, or the ancillary groups. Chirality in the binding site was mentioned above for $pdta^{4-}$ (**1**) and $cdta^{4-}$ (**2**), and a recent example used the 2,2′-dihydroxy-biphenyl unit in the synthesis of a chiral cage [53], but for polynuclear species the only example we know of is the ligand **25** used by Corey [54] to form a trinuclear triple helicate with Ti(IV).

25

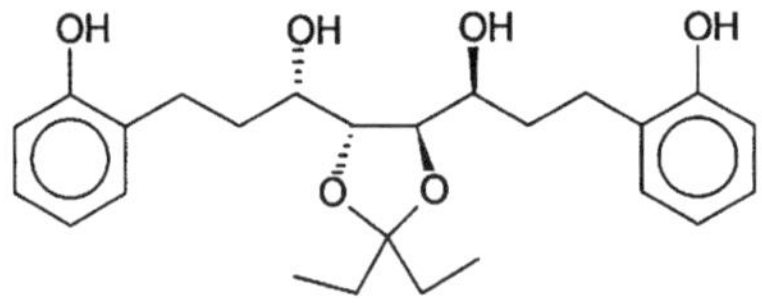

The most efficient route to date involves the chiral podates of the type mentioned previously. Shanzer synthesized a number of chiral tripod ligands such as **26** which showed stereospecific formation of triple helicates [55] which involved interstrand hydrogen bonding between amides, while more recently Siegel [56] has linked mono-, *bis*, and *tris*-bipyridyl units onto a chiral template such as **27** leading to stereospecific syntheses of double helicates.

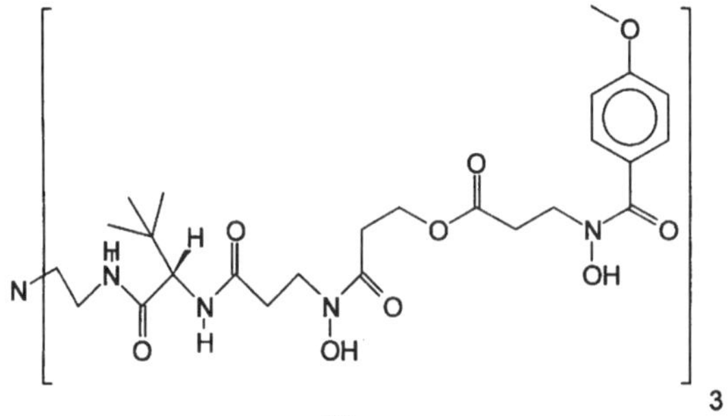

26

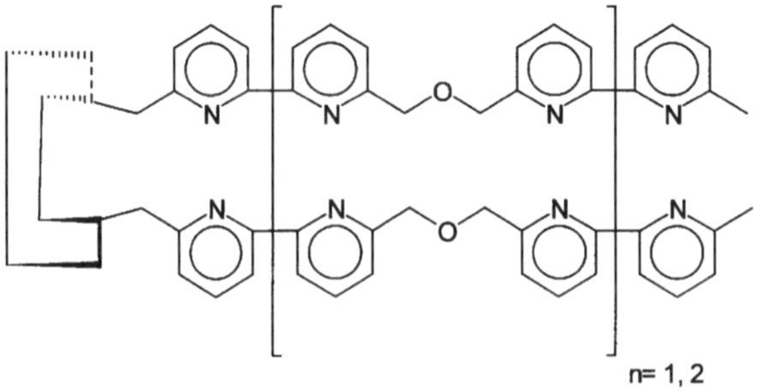

27

Lehn was the first to attempt the stereoselective synthesis of non-podal helicates [57], and introduced a stereogenic centre in the bridge between binding sites as in **28**:

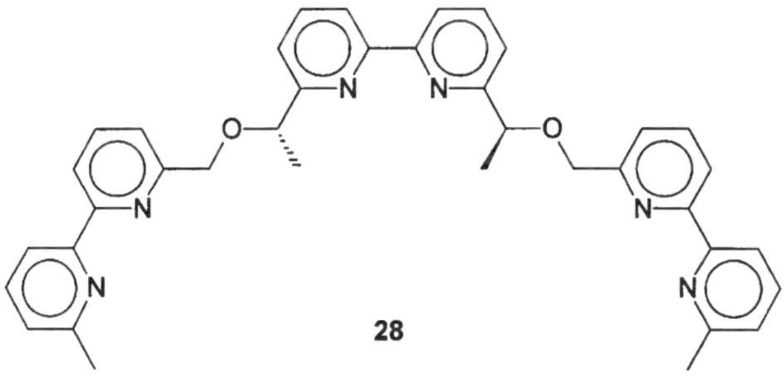

28

NMR showed only one diastereomer to be formed. The bridge was also used as a stereogenic centre by Stack [58] with a *bis*-catechol ligand **29**:

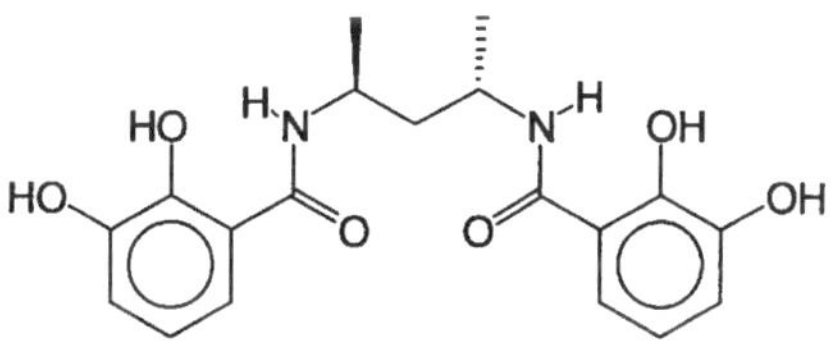

29

The ligand (*R*,*R*-**29**) forms specifically the *M* triple helicate as established by X-ray crystallography. Interestingly, however, the use of a mixture of *R*,*R*-**29** and *S*,*S*-**29** leads to formation of a mixture of homochiral (e.g. $[Ga_2(R,R\text{-}\mathbf{29})_3]^{6-}$) and heterochiral (e.g. $[Ga_2(R,R\text{-}\mathbf{29})(S,S\text{-}\mathbf{29})_2]^{6-}$) complexes in a ratio of 1 : 6.

Somewhat surprisingly, the obvious strategy of grafting a bulky chiral ancillary group onto the helicating ligand was the last to be tried, but there are now several examples. Albrecht has grafted a chiral ancillary group onto a catechol unit (**30**, [59]) while Constable has prepared chiral polypyridyl ligands using bornyl groups [60] or the pinene group (**31**, [61]) developed in another context by von Zelewsky and coworkers [62]. Interestingly the chiral polypyridines show diastereomeric excesses less than 100%. Oxazoline ligands such as **4** may be regarded as chiral modifications of **23** and have been shown to lead to selective formation of one enantiomer of the silver double helicate analogous to $[Cu_2(\mathbf{23})_2]^{2+}$ [63].

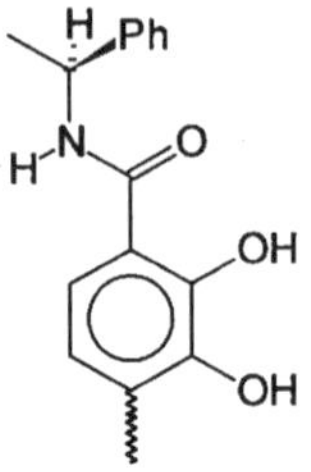

30

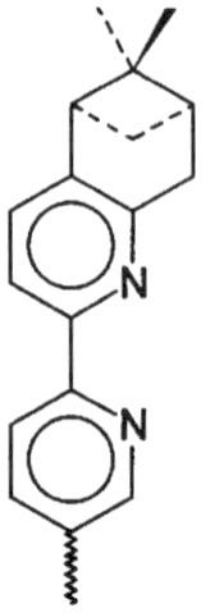

31

4.2 Circular Helicates

A recent development in this field has been the discovery of circular helicates. The geometrical description of these systems is quite simple: the metal ions are no longer aligned along a helical axis, but are spaced at equal intervals around a circle, and the ligand strands twist around this circle. The first example was discovered by Lehn [64] upon reaction of $FeCl_2$ with the ligand **32** which had previously been shown to form a linear triple helicate with Ni(II) [38].

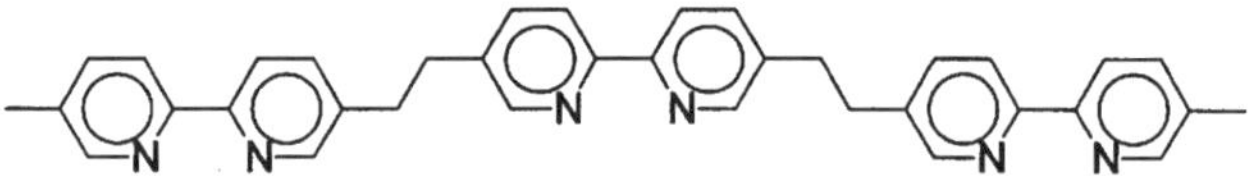

32

With $FeCl_2$, **32** gave a pentameric species shown schematically in Figure 16, in which a chloride ion is strongly bound inside the cavity. It will be noticed that the ligand strands are displaced relative to each other, so that a given pair of adjacent metal ions is linked only by two ligand strands. All five metal centres have identical chirality, Δ in the enantiomer shown.

The factors which favour the formation of a circular helicate with Fe(II) rather than a linear helicate with Ni(II) are unclear, but presumably reflect subtle differences in the coordination preferences. The chloride ion plays a role in the formation of the circular helicate, since the use of $FeSO_4$ instead of $FeCl_2$ leads to formation of a hexagonal helicate with the central cavity apparently unoccupied. $FeBr_2$ gives a mixture of $[Fe_5(\mathbf{32})_5Br]^{9+}$ and $[Fe_6(\mathbf{32})_6]^{12+}$ [65]. The flexibility of the ligand is also a factor of importance. If the bridge linking the bipyridyl units is made more flexible as in **33** then reaction with $FeCl_2$ leads to the tetranuclear cyclic helicate $[Fe_4(\mathbf{33})_4]^{8+}$ [65].

33

34

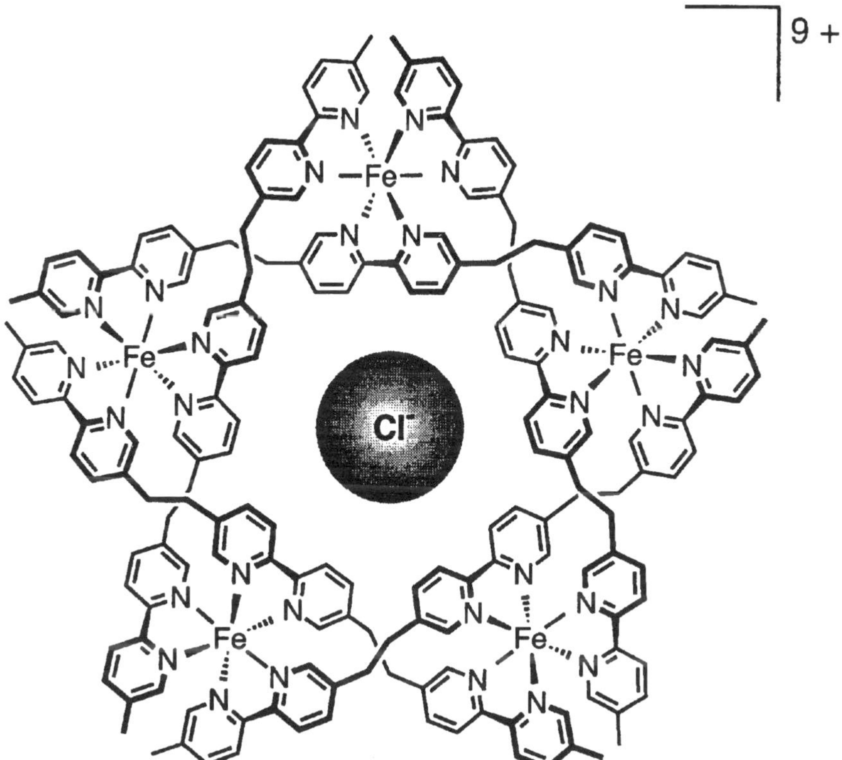

Figure 16 Schematic view of the complex $[Fe_5(\mathbf{32})_5Cl]^{9+}$. Reproduced with permission from reference 64.

Clearly many factors combine to determine what is the final product, and an interesting illustration of this has been given for tetrahedral systems with ligand **34** which reacts with Cu(I) to give a mixture of double helical $[Cu_2(\mathbf{34})_2]^{2+}$, the circular helicate $[Cu_3(\mathbf{34})_3]^{3+}$, and the (presumably) achiral grid $[Cu_4(\mathbf{34})_4]^{4+}$. Crystallization from the solution results in the exclusive isolation of the double helicate salt which redissolves to give a mixture of the three complexes [66]. A similar selective crystallization is observed for the enantiomerically pure complex $[Ag_3(R,R\text{-}\mathbf{4})_3]^{3+}$ where R = Ph [63], where the circular helicate shown in Figure 17 is obtained in the solid state even though NMR suggests the major species in solution to be the double

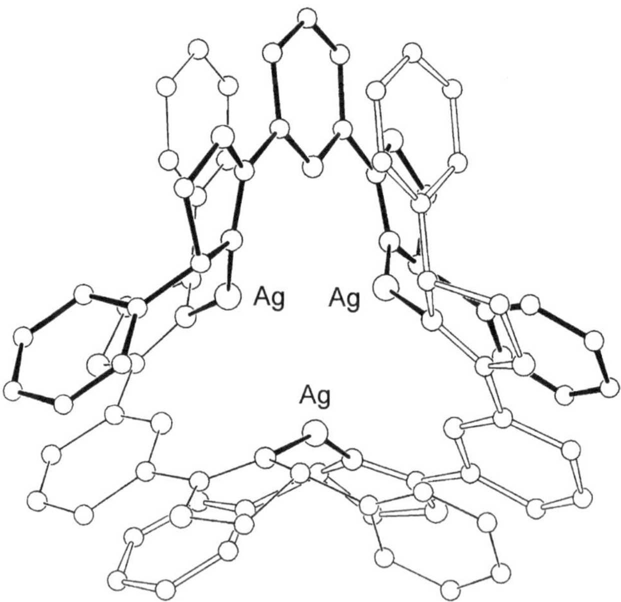

Figure 17 View of the circular helicate $[Ag_3(R, R\text{-}\mathbf{4})_3]^{3+}$ along the threefold axis. The three strands are differentiated by shading of the bonds as full, open or thin lines respectively.

N N N N

35

36

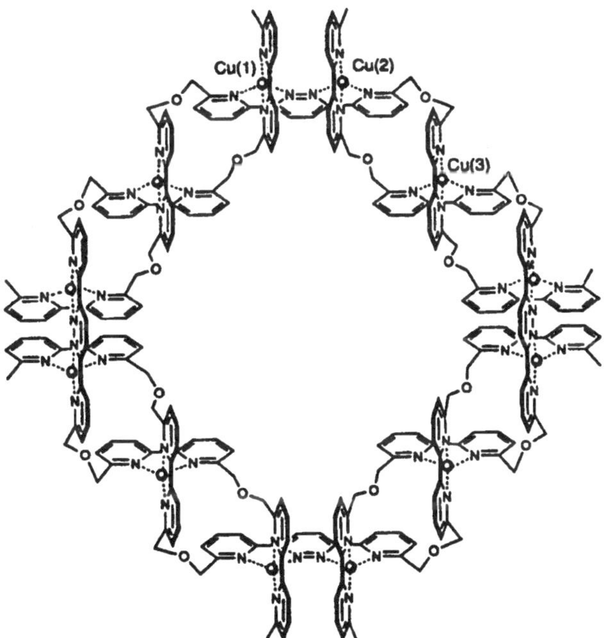

Figure 18 Schematic view of the structure of $[Cu_{12}(\mathbf{36})_4]^{12+}$. Reproduced with permission from reference 68.

37

helicate $[Ag_2(R,R\text{-}\mathbf{4})_2]^{2+}$ [9]. In this case the stacking interaction between the phenyl substituents and pyridines seems to favour the crystallization of the circular helicate.

A second example of a stereospecific circular helicate has been reported very recently with the chiral ligand **35** which links together tetrahedrally coordinated Ag(I) ions to give the hexanuclear species $[Ag_6(\mathbf{35})_6]^{6+}$ [67].

All circular helicates reported to date are homochiral, and they are generally obtained in high yield. The beauty of the structures should not blind us to the current lack of understanding of why they are formed in preference to a linear helicate, and why a particular nuclearity is chosen.

Two related circular structures have been described recently by Lehn [68] and McCleverty and Ward [69]. The segmental ligand **36** reacts with copper(I) to form a circular dodecanuclear species $[Cu_{12}(\mathbf{36})_4]^{12+}$ shown in Figure 18 [68]. The metal

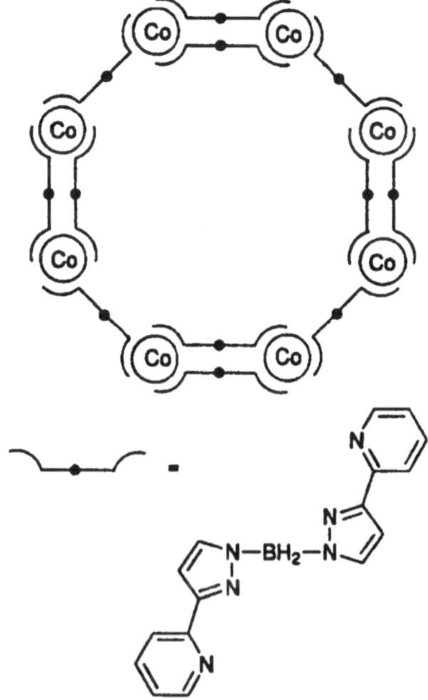

Figure 19 The structure of $[Co_8(\mathbf{37})_{12}]^{4+}$. Reproduced with permission from reference 69.

centres may be grouped as alternating homochiral triplets -(ΛΛΛ)-(ΔΔΔ)-, each half of ligand **36** having the same chirality.

The ligand **37** used by McCleverty and Ward forms an octanuclear species $[Co_8(\mathbf{37})_{12}]^{4+}$ shown schematically in Figure 19 in which each metal has a *tris*-bidentate coordination, linked by one ligand to one neighbour, and two ligands to the other. In this species, however, the metal centres are homochiral [69].

It is interesting to note that the rather rigid pyridazine bridge of **36** results in an inversion of chirality for the two metal centres bound to it, whereas the more flexible bridges in the other parts of the molecule and in **37** lead to homochirality.

4.3 Dendrimers

Dendrimers, also called arborols or cascade molecules, are highly branched compounds synthesized stepwise from a central core and they can be of nanometre size. Since the first report of an organic dendrimer in 1978 [70], a lot of work has been done on this type of molecule. A particularly interesting class of dendrimers is that containing one or more metals; such metallodendrimers are interesting as novel magnetic, electronic or photo-optical materials. Contrary to organic dendrimers, the junctions in metallodendrimers may be based either on covalent bond or on metal-ligand interactions, and this offers more possibilities to chemists to build compounds with particularly symmetrical three-dimensional structures. Using monodentate, bidentate or tridentate rigid ligands with tetra- or hexa-coordinated metal cations provides the bricks necessary to build well-defined three-dimensional ramifications.

The first report of chiral dendrimers dates back to 1979, but concerned a purely organic molecule based on the amino acid lysine [71]. In the past two decades, many publications have appeared dealing with chiral dendrimers, although there are very few reports on chiral metallodendrimers. Several authors have tried to classify the various possibilities for introducing chirality into organic dendritic architectures [72, 73], and we may consider the different types of chirality and the ways in which it may be introduced into metallodendrimers. Chiral metallodendrimers may be based on (1) the chirality of the core only (either organic or inorganic), (2) the chirality of the metal centre (Δ or Λ), (3) chirality of the ligand only, (4) chirality of terminal group only, (5) a combination of some of the properties cited above. All the metallodendrimers in which chirality has been introduced or established are from classes 1 and 2.

Some metallodendrimers with one or more stereogenic centers have been prepared without control of the chirality. Vögtle and Balzani [74] have tried several strategies to prepare dendrimers in which a ruthenium cation is the core of the final compound. In these compounds, the only centre of chirality is that of the metal, but as it was not controlled racemic mixtures were obtained. Controlling the stereochemistry of the starting complex would have allowed the authors to prepare a optically pure metallodendrimer. Denti, Campagna, Balzani, and their co-workers have studied polymetallic dendrimers based on bipyridine and 2,3-*bis*-(2-pyridyl)pyrazine (2,3-

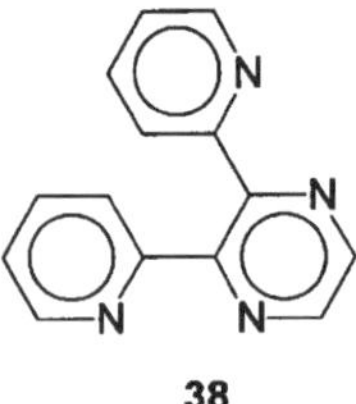

dpp) **38**, prepared by the 'complexes-as-metals and complexes-as-ligands' synthetic strategy [75]. These compounds (Figure 20), containing up to 22 metal centres, are a mixture of several diastereoisomeric species. In fact, the $[Ru(2,3\text{-dpp})_3]^{2+}$ core is a mixture of the *mer* and *fac* isomers in which the *mer* isomer predominates (92%) and each octahedral metal center may have the Δ or Λ stereochemistry. A further complication is the kinetic inertness of the Ru(II) centre, which prohibits extensive rearrangement to the thermodynamically stable product.

Recently, different strategies have been studied to prepare enantiopure metallodendrimers. Constable [76] proposed a divergent strategy starting from a chiral

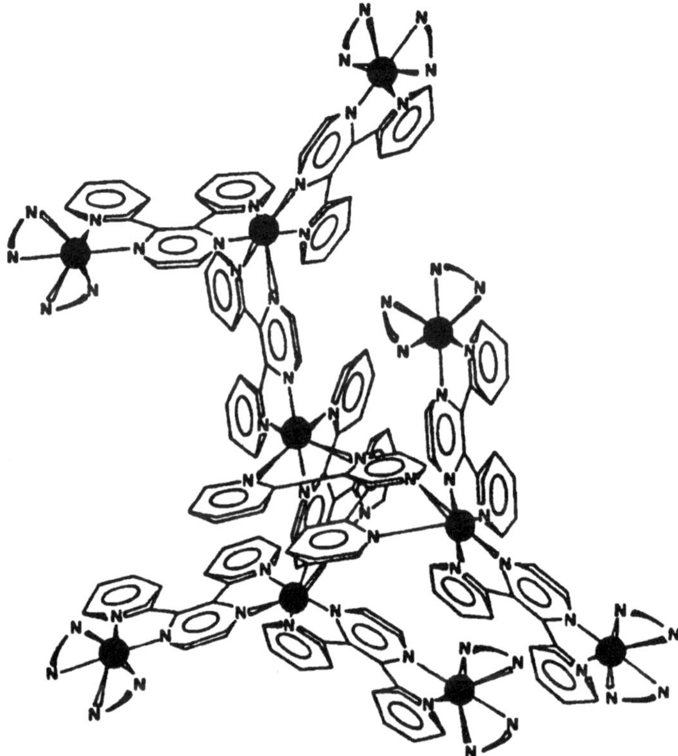

Figure 20 Schematic view of the structure of a decanuclear compound. Reproduced with permission from reference 75.

$[Ru(bipy)_3]^{2+}$ centre, and expanding the dendrimer using achiral $Ru(terpy)_2$ units to give a heptanuclear metallodendrimer (Figure 21), thereby avoiding the problem of a plethora of diastereomers. Unfortunately this was unsuccessful, but the heptanuclear species could be generated by assembly around a labile metal ion (Fe(II) or Co(II)) of the dinuclear ruthenium units.

The first enantiopure metallodendrimer is the tetranuclear ruthenium species of Bodige *et al.* [77] shown in Figure 22. Four diastereomers ($\Delta\Delta_3$, $\Lambda\Delta_3$, $\Delta\Lambda_3$ and $\Lambda\Lambda_3$) have been prepared using a different strategy from Constable and involving the preparation of four optically pure ruthenium complexes which were then linked by ligand-ligand reactions. The different isomers have been clearly identified by NMR spectroscopy and circular dichroism. The CD spectra show the expected mirror image relationship between the two pairs of enantiomers $\Delta\Delta_3$, $\Lambda\Lambda_3$ and $\Lambda\Delta_3$, $\Delta\Lambda_3$. An approximately additive relationship between the number and type of chromophores and the magnitude of the molar ellipticity is observed in the CD.

To our knowledge this the only example of a stereoselective metallodendrimer synthesis. Chiral ligands have not yet been used, but some chiral complexes prepared

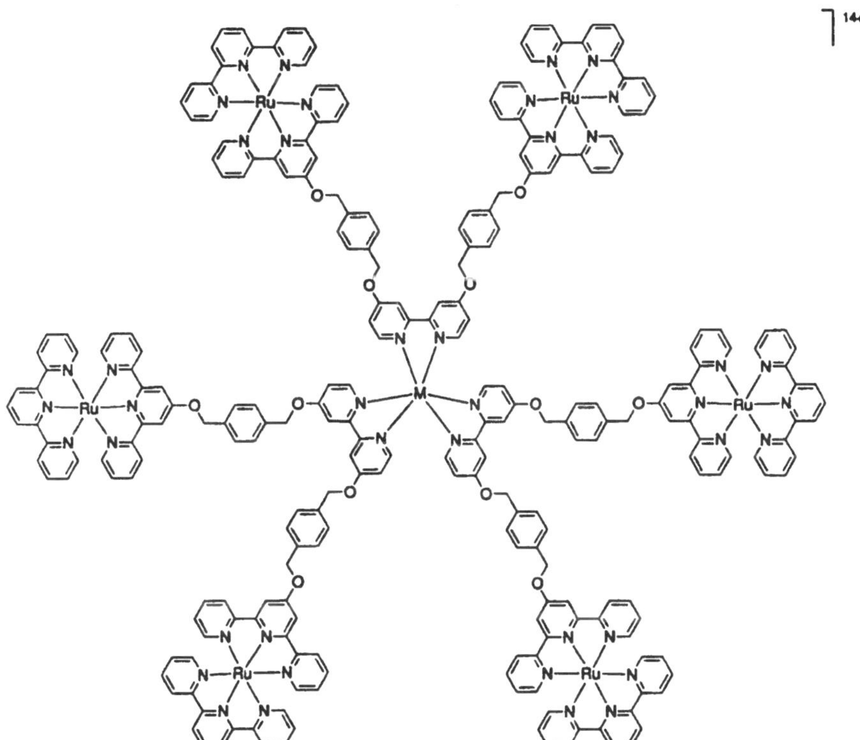

Figure 21 A heptanuclear dendrimer, formed by complexation around the central metal ion. Reproduced with permission from reference 76.

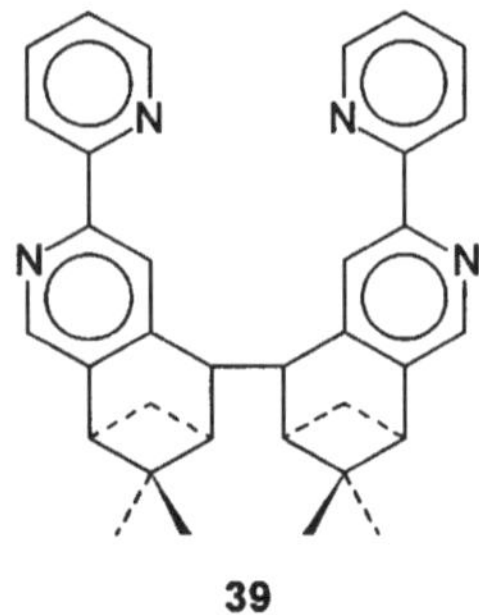

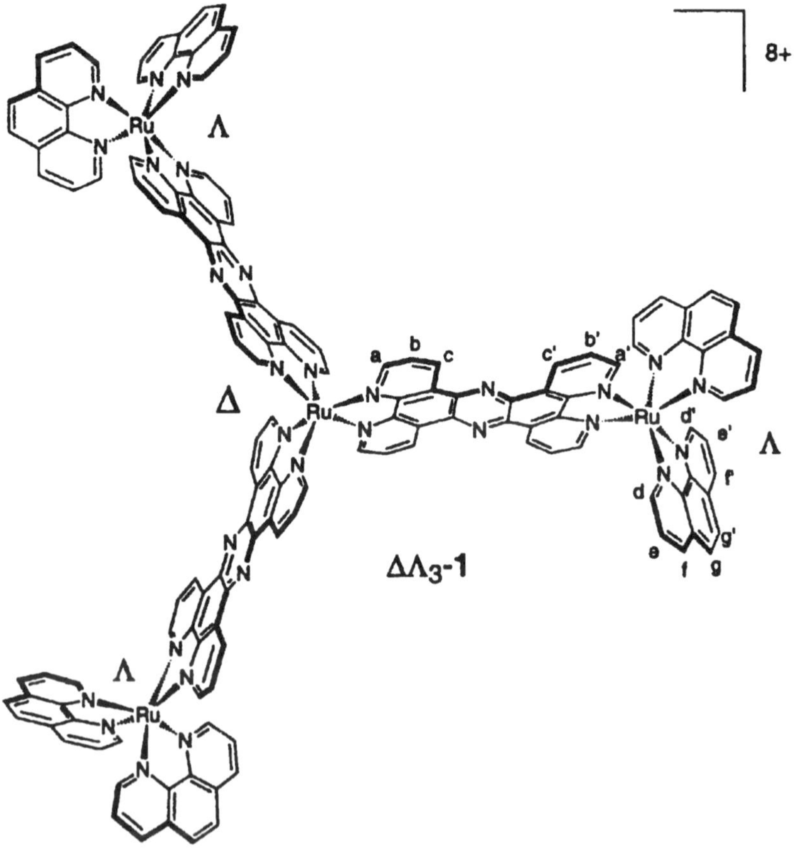

Figure 22 The enantiopure metallodendrimer of Bodige *et al.* Different diastereomers may be prepared by varing the chirality of the components. Reproduced with permission from reference 77.

with 'chiragen' ligand **39** might be interesting building blocks to prepare chiral dendrimers. The trimeric species of von Zelewsky [78], made of three ruthenium cations and four 'chiragen' ligands, is one of these potentially interesting building blocks. The use of chiral ligands seems to be logically one of the next steps in the quest for chiral metallodendrimers.

4.4 Boxes

Over the last few years, self-assembly has emerged as a very promising approach to the generation of compounds with large molecular-sized cavities. Such two-dimensional macrocyclic compounds are called boxes. Three-dimensional macropolycyclic compounds are more often called cages and will be discussed in the next section. The use of non-covalent interactions between metal atoms and organic ligands to generate macrocyclic structures, is particularly attractive since the cyclization is under thermodynamic control. As a result, if the components are well designed, polymerization will not take place. Boxes are prepared by combining angular and linear units and they have the shape of polygons. Depending on the turning angle of the angular component, the resulting macrocyclic compound can be a triangle, a square, a parallelogram, a pentagon or a hexagon [79].

The first macrocyclic square box was reported in 1990 by Fujita *et al.* [80]. They used the square planar geometry of four palladium(II) atoms and four 4,4′-bipyridine ligands to generate a large macrocyclic cavity. Since then they have studied other boxes but they have not discussed the problem of chirality. Stang drew up a list of the ways of introducing chirality when using transition metals and organic ligands as building blocks [79]. He proposed five ways of creating chiral supramolecular species via spontaneous self-assembly: (1) use of a chiral ligand coordinated to a metal; (2) use of an inherently chiral octahedral metal complex; (3) use of an optically active atropoisomeric diaza-bisheterocycle as linker ligand; (4) helicity or twist due to the use of ligands which lack a rotation symmetry about their linkage axis; and (5) a combination of the above methods. We might add a sixth possibility which is the presence of a chiral conformation of the ligand.

Up till now, the chirality in boxes has mostly been introduced via the ligand. A chiral β-ketosulphoxide has been used to prepare an enantiomerically pure palladacycle [81]. The compound was characterized both in the solid state and by NMR spectroscopy in solution. This box is composed of three palladium(II) atoms and three ligands. A section of the structure is shown in Figure 23. We can see the coordination sphere of the palladium atoms and the overall structure has the shape of a triangle.

Stang *et al.* have used several strategies to prepare chiral boxes. The first one was the use of ligands, such as **40**, which lack a rotation symmetry about their linkage axis. By reacting **40** with the achiral metal bisphosphine **41**, they prepared a diastereoisomeric mixture of products [82] which were characterized by ^{31}P and

^{1}H NMR and shown to be a mixture of the six isomers shown in Figure 24 and of linear oligomers.

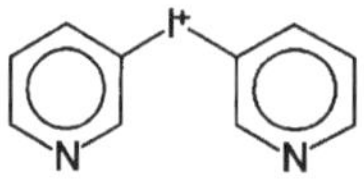

40 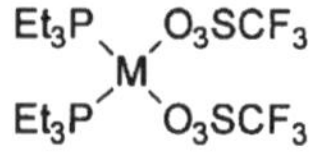 41 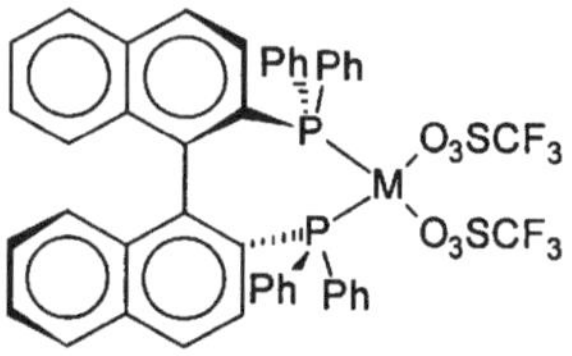 42

M = Pd, Pt

A chiral bisphosphine such as 2,2′-*bis*-(diphenylphosphino)-1,1′-binaphthyl (BINAP) has been extensively used as a chiral chelator in asymmetric catalysis. When Stang *et al.* reacted the chiral metal complex **42** with **40**, they synthesized a square box (Figure 25) and asymmetric induction was observed [79,82] with the formation of an excess of one of the preferred diastereoisomers as measured by NMR spectroscopy. The same reaction has been carried out with **42** and *bis*-4-(4′-pyridyl)phenyl)iodonium triflate, but in this case the diaza ligands of the iodonium species possess rotational symmetry about their linkages. Consequently, the optical activity of the molecular squares obtained is due exclusively to the chiral transition metal auxiliary BINAP.

A second strategy investigated by Stang *et al.* [79,83] is the use of a C_{2h} diaza-bisheterocycle as linker ligand, such as 2,6-diazaanthracene (DAA) **43** and 2,6-diazaanthracene-9,10-dione (DAAD) **44**, as opposed to the D_{2d} or D_{2h} symmetrical ligands commonly used in molecular squares.

Upon formation of the conformationally rigid square box, these ligands are restricted in rotation around the metal–nitrogen bond and remain orthogonal to the transition metal coordination plane. Reaction of such ligands with metal complexes such as **41** can lead to six possible diastereoisomers (Figure 26).

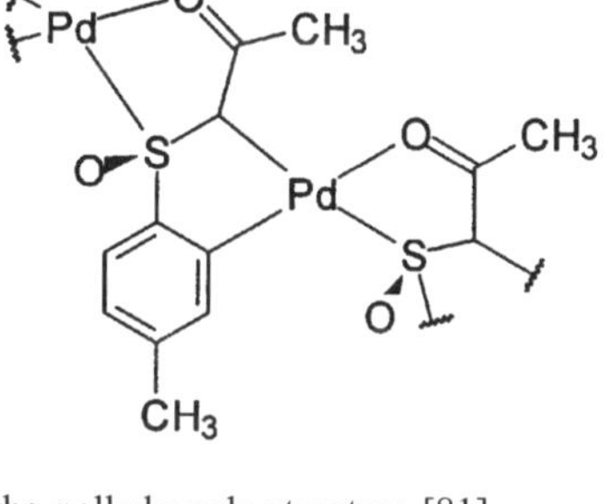

Figure 23 A section of the palladacycle structure [81].

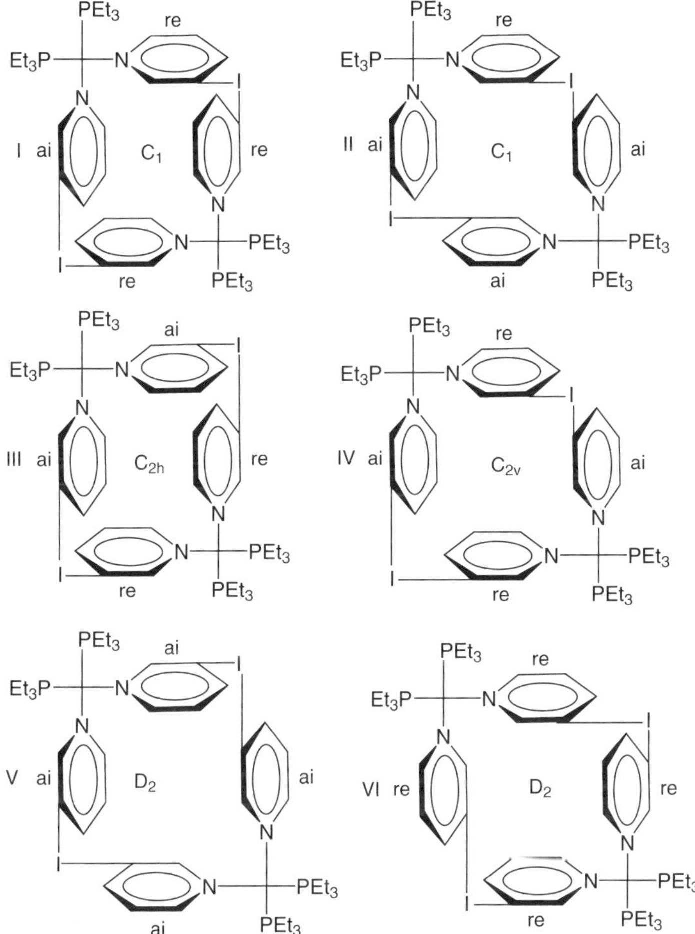

Figure 24 The six possible isomers of a box prepared with two ligands **40** and two square planar coordinated metals. Reproduced with permission from reference 82.

When the chiral complex **42** was mixed with DAA (**43**), the formation of a single diastereomer was observed by ^{31}P NMR, and was shown by NMR and X-ray crystallography to be isomer VI of Figure 26, shown in more detail in Figure 27. When the same experiment was carried out with DAAD (**44**) one diastereomer was the major product (diastereomeric excess in the range of 70% to 80%) but traces of other products were observed by NMR [79,83].

A rather different chiral box is the trimeric cyclometallated compound reported by Rüttiman *et al.* [84]. This triangular box (Figure 28) has a hydrophobic cavity in which a guest molecule such as acetonitrile fits perfectly. The chirality of this box

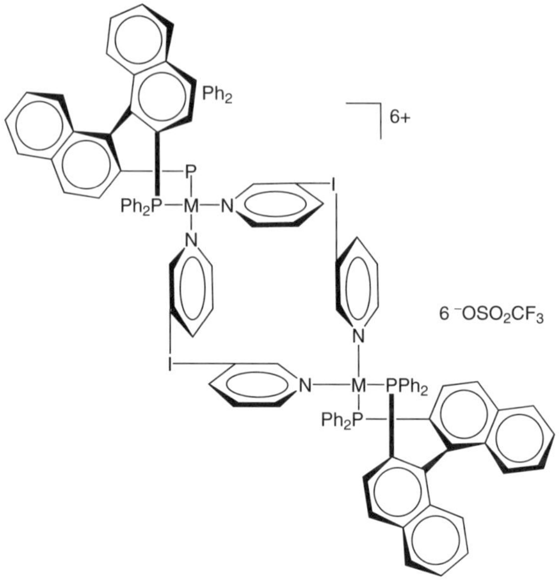

Figure 25 The square chiral box prepared with **40** and **42**. Reproduced with permission from reference 79.

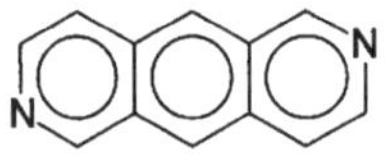

43

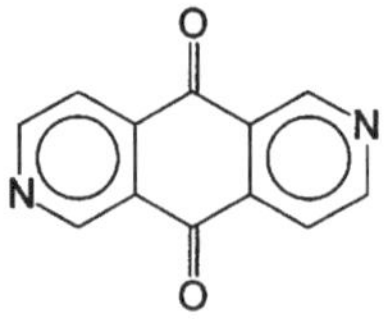

44

results from the locking of the ligands into a chiral conformation upon formation of the macrocycle. In contrast to the previous examples the metal centre no longer has a twofold symmetry axis relating the two bridging ligands. Somewhat surprisingly, the use of a chiral metal centre such as a *cis-bis*(bidentate) octahedral unit has not been reported up to now, but in view of the interest in the use of these boxes as chiral receptors, it seems reasonable to assume that this will soon be tried.

4.5 Cages

If helicates and boxes may be regarded as the organization of metal ions in one and two-dimensional space respectively, then cages may be regarded as their organization in three dimensions. The simplest three-dimensional structure is the tetrahedron, and we may envisage two ways to organize the metal ions into a tetrahedron, by

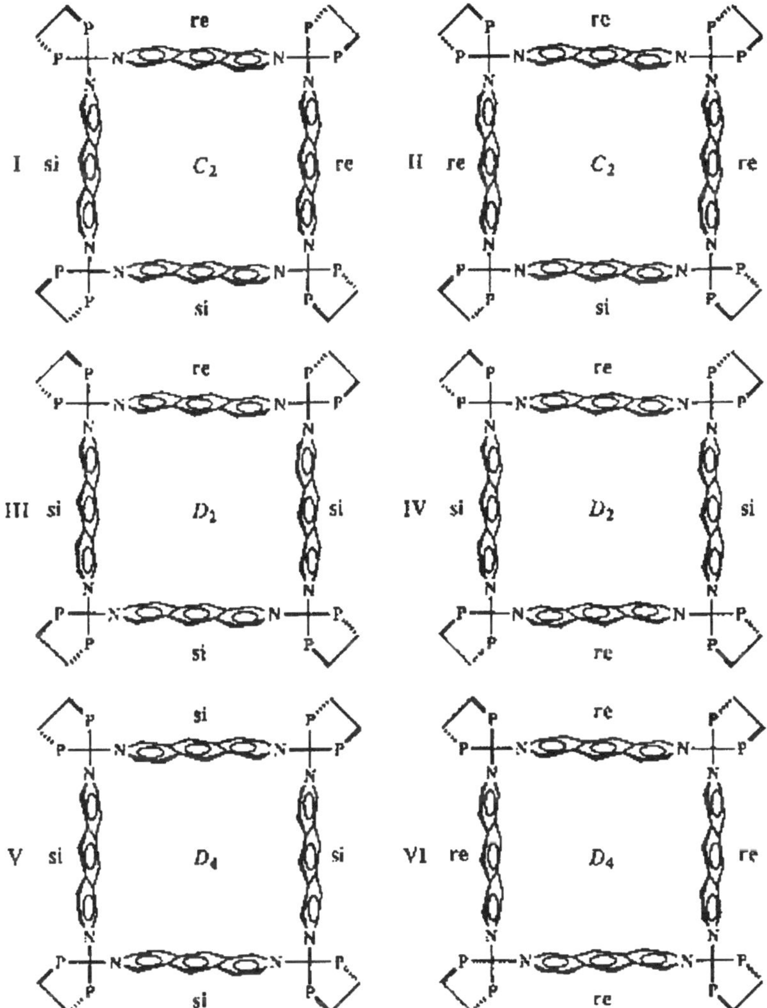

Figure 26 Schematic representation of the six possible diastereoisomers of the cyclic tetranuclear complexes with a C_{2h} symmetric bridging ligand. Reproduced with permission from reference 83.

edge-bridging between two metal ions, and by face-bridging between three metals (Figure 29).

The edge-bridging structure leading to a stoichiometry M_4L_6 is now well established and is frequently referred to as the adamantyl structure. Saalfrank has studied this structure type using ligands of the general type **45**:

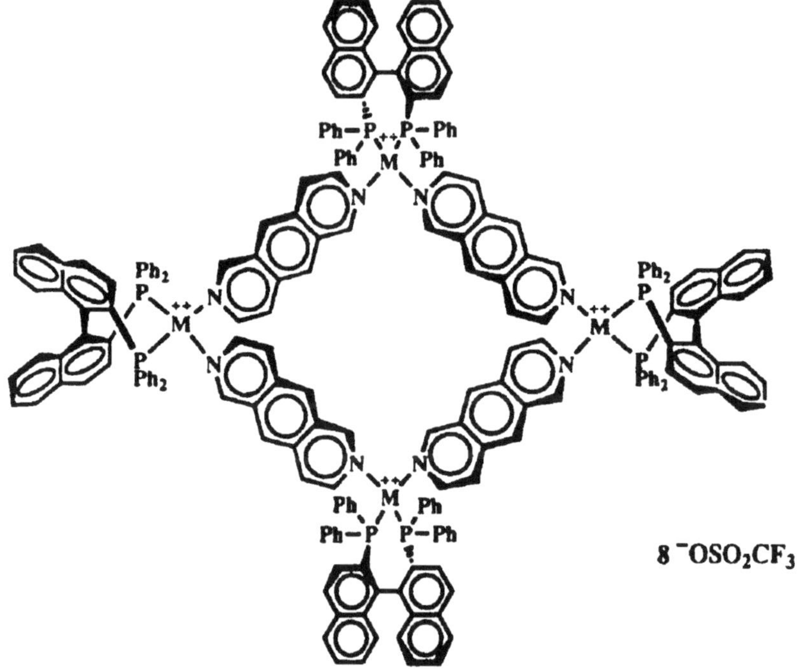

Figure 27 The square chiral box prepared with **42** and **43**. Reproduced with permission from reference 83.

The M_4L_6 structure is observed with M^{II} ions such as Mg, Mn, Co, Ni, and Zn [85], Fe(III) [86], and a mixed valence species with three Fe(III), and one Fe(II) which includes an ammonium ion to balance the charge [87]. The vertices of these structures are octahedral metal ions complexed by three bidentate ligands, and are consequently chiral. X-ray crystallography has shown the existence of two structure types, one with approximate T symmetry in which the four metal centres are homochiral [85,87], and another in which the cluster has crystallographic S_4 symmetry [86], resulting in a *meso* complex with two Δ and two Λ coordination spheres. Raymond [88] has studied a similar system using the hydroxamic acid ligand **46** which forms neutral M_4L_6 complexes with Fe(III) and Ga(III). The crystal structure of the Ga(III) complex shows the S_4 *meso* structure, but the ^{1}H NMR spectrum shows only six proton signals, suggesting a higher symmetry in solution. Raymond has analysed the tendency of this *bis*-bidentate ligand to form tetranuclear M_4L_6 complexes rather than triple helicates in terms of the structure of the bridging unit and the relative orientation of the bridging groups.

The only example of a face-bridging tetrahedron of which we are aware is the complex $[Mn_4(\mathbf{47})_4]^{4+}$ in which the Mn(II) ions have a distorted octahedral

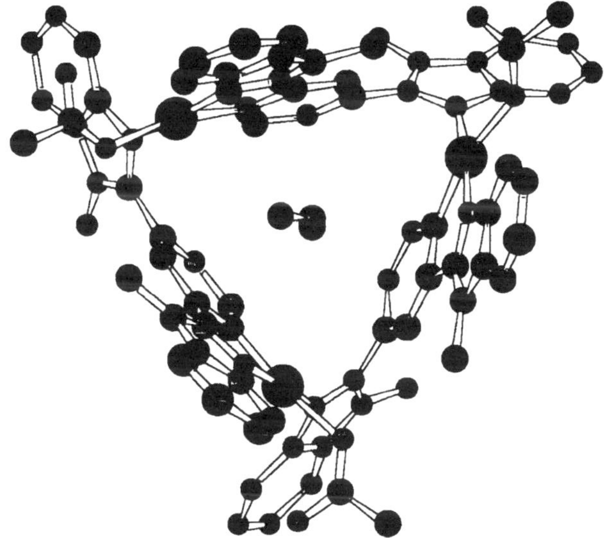

Figure 28 X-ray structure of the triangular box of Rüttiman *et al.* [84]. A molecule of acetonitrile is included in the crystal.

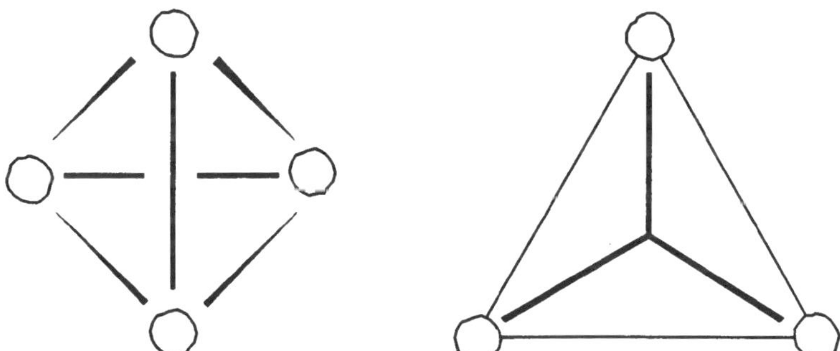

Figure 29 Connectivities of tetrahedral clusters: the edge-bridging structure (left) and the face-bridging structure (right)—only one ligand is shown and the fourth metal ion lies behind the centre of the ligand.

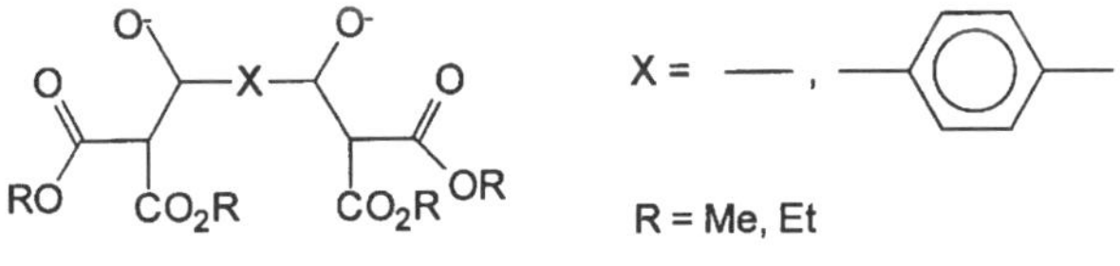

O⁻ X O⁻ O O RO CO_2R CO_2R OR

X = —, (p-phenylene)

R = Me, Et

45

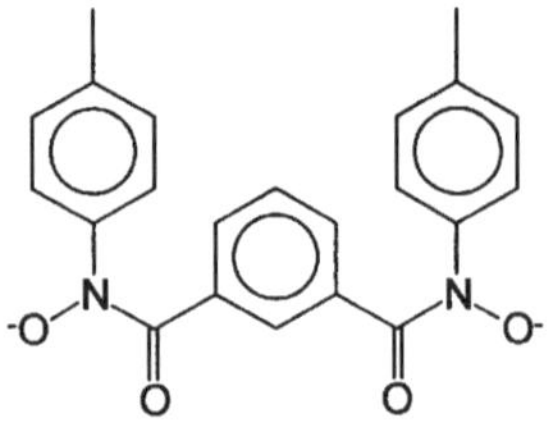

46

coordination by three pyridyl-pyrazole units [89]. The chiralities of the four metal centres are identical, indicating approximate T symmetry.

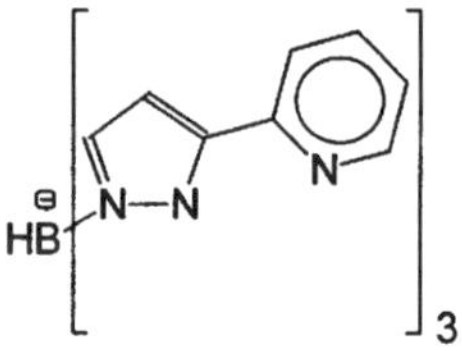

47

An octahedron may be generated by six metal ions at the vertices linked by four face-bridging ligands (Figure 30), and this was first reported by Fujita using ligands such as **48** to link six $[M(en)]^{2+}$ units (M = Pt, Pd; en = 1,2-diaminoethane) [90]. The highest symmetry possible for this structure is T, which requires the six metal centres to be homochiral. Very recently, Stang [91] has replaced the achiral en ligand by chiral *R*-BINAP and using a slightly different ligand from **48** has shown that an octahedral complex of T symmetry is also formed.

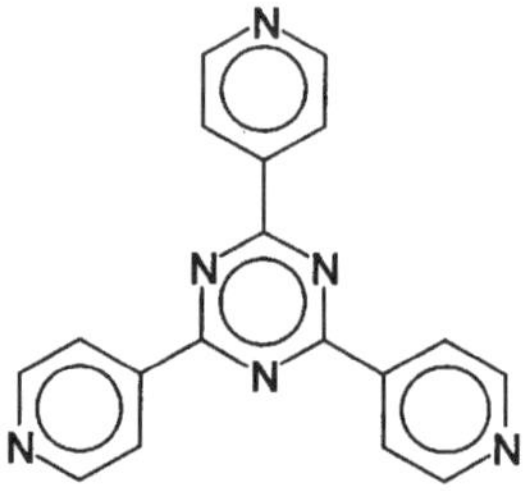

48

4.6 Topologically Complex Molecules

Sauvage and Dietrich-Buchecker have developed a beautiful chemistry of topologically complex molecules in which complexation reactions are designed to favour the threading and interweaving necessary to produce the topological complexity [92]. In

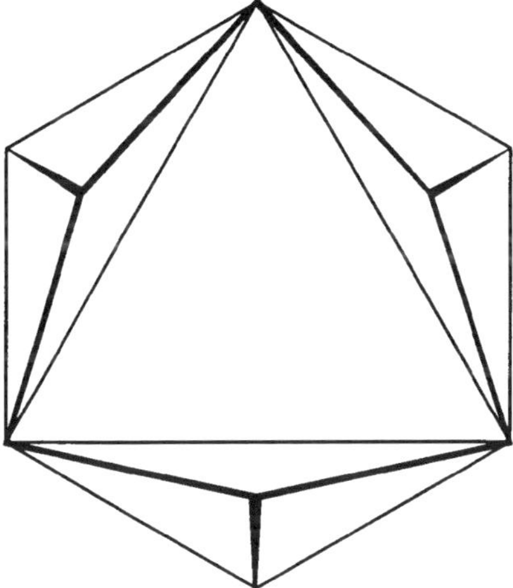

Figure 30 An octahedral cage formed by six metal centres linked by four face-bridging ligands of which only three are shown in the drawing.

general the metal centres used have been either *bis*-bidentate tetrahedral centres, typically with Cu(I) or Ag(I), or *mer-bis*-tridentate octahedral centres associated with d^6 ions such as Fe(II) or Ru(II). The classical catenate synthesis assembles two interlocking ligands around a tetrahedral Cu(I) centre (Figure 31), and is achiral if the ligand is symmetric, but asymmetric substitution of the ligand results in two enantiomers which can be distinguished by NMR in a chiral environment [93].

Chirality may also be introduced into a simple catenate if the reaction used to close the cycles involves the formation of a stereogenic centre. This is the case with ligand **49** where reaction with Fe(II) leads to the formation of the complex $[Fe(\mathbf{49})_2]^{2+}$; the pendant bidentate groups can then be bound together to close the cycle by addition of Ag(I). However, the tetrahedral *bis*-bidentate unit is chiral (see Figure 4), and consequently closing the loops can lead to hetero- or homochiral arrangements (Figure 32).

The product of this reaction shows two different crystal forms which were shown by X-ray crystallography to be the homochiral and heterochiral compounds, respectively [94]. Sauvage has recently used a similar system composed of a

N N N N N N N N N

49

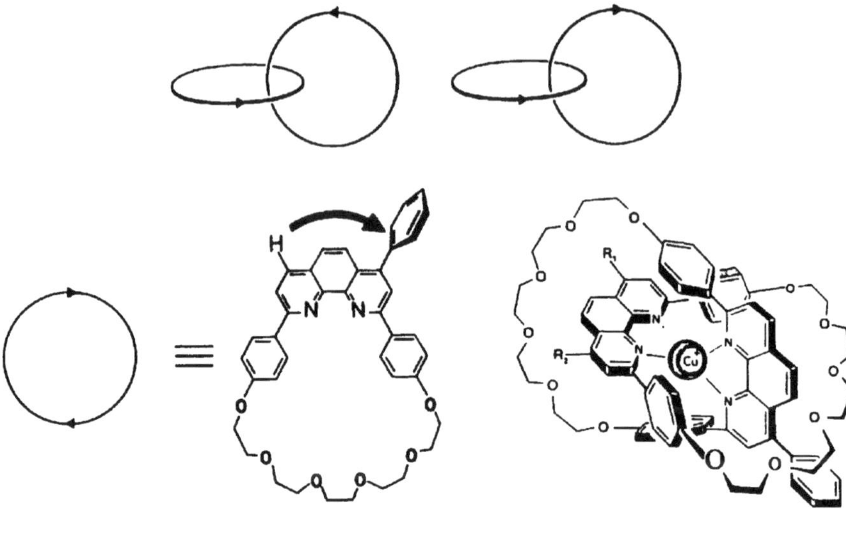

Figure 31 A catenate formed by interlocking of rings around a Cu(I) centre. The system is achiral if the ligand is symmetric, but asymmetric substitution of the ligand ($R_1 = H$; $R_2 = Ph$) results in a directionality in the ligand (indicated by an arrow following the shortest path from R_1 to R_2). The resulting catenate exists as two enantiomeric forms. Reproduced with permission from reference 92.

tridentate-bidentate-tridentate ligand to assemble a catenate around Cu(I), 'clipping' the cycles closed with octahedral metals which bind to the tridentate site [95]. While simple singly interlocked catenates are achiral, doubly interlocked catenates are intrinsically chiral. Figure 33 shows the strategy used to synthesize the doubly interlocked [2]-catenane. It will be noted that the necessary chirality for the synthesis is incorporated via a trinuclear double-helical structure [96].

The use of dinuclear double helicates as a starting point for the synthesis of a trefoil knot has already been mentioned [35,36]. Dietrich-Buchecker and Sauvage have published a general review of the synthesis of molecular knots [97]. The resolution of the synthetic trefoil knot was reported recently [98] and the chiroptical properties of both the metallated and the demetallated knot reported: the α_D values were high and comparable to helicenes. A recent development in this field has been the synthesis of composite knots involving the linking of two chiral precursor complexes [99]. Figure 34 shows the strategy for the synthesis of a simple knot and its extension to a composite knot. Figure 35 shows how the strategy was developed in the laboratory, and illustrates the problems arising from the existence of the precursor in two forms (the preknot, p-K, and the premacrocycle p-M) both of which

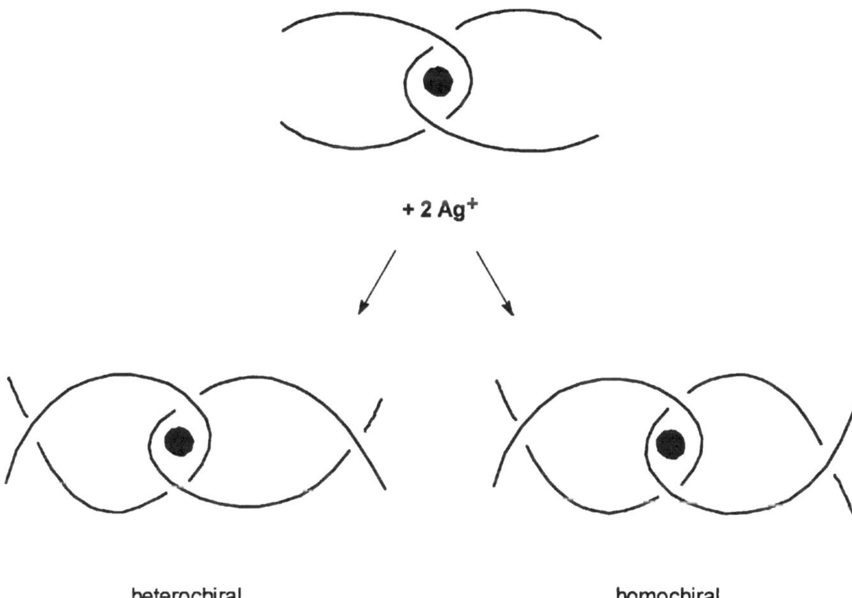

Figure 32 The two different diastereomers arising from reaction of $[Fe(\mathbf{49})_2]^{2+}$ with two equivalents of Ag^+.

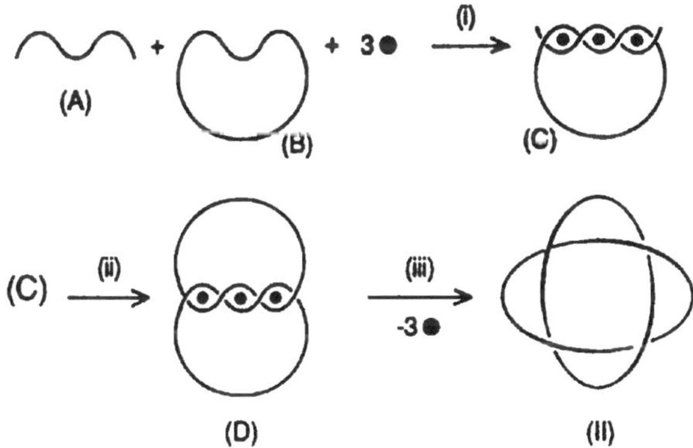

Figure 33 Strategy for the synthesis of a doubly interlocked catenane. Reproduced with permission from reference 96.

exist as enantiomeric pairs. A total of 10 compounds may be anticipated from the linking of these fragments, and their identification and analysis is a complicated problem. Surprisingly the yield of the composite knot formed by linking of two p-K fragments was significantly higher than the statistical value; furthermore, a signifi-

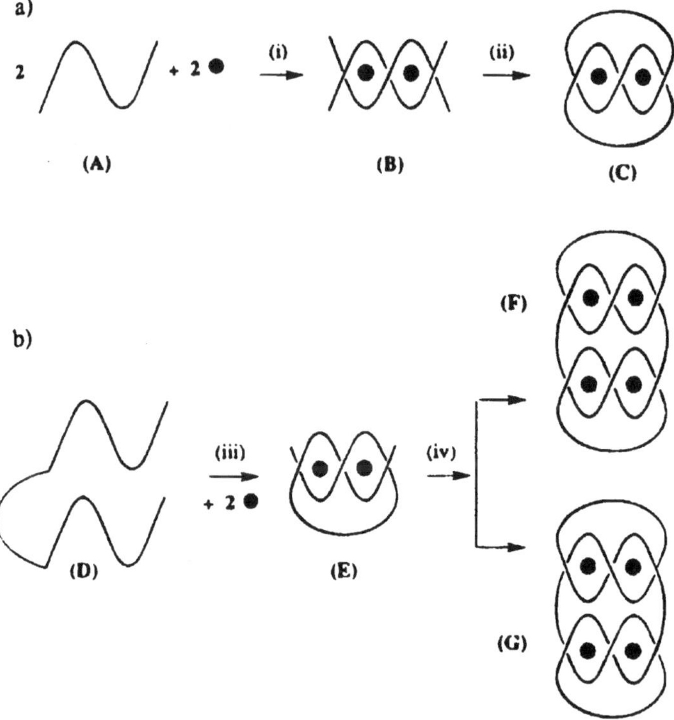

Figure 34 Strategies for the synthesis of simple (a) and composite (b) knots. Complex B is a dinuclear double helicate which is then cyclized to yield the simple knot. Ligand D has sufficient binding sites to form the pre-knot E which contains the double-helicate unit which is then coupled to give the composite knots F and the *meso*- form G. Reproduced with permission from reference 99.

cantly higher amount of the *meso* K-K dimer was observed than predicted by statistical calculations

4.7 Other Structures

In the previous sections we have concentrated on complexes where the ligands were highly structured, and thereby enabled a degree of control over the formation of the polynuclear species. However, many polynuclear complexes may be formed upon condensation of very simple ligands, and in this section we will discuss some examples of these. One of the oldest examples is to be found in the acetylacetonates. Dehydration of the octahedral complex $[Ni(acac)_2(OH_2)_2]$ leads to $[Ni(acac)_2]_3$. The structure of the trimer [100] is shown in Figure 36.

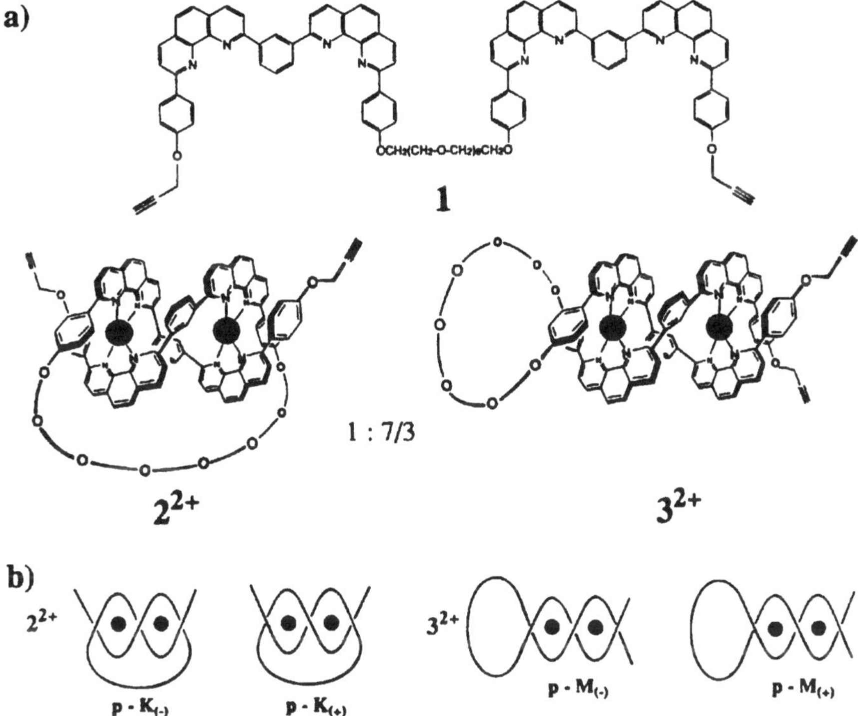

Figure 35 The system used to apply the strategy defined in Figure 34. The precursor ligand binds two copper ions to give either the preknot (p-K, $\mathbf{2}^{2+}$) or the premacrocyle (p-M, $\mathbf{3}^{2+}$) (a), both species existing in two enantiomeric forms (b). Reproduced with permission from reference 99.

Six of the oxygen atoms of the acac ligands are bound to two nickel atoms. The external nickel sites have *tris*(bidentate) coordination, one with Δ configuration, one with Λ, while the central atom has an achiral coordination sphere with two acac ligands in a plane, and the fifth and sixth coordination sites occupied by oxygens from acac ligands bound to the external Ni atoms. The three NiO_6 coordination spheres may be regarded as face-sharing octahedra. Somewhat surprisingly the analogous Co(II) complex forms a tetramer $[Co(acac)_2]_4$ with a completely different structure, composed of two pairs of face-sharing octahedra with identical conformations ($\Delta\Delta$ and $\Lambda\Lambda$), which share a common edge, the molecule as a whole having a plane of symmetry. Cotton has further shown how $[Co(acac)_2]_4$ may be progressively 'cut open' by hydration to $[Co_3(acac)_6(OH_2)]$ then $[Co_2(acac)_4(OH_2)_2]$ [102], without change of configuration at the cobalt. A recent example of this type of condensation was reported by Vahrenkamp [103] in which the bidentate ligand **50**

N O$^-$

50

reacted with $ZnCl_2$ to give the complex $[Zn_7(\mathbf{50})_{12}]^{2+}$ whose structure is shown in Figure 37.

The central core is quite reminiscent of the complex reported by Lipppard *et al.* (Figure 1) and may be described as a circular arrangement of six octahedra sharing edges, with a seventh octahedral coordination site at the centre. The chirality of the external zinc atoms alternates on moving round the cycle to give an overall D_{3d} symmetry to the cluster. This alternation is essential to generate the seventh octahedral site at the inside of the cluster. The observation of D_{nd} symmetry in these cyclic systems is, however, quite frequent, and we may note D_{5d} symmetry for the molecular 'ferric' wheel of Lippard *et al.* [104], D_{4d} in a cyclic $[Co_8(CH_3CO_2)_8$-

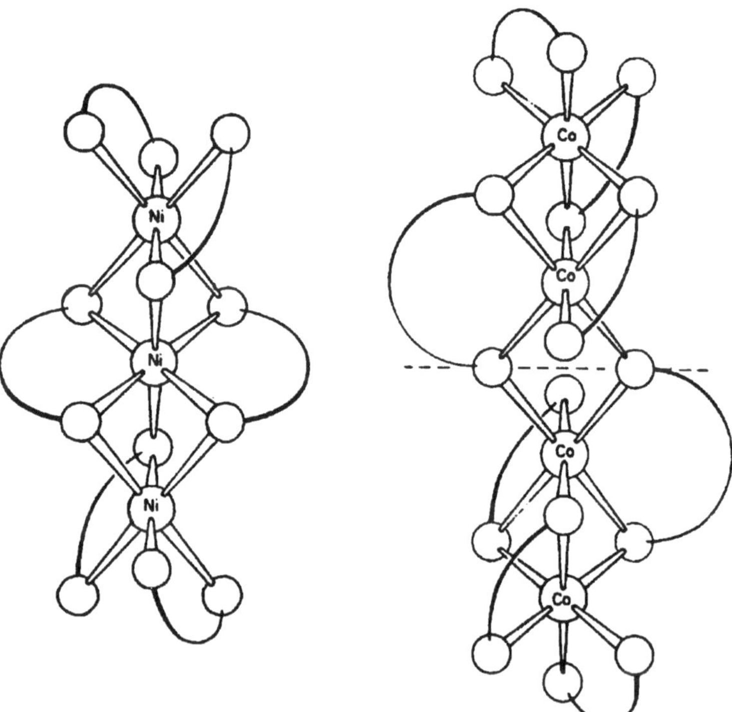

Figure 36 Schematic view of $[Ni(acac)_2]_3$[100] (left) and $[Co(acac)_2]_4$ [101] (right). Reproduced with permission from reference 101.

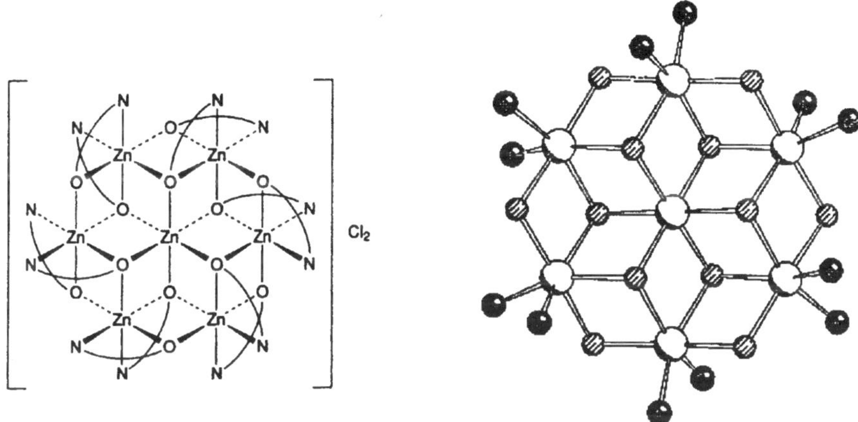

Figure 37 Structure of $[Zn_7(\mathbf{50})_{12}]^{2+}$ shown schematically (left) and illustrating the alternating chirality of the six external zinc sites and (right) the $Zn_7O_{12}N_{12}$ core (Zn colourless, O striped, N dark). Reproduced with permission from reference 103.

$(CH_3O)_{16}]$ [105], and D_{3d} again in the spectacular 18-membered ring recently reported by Lippard's group [106]. All these species contain coordination centres linked by bridging alkoxides or carboxylates.

4.8 Extended Structures

We conclude this survey with extended systems, where the assembled structures show the translational symmetry characteristic of the crystalline state. The packing of chiral units into a crystal lattice will inevitably involve some type of diastereoselectivity, either homochiral or heterochiral, although this is frequently not discussed in crystal stucture reports. If the associations are all homochiral, then an enantiomerically pure crystal will be obtained, and a solution of the racemate will yield a racemic mixture (see section 3). If, on the other hand, heterochiral association (often related by a centre of inversion or glide plane) is favoured, a racemic compound will crystallize. The double helicate $[Cu_2(\mathbf{51})_2]^{2+}$ crystallizes with a homochiral association of complexes along the helical axis (Figure 38). The homochiral columns are then arranged in pseudohexagonal arrays with the chirality

51

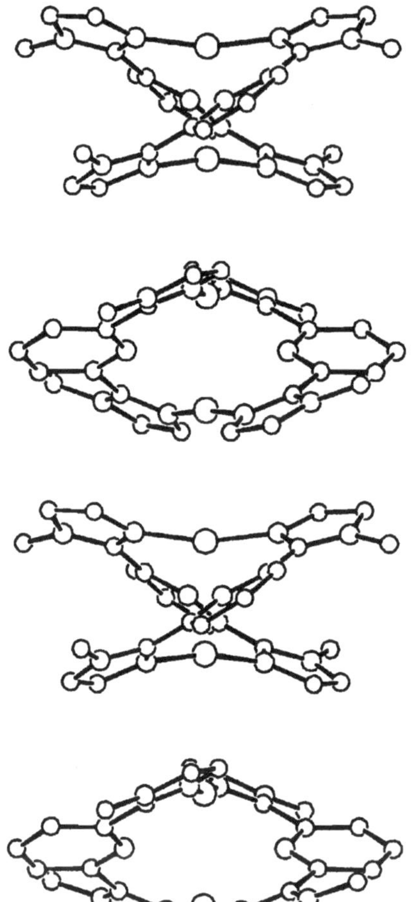

Figure 38 Double helicates $[Cu_2(\mathbf{51})_2]^{2+}$ stack on top of each other in a homochiral manner to give an infinite double-helical column in the crystal.

alternating, to give a racemic compound [107]. The crystal packing in helicates has been discussed in some detail in reference 23.

Homochiral association is also seen in the compound $[Co(\mathbf{52}H)_3]\cdot 10\ H_2O$ [108]. Protonation of the complex $[Co(\mathbf{52})_3]^{3-}$ leads to the structure shown at the left of Figure 39 in which each unit possesses three H-bond donor groups (–COOH) and 3 H-bond acceptor groups ($-COO^-$) and can consequently form six hydrogen bonds.

52

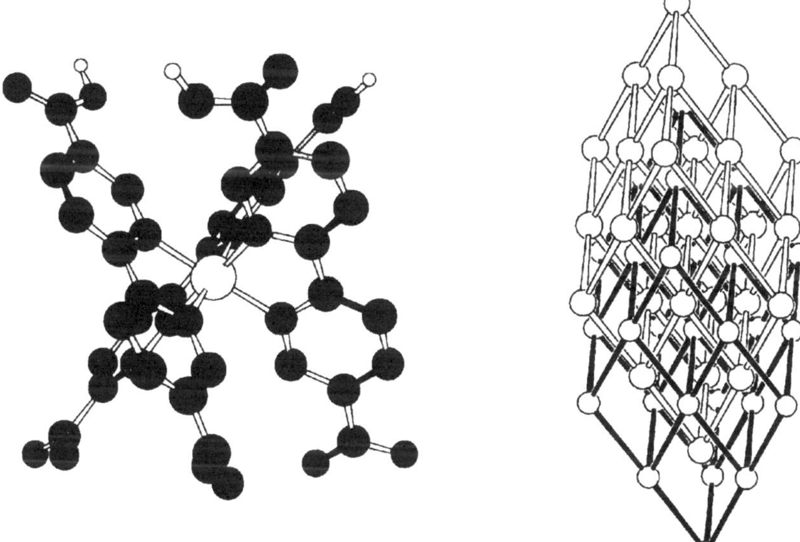

Figure 39 Left: the complex [Co(**52**H)$_3$] carries three H-bond donor groups (–COOH, top) and 3 H-bond acceptor groups (–COO$^-$, below). Right: the [Co(**52**H)$_3$] units are hydrogen bonded into a homochiral pseudorhombohedral lattice (black) which is interpenetrated by an identical lattice of opposite chirality (white). Reproduced with permission from reference 108.

The pseudo threefold symmetry results in a rhombohedral lattice involving only complexes of identical chirality. The resulting network has large voids which are filled by a second network of opposite chirality and molecules of water.

The use of enantiomerically pure ligands automatically excludes the presence of certain elements of symmetry, and, as mentioned for ligand **3** in section 2, can lead to the development of extended structures based on helical symmetry. Another example has been given by Yamauchi using complexes of copper(II) with arginine in which the infinite helices are stitched together by hydrogen bonding between arginines and *meta*-phthalate or dipicolinate anions [109]. Hydrogen bonding is thus a powerful tool for mediating the interaction between two complexes, but if suitable binding groups are present, metal ions may also be used: an interesting recent example showed how three copper(I) ions bridged two [Ru(**53**)$_3$]$^-$ units giving selectively the homochiral biruthenium species [110].

N N N-

53

There is currently considerable interest in coordination polymers, in which the extended structure is built up exclusively with coordinate bonds. In most cases studied to date, the metal centres and bridging ligands are achiral, and we shall concentrate here on the examples where chirality plays an important role. The efficiency of packing of helical structures, especially for one-dimensional systems, results in the frequent observation of extended helical structures and a number have been reported [111–113], although generally helices of both chiralities are present in the crystal. Even quite simple ligands can give rise to complicated networks such as $[Ag_2(2,3\text{-}Me_2pyz)_3][SbF_6]_2$ (2,3-Me_2pyz = 2,3-dimethylpyrazine) which crystallises as two interpenetrating $SrSi_2$ type lattices of opposite chirality [114]. Interpenetration on a grand scale is observed in the compound $ZnAu_2(CN)_4$ [115], in which no less than six quartz-like networks interpenetrate; the zinc atoms show tetrahedral coordination, while the linear $[NCAuCN]^-$ units act as the equivalent of the –O– units in quartz. In this case all networks have the same chirality.

The best studied examples, however, concern the oxalates. The oxalate anion can act as a bridging anion, and has the great advantage of acting as an efficient magnetic bridge between paramagnetic centres [116]. The $[M(C_2O_4)_3]^{n-}$ unit can bridge to three other metal ions, and this connecting unit may be used to build up a network. The work of Decurtins and colleagues has shown how the nature of the network depends strongly on chirality: the unit itself is chiral and consequently the linking to another unit can be homochiral or heterochiral.

Heterochiral pairing (Figure 40) results in a two-dimensional honeycomb lattice (Figure 41) of general composition $[MM'(C_2O_4)_3]^{n-}$ where M and M′ are octahedral metal centres, and each M is linked by bridging oxalates to three M′ ions and vice versa. For the systems studied by Decurtins [117,118] the ions are bivalent and trivalent respectively, and the negative charge of the lattice is balanced by a large

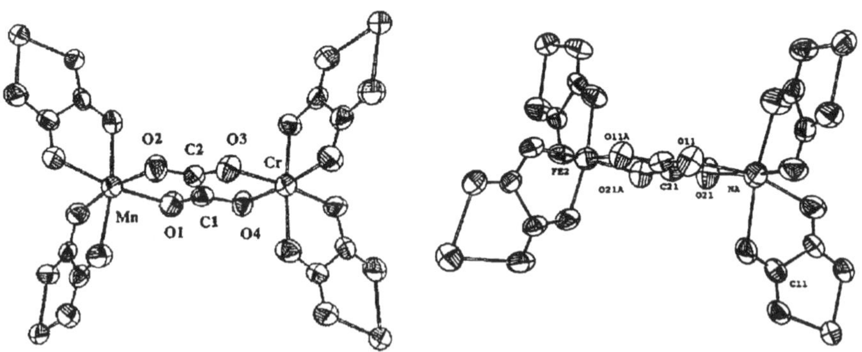

Figure 40 The two modes of linkage for the $[M(C_2O_4)_3]^{n-}$ unit: heterochiral (a) in which the left-hand ion (Mn) has Δ configuration and the right hand (Cr) has Λ configuration, and homochiral (b) in which both ions have Δ configuration. Reproduced with permission from references 118 and 122.

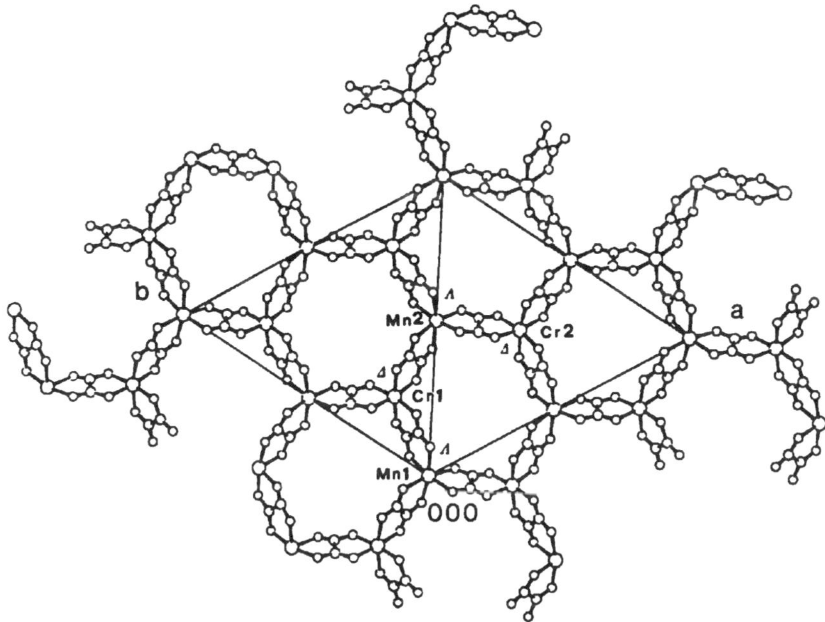

Figure 41 The hexagonal lattice formed by homochiral pairing in $[MnCr(C_2O_4)_3]^-$. Reproduced with permission from reference 117.

cation such as $PPh_4{}^+$ which is partially included in the hexagonal holes of the lattice, the remainder lying in between the anionic layers. Similar structures have been reported in which the counter ion is decamethylferrocenium [119], and an example of a $[NaCr(C_2O_4)_3]^{2-}$ lattice has been reported [120].

Homochiral pairing leads to a totally different structure in which the $[M(C_2O_4)_3]$ units form a chiral three-dimensional network which may be described as a [10,3] net in which each point has a connectivity of 3 and the shortest circuit from one point without retracing one's steps will pass through 10 links (Figure 42). The system initially discovered by Decurtins [121] was $[Fe(bipy)_3][Fe_2(C_2O_4)_3]$, and the $[Fe(bipy)_3]^{2+}$ units fill the vacancies in the network. The resulting structure has cubic symmetry and appears to be remarkably stable: the anions may be replaced by $[M^IM^{III}(C_2O_4)_3]^{2-}$ units and the cations by other $[M(bipy)_3]^{2+}$ complexes [122], or even by $[M(bipy)_3]^{3+}$ with a counter anion [123]. There is strict matching between the chirality of the $[M(bipy)_3]$ unit (Δ or Λ) and that of the $M(C_2O_4)_3$ units of the lattice. Crystals grow from a racemic solution as well-formed enantiomerically pure octahedra.

The oxalate ion offers almost limitless possibilities for bridging extended structures, and an interesting example showed recently how the complex $[Cr(bipy)(C_2O_4)_2]^-$ forms extended enantiomerically pure chains in [Mn{Cr(bipy)-

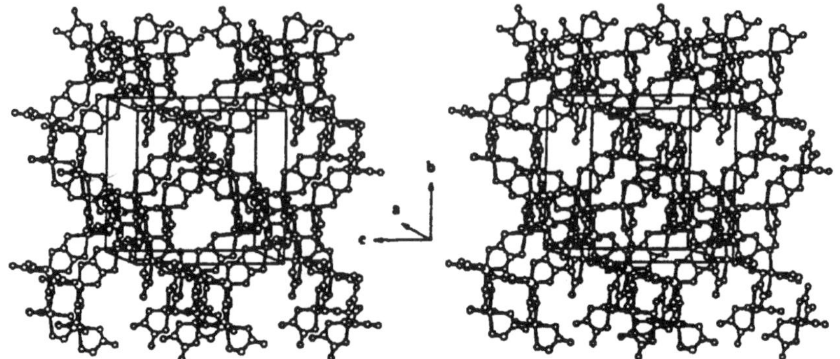

Figure 42 A stereoview of the three-dimensional lattice of the $[M^{I}M^{III}(C_2O_4)_3]^{2-}$ type. Reproduced with permission from reference 122.

$(C_2O_4)_2\}_2]_n$ where the manganese ion is eight-coordinate, two $[Cr(bipy)(C_2O_4)_2]^-$ units bridging between two adjacent manganese ions [124].

5 CONCLUSIONS

We have tried to draw attention to the many stereogenic centres present in self-assembled polynuclear complexes. A full description of the structures of these compounds requires the analysis of these centres and the identification of any diastereoselectivity present. In preparing this manuscript we were surprised by how often this aspect is ignored, and we were frequently reduced to careful examination of plots of the crystal structure and of the reported space group. Indeed it is only in the helicates, where even the most obtuse chemist could scarcely fail to recognize the chirality, that the subject is systematically discussed. There are, however, many aspects which we believe should encourage chemists to turn their attention to this question. The introduction of one or two elements of chirality among the components can dramatically influence the structure. Diastereomeric excesses are high and statistical mixtures of diastereomers are the exception rather than the rule. The use of enantiomerically pure ligands (or metal complexes when these are used as building bricks) excludes the presence of certain elements of symmetry and can thereby determine which of several possible structures is adopted. The search for novel polar or non-linear optical materials favours the use of enantiomerically pure systems since centrosymmetric space groups are automatically excluded. The use of enantiomerically pure ligands clearly offers interesting prospects for the future development of this field. It would also be useful to develop efficient methods which can predict the diastereoselectivity observed in these systems, and modern molecular mechanics programmes could surely be developed to study this.

Even if chirality is not a primary interest, the chemist should be aware of its presence, notably in the interpretation of NMR data for structure determination. In kinetically labile systems formed by self-assembly, the study of racemization kinetics offers vital information on the robustness of the system, and the possible mechanisms of decomposition or fragmentation. In short, what appears at first to be an extra complication in an already complex system does in fact offer the chemist new means of investigation and control over the properties of these compounds.

ACKNOWLEDGEMENT

The authors' interest in this subject was stimulated by the Chiral2 programme of the Swiss National Science Foundation, and we would like to acknowledge the encouragement and support given by this programme.

REFERENCES

1. A. Werner, *Z. Anorg. Chem.*, **3**, 267 (1893).
2. A. von Zelewsky, *Stereochemistry of Coordination Compounds*, J. Wiley, Chichester, 1996. This book is strongly recommended for discussion of general aspects of chirality in coordination chemistry.
3. A. Werner, *Chem. Ber.*, **47**, 3087 (1914).
4. J. S. Lindsey, *New J. Chem.*, **15**, 153 (1991).
5. A. Caneschi, A. Cornia and S. J. Lippard, *Angew. Chem. Int. Ed. Engl.*, **34**, 467 (1995).
6. International Union of Pure and Applied Chemistry, *Nomenclature of Inorganic Chemistry, Recommendations 1990*, Blackwell Scientific Publications, Oxford 1990; B.P. Block, W.H. Powell, W.C. Fernelius, *Inorganic Chemical Nomenclature*, American Chemical Society, Washington, 1990.
7. F. P. Dwyer and F. L. Garvan, *J. Am. Chem. Soc.*, **83**, 2610 (1961).
8. T. Suzuki, H. Kotsuki, K. Isobe, N. Moriya, Y. Nakagawa and M. Ochi, *Inorg. Chem.*, **34**, 530 (1995).
9. C. Provent, S. Hewage, A. F. Williams and G. Bernardinelli, unpublished observations.
10. J. R. Pollack and J. B. Neilands, *Biochim. Biophys. Res. Commun.*, **38**, 989 (1970); I. G. O'Brien and F. Gibson, *Biochim. Biophys. Acta*, **21**, 393 (1970).
11. P. N. W. Baxter, J.-M. Lehn and K. Rissanen, *Chem. Commun.*, 1997, 1323.
12. A good discussion of this subject is given in chapter 14 of J.P. Glusker, M. Lewis and M. Rossi, *Crystal Structure Analysis for Chemists and Biologists*, VCh Weinheim, 1994.
13. H. D. Flack, *Acta Cryst.* **A39**, 876 (1983); H.D. Flack and G. Bernardinelli, *Acta Cryst.*, **A41**, 500 (1985).
14. L. J. Charbonnière, A. F. Williams, C. Piguet, Gernardinelli and E. Rivara-Minten, *Chem. Eur. J.*, **4**, 481 (1998).
15. A. von Zelewsky and J. Nachbaur, personal communication. See also ref. 2, p. 193 ff.
16. W. H. Pirkle, D. L. Sikkenga and M. S. Pavlin, *J. Org. Chem.* **42**, 384 (1977).
17. J.-M. Lehn, A. Rigault, J. Siegel, J. Harrowfield, B. Chevrier and D. Moras, *Proc. Natl Acad. Sci. USA*, **84**, 2565 (1987).
18. J. Lacour, C. Ginglinger, C. Grivet and G. Bernardinelli, *Angew. Chem. Int. Ed. Engl.*,

36, 608 (1997); J. Lacour, C. Ginglinger, F. Favarger, S. Torche-Haldimann, *Chem. Commun.*, 2285 (1997).
19. Y. Yoshikawa and K. Yamasaki, *Coord. Chem. Rev.*, **28**, 205 (1979); H. Yoneda and K. Miyoshi, in ACS Symposium, Series, **565**, *Coordination Chemistry, A Century of Progress*, G.B. Kaufmann Ed, ACS Washington, 1994, chap. 26.
20. K. Nakanishi, N. Berova, R. W. Woody, *Circular Dichroism. Principles and Applications*, VCh Weinheim, 1994.
21. J.-M. Lehn, J.-P. Sauvage, J. Simon, R. Ziessel, C. Piccinni-Lombard, G. Germain, J.-P. Declercq and M. van Meersche, *Nouv. J. Chim.*, **7**, 413 (1983).
22. C. Piguet, G. Bernardinelli, B. Bocquet, O. Schaad and A.F. Williams, *Inorg. Chem.*, **33**, 4112 (1994).
23. C. Piguet, G. Bernardinelli and G. Hopfgartner, *Chem. Rev.*, **97**, 2005 (1997).
24. E. C. Constable in *Comprehensive Supramolecular Chemistry, Vol. 9*, J.-P. Sauvage, M. W. Hosseini Eds, Pergamon, Oxford, 1996, Chapter 6.
25. A. F. Williams, *Chem. Eur. J.*, **3**, 9 (1997).
26. R. S. Cahn, C. Ingold and V. Prelog, *Angew. Chem. Int. Ed. Engl.*, **5**, 383 (1966).
27. M. Albrecht and S. Kotila, *Angew. Chem. Int. Ed. Engl.*, **34**, 2134 (1995).
28. M. Albrecht and C. Riether, *Chem. Ber.*, **129**, 829 (1996).
29. D. Zurita, P. Baret and J.-L. Pierre, *New J. Chem.*, **18**, 1143 (1994).
30. B. R. Serr, K. A. Andersen, C. M. Elliott and O. P. Anderson, *Inorg. Chem.*, **27**, 4499 (1988).
31. R. W. Saalfrank, A. Dresel, V. Seitz, S. Trummer, F. Hampel, M. Teichert, D. Stalke, C. Stadler, J. Daub, V. Schënemann and A. X. Trautwein, *Chem. Eur. J.*, **3**, 2058 (1997).
32. A. L. Airey, G. F. Swiegers, A. C. Willis and S. B. Wild, *Inorg. Chem.*, **36**, 1588 (1997).
33. A. Bilyk and M. M. Harding, *J. Chem. Soc. Dalton Trans*, 77 (1994).
34. A. Bilyk and M. M. Harding, *J. Chem. Soc. Chem. Commun.*, 1697 (1995).
35. C. O. Dietrich-Buchecker, J.-P. Sauvage, J.-P. Kintzinger, P. Malthèse, C. Pascard and J. Guilhem, *New. J. Chem.*, **16**, 931 (1992).
36. C. O. Dietrich-Buchecker, J.-P. Sauvage, A. De Cian and J. Fischer, *J. Chem. Soc. Chem. Commun.*, 2231 (1997).
37. R. C. Scarrow, D. L. White and K. N. Raymond, *J. Am. Chem. Soc.*, **107**, 6540 (1985).
38. R. Krämer, J.-M. Lehn, A. De Cian and J. Fischer, *Angew. Chem. Int. Ed. Engl.*, **32**, 703 (1993).
39. L. J. Charbonnière, G. Bernardinelli, C. Piguet, A. M. Sargeson and A. F. Williams, *J. Chem. Soc. Chem. Commun.*, 1419 (1994).
40. B. Hasenknopf and J.-M. Lehn, *Helv. Chim. Acta*, **79**, 1643 (1996).
41. A review of helicenes together with a compilation of optical rotations may be found in K. P. Meurer and F. Vögtle, *Topics Curr. Chem.*, **127**, 1 (1985).
42. L. J. Charbonnière, M.-F. Gilet, K. Bernauer and A. F. Williams, *J. Chem. Soc. Chem. Commun.*, **39** (1996).
43. A. F. Williams, C. Piguet and R. Carina in *Transition Metals in Supramolecular Chemistry, NATO ASI Series C* Vol. 448, L. Fabbrizzi and A. Poggi, Eds, Kluwer, Dordrecht, 1994, p. 409.
44. L. J. Charbonnière, A. F. Williams, U. Frey, A. E. Merbach, P. Kamalaprija and O. Schaad, *J. Am. Chem. Soc.*, **119**, 2488 (1997).
45. B. Kersting, M. Meyer, R. E. Powers and K. N. Raymond, *J. Am. Chem. Soc.*, **118**, 7221 (1996); M. Meyer, B. Kersting, R. E. Powers and K. N. Raymond, *Inorg. Chem.*, **36**, 5179 (1997).
46. B. Kersting, J. R. Telford, M. Meyer and K. N. Raymond, *J. Am. Chem. Soc.*, **118**, 5712 (1996).
47. M. Albrecht and S. Kotila, *Angew. Chem. Int. Ed. Engl.*, **35**, 1208 (1996).
48. M. Albrecht and M. Schneider, *J. Chem. Soc. Chem. Commun.*, **137** (1998).

49. C. Piguet, G. Bernardinelli and A. F. Williams, *Inorg. Chem.*, **28**, 2920 (1989).
50. S. Rüttimann, C. Piguet, B. Bocquet, G. Bernardinelli and A. F. Williams, *J. Am. Chem. Soc.*, **114**, 4230 (1992).
51. R. Carina, C. Piguet and A. F. Williams, *Helv. Chim. Acta.*, **81**, 548 (1998).
52. A. Marquis-Rigault, A. Dupont-Gervais, A. Van Dorsselaer and J.-M. Lehn, *Chem. Eur. J.*, **2**, 1395 (1996).
53. P. Baret, V. Beaujolais, D. Gaude, C. Coulombeau and J.-L. Pierre, *Chem. Eur. J.*, **3**, 969 (1997).
54. E. J. Corey, C. L. Cywin and M. C. Noe, *Tet. Lett.*, **35**, 69 (1994).
55. J. Libmann, Y. Tor and A. Shanzer, *J. Am. Chem. Soc.*, **109**, 5880 (1987).
56. C. R. Woods, M. Benaglia, F. Cozzi and J. S. Siegel, *Angew. Chem. Int. Ed. Engl.* **35**, 1830 (1996).
57. W. Zarges, J. Hall, J.-M. Lehn and C. Bolm, *Helv. Chim. Acta*, **74**, 1843 (1991).
58. E. J. Enemark and T. D. P. Stack, *Angew. Chem. Int. Ed. Engl.*, **34**, 996 (1995).
59. M. Albrecht, *Synlett*, 565 (1966).
60. E. C. Constable, T. Kulke, M. Neuburger and M. Zehnder, *Chem. Commun.*, 489 (1997).
61. G. Baum, E. C. Constable, D. Fenske and T. Kulke, *Chem. Commun.*, 2043 (1997)
62. P. Hayoz, H. Stoeckli-Evans and A. von Zelewsky, *J. Am. Chem. Soc.*, **115**, 5111 (1993); H. Mürner, P. Belser and A. von Zelewsky, *J. Am. Chem. Soc.*, **118**, 7989 (1996).
63. C. Provent, S. Hewage, G. Brand, G. Bernardinelli, L. J. Charbonnière and A. F. Williams, *Angew. Chem. Int. Ed. Engl.*, **36**, 1287 (1997).
64. B. Hasenknopf, J.-M. Lehn, B. O. Kneisel, G. Baum and D. Fenske, *Angew. Chem. Int. Ed. Engl.*, **35**, 1838 (1996).
65. B. Hasenknopf, J.-M. Lehn, N. Boumediene, A.- Dupont-Gervais, A. Van Dorsselaer, B. O. Kneisel and D. Fenske, *J. Am. Chem. Soc.*, **119**, 10956 (1997).
66. P. N. W. Baxter, J.-M. Lehn and K. Rissanen, *Chem. Commun.*, 1323 (1997).
67. O. Mamula, A. von Zelewsky and G. Bernardinelli, *Angew. Chem. Int. Ed. Engl.*, **37**, 290 (1998).
68. D. P. Funeriu, J.-M. Lehn, G. Baum and D. Fenske, *Chem. Eur. J.*, **3**, 99 (1997).
69. P. L. Jones, K. J. Byrom, J. C. Jeffery, J. A. McCleverty and M. D. Ward, *Chem. Commun.*, 1361 (1997).
70. E. Buhleier, W. Wehner and F. Vögtle, *Synthesis*, 155 (1978).
71. R. G. Denkewalter, J. F. Kolc and W. J. Lukasavage, *U.S. Pat.* 4,410,688 (1979)
72. D. Seebach, J.-M. Lapierre, K. Skobridis and G. Greiveldinger, *Angew. Chem. Int. Ed. Engl.*, **33**, 440 (1994).
73. H. W. I. Peerlings and E. W. Meijer, *Chem. Eur. J.*, **3**, 1563 (1997).
74. J. Issberner, F. Vögtle, L. De Cola and V. Balzani, *Chem. Eur. J.*, **3**, 706 (1997).
75. G. Denti, S. Campagna, S. Serroni, M. Ciano and V. Balzani, *J. Am. Chem. Soc.*, **114**, 2944 (1992); S. Campagna, G. Denti, S. Serroni, A. Juris, M. Venturi, V. Ricevuto and V. Balzani, *Chem. Eur. J.*, **1**, 211 (1995); S. Campagna, A. Giannetto, S. Serroni, G. Denti, S. Trusso, F. Mallamace and N. Micali, *J. Am. Chem. Soc.*, **117**, 1754 (1995); V. Balzani, S. Campagna, G. Denti, A. Juris, S. Serroni and M. Venturi, *Acc. Chem. Res.*, **31**, 26 (1998).
76. E. C. Constable, *Chem. Commun.*, 1073 (1997).
77. S. Bodige, A. S. Torres, D. J. Maloney, D. Tate, G. R. Kinsel, A. K. Walker and F. M. MacDonnell, *J. Am. Chem. Soc.*, **119**, 10364 (1997).
78. N. C. Fletcher, F. R. Keene, H. Viebrock and A. von Zelewsky, *Inorg. Chem.*, **36**, 1113 (1997).
79. P. J. Stang and B. Olenyuk, *Acc. Chem. Res.*, **30**, 502 (1997).
80. M. Fujita, J. Yazaki and K. Ogura, *J. Am. Chem. Soc.*, **112**, 5645 (1990).
81. J. L. Garcia-Ruano, A. M. Gonzalez, I. Lopez-Solera, J. R. Masaguer, C. Navarro-

Ranninger, P. R. Raithby and J. H. Rodriguez, *Angew. Chem. Int. Ed. Engl.*, **34**, 1351 (1995).
82. B. Olenyuk and J. A. Whiteford, P. J. Stang, *J. Am. Chem. Soc.*, **118**, 8221 (1996).
83. P. J. Stang and B. Olenyuk, *Angew. Chem. Int. Ed. Engl.*, **35**, 732 (1996).
84. S. Rüttiman, G. Bernardinelli and A. F. Williams, *Angew. Chem. Int. Ed. Engl.*, **32**, 392 (1993).
85. R. W. Saalfrank, A. Stark, K. Peters and H. G. von Schnering, *Angew. Chem. Int. Ed. Engl.*, **27**, 851 (1988).
86. R. W. Saalfrank, B. Hörner, D. Stalke and J. Stalbeck, *Angew. Chem. Int. Ed. Engl.*, **32**, 1179 (1993).
87. R. W. Saalfrank, R. Burak, A. Breit, D. Stalke, R. Herbst-Ilmer, J. Daub, M. Porsch, E. Bill, M. Müther and A. X. Trautwein, *Angew. Chem. Int. Ed. Engl.*, **33**, 1621 (1994).
88. T. Beissel, R. E. Powers and K. N. Raymond, *Angew. Chem. Int. Ed. Engl.*, **35**, 1084 (1996).
89. A. J. Amoroso, J. C. Jeffery, P. L. Jones, J. A. McCleverty, P. Thornton and M. D. Ward, *Angew. Chem. Int. Ed. Engl.*, **34**, 1443 (1995).
90. M. Fujita, D. Oguro, M. Miyazawa, H. Oka, K. Yamaguchi and K. Ogura, *Nature*, **378**, 469 (1995).
91. P. J. Stang, B. Olenyuk, D. C. Muddiman and R. D. Smith, *Organometallics*, **16**, 3094 (1997).
92. J.-P. Sauvage, *Acc. Chem. Res.*, **23**, 320 (1990).
93. J.-Cl. Chambron, D. K. Mitchell and J.-P. Sauvage, *J. Am. Chem. Soc.*, **114**, 4625 (1992).
94. C. Piguet, G. Bernardinelli, A. F. Williams and B. Bocquet, *Angew. Chem. Int. Ed. Engl.*, **34**, 582 (1995).
95. D. J. Cardenas and J.-P. Sauvage, *Inorg. Chem.*, **36**, 2777 (1997).
96. J.-F. Nierengarten, C. O. Dietrich-Buchecker and J.-P. Sauvage, *J. Am. Chem. Soc.*, **116**, 375 (1994).
97. C. O. Dietrich-Buchecker and J.-P. Sauvage, *New J. Chem.*, **16**, 277 (1992).
98. G. Rapenne, C. O. Dietrich-Buchecker and J.-P. Sauvage, *J. Am. Chem. Soc.*, **118**, 10932 (1996).
99. R. F. Carina, C. O. Dietrich-Buchecker and J.-P. Sauvage, *J. Am. Chem. Soc.*, **118**, 9110 (1996).
100. G. J. Bullen, R. Mason and P. Pauling, *Inorg. Chem*, **4**, 456 (1965).
101. F. A. Cotton and R. C. Elder, *Inorg, Chem.*, **4**, 1145, (1965).
102. F. A. Cotton and R. Eiss, *J. Am. Chem. Soc.*, **90**, 38 (1968).
103. M. Tesmer, B. Müller and H. Vahrenkamp, *Chem. Commun.*, 721 (1997).
104. K. L. Taft, C. D. Delfs, G. C. Papaefthymiou, S. Foner, D. Gatteschi and S. J. Lippard, *J. Am. Chem. Soc.*, **116**, 823 (1994).
105. J. K. Beattie, T. W. Hambley, J. A. Klepetko, A. F. Masters and P. Turner, *Chem. Commun.*, 45 (1998).
106. S. P. Watton, P. Fuhrmann, L. E. Pence, A. Caneschi, A. Cornia, G. L. Abbati and S. J. Lippard, *Angew. Chem. Int. Ed. Engl.*, **36**, 2774 (1997).
107. R. F. Carina, G. Bernardinelli and A. F. Williams, *Angew. Chem. Int. Ed. Engl.*, **32**, 1463 (1993).
108. P. G. Desmartin, G. Bernardinelli and A. F. Williams, *New J. Chem.*, **19**, 1109 (1995).
109. N. Ohata, H. Masuda and O. Yamauchi, *Angew. Chem. Int. Ed. Engl.*, **35**, 531 (1996).
110. M. H. W. Lam, S. T. C. Cheung, K.-M. Fung and W.-T. Wong, *Inorg. Chem.*, **36**, 4618 (1997).
111. I. G. Dance, L. J. Fitzpatrick, A. D. Rae and M. L. Scudder, *Inorg. Chem.*, **22**, 3785 (1983).

112. M. Munakata, L. P. Wu, T. Kuroda-Sowa, M. Maekawa, K. Moriwaki and S. Kitagawa, *Inorg. Chem.*, **36**, 5416 (1997).
113. L. Carlucci, G. Ciani, D. W. v. Gudenberg and D. M. Proserpio, *Inorg. Chem.*, **36**, 3812 (1997).
114. L. Carlucci, G. Ciani, D. M. Proserpio and A. Sironi, *Chem. Commun.*, 1396 (1996).
115. B. F. Hoskins, R. Robson and N. V. Y. Scarlett, *Angew. Chem. Int. Ed. Engl.*, **34**, 1203 (1995).
116. H. Tamaki, Z. J. Zhong, N. Matsumoto, S. Kida, M. Koikawa, N. Achiwa, Y. Hashimoto and H. Okawa, *J. Am. Chem. Soc.*, **114**, 6974 (1992).
117. S. Decurtins, H. W. Schmalle, H. R. Oswald, A. Linden, J. Ensling, P. Gütlich and A. Hauser, *Inorg. Chim. Acta*, **216**, 65 (1994).
118. R. Pellaux, H. W. Schmalle, R. Huber, P. Fischer, T. Hauss, B. Ouladdiaf and S. Decurtins, *Inorg. Chem.*, **36**, 2301 (1997).
119. M. Clemente-Léon, E. Coronado, J.-R. Galán-Mascarós and C. J. Gómez-Garcia, *Chem. Commun.*, 1727 (1997).
120. R. P. Farrell, T. W. Hambley and P. A. Lay, *Inorg. Chem.*, **34**, 757 (1995).
121. S. Decurtins, H. W. Schmalle, P. Schneuwly and H. R. Oswald, *Inorg. Chem.*, **32**, 1888 (1993).
122. S. Decurtins, H. W. Schmalle, P. Schneuwly, J. Ensling and P. Gütlich, *J. Am. Chem. Soc.*, **116**, 9521 (1994).
123. S. Decurtins, H. W. Schmalle, R. Pellaux, P. Schneuwly and A, Hauser, *Inorg. Chem.*, **35**, 1451 (1996).
124. F. D. Rochon, R. Melanson and M. Andruh, *Inorg. Chem.*, **36**, 6086 (1996).

Chapter 5

Design and Serendipity in the Synthesis of Polymetallic Complexes of the 3d-Metals

RICHARD E. P. WINPENNY

University of Edinburgh, Scotland, UK

When all the involved calculations prove false, and the philosophers themselves have nothing more to tell us, it is excusable to turn to the random twitter of birds, or toward the distant mechanism of the stars.

Marguerite Yourcenar, *Memoirs of Hadrian*

1 INTRODUCTION

The success of organic chemists in establishing methods for making large, complicated molecules in a systematic and controlled manner is one of the great collective achievements of twentieth-century science. The attitude of mind involved in such as process has become inculcated in all synthetic chemists, where we look for molecules to be built step by step, with the overall design becoming apparent as fragment after fragment is attached to the growing molecule. By comparison transition metal chemists have made little progress in discovering general approaches to making compounds containing large numbers of metal centres. This is because until recently means for routine characterization of such compounds did not exist, and because the obvious biological relevance and commercial applications of large organic molecules were not matched by properties of polynuclear metal complexes.

Despite the paradigm represented by organic synthesis, no scientist working on the synthesis of polymetallic cages adopts a similar approach. For large organic

Transition Metals in Supramolecular Chemistry, edited by J. P. Sauvage.

molecules convergent synthetic methods are used, where three or four different sections of a molecule are made separately before the entirety is assembled in two or three final steps. To make large metal cages we rely on precursor fragments and little apparent control of the product. The lack of control has led to the neologism 'self-assembly' being coined, however, it is unclear whether we have understood the change in philosophy this represents, or whether we have merely named it.

At the time of writing several groups are introducing an element of design into the assembly process by choosing rigid ligands which have strong preferences for specific bonding modes. This 'designed assembly' approach has produced many beautiful molecules, ranging as large as the nonanuclear silver grid described by Lehn and co-workers [1]. Typically, one element of such ligands is a bidentate head-group disposed so that coordination will produce a five-membered chelate ring. The structures of such compounds are moderately predictable, based on the preferred coordination geometry of the metal, the number of bidentate donor sites presented by the ligand, and on the limited flexibility of the ligand. The resulting compounds tend to be homoleptic.

Other researchers use much less well-behaved ligands, typically 1,3-bridging ligands where chelation would produce the thermodynamically disfavoured four-membered chelate ring. Once formation of a five- or six-membered chelate ring is excluded, the coordinative flexibility of any polydentate ligand increases enormously. This flexibility in turn allows stabilization of many unpredictable structures, almost invariably incorporating further ligands such as oxide, hydroxide or alkoxides. Compared with designed assembly using pre-organized ligands this approach, which I will term 'serendipitous assembly', produces heteroleptic complexes, appears inelegant, and this approach is at least aesthetically less appealing.

The advantages of serendipitous assembly are considerable. At the moment designed assembly requires similar coordination at each metal site, and this restricts accessible structures to Platonic solids, rings and grids; serendipitous assembly often creates cages with lower symmetry and also metal-centered cages. This vastly increases the range of compounds available for study, at least in part because no one would chose as a target a cage with little or no symmetry. These unusual structures can lead to novel properties.

For example, the Mn_{12} cage first described by Lis [2] and rediscovered by Christou [3], involves a central Mn_4O_4 cubane surrounded by an Mn_8 ring. This cage shows quantum tunnelling of its magnetization at low temperature [4], a phenomenon which may have profound implications for studies at the quantum-classical interface. No one would have dreamt of deliberately making a molecule with this structure, therefore it could never have been made by any designed synthesis.

As the field develops the boundary between 'designed' and 'serendipitous' assembly is becoming blurred. Already polypyridyl-like molecules containing flexible linking groups are appearing [5, 6], which impose specific coordination geometries on metal centres without controlling the overall structure, thus reducing the amount of control in a designed assembly. On the other side, attempts to link cages into larger fragments, and to control structure by inclusion of templates introduce directing factors into serendipitous assembly. It is also the case that as coordinatively flexible ligands are further studied we will begin to understand their reactivity. It is like playing poker; if you play long enough you begin to make decisions based on probability, and if fortunate you may recognize how the cards are marked.

A problem in recognizing trends in these cages is the asymmetry and complexity of the structures. Superficial differences may hide fundamental similarities between cages, and a systematic classification of cages into different families is necessary as a first step to understanding probabilities. We could classify polynuclear structures in at least three ways:

(a) by relation to extended lattices, e.g. metal oxides or hydroxides;
(b) by relation to regular polyhedra, e.g. icosahedra, trigonal prism;
(c) by reference to constituent fragments from which the cage is constructed, e.g. linked tetrahedra or triangles.

In what follows I will concentrate on approach (c), referring to the other two methods where they are clearly superior for the specific structure involved. This approach is only valid if a large number of structures can be built from a small selection of building blocks. The two to be used here are: a metal tetrahedron (which may occur with O atoms on the faces or edges or at the centre of the tetrahedron) and an M_4O_2 butterfly (i.e. two M_3O triangles sharing an edge) (Figure 1). Both blocks can be derived from the triangle, however, they are so common that they deserve to be considered as structural components in their own right. The butterfly description used here is imprecise as the dihedral angle between the two triangles within the butterfly may vary.

In a tetrahedron all vertices are equivalent. For the M_4O_2 block there are two distinct metal sites; the metal sites in the edge shared by the two component triangles, which will be called 'body' sites, and the two 'wing-tip' sites. Quite distinct structures result depending on the type of vertex shared. Below I have attempted to

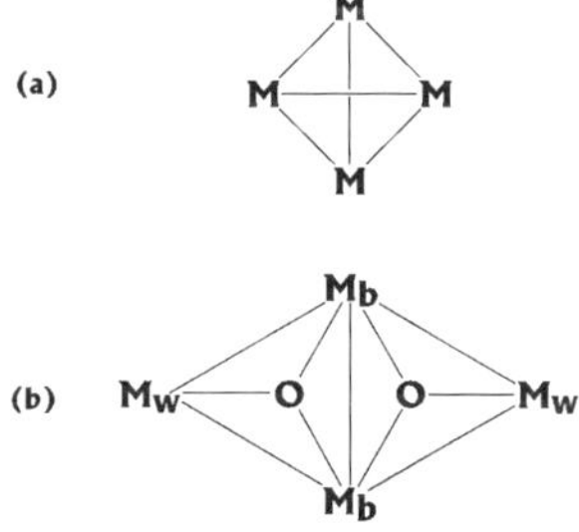

Figure 1 The two building blocks used to construct polynuclear cages: (a) tetrahedra; (b) M_4O_2 butterfly, M_w = wing-tip site, M_b = body site.

classify structures depending on whether tetrahedra or M_4O_2 blocks are involved, and have further subdivided the structures depending on how the building blocks are linked.

Partly for reasons of space, the discussion is not encyclopedic but is restricted to cages featuring seven or more metal centres, and the elements chromium to nickel. Smaller cages are legion, and are normally easily described. Earlier 3d-elements tend to form quite different polyoxometallate structures, and therefore require separate consideration. Copper is excluded here because the distorted geometries adopted by Cu(II) in most complexes lead to a unique polymetallic chemistry. The final restriction is on the choice of ligands. I have only included O and N donors, again because the proliferation of structures involving sulphur donors would render the review unwieldy, and others are far better qualified to survey such science.

2 CLASSIFICATION OF STRUCTURES

2.1 Structures Built with M_4O_2 Blocks

2.1.1 Structures involving discrete M_4O_2 blocks

There are several examples where M_4O_2 units are either linked by organic or inorganic fragments, or which stack. All but one of these compounds involves manganese. In $[Mn_8(O)_4(O_2CPh)_{12}(H_2O)_2(dem)_2]$ **1** (dem = 2,2-diethylmalonate) the two M_4O_2 blocks stack, sharing two of the bridging oxide ligands, with the blocks also bridged by the malonate ligands [7]. In $[Mn_8(O)_6O_2(CMe)_{12}(pic)_4]$ **2** pic = picolinate [8] and in $[Mn_8(O)_4(O_2CPh)_6(dbm)_4(bpe)_2]$ **3** trans-1,2-bis(4-pyridyl)ethene ligands (= bpe) [9] link two such planes. In $[Mn_9(O)_4(O_2CPh)_8(py)_2(sal)_4(salH)_2]$ **4** ($salH_2$ = salicylic acid) [10] the two butterflies are linked by a 'complex' ligand $[Mn(sal)_4]^{6-}$ which involves an eight-coordinate Mn(II) centre chelated by the carboxylate groups from four salicylates, with the phenol oxygen and carboxylate oxygens binding additionally to the butterflies (Figure 2). The Mn $\cdots$ Mn contacts within the individual butterflies in **1**–**4** are bridged by carboxylates which are generally 1,3-bridging. In **2** and **3** the synthesis of the octanuclear cages is from reaction of Mn_4O_2 butterflies with either picolinate or bis(4-pyridyl)ethene.

The fifth example is more complicated, and involves a decanuclear iron cage $[Fe_{10}Na_2(O)_6(OH)_4(O_2CPh)_{10}(chp)_6(H_2O)_2(Me_2CO)_2]$ **5** [11] (chp = 6-chloro-2-pyridonate). Here two Fe centres are sandwiched between two bent M_4O_2 blocks (Figure 3). The blocks are attached to these Fe centres through μ_3-oxide and hydroxide ligands, and the exterior of the cage is coated with 1,3-bridging benzoate and μ_2-O donors from pyridonate ligands. Recognizing that the iron oxide core is a fragment of the rock salt structure is a much more immediately recognizable description of this individual structure. The classification based on M_4O_2 blocks

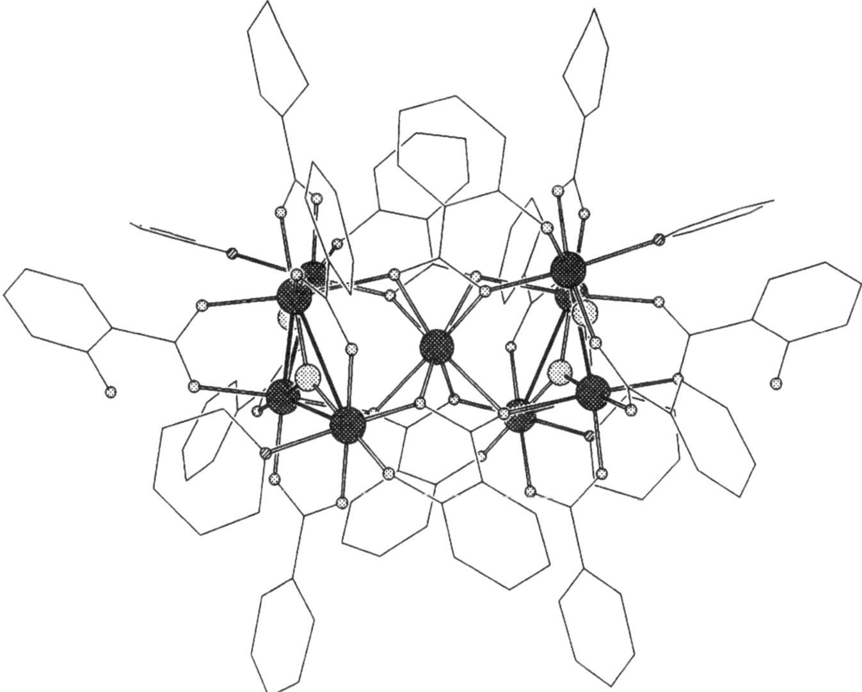

Figure 2 The structure of $[Mn_9(O)_4(O_2CPh)_8(py)_2(sal)_4(salH)_2]$ **4** showing the two Mn_4O_2 blocks linked by a $Mn(sal)_4$ complex ligand. In this and the following figures the metal sites are shown as large cross-hatched circles, oxygen atoms which form part of M_4O_2 butterflies or tetrahedra are shown as large circles with regular dot patterns. All other non-C atoms are shown as small circles with C atoms shown as lines. H Atoms are excluded for clarity. In this and the following figures the M_4O_2 butterflies are shown by full lines unless otherwise stated.

will allow us to show that this structure belongs to a family with other iron-oxo cages which do not have a rock salt core.

2.1.2 Structures involving wing-tip sharing M_4O_2 blocks

There are a couple of straightforward examples of these structures, and several more complicated ones. Two Mn_7 cages, $(NEt_4)[Mn_7(O_2CMe)_{10}(dbm)_4]$ **6** (Hdbm = dibenzoylmethane) [12] and $[Mn_7(O)_4(O_2Cme)_8(trien)_2(dien)_2]$ **7** (trien = triethylenetetramine, dien = diethylenetriamine) [13], represent the simple cases, and the metal cores of these heptanuclear cages are easily visualized. In both complexes there are eight 1,3-bridging carboxylates, with two additional μ_3-O_2CMe and two chelating dbm ligands in **6** and chelating tetra- and tri-amines in **7**.

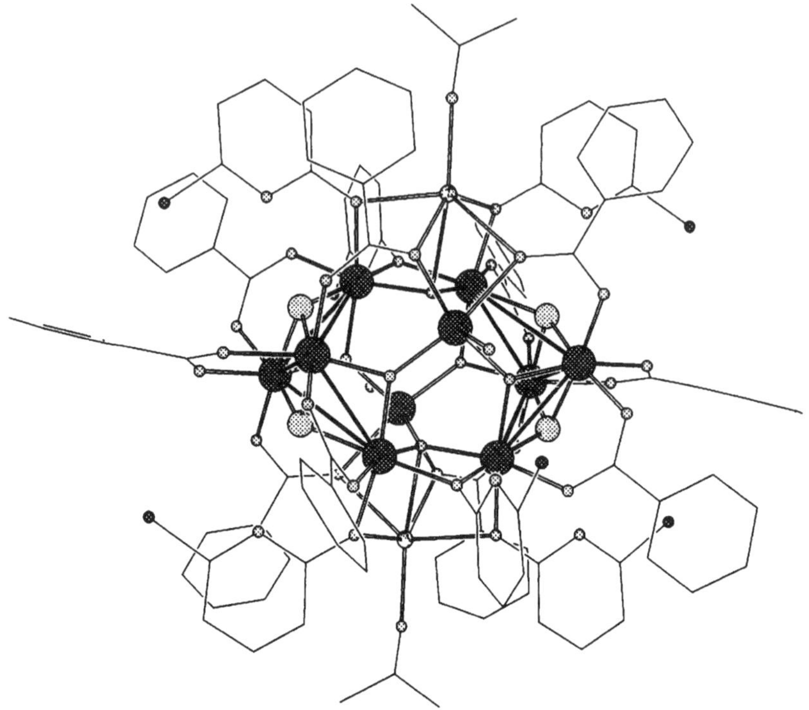

Figure 3 The structure of $[Fe_{10}Na_2(O)_6(OH)_4(O_2CPh)_{10}(chp)_6(H_2O)_2(Me_2CO)_2]$ **5** showing two central Fe centres surrounded by two Fe_4O_2 units.

The complex cases are the more common. One such group is normally described as tricapped trigonal prisms, and examples exist for Cr, Fe, Co and Ni. These tricapped trigonal prisms can be imagined as containing three M_4O_2 units sharing their wing-tip vertices (Figure 4). We have made a nonanuclear cobalt cage $[Co_9(CO_3)_2(phth)_6]^{2+}$ **8** which has this precise structure [14]. More commonly the triangular faces produced at the top and bottom of the cage are capped, either externally—as in the undecanuclear cage $[Fe_{11}(O)_6(OH)_6(O_2CPh)_{15}]$ **9**, reported by Lippard [15]—internally, as in $[Co_{10}(OH)_6(mhp)_6(O_2CPh)_7(Hmhp)_3Cl(MeCN)]$ **10** [16] (mhp = 6-methyl-2-pyridonate), or both externally and internally—as in $[Cr_{12}(O)_{12}(O_2CCMe_3)_{15}]$ **11** and $[Co_{12}(OH)_6(O_2CMe)_6(mhp)_{12}]$ **12** reported by Batsanov *et al.* [17] and the Garner group [18], respectively, and in several undecanuclear nickel cages **13–15** we have reported [16, 19]. In each case the additional capping atoms produce M_4 tetrahedra (Figure 4). In **9** and **11** the oxygens of the M_4O_2 butterflies come from hydroxide or oxide while in **10**, **12–15** they come from pyridonate ligands. It is striking that the metal centres on the trigonal axis of the prisms appear to be optional, but that those centres which comprise the M_4O_2 butterflies are essential.

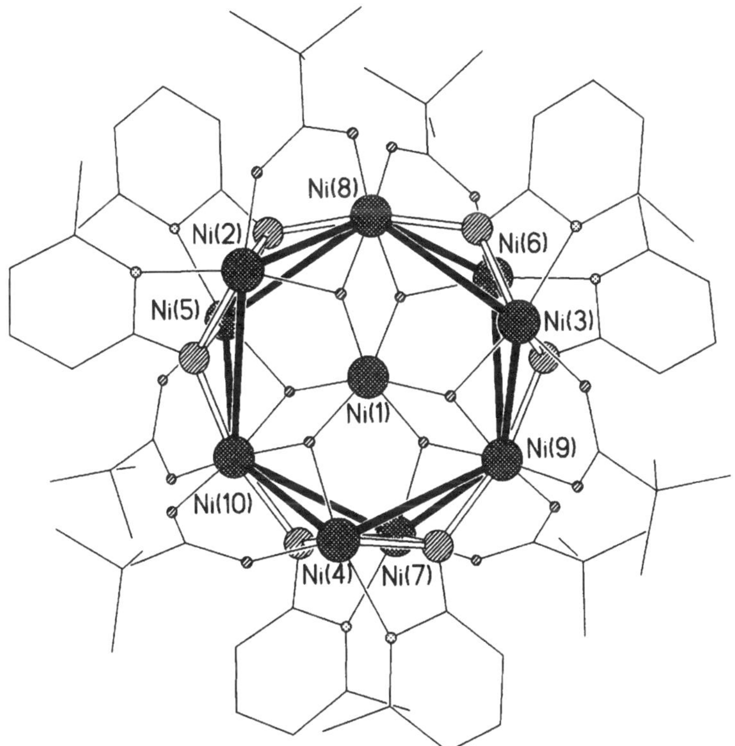

Figure 4 The construction of a tricapped trigonal prism from three vertex-sharing M_4O_2 butterflies, shown for $[Ni_{10}(OH)_6(O_2CCPh)_6(chp)_6(Cl)(H_2O)_2(MeOH)_2]$ **15**. The butterflies can be seen by considering Ni(10), Ni(2), Ni(5) and Ni(8), with Ni(10) and Ni(8) as shared vertices. The 'interior' tetrahedra would involve Ni(1), Ni(2), Ni(3) and Ni(4), or Ni(1), Ni(5), Ni(6) and Ni(7).

Initially this appears a long-winded means to show congruence between structures which are self-evidently related, however, we can extend this family further by this description. Manganese is currently absent from this category if we restrict ourselves to tricapped trigonal prisms. If we consider the structures in terms of M_4O_2 blocks there are several Mn structures within the family.

The most famous is $[Mn_{12}(O)_{12}(O_2CR)_{16}(H_2O)_4]$, initially reported by Lis (where R = Me) **16** [2], and then synthesized by Christou's group using benzoate **17** [3] and propionate **18** [20] in place of acetate. The structure involves four M_4O_2 units sharing wing-tips (Figure 5). Unlike in the trigonal prisms the M_4O_2 fragments are tilted with respect to the plane passing through them, and the innermost metal sites then form the vertices of a central tetrahedron. The carboxylates are all 1,3-bridging, with 12 bridging wing-tip and body vertices, and four spanning body-body contacts.

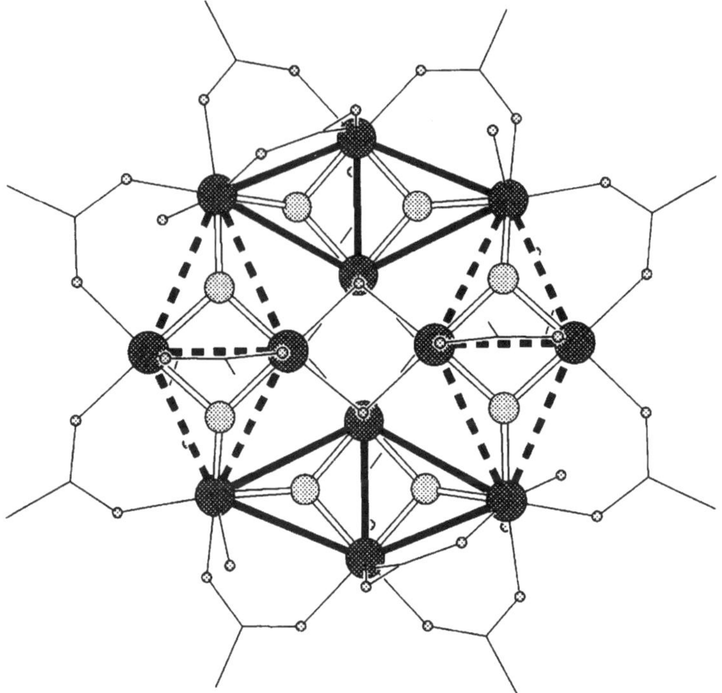

Figure 5 The structure of $[Mn_{12}(O)_{12}(O_2CR)_{16}(H_2O)_4]$ drawn to illustrate the construction of the cage from four vertex-sharing M_4O_2 butterflies. Alternate butterflies are shown with full or dashed lines.

There are now five examples of this metal polyhedron in the literature [2, 3, 20, 21], all containing manganese, with one example where the cental cubane contains Fe surrounded by an Mn ring. There is therefore a contrast between Mn and the other elements between Cr and Ni, but the polyhedra are related when the constituent fragments are considered.

A further Mn_{12} cage, $[Mn_{12}(L1)_6(O_2CMe)_2$ **21** synthesized by Tuchagues and co-workers [22], involving a more complicated ligand, 2,6-bis(o-hydroxyphenyl)-iminomethyl-p-cresol (=L1), also belongs to this class of structure. It is built of four M_4O_2 sharing wing-tip vertices, but here the M_4O_2 units are again perpendicular to the plane passing through them, resulting in a belt-like structure (Figure 6). The pentadentate Schiff base ligands bind to two Mn(II) centres, and the phenolic oxygens then bridge to neighbouring dimers creating this dodecanuclear structure.

Three further cages belong here. The $Fe_{16}M$ cages of general formula $[Fe_{16}M(O)_{10}(OH)_{10}(O_2CPh)_{20}]$ (where M = Mn **22**, Fe **23**, Co **24**) made by the Lippard group [23] contain a belt of four M_4O_2 blocks, but here five further metal

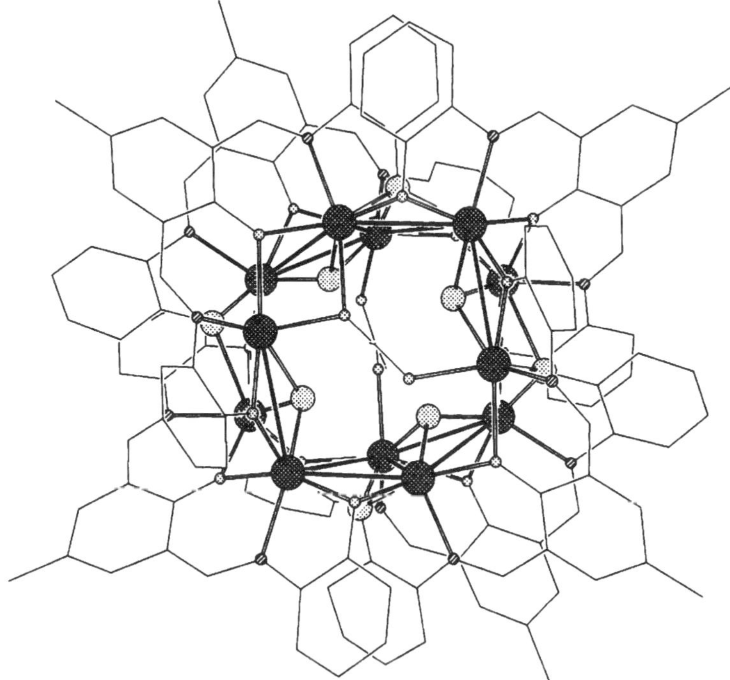

Figure 6 The structure of $[Mn_{12}(L1)_6(OH)_4(O_2CMe)_2]$ **21** showing the construction of the cage from four vertex-sharing M_4O_2 butterflies. Each of the four butterflies from one edge of the 'square' formed by the molecule.

centres are contained within the belt, and form a vertex-sharing bitetrahedron (Figure 7). These structures have been related to two face-sharing trigonal prisms by Lippard [23], which allows the relationship with **9** to be seen. If we break the structure down further we can also show it is related to the decanuclear cage **5** discussed in section 2.1.1 which, compared with the $Fe_{16}M$ cages lacks the backbones of the two central M_4O_2 fragments. The relationship between **5** and **22–24** is shown schematically in Figure 8. Again, this illustrates the advantage of the building-block approach—apparently dissimilar structures can be related by considering the fragments from which they are constructed.

The large number of cages between 10 and 12 metal centers which belong to this classification requires some explanation. The structures are not obviously fragments of a common mineral archetype, therefore a rationalization may be that they grow by a similar pathway, despite the very diverse synthetic methods reported. For example, the iron cages **9** and **22–24** are synthesized from trinuclear precursors, the chromium cage **11** from an uncharacterized polymer, and the cobalt, nickel and manganese cages from simple metal salts. MeCN is the most frequently used solvent for this

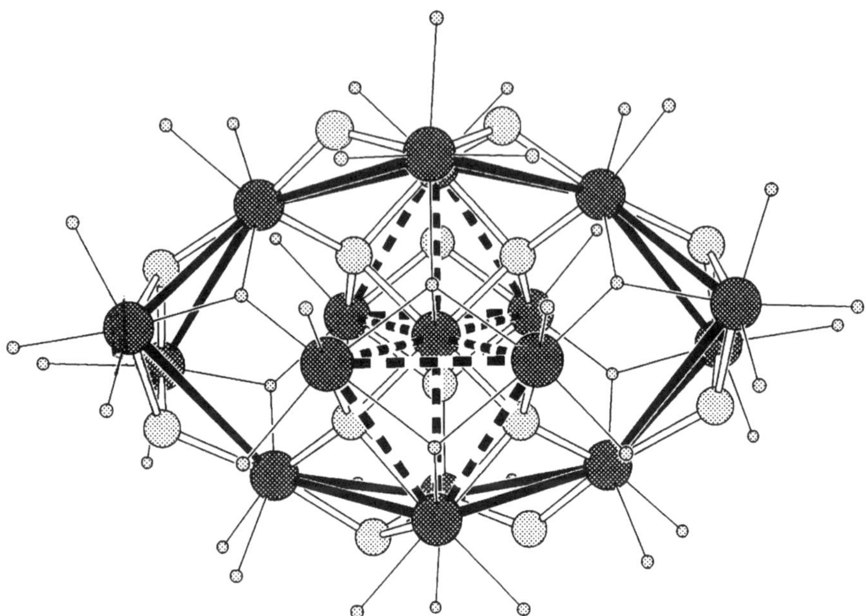

Figure 7 The structure of $[Fe_{16}M(O)_{10}(OH)_{10}(O_2CPh)_{20}]$ showing the four M_4O_2 butterflies with full lines, and the internal vertex-sharing bitetrahedron of metals as dashed lines.

chemistry, with Mn, Fe, Co and Ni cages prepared from this solvent, however, CH_2Cl_2 has also been used. A plausible explanation may be that trinuclear triangles or butterflies are initially formed, perhaps stabilized by weakly coordinating solvents, before oligomerizing into larger cages as crystallization occurs with loss of any coordinated solvent during this step.

2.1.3 Structures involving M_4O_2 blocks sharing body vertices

Cages within this class are less common than the former class. There are four Mn cages within the group and a heterometallic compound involving cobalt.

The simplest example is $[Mn_7(hmp)_{12}]$ **25** (hmp = 2-hydroxymethylpyridine), reported by Christou [24], which could alternately be described as a manganese hexagon containing a seventh metal centre. Combining two M_4O_2 units in this way creates two new M_3 triangles and in both $[Mn_9(O)_7(O_2CPh)_{13}(bpy)]$ **26** (bpy = 2,2′-bipyridyl) [25] and $[Mn_9Na_2(O)_7(O_2CPh)_{15}(MeCN)_2]$ **27** [26] these new triangles are capped by further metals, creating tetrahedra of metals upon the heptanuclear fragment formed from two M_4O_2 blocks. This ‘growth’ sequence is illustrated in Figure 9. An alternative description of **27** is given below in section 2.5.

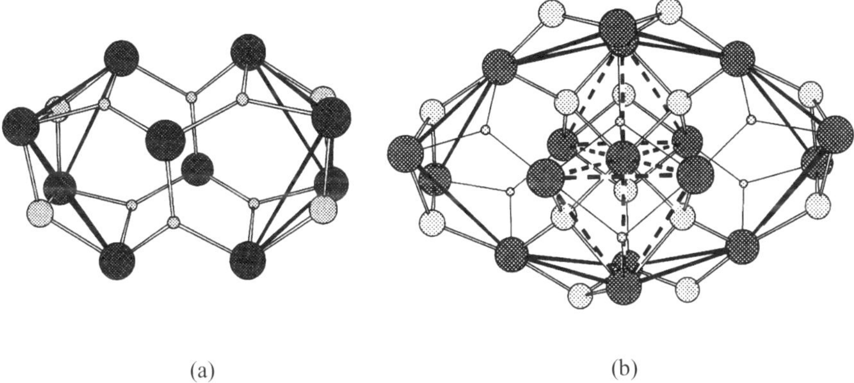

Figure 8 The relationship between (a) **5** and (b) **22–24** illustrated schematically. In both cases two external M_4O_2 butterflies encapsulate the remaining metal centres, either two or nine metals, respectively.

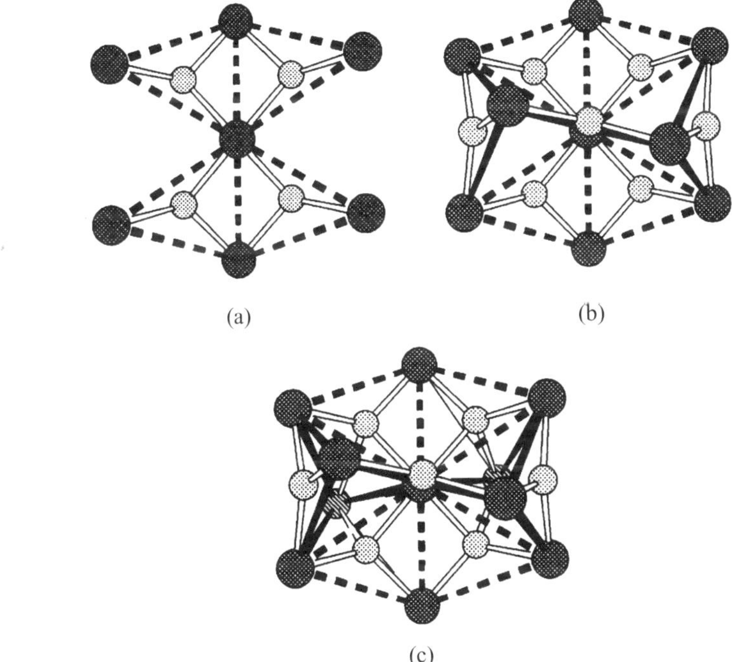

Figure 9 The relationship between hepta-, nona- and undecanuclear cages featuring M_4O_2 units sharing body vertices, shown as dashed lines. The tetrahedra formed by capping the triangular faces are shown as full lines. The polyhedra are taken from (a) and (b) $Mn_9(O)_7(O_2CPh)_{13}(bpy)]$ **26** and (c) $[Mn_9Na_2(O)_7(O_2CPh)_{15}(MeCN)_2]$ **27**.

A heterometallic complex $[Co_7Cu_2(OH)_2(chp)_{10}(O_2CMe)_6]$ **28** [27] also belongs to this category. It is an irregular structure, however, a Co_7O_4 fragment can be recognized at the centre of the molecule, with the oxygen atoms deriving from two hydroxides and two pyridonate ligands (Figure 10). The two Cu sites in the molecule cap edges of this core.

2.1.4 Structures involving edge-sharing M_4O_2 blocks

Most of the cages within this class are products of hydrolysis or oxidation. $[Fe_8(O)_2(tacn)_6]^{8+}$ **29** (tacn = 1,4,9-triazacyclononane) [28] is the oldest member of the group. It is the hydrolysis product from a dinuclear iron complex of this ligand, and contains six Fe centres in an M_6O_2 fragment where μ_3-oxo ligands are missing from two of triangles. The additional two Fe centres describe further triangles about the M_6O_2 core. A decanuclear manganese cage, $[Mn_{10}(O)_{14}\{N(CH_2CH_2NH_{12})_3\}_6]^{8+}$ **30** [29], is a larger version of the same

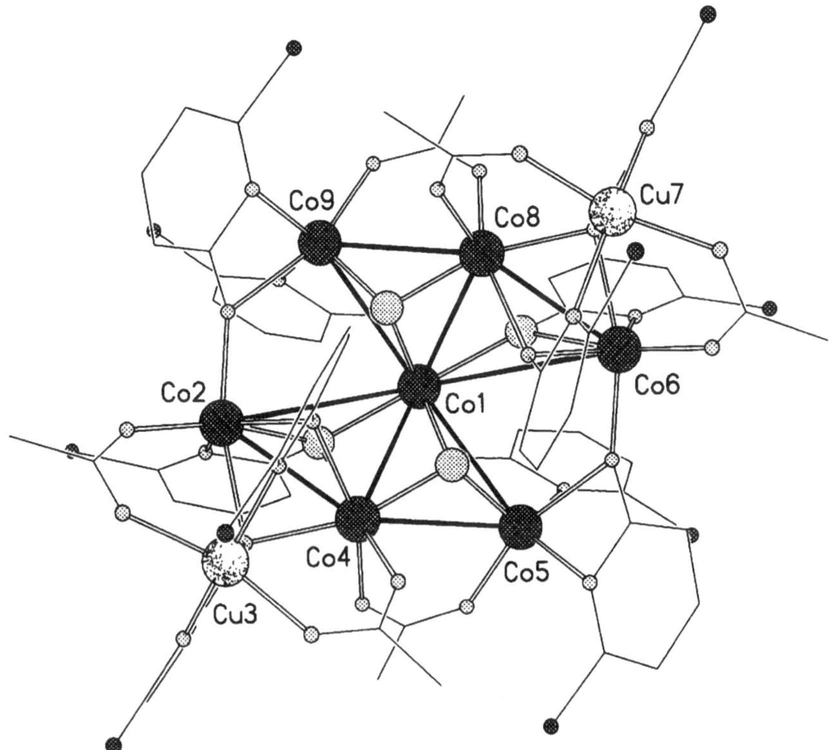

Figure 10 The structure of $[Co_7Cu_2(OH)_2(chp)_{10}(O_2CMe)_6]$ **28**.

structure, with all external edges of the central M_6 fragment capped by further metals, and uses a tripodal amine (Figure 11). The synthesis of **30** involves deliberate air oxidation of a solution of Mn(II) and the tripodal amine. A similar structure is found for nickel where a cyclic polydentate siloxane ligand is used to trap a layer of nickel oxide. The octanuclear cage $[Ni_8(O)_2(\{O_3SiPh\}_6)_2]$ **31** [30] is the only member of this class which is not produced by reaction of a smaller precursor with water or air. **31** is one of a number of cages reported using this polysiloxane ligand.

Three much larger cages are clearly expanded versions of this later structure. The earlier two reported are the cages $[Fe_{19}(O)_6(OH)_{14}(heidi)_{10}(H_2O)_{12}]^+$ **32** and $[Fe_{17}(O)(OH)_{16}(heidi)_8(H_2O)_{12}]^{3+}$ **33** synthesized by Powell and co-workers [31], which uses the tripodal ligand $\{N(CH_2COOH)_2(CH_2CH_2OH)\}$ (=H_3heidi) (Figure 12), and the latest is $[Co_{24}(OH)_{18}(OMe)_2(Cl)_6(mhp)_{22}]$ **34** made in our laboratories [32] using the 1,3-bridging ligand 6-chloro-2-pyridonate (Figure 13). The M_4O_2 fragments are easily recognized in both cases and build up a layer structure which is terminated either by a tripodal ligand (in **32** and **33**) or by pyridonates (in **34**). All three cages are made by hydrolysis reaction, **32**, and **33** by adding base to an

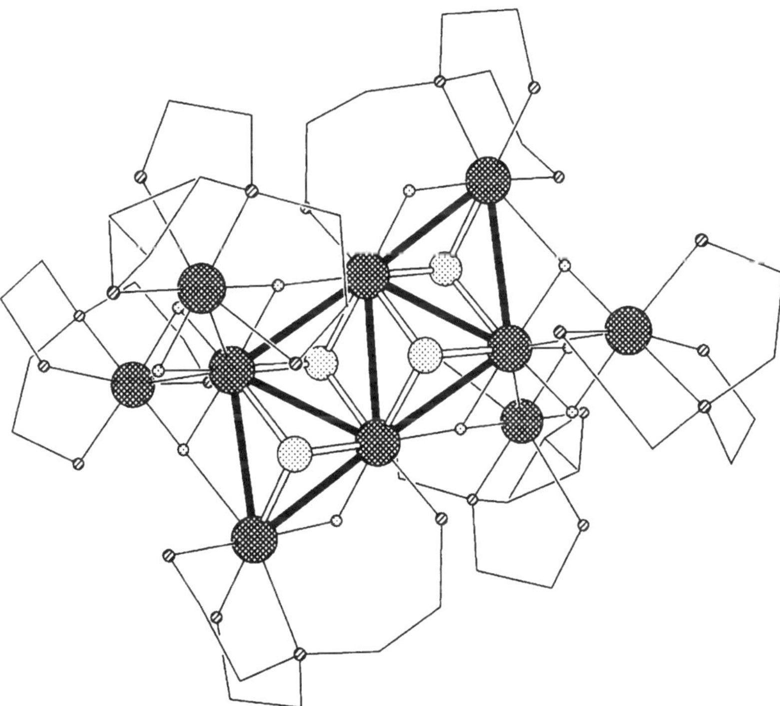

Figure 11 The structure of $[Mn_{10}(O)_{14}\{N(CH_2CH_2NH_2)_3\}_6]^{8+}$ **30**.

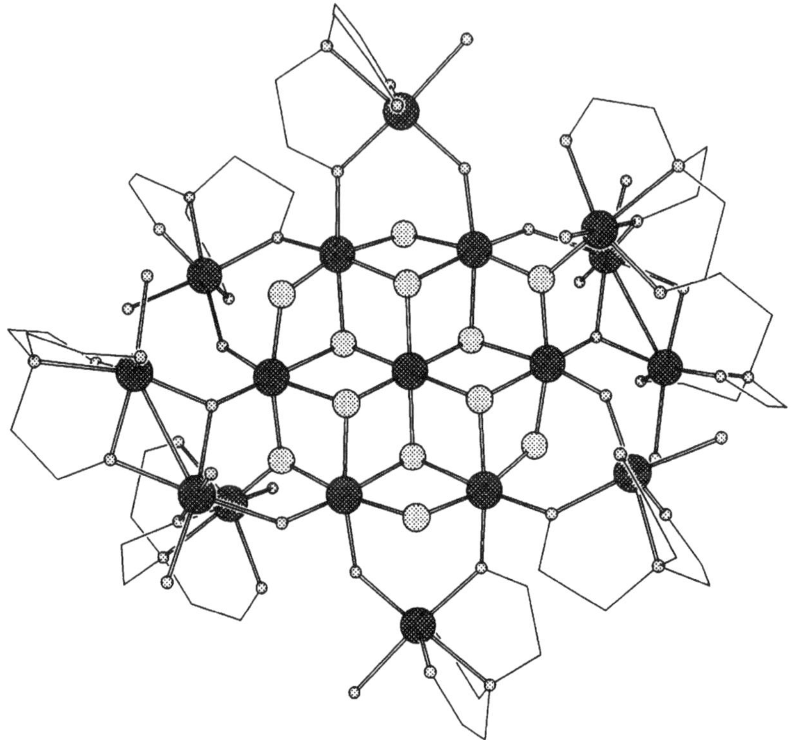

Figure 12 The structure of $[Fe_{19}(O)_6(OH)_4(heidi)_{10}(H_2O)_{12}]^+$ **32**.

aqueous solution of iron and H_3heidi, **34** by judicious use of damp ethyl acetate as a solvent. This Co_{24} cage is the only one within this category which does not feature a polydentate ligand. Powell has also reported an Al_{13} cage which has a similar structure and uses the tripodal ligand heidi [33].

2.2 Structures Built with Tetrahedra

2.2.1 Structures involving discrete M_4 tetrahedra

The bridges between tetrahedra in these structures come from a variety of sources.

In the octanuclear iron cage $[Fe_8(O)_2(O_2CNPr_2)_{12}]$ **35** [34] the two oxo-centred Fe tetrahedra are bridged by four oxygens derived from carbamate ligands (Figure 14). In $[Ni_8(OH)_2(cit)_6(H_2O)_2]$ **36** [35] (cit = citrate) two oxygens from citrate carboxylate link the two tetrahedra. In the latter complex three of the faces of each tetrahedra are capped by an oxygen from citrate. Related to these two structures is a

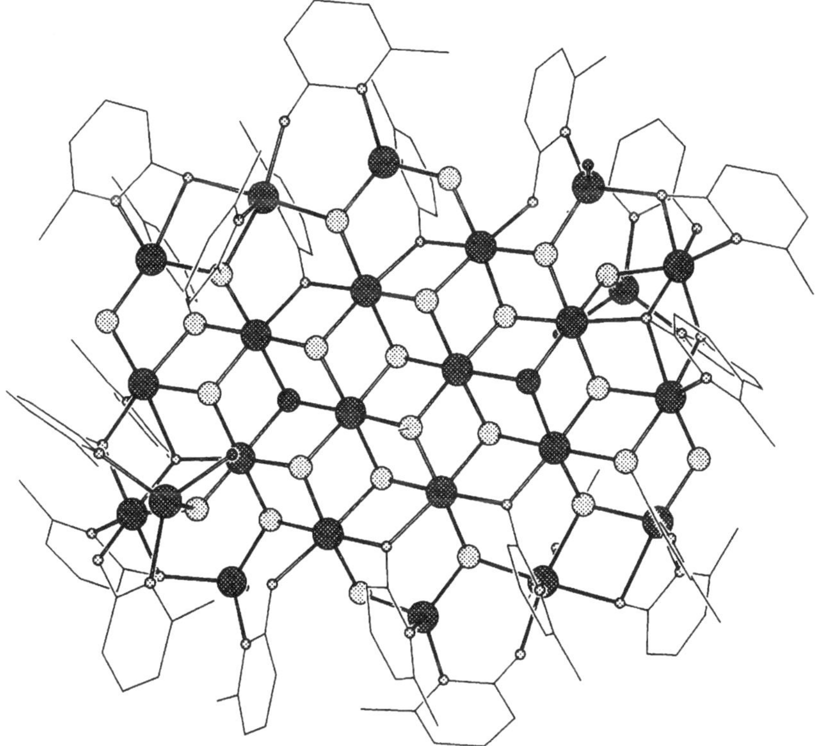

Figure 13 The structure of $[Co_{24}(OH)_{18}(OMe)_2(Cl)_6(mhp)_{22}]$ **34**.

heptanuclear cage $[Co_7(OH)_2(O_2CPh)_4(chp)_8(MeCN)]$ **37** [36], (chp = 6-chloro-2-pyridonate), however, here only one of the two tetrahedra is complete, so the structure appears as a tetrahedron linked through pyridonate ligands to a hydroxide-centred cobalt triangle (Figure 15).

Metal-containing bridges are found in two closely related cobalt complexes which involve pyridonate ligands. In the dodecanuclear complex $[Co_{12}(chp)_{18}(Cl)_2(OH)_4(Cl)_2(Hchp)_2(MeOH)_2]$ **38** [36] (Hchp = 6-chloro-2-pyridone) two tetrahedra are linked through a central eight-membered ring which involve four Co and four O atoms (Figure 16). The tetrahedra are capped on the faces by three O and one Cl atoms, resulting in distorted $[Co_4O_3Cl]$ cubanes. Two further peripheral Co atoms are also attached to the structure. We have also crystallized weakly diffracting crystals of a related decanuclear cage $[Co_{10}(chp)_{16}(OH)_4(MeCN)_2]$ **39** [13] which contain the same motif of two tetrahedra linked through an eight-membered ring, but which lack the two peripheral Co centers.

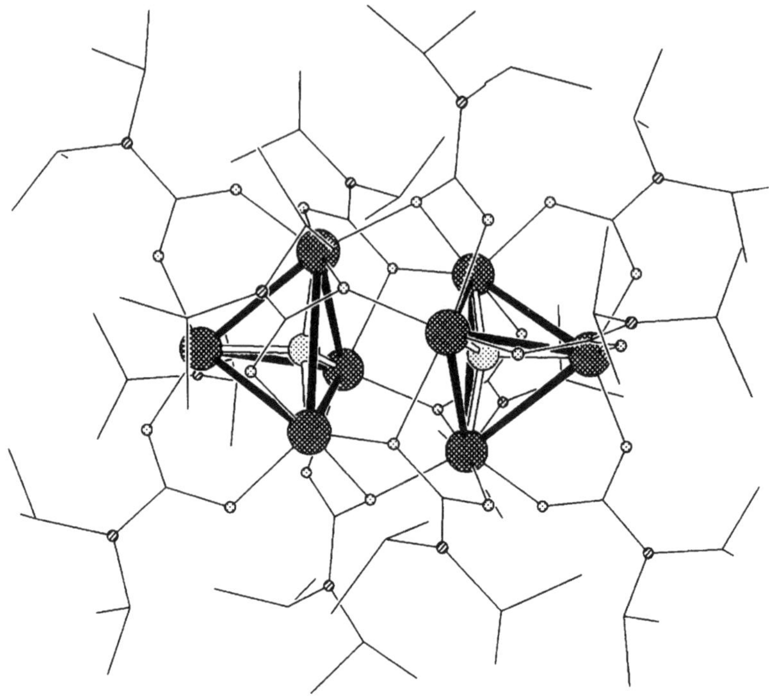

Figure 14 The structure of $[Fe_8(O)(O_2CNPr_2)_{12}]$ **35**.

The largest member of this class is a cage which also involves sodium centres. The complex $[Ni_{16}Na_6(chp)_4(phth)_{10}(OMe)_{10}(OH)_2(MeOH)_{20}]$ **40** [37] consists of four nickel tetrahedra attached to a central Na_6 octahedron through bridging phthalate ligands (Figure 17). The faces of the Ni_4 tetrahedra are capped by OMe groups, giving Ni_4O_4 cubanes. There are eight bridging phthalate ligands in this structure, with each bound to one face of the Na_6 octahedron through a carboxylate group. Therefore in some ways the symmetry of the central polyhedron is directing the structure of the surrounding assembly resulting in a square of cubanes.

2.2.2 Structures involving M_4 tetrahedra sharing vertices, edges or faces

There appear to be many more examples of metal butterflies assembled to form larger cages than large structures built of tetrahedra, possibly distortions of structure caused by linking together polyhedra tend to favour lower symmetry structures.

The octanuclear cage $[Co_8(O)_4(O_2CPh)_{12}(H_2O)(MeCN)_3]$ **41** [38] features a central cobalt tetrahedron capped on each face by a μ_4-oxo unit which binds to a further Co atom above the face (Figure 18). The structure could therefore be

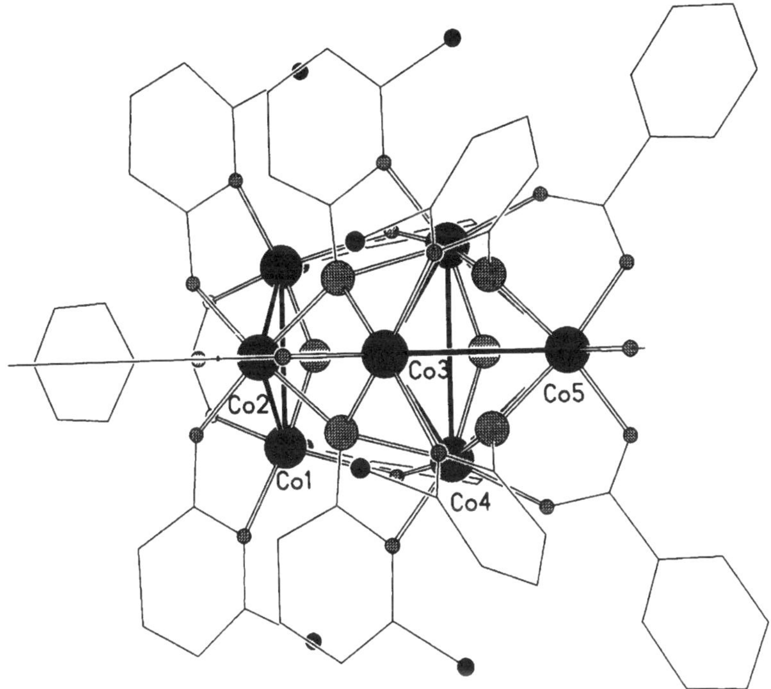

Figure 15 The structure of $[Co_7(OH)_2(O_2CPh)_4(chp)_8(MeCN)]$ **37**. The molecules lies on a mirror plane, and the tetrahedron consists of Co(3), Co(5), Co(4) and the symmetry equivalent of Co(4).

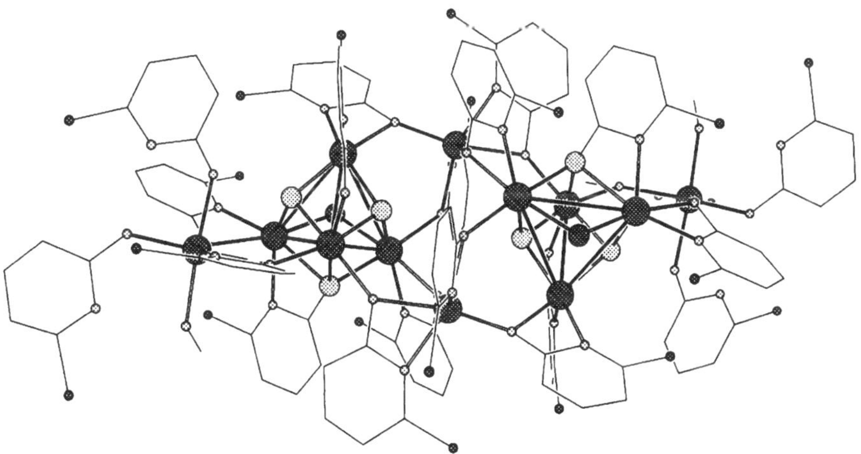

Figure 16 The structure of $[Co_{12}(chp)_{18}(OH)_4(Cl)_2(Hchp)_2(MeOH)_2]$ **38**. The chloride ions which occupy vertices of the cubanes are shown as cross-hatched circles.

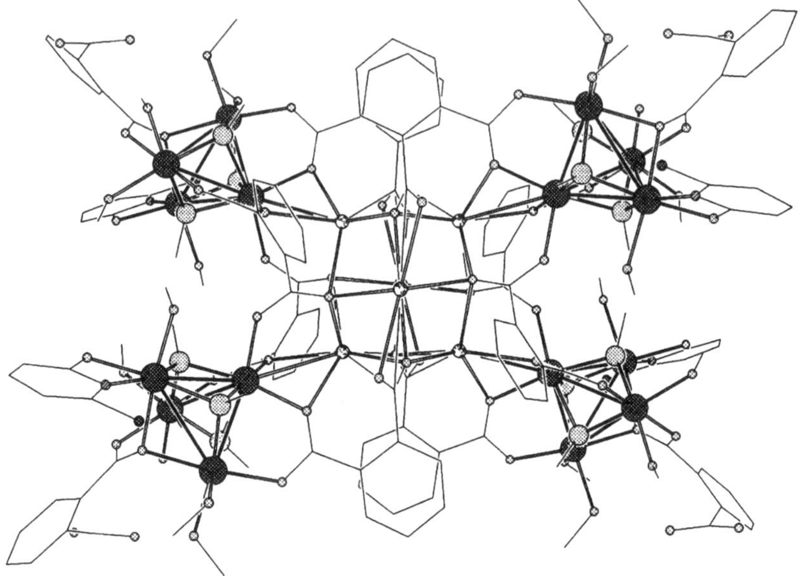

Figure 17 The structure of $[Ni_{16}Na_6(chp)_4(phth)_{10}(OMe)_{10}(OH)_2(MeOH)_{20}]$ **40**.

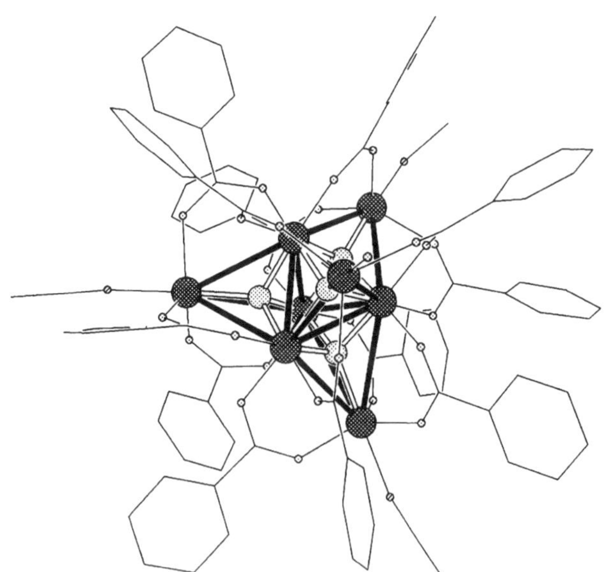

Figure 18 The structure of $[Co_8(O)_4(O_2CPh)_{12}(H_2O)(MeCN)_3]$ **41**.

described as four oxo-centred Co tetrahedron, with each tetrahedron sharing edges with the three other tetrahedra in the structure. In two decanuclear manganese cages $[Mn_{10}(O)_4(X)_{12}(biphen)_4]^{4-}$ (X = Cl **42** or Br **43**; biphen = 2,2′-bis{phenoxide}) [39, 40] four oxo-centred Mn tetrahedra share vertices with each other, resulting in a cage which could alternatively be described as an Mn octahedron capped by further Mn atoms on alternate faces (Figure 19).

In these structures the M_4 tetrahedra are oxo-centered, while for the discrete cases they were, in general, face-capped by oxygens leading to M_4O_4 cubanes. A further series of linked tetrahedra involve cages where the edges of the tetrahedra are bridged, producing adamantane-like structures. These structures have been found for both nickel and cobalt, using pyridonate ligands [41]. The lowest nuclearity cage of this type is $[Ni_7(chp)_{12}(Cl)_2(MeOH)_6]$ **44**, in which two nickel adamantanes share a common vertex. In a related cobalt cage $[Co_9(chp)_{18}]$ **45** there are four adamantanes sharing two faces and one vertex (Figure 20). This latter complex is unique in this area, in that it is homoleptic. The oxygen atoms of the adamantane are derived from

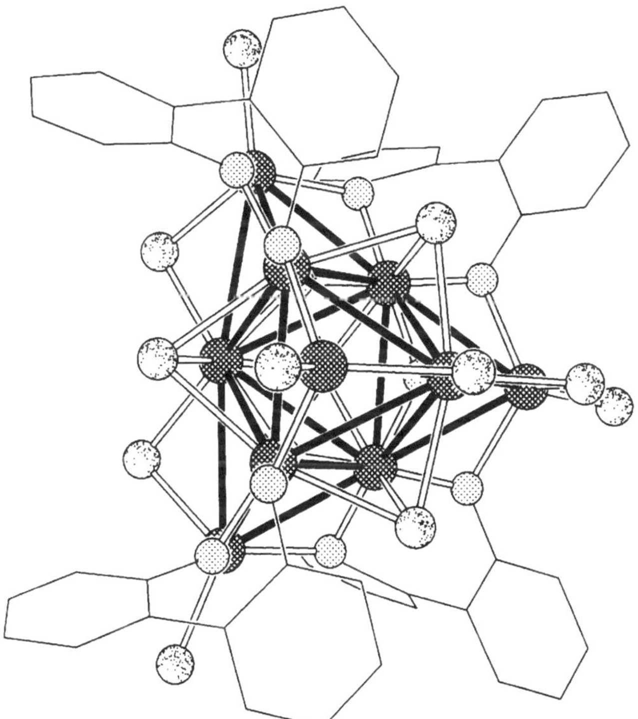

Figure 19 The structure of $[Mn_{10}(O)_4(Cl)_{12}(biphen)_4]^{4-}$ **42**. The chloride ligands involved in the structures shown as irregularly shaded circles.

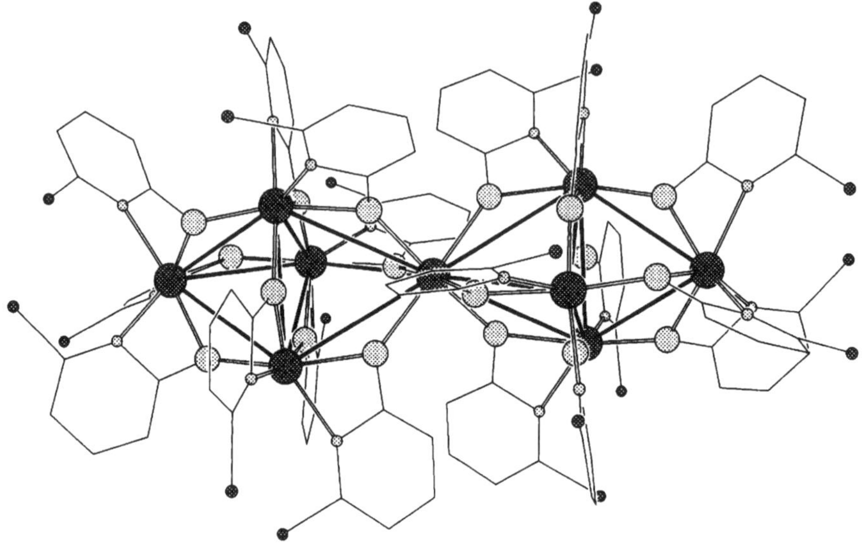

Figure 20 The structure of $[Co_9(chp)_{18}]$ **45**.

the pyridonate ligands, which chelate to one metal through both N and O donors, while the oxygen bridges to a second. Two further nickel cages, one hepta- and one nonanuclear, have a similar array of metal centres, but lack of the perfect M_4O_6 adamantine core as μ_3-hydroxide groups are found on one face of each adamantane [41].

These latter two cages are distorted towards the structure of the tridecanuclear cage $[Co_{13}(chp)_{20}(phth)_2(Cl)_2]$ **46** [36]. Here the central M_7 fragment of two metal tetrahedra sharing a common vertex is extremely distorted, with a carboxylate from a phthalate ligand bridging one $Co \cdots Co$ edge where previously a μ_2-oxygen had been found, and with a μ_3-hydroxide linking the three Co centres of each tetrahedra. The relationship between the cores of **45** and **46** are shown schematically in Figure 21. Additionally in **46** two more Co_3 triangles are found in the structure, which leads to a similarity with $[Co_7(OH)_2(O_2CPh)_4(chp)_8(MeCN)]$ **37** [36] in which one tetrahedra and one triangle were found.

2.3 Structures Which Feature Both Butterflies and Tetrahedra

These structures are among the most complicated yet characterized. $[Fe_{10}(O)_4(OMe)_{16}(dbm)_6]$ **47** contains two stacked M_4O_2 butterflies, linked by two tetrahedra [42] (Figure 22). The tetrahedra each contain one wing-tip iron from one butterfly, two iron centres from the other butterfly and a capping iron. The tetrahedra are both

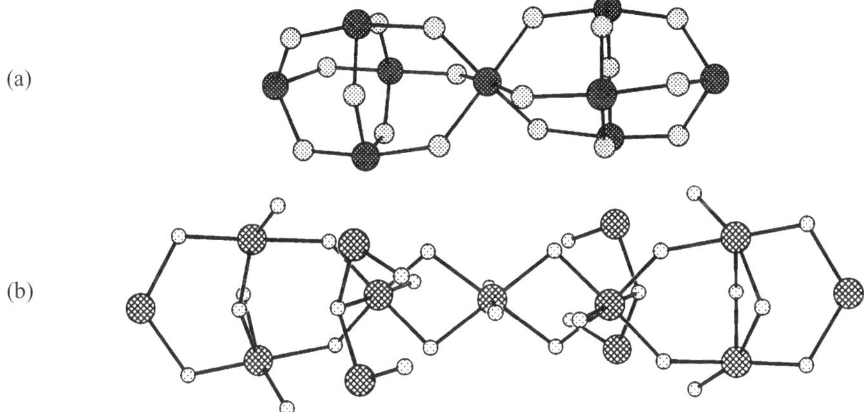

Figure 21 The relationship between structures (a) **45** and (b) **46** illustrated schematically. The similarity is most apparent for the central seven metal centres of both structures.

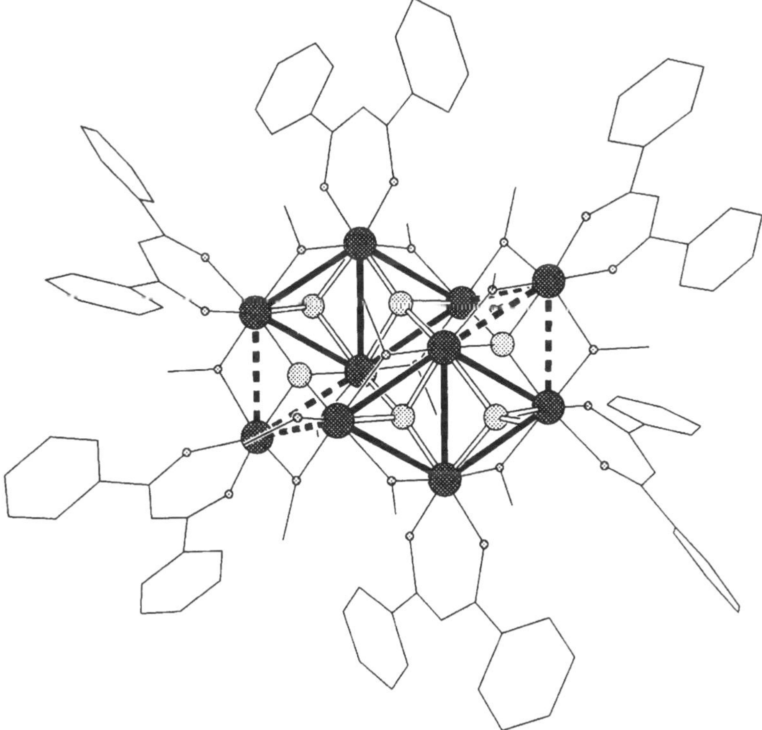

Figure 22 The structure of $[Fe_{10}(O)_4(OMe)_{16}(dbm)_6]$ **47**. The M_4O_2 butterflies are shown as full lines, and the linking tetrahedra as dashed lines.

oxo-centred with three μ_2-methoxides bridging the Fe···Fe edges which involve the capping iron.

A similarly complicated structure is found in $[Fe_{12}(O)_2(OMe)_{18}(O_2CMe)_6$-$(EtOH)_3]$ **48** [43], where there again two stacked butterflies linked by tetrahedra (Figure 23). Here the tetrahedra contain one wing-tip iron from each butterfly, with two further atoms completing the polyhedron. The tetrahedra are now capped on three faces by μ_3-methoxides, with the fourth face occupied by a μ_4-oxide. The butterfly units are distorted because these μ_4-oxides occupy the sites where one of the two μ_3-oxides would be expected, but are displaced towards the tetrahedra.

There are also two examples where M_4O_2 units link tetrahedra. In $[Mn_{11}(O)_{10}(O_2CMe)_{11}(Cl)_2(bpy)_2]$ **49** [44] two tetrahedra are linked through a pair of Mn_4O_2 units which share a body vertex (Figure 24). This central M_7O_4 fragment resembles those found in $[Mn_9(O)_7(O_2CPh)_{13}(bpy)]$ **26** [25] and $[Mn_9Na_2(O)_7(O_2CPh)_{15}(MeCN)_2]$ **27** [26]. The wing-tip metal centres form part of the two tetrahedra. These tetrahedra are capped on the faces by three μ_3-oxides and one μ_3-chloride.

The octadecanuclear cage $[Mn_{18}(O)_{16}(phth)_2(O_2CPh)_{22}]$ **50** [45] follows a similar pattern on a larger scale. Here three Mn_4O_2 units share body vertices to form a central $Mn_{10}O_6$ region which links to tetrahedra, although here no metals are shared by the M_4O_2 butterflies and the tetrahedra (Figure 25). The tetrahedra are capped on

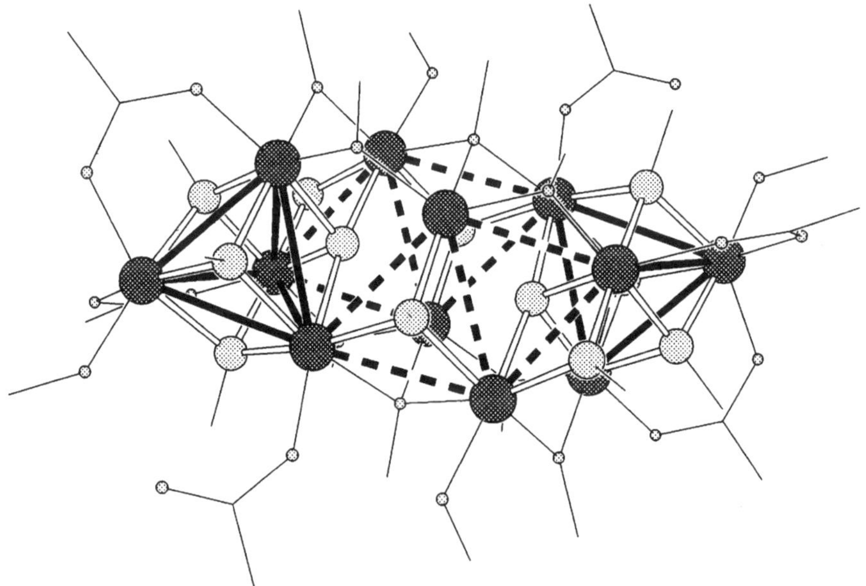

Figure 23 The structure of $[Fe_{12}(O)_2(OMe)_{18}(O_2CMe)_6(EtOH)_3]$ **48**. The M_4O_2 butterflies are shown as dashed lines, and the linking tetrahedra as full lines.

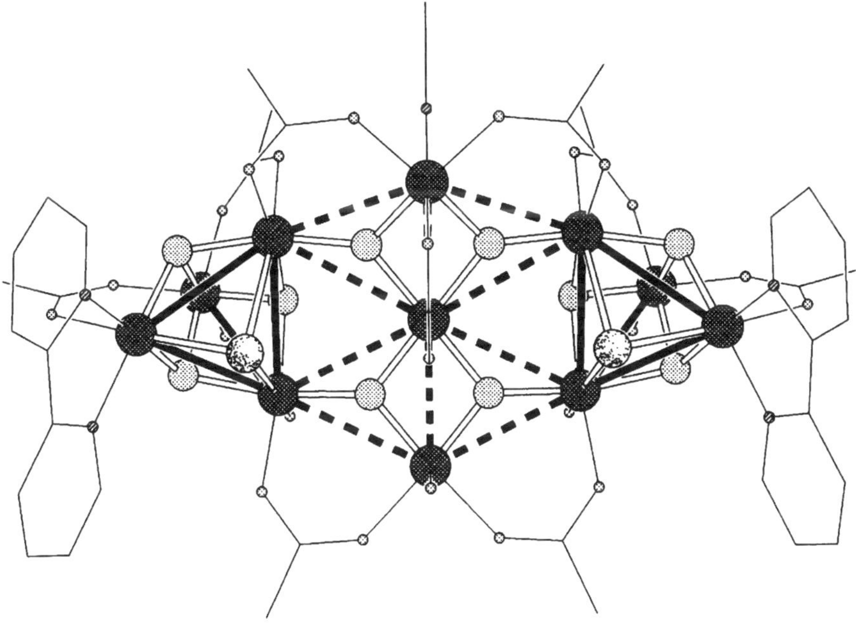

Figure 24 The structure of $[Mn_{11}(O)_{10}(O_2CMe)_{11}(Cl)_2(bipy)_2]$ **49**. The linking M_4O_2 butterflies are shown as dashed lines, and the tetrahedra as full lines.

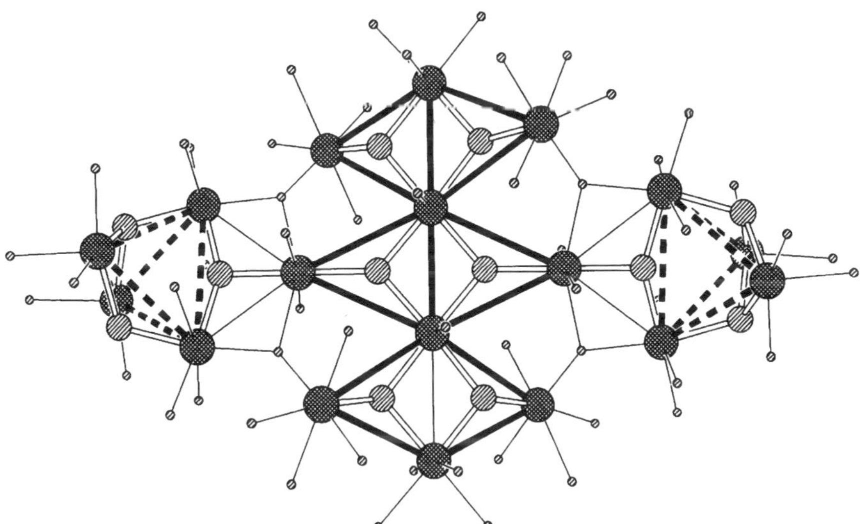

Figure 25 The structure of $[Mn_{18}(O)_{16}(phth)_2(O_2CPh)_{22}]$ **50**. The linking M_4O_2 butterflies are shown as full lines, and the tetrahedra as dashed lines.

two faces and one edge by μ_3-oxides, with μ_3-oxides also serving to link the tetrahedra to the $Mn_{10}O_6$ segment of the structure.

2.4 Wheels

There is a large group of polynuclear cages of 3d-metals in moderate oxidation states which do not fit within the above classifications. These are the 'metal wheels'. There are rings known for the metals from titanium to nickel (with the exception of manganese), and with nuclearities varying from 8 to 12. This is in itself another family of structures.

For the early transition metals, which have not been discussed for the more compact cages, octanuclear structures have been reported: $[Ti(O)(O_2CC_6F_5)_2]_8$ **51** [46], $[V_8(OH)_4(OEt)_8(O_2CMe)_{12}]$ **52** [47] and $[CrF(O_2CCMe_3)_2]_8$ **53** [48]. Iron provides examples of both octanuclear $[FeF(O_2CCMe_3)_2]_8$ **54** [49] and decanuclear rings $[Fe(OMe)_2(O_2CR)]_{10}$ [R = CH_2Cl **55** or CH_3 **56**] [10, 50]. For cobalt and nickel dodecanuclear wheels are known which are isostructural [13, 51]: $[M_{12}(xhp)_{12}(O_2CMe)_{12}(H_2O)_6(THF)_6]$ [M = Ni, xhp = 6-chloro- **57** or 6-bromo-2-pyridonate **58**: M = Co, xhp = 6-chloro-2-pyridonate **59**] (Figure 26). In all these

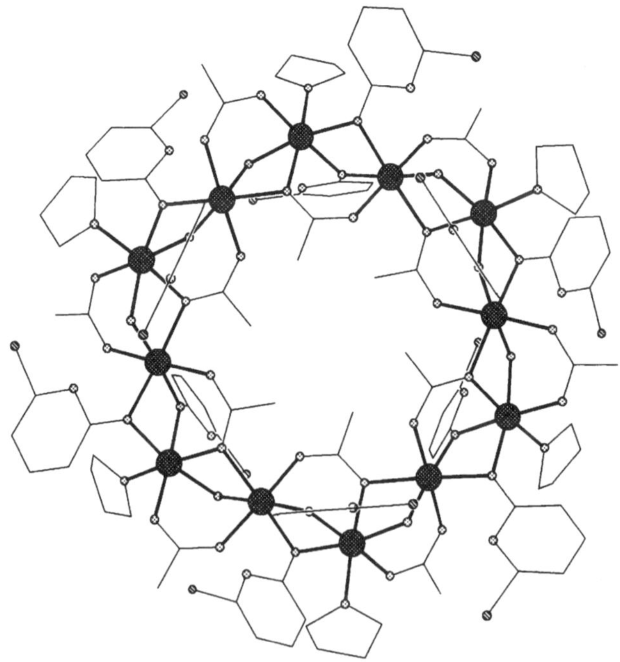

Figure 26 The structure of $[Ni_{12}(chp)_{12}(O_2CMe)_{12}(H_2O)_6(THF)_6]$ **57**.

structures all M···M vectors are bridged by a mixture of 1,3-bridging carboxylates, and μ_2-bridging anions, e.g. hydroxide, fluoride, oxide or a μ_2-oxygen from an alkoxide or pyridonate ligand. This mixture of bridges across each vector seems to cause sufficient curvature in the structure to lead to cyclic structures.

Manganese is presently absent from this family. Our attempts to make an Mn wheel similar to the cobalt and nickel wheels led to a one-dimensional polymer $[Mn_2(chp)_2(O_2CMe)_2(MeOH)_2]_n$ [52]. There is in this structure curvature within each trinuclear unit, however, in neighbouring units the structure curves in opposite directions, leading to a zigzag polymer rather than a wheel. There must be a fine balance between these two alternative structures.

2.5 Other Exceptions

All beautiful theories are doomed when required to explain all ugly facts. There are several structures which do not fit into the above categorization, some of which are the result of our own work.

One of the most beautiful exceptions is the heterometallic cage $[Mn_8Sb_4(O)_4(OEt)_{20}]$ **60** [53], which features Mn_4Sb square-based pyramids. The structure is best described as two Mn_6Sb_2 square anti-prisms sharing a common square face (Figure 27). It is possibly the presence of the p-block metalloid which is leading to this new structural type.

Other exceptions are the two decanuclear cages $[Ni_{10}(OH)_4(mhp)_{10}(O_2CCMe_3)_6$-$(sol)_2]$ (sol = MeOH **61** or H_2O **62**) [54]. The synthesis and stoichiometry of these cages are similar to those for the series of tricapped-trigonal prisms **10**, **13–15** (see Section 1b), however, pivalate was used in place of less sterically demanding carboxylates and this results in a less regular structure. Any attempt to describe this structure in terms of butterflies or tetrahedra is extremely contrived.

It is much more easily described by considering the metal centres to occupy the centre and nine vertices of a 14-vertex tetracosahedron (Figure 28a). An interesting observation is that in changing the ligand, and thus distorting the polyhedron from tricapped-trigonal prism to tetraicosahedron, the triangular faces of the metal cage have been retained, presumably due to the large number of μ_3-oxygen bridges found in these structures. This observation led us to look for the intermediate fully triangulated polyhedra, i.e. those with between nine and 14 vertices. Although none of the complete polyhedra have been structurally characterized, $[Mn_9Na_2(O)_7(O_2CPh)_{15}(MeCN)_2]$ **27** [26] can be shown to be a fragment of a centred icosahedra, i.e. the 12-vertex deltahedron (Figure 28b).

The presence of triangulated faces gives some resemblance to the structural chemistry of boranes, however, the reason for the triangular faces is quite different. In boranes it is due to delocalized bonding between the 2p elements, while in these metal cages it is due to the preference of oxygen donors to act as μ_3-bridges.

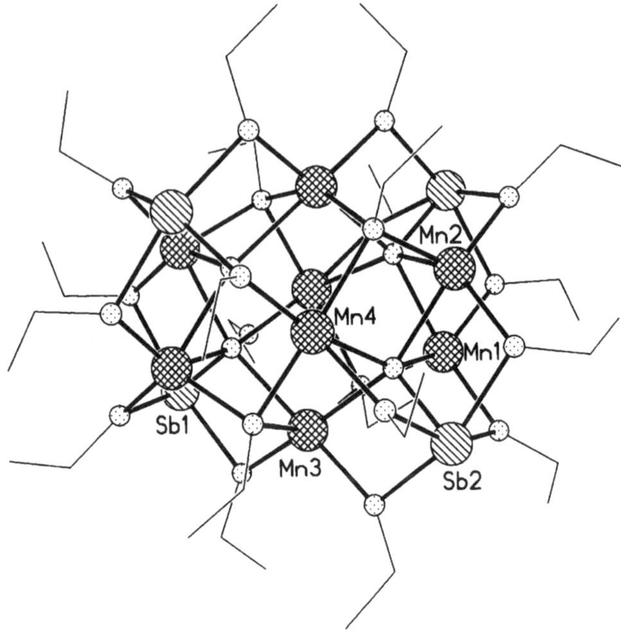

Figure 27 The structure of $[Mn_8Sb_4(O)_4(OEt)_{20}]$ **60**. One of the square-based pyramids from which the structure is constructed is labelled Mn(1), Mn(2), Mn(3), Mn(4) and Sb(2).

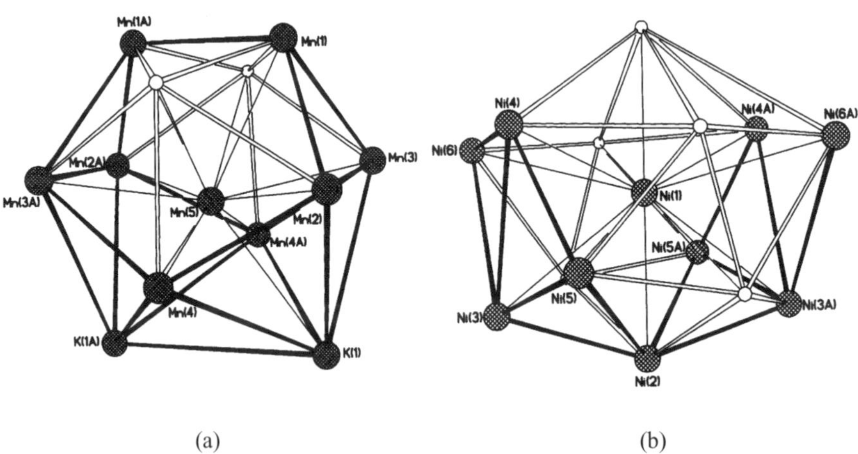

Figure 28 The metal polyhedra in (a) $[Mn_9Na_2(O)_7(O_2CPh)_{15}(MeCN)_2]$ **27** and (b) $[Ni_{10}(OH)_4(mhp)_{10}(O_2CCMe_3)_6(sol)_2]$ **61** illustrating the resemblance to deltahedra. The additional circles required to complete the tetraicosahedron and icosahedron, respectively, are shown as small open circles.

A heptadecanuclear iron cage, $[Fe_{17}(O)_{15}(OH)_6(chp)_{12}(phen)_8(OMe)_3]$ **63** (phen = 1,10-phenanthroline) is an exception to any attempt at classification [55]. The metal core is extremely irregular (Figure 29) and, although M_4O_2 units can be found within the cage, it is impossible to explain as a fragment of a mineral, a common polyhedron or by reference to the simple building blocks used above. The irregularity of the structure is due to the presence of the phenanthroline ligands, which stack along edges of the cage leading to a unique polymetallic cage.

The final two exceptions are the first examples of Ni or Co cages of this size being produced by 'designed' assembly. The first is a beautiful Ni_8 cage, $[Ni_8(Hvi)_{10}(H_2vi)_2]^{6-}$ **64** (where H_3vi = violuric acid) [56], where the eight nickel(II) centres lie at the corners of a cube with the 12 ligands lying on the cube edges. The second is a cyclic octanuclear cobalt cage, $[Co_8(L2)_{12}(ClO_4)]$-$(ClO_4)_3$ **65** [where L2 = bis{3-(2-pyridyl)pyrazol-1-yl}dihydroborate] [6], where the ligand is a flexible 'tris-bypyridyl' analogue.

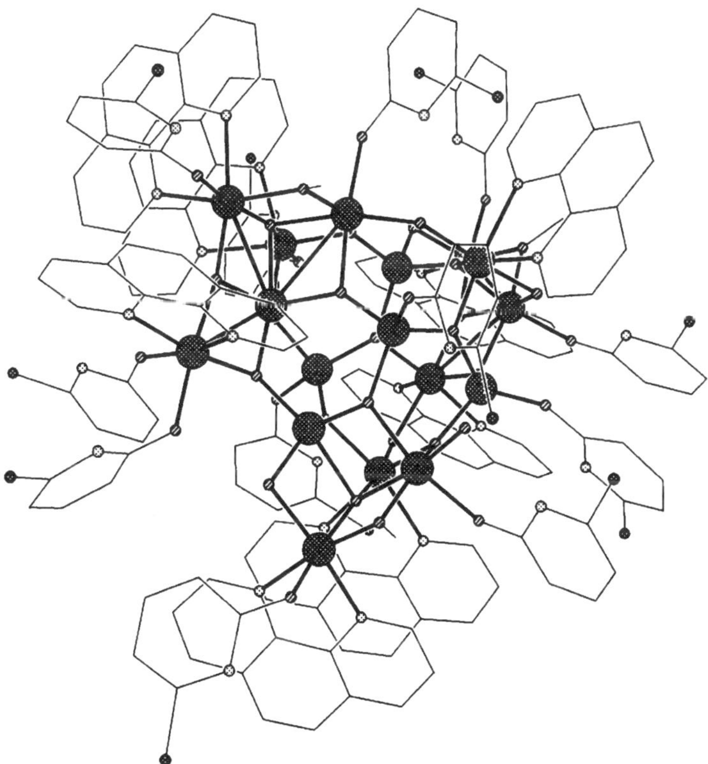

Figure 29 The structure of $[Fe_{17}(O)_{15}(OH)_6(chp)_{12}(phen)_8(OMe)_3]$ **63**.

3 DISCUSSION OF CLASSIFICATION OF STRUCTURES

The procedure of classifying structures based on the blocks from which they are constructed is laborious, and somewhat repetitive. It leads to the observation of relationships between structures which might not otherwise be obvious. This in turn can be related to features of the synthetic procedure. The complicating factor is the fact that all but one of these structures are heteroleptic, and the vast majority contain three or more different ligands, therefore it is difficult to decide how individual ligands within a structure have influenced the shape adopted. It is always necessary to discuss the influence of one ligand in the presence of another, therefore all conclusions are tentative and there are exceptions to most of the 'rules'.

One clear observation is that the presence of chloride or methoxide bridges leads to cages built of tetrahedra rather than M_4O_2 blocks. This may allow an element of design into synthetic procedures—if you wish to make a cage based on tetrahedra, use MeOH as a solvent and add a source of methoxide or chloride. In passing, it is worth noting the presence of chloride as a co-factor in the tetranuclear oxygen-evolving complex. As chloride is a means for making tetrahedra in coordination chemistry, a possible role for the chloride in the enzymatic process is to convert a puckered Mn_4 butterfly towards an Mn_4 tetrahedron. Christou and coworkers have illustrated exactly this transformation by addition of chloride to the butterfly cage $[Mn_4(O)_2(O_2CMe)_6(dbm)_2]$ [57].

The means by which the more complex organic ligands direct structure is less clear. Carboxylates are the most frequently used ligands in this chemistry. They appear to favour structures containing discrete M_4O_2 butterflies, structures comprising M_4O_2 units sharing wing-tip vertices and metal wheels. Carboxylates are rarely found in structures belonging to other categories, which suggests that carboxylates have a directing role towards the butterfly building block. The exceptions are where methoxide or chloride are present.

Dicarboxylates, such as phthalate and citrate, appear quite different, and have thus far only been found in structures where tetrahedra are present. This may reflect the extremely small amount of chemistry reported for such ligands, and the fact that in the relevant complexes methoxide or chloride are also present, and therefore the influence of these tetrahedrally directing ligands may be paramount. Polydentate ligands, e.g. tacn, tren and heidi, seem to favour structures based on edge-sharing butterflies but this may also reflect the synthetic methods used in these cages; all the edge-sharing butterflies are made by hydrolysis or oxidation reactions.

The only homoleptic cage described above is $[Co_9(chp)_{18}]$ **45**, and therefore one might conclude that, in the absence of other influences, pyridonates favour structures based on tetrahedra. However, structures featuring pyridonates are found in virtually all the categories, chiefly due to the coordinative flexibility of the ligands. There is also a marked difference between the coordination chemistry of the 6-chloro- and 6-methyl-2-pyridonate ligands, due to electronic factors which influence the coordinative ability of the ring nitrogen atom. Of all the ligands these seem the most 'ill-behaved' or the least directing.

We can therefore propose a hierarchy of influences. Alkoxides and chlorides are the most directing, and will lead to cages based on tetrahedra. In the absence of these ligands carboxylates direct structure towards cages involving linked M_4O_2 butterflies. Pyridonates are the least directing of the frequently reported ligands, but alone they direct towards structures based on tetrahedra. Even if these rules are correct, however, they give no information about which specific cages will form, only a rough guide to the likely classification.

There has been a very limited range of bridging ligands used in this chemistry. Acetylacetonates, especially dibenzoylmethane, have been used on occasion, and one example of the use of 2,2′-biphenoxide has been reported. Only one high nuclearity cage featuring a bridging carbamate ligand has been reported. The potential of these, and other simple ligands in this chemistry remains to be explored.

4 ACKNOWLEDGEMENTS

The synthetic work from my own group which I have discussed above is due to Euan Brechin, Alasdair Graham, Craig Grant, Mark Murrie, David Nation and Greg Solan with the structural studies performed by Simon Parsons and Steve Harris. To discuss the other structures, and to generate the figures, I used the Cambridge Structural Database and I am grateful for access. This work was supported by the EPSRC(UK) and The Leverhulme Trust.

5 REFERENCES

1. P. N. W. Baxter, J.-M. Lehn, J. Fisher and M.-T. Youinou, *Angew. Chem. Int. Ed. Engl.*, **33**, 2284 (1994).
2. T. Lis, *Acta Crystallogr.*, *Sect. B*, **36**, 2042 (1980).
3. R. Sessoli, H.-L. Tsai, A. R. Schake, S. Wang, J. B. Vincent, K. Folting, D. Gatteschi, G. Christou and D. N. Hendrickson, *J. Amer. Chem. Soc.*, **115**, 1804 (1993).
4. L. Thomas, F. Lionti, R. Ballou, D. Gatteschi, R. Sessoli and B. Barbara, *Nature*, **383**, 145 (1996).
5. P. N. W. Baxter, J.-M. Lehn and K. Rissanen, *Chem. Commun.*, 1323 (1997).
6. P. L. Jones, K. J. Byrom, J. C. Jeffrey, J. A. McCleverty and M. D. Ward, *Chem. Commun.*, 1361 (1997).
7. M. W. Wemple, H.-L. Tsai, W. E. Streib, D. N. Hendrickson and G. Christou, *J. Chem. Soc., Chem. Commun.*, 1031 (1994).
8. E. Libby, K. Folting, J. C. Huffman and G. Christou, *J. Amer. Chem. Soc.*, **112**, 5354 (1990).
9. S. Wang, H.-L. Tsai, K. Folting, J. D. Martin, D. N. Hendrickson and G. Christou, *J. Chem. Soc., Chem. Commun.*, 671 (1994).

10. C. Christmas, J. B. Vincent, H.-R. Chang, J. C. Huffman, G. Christou and D. N. Hendrickson, *J. Amer. Chem. Soc.*, **110**, 823 (1988).
11. C. Benelli, S. Parsons, G. A. Solan and R. E. P. Winpenny, *Angew. Chem., Int. Ed. Engl.*, **35**, 1825 (1996).
12. S. Wang, H.-L. Tsai, W. E. Streib, G. Christou, and D. N. Hendrickson, *J. Chem. Soc., Chem. Commun.*, 677 (1992).
13. R. Bhula and D. C. Weatherburn, *Angew, Chem., Int. Ed. Engl.*, **30**, 688 (1991).
14. K. Brechin, S. G. Harris, S. Parsons and R. E. P. Winpenny, unpublished results.
15. S. M. Gorun, G. C. Papaefthymiou, R. B. Franekl and S. J. Lippard, *J. Amer. Chem. Soc.*, **109**, 3337 (1987).
16. E. K. Brechin, S. Parsons and R. E. P. Winpenny, *J. Chem. Soc., Dalton Trans.*, 3745 (1996).
17. A. S. Batsanoc, G. A. Timko, Y. T. Struchkov, N. V. Gérbéléu and K. M. Indirchan, *Koord. Khim.*, **17**, 662 (1991).
18. W. Clegg, C. D. Garner and M. H. Al-Samman, *Inorg. Chem.*, **22**, 1534 (1983).
19. A. J. Blake, E. K. Brechin, A. Codron, R. O. Gould, C. M. Grant, S. Parsons, J. M. Rawson and R. E. P. Winpenny, *J. Chem. Soc., Chem. Commun.*, 1983 (1995).
20. H. J. Eppley, H.-L. Tsai, N. De Vries, K. Folting, G. Christou and D. N. Hendrickson, *J. Amer. Chem. Soc.*, **117**, 301 (1995).
21. A. R. Schake, H.-L, Tsai, R. J. Webb, K. Folting, G. Christou and D. N. Hendrickson, *Inorg. Chem.*, **33**, 6020 (1994).
22. D. Luneau, J.-M. Savariault and J.-P. Tuchagues, *Inorg. Chem.*, **27**, 3912 (1988).
23. W. Michlitz, V. McKee, R. L. Rardin, L. E. Pence, G. C. Papaethymiou, S. G. Bott and S. J. Lippard, *J. Amer. Chem. Soc.*, **116**, 8061 (1994).
24. M. A. Bolcar, S, M. J. Aubin, K. Folting, D. N. Hendrickson and G. Christou, *Chem. Commun.*, 1485 (1997).
25. D. W. Low, D. M. Eichhorn, A. Draganescu and W. H. Armstrong, *Inorg. Chem.*, 330, 877 (1991).
26. H.-L. Tsai, S. Wang, K. Folting, W. E. Streib, D. N. Hendrickson and G. Christou, *J. Amer. Chem. Soc.*, **117**, 2503 (1995).
27. E. K. Brechin, S. G. Harris, S. Parsons and R. E. P. Winpenny, *J. Chem. Soc., Dalton Trans.*, 3403 (1997).
28. K. Wieghardt, K. Pohl, I. Jibril and G. Huttner, *Angew. Chem., Int. Ed. Engl.*, **23**, 77 (1984).
29. K. S. Hagen, W. H. Armstrong and M. M. Olmstead, *J. Amer. Chem. Soc.*, **111**, 774 (1989).
30. M. M. Levitsky, O. I. Schegolikhina, A. A. Zhdanov, V. A. Igonin, Y. E. Ovchinnikov, V. E. Shklover and Y. T. Struchov, *J. Organomet. Chem.*, **401**, 199 (1991).
31. A. K. Powell, S. L. Heath, D. Gatteschi, L. Pardi, R. Sessoli, G. Spina , F. Del Giallo and F. Pieralli, *J. Amer. Chem. Soc.*, **117**, 2491 (1995).
32. E. K. Brechin, S. G. Harris, A. Harrison, S. Parsons, A. G. Whittaker and R. E. P. Winpenny, *Chem. Commun.*, 653 (1997).
33. S. L. Heath, P. A. Jordan, I. D. Johnson, G. R. Moore, A. K. Powell and M. Helliwell, *J. Inorg. Biochem.*, **59**, 785 (1995).
34. D. B. Dell'Amico, F. Calderozzo, L. Labella, C. Maichle-Mossmer and J. Strahle, *J. Chem. Soc., Chem. Commun.*, 1555 (1994).
35. J. Strouse, S. W. Layten and C. E. Stouse, *J. Amer. Chem. Soc.*, **99**, 562 (1977).
36. E. K. Brechin, S. G. Harris, S. Parsons and R. E. P. Winpenny, *J. Amer. Chem. Soc.*, **118**, 11293 (1996).
37. K. Dimitrou, K. Folting, W. E. Streib and G. Christou, *J. Chem. Soc., Chem. Commun.*, 1385 (1994).

38. D. P. Goldberg, A. Caneschi and S. J. Lippard, **115**, 9299 (1993).
39. D. P. Goldberg, A. Caneschi, C. D. Delfs, R. Sessoli and S. J. Lippard, *J. Amer. Chem. Soc.*, **117**, 5789 (1995).
40. E. K. Brechin, S. G. Harris, S. Parsons and R. E. P. Winpenny, *Angew. Chem., Int. Ed. Engl.*, **36**, 1967 (1997).
41. A. Caneschi, A. Cornia, A. C. Fabretti and D. Gatteschi, *Angew. Chem., Int. Ed. Engl.*, **34**, 2716 (1995).
42. K. L. Taft, G. C. Papaehtymiou and S. J. Lippard, *Inorg. Chem.*, **33**, 1510 (1994).
43. S. P. Perlepes, J. C. Huffman and G. Christou, *J. Chem. Soc., Chem. Commun.*, 1657 (1991).
44. R. C. Squire, S. M. J. Aubin, K. Folting, W. E. Streib, D. N. Hendrickson and G. Christou, *Angew Chem., Int. Ed. Engl.*, **34**, 887 (1995).
45. H. Barrow, D. A. Brown, N. W. Alcock, H. H. Clase and M. G. H. Wallbridge, *J. Chem. Soc., Chem. Commun.*, 1231 (1995).
46. H. Kumahai and S. Kitagawa, *Chem. Lett.*, 471 (1996).
47. N. V. Gerbeleu, Y. T. Struchkov, F. A. Timko, A. S. Batsanov, K. M. Indrichan and G. A. Popovich, *Dokl. Akad. Nauk. SSSR.*, **313**, 1459 (1990).
48. E. V. Gerbeleu, Y. T. Struchkov, O. S. Manole, G. A. Timko and A. S. Batsanov, *Dokl. Akad. Nauk. SSSR.*, **331**, 184 (1993).
49. K. L. Taft, C. D. Delfs, G. C. Papaefthymiou, S. Foner, D. Gatteschi and S. J. Lippard, *J. Amer. Chem. Soc.*, **116**, 824 (1994).
50. A. J. Blake, C. M. Grant, S. Parsons, J. M. Rawson and R. E. P. Winpenny, *J. Chem. Soc., Chem. Commun.*, 2363 (1994).
51. A. J. Blake, C. M. Grant, S. Parsons, J. M. Rawson, G. A. Solan and R. E. P. Winpenny, *J. Chem. Soc., Dalton Trans.*, 2311 (1995).
52. U. Bemm, R. Norrestam, M. Nygren and G. Westin, *Inorg. Chem.*, **34**, 2367 (1995).
53. E. K. Brechin, A. D. Graham, S. G. Harris, S. Parsons and R. E. P. Winpenny, *J. Chem. Soc., Dalton Trans.*, 3405 (1997).
54. S. Parsons, G. A. Solan and R. E. P. Winpenny, *J. Chem. Soc., Chem. Commun.*, 1987 (1995).
55. J. Faus, F. Lloret, M. Julve, J. M. Clemente-Juan, C. Munoz, X. Solans and M. Font-Bardia, *Angew. Chem., Int. Ed. Engl.*, **35**, 1485 (1996).
56. S. Wang, K. Folting, W. E. Streib, E. A. Schmitt, J. K. McKusker, D. N. Hendrickson and G. Christou, *Angew Chem., Int. Ed. Engl.*, **30**, 305 (1991).

Chapter 6

Rotaxanes: From Random to Transition Metal-Templated Threading of Rings at the Molecular Level

JEAN-CLAUDE CHAMBRON

Université Louis Pasteur, France

INTRODUCTION AND HISTORICAL ASPECTS

Supramolecular chemistry is concerned with assemblies of molecules held together by non-covalent forces, such as hydrogen bonds, donor–acceptor interactions between aromatic stacks, ion–ion, ion–dipole, dipole–dipole and van der Waals attractions [1, 2]. Molecules can also be gathered around a metal: this is the realm of coordination chemistry. Last, molecules can be linked together without the need for any chemical bond: the so-called physical or *mechanical* bond is found in catenanes (species formed of interlocked rings) and rotaxanes (Figure 1).

The simplest rotaxanes are molecules made from a macrocycle threaded onto a dumbbell-shaped molecule consisting of an axle bearing bulky end groups (or stoppers) to prevent macrocycle unthreading. The name 'rotaxane' was coined in 1967 by Schill and Zollenkopf [3] and supplanted the term 'hooplane' proposed the same year by Harrison and Harrison [4]. As a matter of fact, these authors described quite simultaneously two very different rotaxane syntheses.

If mechanical bonds are formed both in catenanes and rotaxanes, the former differ from the latter by their topological properties [5–7]. Whereas catenanes cannot be dissociated into the separate rings without bond breaking and reforming, rotaxanes

Transition Metals in Supramolecular Chemistry, edited by J. P. Sauvage.

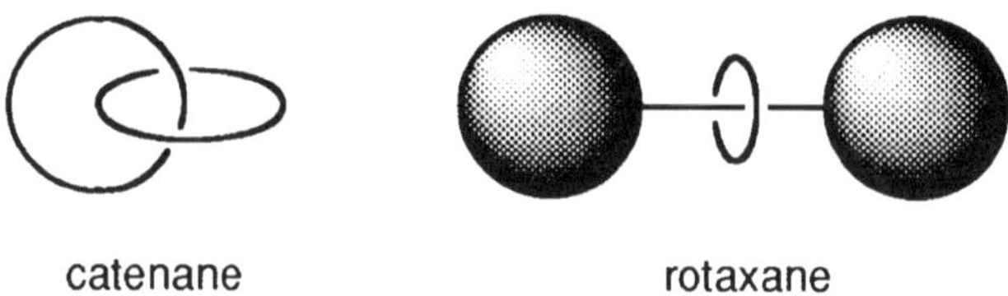

Figure 1 Catenane and rotaxane structures.

owe their existence to the larger size of the stoppers as compared to the ring diameter, that is, to steric factors. Such a difference between the physical bonding in catenanes and rotaxanes has been correctly expressed by calling 'topological bond' the mechanical bond of catenanes. The very different topological nature of catenanes and rotaxanes was established in 1961 by Frisch and Wasserman [5], who provided a definition and a sketch of what would be coined 'rotaxane' six years later.

In fact, according to a paper published in 1958 by Lüttringhaus, Cramer, Prinzbach and Henglein [8], and which describes the first attempted synthesis of a catenane by cyclization of long-chain dithiols threaded by a cyclodextrin (see Figure 2: **1** is the threaded precursor to catenane **2**), as early as 1950, Freudenberg and Cramer 'attempted to form inclusion compounds between cyclodextrin and long chain aldehydes in aqueous solution and react them with hydrazine derivatives that would play the role of a barrier, e.g. napthylhydrazine'. Their idea was that 'if the terminal aldehyde groups protruded out of the cyclodextrin and reacted with the hydrazines, compounds should have formed in which the aldehyde derivatives would no longer be able to come out, although they would not be chemically linked to the cyclodextrin.' However, this 'riveting reaction' as these authors called it, was unsuccessful. In fact, cyclodextrin-based rotaxanes [9] and catenanes [10] were to

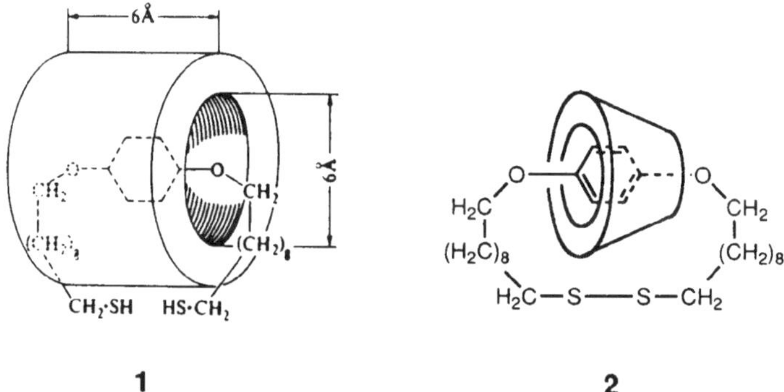

Figure 2 (a) Original drawing, taken from the paper of Lüttringhaus, Cramer, Prinzbach, and Henglein [8], of a cyclodextrin threaded onto a long chain dithiol (**1**). (b) The target catenane (**2**), which should have been obtained by oxidative coupling of the thiol functions of (**1**).

be prepared a few decades later, proving the feasibility of the ideas of these German chemists.

The importance of this 1958 paper is twofold: it shows that (i) as early as 1950 chemists were thinking of and trying to make molecules consisting of mechanically bound molecular components; (ii) advantage was taken of non-covalent molecular interactions for the purpose of synthesizing rotaxanes and catenanes.

Actually, the so-called templating technique will be used successfully not less than 25 years later, in the early 1980s, by Sauvage and Dietrich-Buchecker [11–16], who developed a transition metal-templated route to catenanes, rotaxanes and knots, and Ogino [9, 17, 18], who succeeded in preparing the first cyclodextrin-based rotaxane. In the late 1980s and early 1990s, Stoddart and coworkers introduced the use of π-donor/π-acceptor interactions between electron-rich and electron-poor aromatics for making mechanically bound molecular species [19–21], and Gibson and coworkers gave a large development to the field of polyrotaxanes [22–24]. Since 1991, the chemistry of rotaxanes literally exploded: novel methods such as the amide templates of Hunter [25] and Vögtle [26, 27] were explored, and rotaxanes were envisaged from the viewpoint of their functions, the most specific being the *controlled* motions of the ring along the dumbbell [28–31].

The very first rotaxanes were, however, not prepared via the template technique. In 1967, Schill and Zollenkopf [3] described a directed rotaxane synthesis (i.e. by the use of temporary, cleverly disposed covalent bonds), which was traced from the first directed catenane synthesis developed by Schill and Lüttringhaus in 1964 [32]. Quite simultaneously, Harrison and Harrison published the first rotaxane synthesis by statistical threading [4].

Whereas the directed synthesis was an isolated adventure, the statistical technique was subjected to scrutiny from time to time [33–39] and was finally borrowed by polymer chemists for making polyrotaxanes [22, 40, 41].

Several reviews on rotaxanes and catenanes have appeared in the literature [13–16, 20–24, 27, 42–46]. The present review, mainly devoted to synthetic aspects, will be divided into two main topics. In the first part the synthetic methodology of rotaxanes will be presented and discussed, and illustrated with selected examples from the literature. In the second part, we shall focus on the transition metal-template technique and its relevance to the preparation of rotaxanes.

PART I: SYNTHETIC METHODOLOGY OF ROTAXANES

1 Overview of the Synthetic Methodology of Rotaxanes

Rotaxanes are synthesized by two basically different methods (Figure 3): threading (or 'slippage'), and 'clipping' [47]. In the threading strategy a macrocycle is threaded onto a molecular axle at first; then the stoppers are attached to the ends of the axle. In a variant, the molecular axle is end-blocked at one extremity prior to

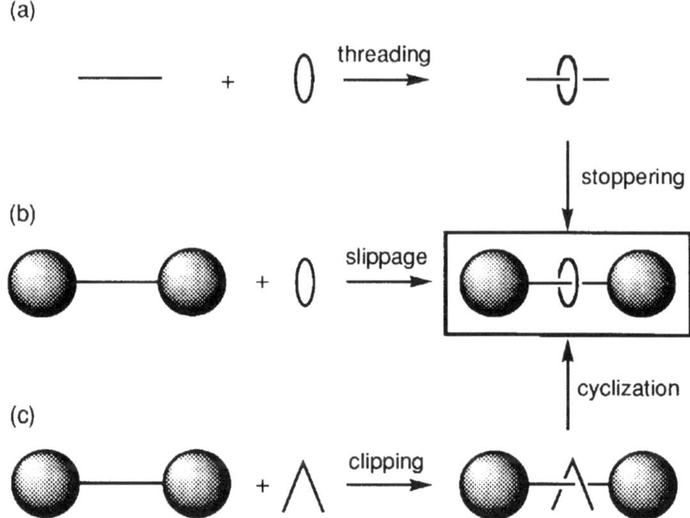

Figure 3 The three methods of rotaxane construction: (a) threading and stoppering; (b) slippage; (c) clipping and cyclization.

threading. This is particularly useful when a rotaxane with two different stoppers is wanted. The slippage method is in fact a particular case of threading, except that the ring is threaded onto the dumbbell component itself, instead of the open molecular axle. This is made possible by a transitory increase of the ring diameter. Finally, in the clipping strategy the macrocycle is constructed around the dumbbell component.

Another way to discuss the strategies outlined above is to view them as: assembling the presynthesized ring and dumbbell (slippage); or constructing the dumbbell inside the ring (threading); or constructing the ring around the dumbbell (clipping). Several of the synthetic schemes presented below were set up by Lüttringhaus and coworkers in the 1958 paper [8].

1.1 Threading and slippage

As first described by Gibson [24, 41, 48], threading a macrocycle onto a molecular axle can be represented by the equilibrium of scheme 1:

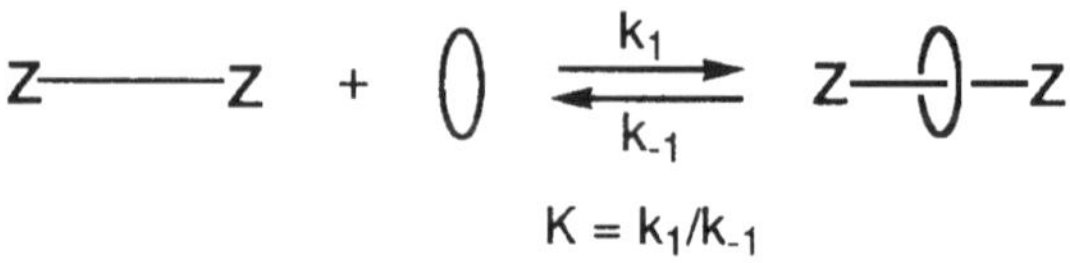

$$\text{Z—Z} + \text{O} \underset{k_{-1}}{\overset{k_1}{\rightleftharpoons}} \text{Z—O—Z}$$

$$K = k_1/k_{-1}$$

Scheme 1

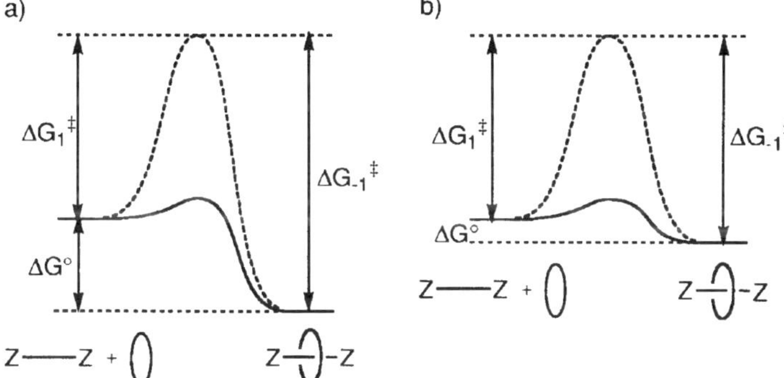

Figure 4 Thermodynamic profile of the reaction equilibrium in the case of threading (full line) and slippage (dashed line). (a) Strongly negative ΔG°; (b) weakly negative ΔG°.

In this scheme and the following, Z is a precursor to the stoppers (case of threading *stricto sensu*), or the stoppers themselves (case of slippage). The equilibrium of scheme 1 can be described by the following equations:

$$\ln K = -\Delta G^{\circ}/RT = -\Delta H^{\circ}/RT + \Delta S^{\circ}/R \tag{1}$$

$$k_1 = A_1 \exp(-\Delta G_1{}^{\ddagger}/RT) \tag{2}$$

$$k_{-1} = A_{-1} \exp(-\Delta G_{-1}{}^{\ddagger}/RT) \tag{3}$$

$$K = k_1/k_{-1} \tag{4}$$

The relevant thermodynamic and kinetic parameters are displayed diagrammatically in Figure 4. From the point of view of kinetics, $\Delta G^{\ddagger}$ is significant only in the case of slippage. From the point of view of thermodynamics, if $\Delta H^{\circ} \approx 0$, the threading is controlled by entropy: this is the case of the so-called 'statistical' threading. If $\Delta H^{\circ} < 0$, the threading is controlled by enthalpy: the threaded species is thermodynamically much more stable than the separated components. In this latter case, the equilibrium of scheme 1 can be rewritten as shown in scheme 2:

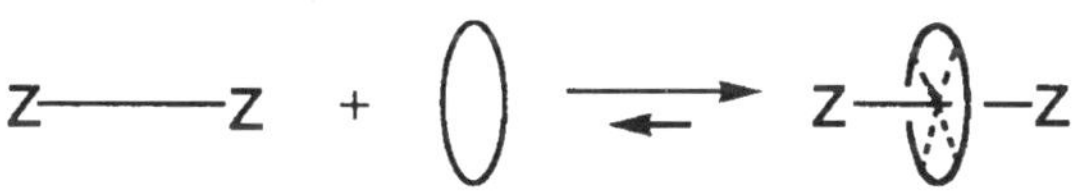

Z——Z + ⇌ Z–Z

Scheme 2

The dotted lines indicate that some kind of non-covalent bond between the molecular thread and the macrocycle is established. As shown in the examples developed later on, these interactions are either hydrophobic in nature, or π-donor/π-acceptor interactions between aromatic nuclei, or hydrogen bonds or metal-ligand coordination bonds. Their effect is to shift the equilibrium of scheme 2 towards the formation of the threaded species, which can thus be obtained quantitatively from the separated

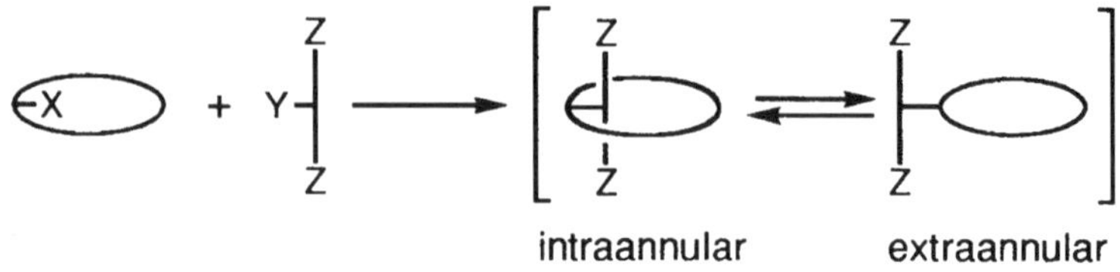

intraannular extraannular

Scheme 3

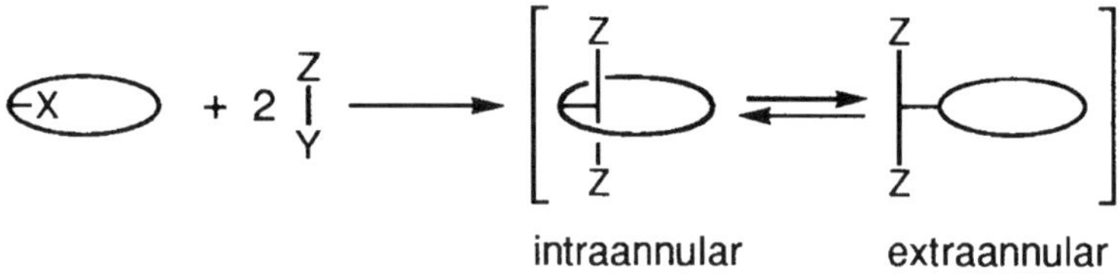

intraannular extraannular

Scheme 4

components in many cases, and be handled without unthreading. The threaded complex is sometimes called 'pseudorotaxane' [49–51]. It is nothing but a host–guest complex, the macrocycle acting as a receptor for the thread component [21].

In another, very different approach, it was imagined that the threaded species could be formed by the means of covalent bonds, either by reaction of a functionalized macrocycle with the appropriate functionalized molecular thread (scheme 3) or by construction of the molecular thread inside the functionalized macrocycle (scheme 4) [8]. In these schemes and the following, X and Y represent complementary reactive functions.

This technique, called 'semi statistical' [52], was faced with the problem of the coexistence of two conformers, one with the thread inside the macrocycle (intraannular conformer) and the other with the thread outside the macrocycle (extraannular conformer). The problem could be overcome, in principle, by a strict control of the stereochemistry of the macrocycle. The so-called 'directed synthesis' (see below) solved elegantly this problem.

1.2 Clipping

In the clipping method the macrocycle is constructed around the linear or dumbbell component. Obviously the precursor to the macrocycle must be bound to the linear component prior to the cyclization step. This may be achieved either covalently (schemes 5 and 6) or non-covalently (scheme 7).

In the first case, since the linear component is anchored twice to the macrocycle, the unwanted equilibrium between intra- and extraannular conformations should be avoided. However, if again the stereochemistry of the precursor of scheme 5 is not

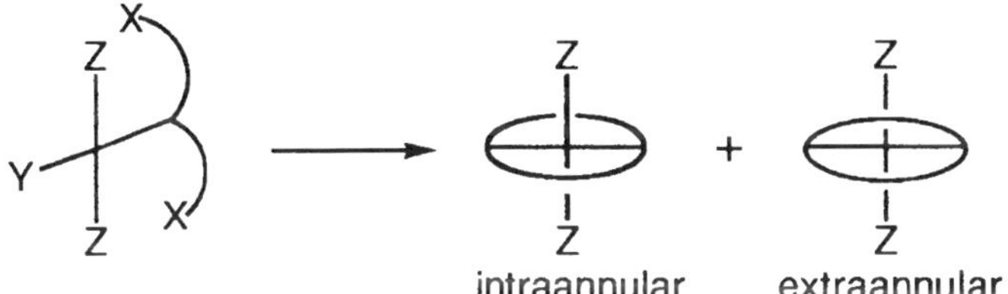

Scheme 5

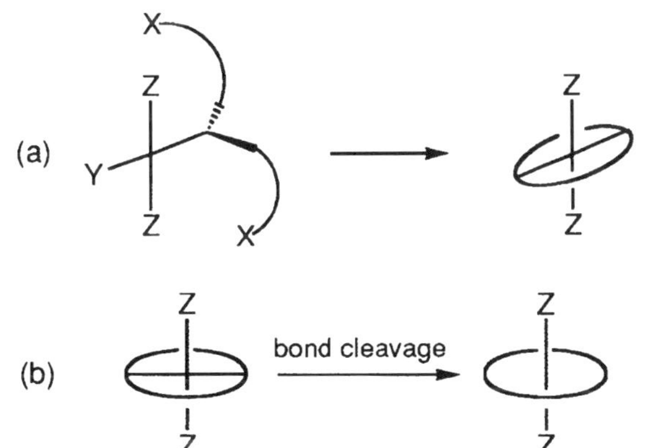

Scheme 6

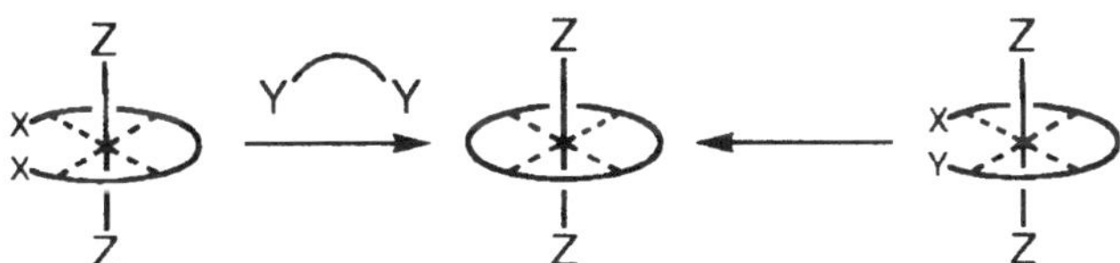

Scheme 7

strictly controlled, the possibility arises that the macrocycle does not form around the linear component. The directed synthesis, shown in scheme 6, circumvents this possibility by maintaining rigidly at right angles the precursor to the macrocycle and the linear component [3, 32]. The rotaxane itself is obtained by cleaving the covalent bonds linking the macrocycle and the dumbbell.

In the second case, non-covalent interactions are used to create a complex between the precursor to the macrocycle and the dumbbell component (scheme 7). Therefore the latter templates the macrocyclization reaction which may occur either intramolecularly, if the precursor bears two complementary functions, or intermolecularly if a third molecular fragment (i.e. Y–Y) is involved.

As underlined in the introduction to this chapter, both rotaxanes and catenanes involve mechanical bonds. Obviously, they are also related molecules in terms of

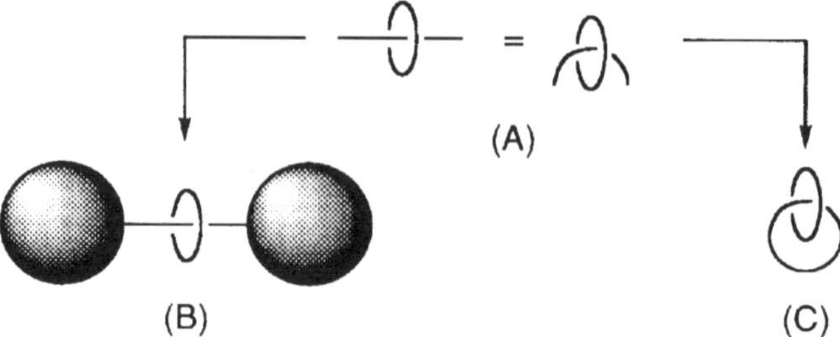

Figure 5 A threaded species (A) is a common precursor to a rotaxane (B) or a catenane (C).

synthetic methodology, since they can be viewed as arising from a common precursor, the threaded complex (Figure 5). The rotaxane is obtained by a stoppering reaction, whereas the catenane is obtained by a cyclization reaction. In most cases, a method developed for making rotaxanes can be successfully transposed to the case of catenanes and vice versa. Actually, no catenane, with one exception, was synthesized by a statistical route [53]. However, rotaxanes prepared by the statistical method were used as stable intermediates for the synthesis of interlocked species [36, 54, 55]. Lastly, since catenanes were always considered as more fascinating synthetic targets, the synthetic methods were generally developed at first for making catenanes and then extended to the case of rotaxanes.

2 Nomenclature, Morphologies and Constitution of Rotaxanes

Until now, we considered the simplest rotaxane morphology, as defined in the introduction and represented in Figure 1. This molecular system is noted as [2]-

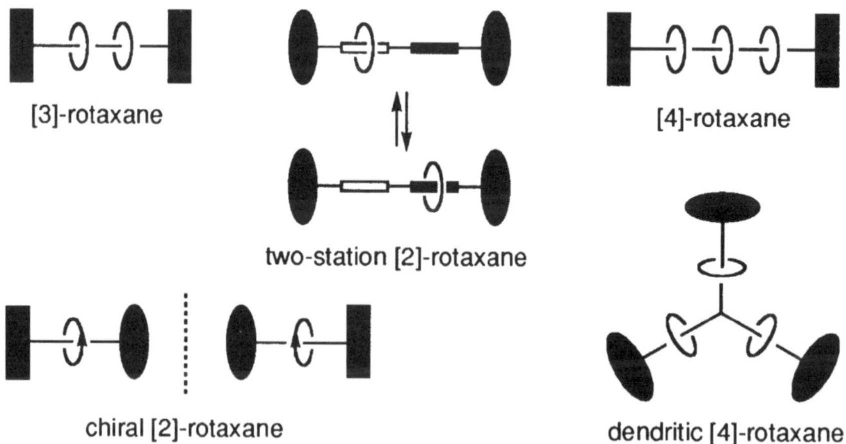

Figure 6 Selected rotaxane morphologies.

rotaxane to express that it is made up with two components: a ring and a dumbbell [42]. More generally, an [*n*]-rotaxane contains $n - 1$ rings threaded onto a dumbbell component. Figure 6 shows several rotaxane morphologies that are relevant to the present study, and that have been realized.

Two higher homologues of the [2]-rotaxane are represented: the [3]-rotaxane and the [4]-rotaxane. Whereas [3]-rotaxanes are relatively common [56–60], this is not the case for the linear [4]-rotaxane [61] and its branched or dendritic analogue [62]. Very important in the recent developments of rotaxane chemistry is the [2]-rotaxane whose dumbbell contains two separated sites for interaction of the threaded macrocycle. In this molecule, the macrocycle may be moved between two sites of the dumbbell, in a degenerate manner, if the sites are identical [63–65], or in a controlled manner, thanks to an external trigger, if the sites are different [28–31]. A chiral [2]-rotaxane, whose chirality arises from the mechanical bonding, is obtained if the macrocycle is oriented (by an appropriate substitution pattern) and the dumbbell made non-symmetrical, for example by attaching two different stoppers [66].

The chemical constitution of rotaxanes will depend, of course, on the methodology used for their elaboration. The statistical synthesis is not demanding in terms of functionality. On the contrary, template methods require that the components coming in interaction (e.g. the thread and the macrocycle in the case of the threading route) contain all the necessary complementary chemical information: for example chelates subunits or functional groups able to form hydrogen bonds. Cyclodextrins, cationic, or amide-based cyclophanes or electron-rich benzo crown ethers or even simple crown ethers are extensively used macrocycles. The most common stoppers are derived from the triphenylmethane (trityl) group. Metal complex fragments have also been used. The last sections of this chapter will focus on the implementation and use of porphyrinic stoppers. As developed farther, the relative size of the components must be carefully controlled: the cross-section of the macrocycle must be large enough to allow a free motion along the thread and small enough to ensure efficient stoppering [24]. The thread must protrude sufficiently from the rims of the macrocycle to prevent steric hindrance of the stoppering reaction [17].

3 Threading and Slippage

3.1 Statistical threading

One of the earliest rotaxane syntheses involved statistical threading. Harrison and Harrison [4] coupled the 30-membered macrocycle (**3**) bearing a pendent carboxylic group to a Merrifield's peptide resin, forming the resin adduct (**4**) (Figure 7). Next, a column was charged with this modified resin and treated with a solution of decane-1, 10-diol (**5**) and triphenylmethyl chloride (**6**) in a mixture of pyridine, dimethylformamide and toluene. The process was repeated 70 times. After washing the column,

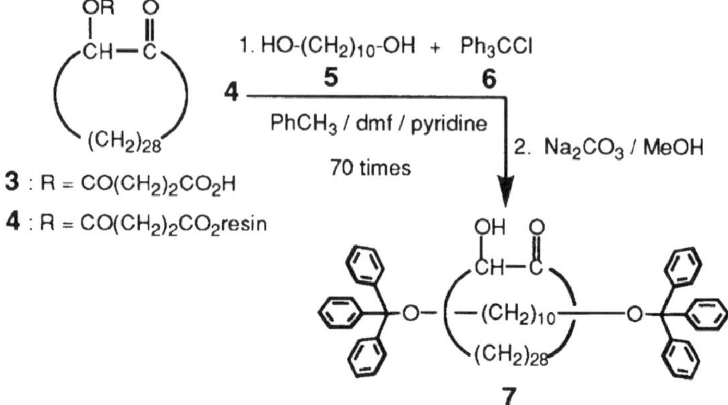

Figure 7 Statistical threading. Preparation of [2]-rotaxane (**7**) [4].

treatment of the resin with sodium carbonate liberated the macrocycle-containing species out of which rotaxane **7** was isolated by chromatography in 6% yield. Its structure was supported by chemical degradation, i.e. selective cleavage of either macrocycle or stoppers, depending on the reagent used. In summary, a stable rotaxane could be obtained with a 30-membered macrocycle and a decamethylene chain end-blocked with triphenylmethyl (trityl) groups.

Harrison [33, 34], and later, Schill and coworkers [38] refined this result. They studied threading of methylene rings onto open methylene chains or temperature-controlled slippage of the same rings onto methylene chains end-blocked by trityl groups. Their conclusions are the following: the limiting size for a methylene ring to be threaded onto a methylene chain (diameter: 4.5 Å) is 21–22 atoms. Rotaxanes containing 30-membered rings or more are not stable with respect to dethreading from a trityl-blocked dumbbell. The length of the methylene chain is also important: the longer the chain, the better the threading. Schill and coworkers [38] established the optimum conditions for slippage and could synthesize in 11% yield 150 mg of rotaxane (**10**). As shown in Figure 8, macrocycle (**9**) was forced to thread onto the dumbbell-shaped linear component (**8**) by heating a solvent-free mixture of the compounds. Upon cooling of the reaction mixture the [2]-rotaxane species (**10**) was trapped and could thus be isolated.

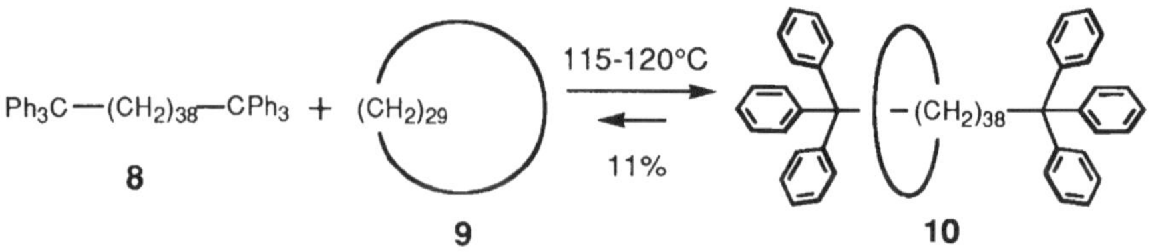

$Ph_3C—(CH_2)_{38}—CPh_3$ (**8**) + $(CH_2)_{29}$ (**9**) ⇌ (115-120°C, 11%) $—(CH_2)_{38}—$ (**10**)

Figure 8 Statistical threading (temperature-controlled slippage). Preparation of [2]-rotaxane (**10**) [38].

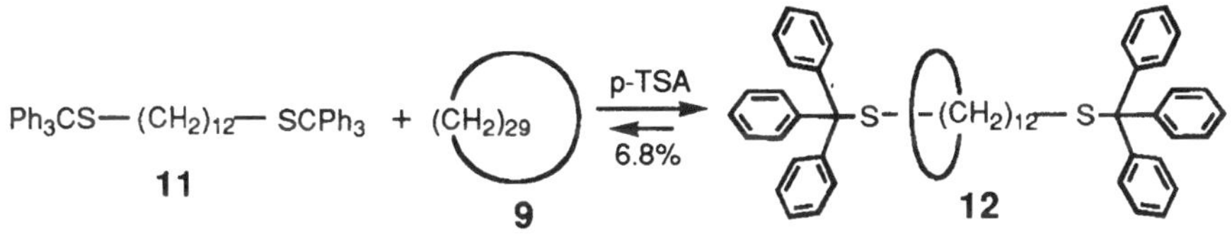

$Ph_3CS—(CH_2)_{12}—SCPh_3$ + $(CH_2)_{29}$ ⇌ (p-TSA, 6.8%) S–$(CH_2)_{12}$–S

11 9 12

Figure 9 Statistical threading and stoppering *in situ*. Preparation of [2]-rotaxane (**12**) [38].

Threading onto open methylene chains was examined using stoppers cleavable *in situ*, e.g. trityl ether or thioether end groups (Figure 9) [38]. In acidic medium, the thioether linkages of the trityl stoppers of molecular dumbbell (**11**) are cleaved, allowing statistical threading of macrocycle (**9**) onto the open dodecamethylene molecular thread. Quenching of the acid by a base traps the threaded species as rotaxane (**12**) by allowing the thioether bonds of the stoppers to be restored.

This threading and stoppering technique was used by Schill and coworkers for a very elegant synthesis of a 'hydrocarbon' catenane [54, 55]. The intermediate rotaxane was prepared in gram scale amounts, exemplifying the preparative value of the statistical method.

The results presented above were obtained for 'hydrocarbon' rotaxanes, since the thread and the ring component were made from oligomethylene chains. Since the discovery of their complexing ability for alkali metal cations [67–69], crown ethers form an extensively used class of macrocyclic compounds. The introduction of crown ethers in rotaxane chemistry was due to Zilkha and coworkers [35, 36]. These authors studied the threading of dibenzo crown ethers onto polyethyleneglycol threads. They observed essentially that the larger the ring, the greater the extent of threading, and that each ring size had an optimal chain length for maximum threading (i.e. 2.5–3 times the ring diameter). The maximum extent of threading was 63%. These results match qualitatively those obtained for 'hydrocarbon' rotaxanes.

Next, Zilkha and coworkers prepared a rotaxane from a *c*. 44-membered dibenzo crown ether and PEG 400 (containing *c*. 9 ethyleneoxy units) using trityl groups as stoppers: after equilibration of the equimolecular amounts of macrocycle and PEG 400 at 120°C during 30 min, solid Ph_3CCl and pyridine were added. The rotaxane was isolated in 15% yield. It was stable with respect to dethreading, despite the fact that the diameter of the extended ring was about twice that of the effective diameter of the trityl group: at 150°C, only 52% of the rotaxane had decomposed. To account for this unexpected stability, the authors invoked dipole–dipole interactions between the ring and the chain. Commenting on this work and the high yields of ring threading observed, Gibson argued about the fact that threading was probably not purely statistical, and suggested that it involved enthalpic contributions, for example hydrogen bonding between the hydroxyl end groups of the open threads and the oxygen atoms of the polyether macrocycles [24].

Gibson and his coworkers made an extensive use of this ‘pseudo-statistical’ threading for preparing polyrotaxanes [23, 24]. Polyrotaxanes are usually not prepared by threading macrocycles onto preformed polymers, even of low molecular weights, because of limited chain-end concentrations and slow diffusion processes of the rings onto the chains. Rather, the monomeric, open thread is mixed with the macrocycle and once the equilibrium is reached, the polymerization reaction is initiated. Le Châtelier’s principle is used as a golden rule to shift the equilibrium towards the formation of the threaded species. Therefore, to achieve a high concentration of reactants, the macrocycles are used as solvents. An example is given in Figure 10 [41]. At first decane-1,10-diol (**5**) was allowed to equilibrate with macrocycle 30-crown-10 (**13**) at 90°C. Then sebacoyl dichloride (**14**) was added to the reaction mixture, and the polymerization allowed to go for 40 h. Finally, it was quenched by addition of Ph_3CCH_2COCl (**15**), which reacted with the free ends of the polymer chains to form the stoppers of the polyrotaxane (**16**). The extent of macrocycle incorporation was found to be nearly seven per repeating unit.

The observations made by Zilkha and coworkers were confirmed and supplemented, in particular the major role played by the ring size in the threading efficiency. For example, crown ethers of various sizes were allowed to thread onto tetraethylene glycol by heating the mixture at 90°C [48]. Subsequently, bis(p-isocyanatophenyl)methane was added and the polymerization reaction was conducted for 24 h. The equilibrium constants of the threading step were determined according to the macrocycle content of the polydiisocyanate obtained thereof. Log *K* was shown to increase linearly with ring size (from 36-crown-12 to 60-crown-20). This correlation allowed to extrapolate that *K* approached zero at a ring size of *c.* 24 atoms, which is consistent with the experimental findings of Harrison and Schill for oligomethylene analogues.

Importantly, it was observed that in the case of polymers lacking end-blocking groups, dethreading of macrocycles was not significant. This result was interpreted in terms of chain-macrocycle entanglements in the polyrotaxane, the macrocycle being trapped in random coils of the backbone [40].

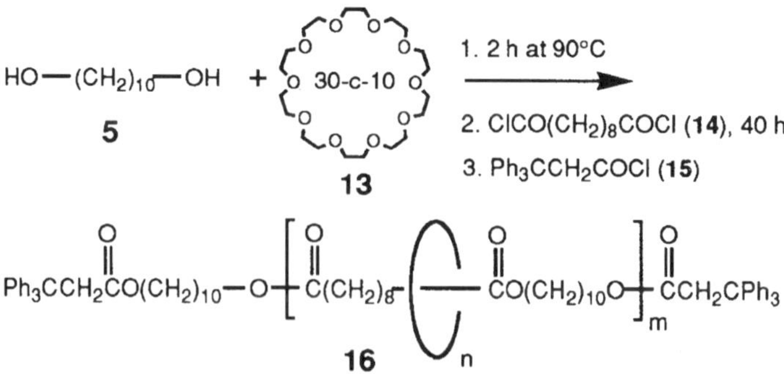

Figure 10 Preparation of polyester rotaxane (**16**) [41].

3.2 Threading by non-covalent interactions

Hydrophobic interactions Cyclodextrins (sometimes abbreviated CDX) are cyclic oligosaccharides constructed from glucose units (Figure 11) [70]. The most common are the hexa-, hepta- and octamer species, called respectively, α-, β- and γ-cyclodextrins. In 1948, Freudenberg and Cramer discovered that cyclodextrins could form inclusion compounds [71]. The formation of these host–guest complexes was studied systematically by Cramer and coworkers [72].

Cyclodextrins are hollow, cone-shaped molecules. Their length is 7.9 Å and their diameter is 5.7 Å (α-CDX), 7.8 Å (β-CDX) and 9.5 Å (γ-CDX). The primary hydroxyl groups lie on the wide rim and the secondary hydroxyl groups are located on the narrow rim. The interior is a non-polar, hydrophobic cavity. Therefore cyclodextrins normally show their complexing properties for hydrocarbon molecules in very polar solvents, such as water, or dmso.

As disclosed in the introduction, Freudenberg and Cramer realized quite early that the inclusion compounds of cyclodextrins with guests long enough to extend beyond the host could be, in principle, 'locked' with bulky end groups. However, in the late 1950s, Lüttringhaus, Cramer and coworkers [8, 73] were probably more interested in making catenanes, but were unsuccessful, and it is only many years later, in 1981, that Ogino eventually prepared the first cyclodextrin-based rotaxane [9]. The idea was to use kinetically inert cobalt(III) complex fragments as stoppers. Therefore, the molecular thread had to be functionalized with coordinating end groups. One of the synthetic routes developed is depicted in Figure 12.

Diamino-1,12-dodecane (**17**$_{12}$) was first allowed to equilibrate with β-CDX in dmso, allowing for the formation of the threaded species (**18**$_{12}$). Two equivalents of $[CoCl_2en]^+$ were then added to the reaction mixture. The desired [2]-rotaxane (**19**$_{12}$) could be isolated in *c.* 7% yield after chromatography on Sephadex. Next, Ogino and

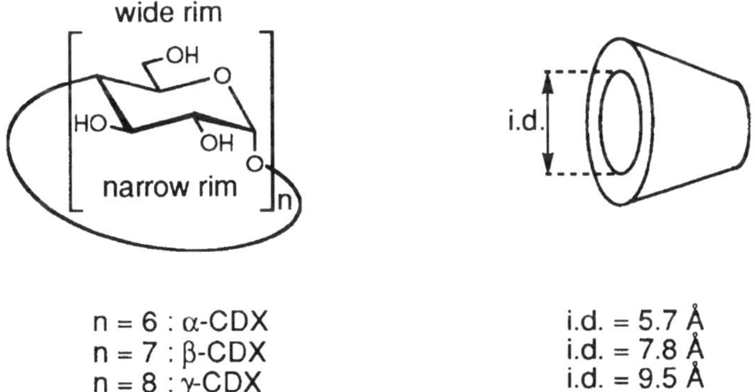

Figure 11 Chemical structures and molecular dimensions of the most common cyclodextrins, abbreviated α, β, and γ-CDX.

ring	n	yield of $\mathbf{16_n}$ (%)
α-CDX	10	5.7
	12	19
	14	12
β-CDX	10	2.4
	12	6.7
	14	1.9

Figure 12 Threading by hydrophobic interactions. Preparation of the cyclodextrin-based [2]-rotaxanes ($\mathbf{19_n}$) [9, 17].

Ohata showed that the yield of rotaxane formation could be increased up to 19% when the smaller α-CDX was used [17]. Employing shorter (C_8 and C_{10}) or longer (C_{14}) methylene chains lowered the yields: stoppering was hampered for steric reasons. When long chains were used, steric crowding was the consequence of threading of more than one CDX unit.

Working on similar rotaxanes, Yamanari and Shimura [74] studied the stereoselectivity of the stoppering reaction induced by the natural chirality of the cyclodextrin. When α-CDX and a racemic mixture of Co(III) precursor complexes were used, the (Δ,Δ) configuration of the stoppers was predominant. Isnin and Kaifer [75] were also interested in stereochemical issues. They threaded α-CDX onto alkyl chains closed at one end by a ferrocenyl group and stoppered the other extremity with a different stopper, a naphtalene sulphonate group. Since the cyclodextrin is cone shaped, two CDX-positional rotaxane isomers were obtained, as a nearly equimolecular mixture. Another remarkable example of a cyclodextrin-based rotaxane whose end groups are metal complex fragments is the cobalamin-stoppered rotaxane described by Kraütler and coworkers [76]. Comparison of the rotaxane and the isolated dumbbell component in aqueous solution, shows that the dodecamethylene chain of the thread is extended in the former and has collapsed in the latter, due to hydrophobic effects.

Wenz and coworkers [77, 78], demonstrated that it was possible to thread lipophilic β-CDXs onto cationic linear components blocked at one extremity with a trityl group, in organic solvents. Remarkably, the resulting rotaxanes, obtained in 20–37% yields, were generated from prerotaxanes in which the CDX rings had threaded with the same orientation: the one insuring the less steric crowding to the free end of the thread.

Aromatic and conjugated rods were utilized by Anderson and Anderson. Threading α-CDX or β-CDX onto the *para*-diazonium salt of azo-benzene and stoppering by diazonium coupling with β-hydroxynaphtalene produced the corresponding rotaxanes in 12–15% yield [79]. The dumbbell unit of these rotaxanes is a dye: rotaxane formation irreversibly encapsulates the dye which is also protected inside the cavity of the CDX. The same authors also introduced water-soluble cyclophanes in the field of rotaxanes, as macrocycles displaying complexing properties similar to those of cyclodextrins [58]. This improved substantially the yields of rotaxane formation. For example, the rotaxane based on the azo-dumbbell mentioned above was obtained in 46% yield. A [3]-rotaxane (**24**) imprisoning two molecules of cyclophane (**20**) (Figure 13) could be prepared in 21% yield by Glaser coupling of phenylacetylene-based rods (e.g. (**21**)), terminated at one end with *bis*-N-methylpyridinium carbinol stoppers. The [2]-rotaxane homologue (**23**) and the free dumbbell component were also isolated from the reaction mixture, in 27% and 9% yields, respectively.

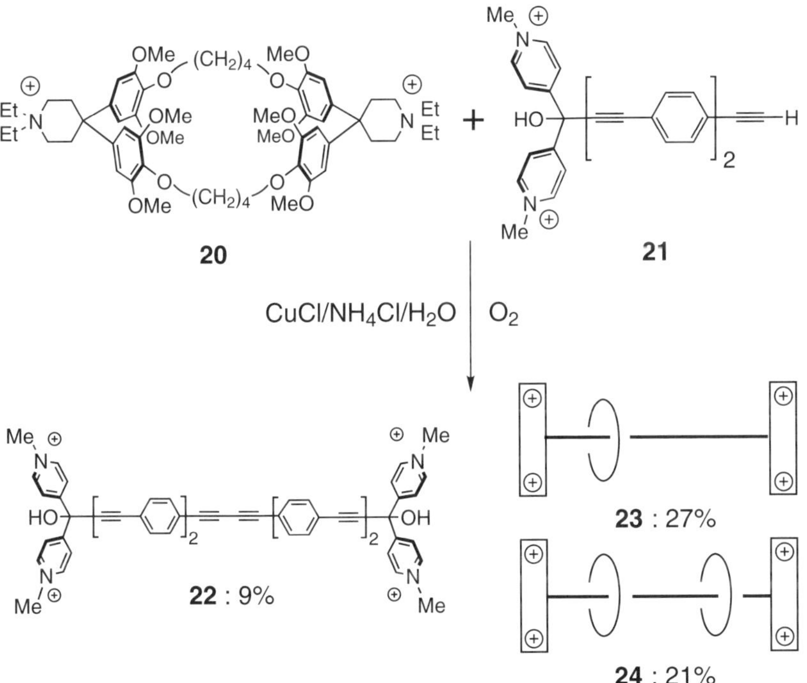

Figure 13 Threading by hydrophobic interactions. Preparation of [2]- and [3]-rotaxanes (**23**) and (**24**) [58].

Cyclodextrins were also threaded onto *polymeric* threads. Wenz and Keller used poly(methylene) chains [80]. Equilibration of α-CDX with poly(iminotrimethylene-iminodecamethylene) was reached in *c.* 9 days. The dethreading by dialysis was not even complete after two weeks. To analyse the degree of threading, the CDX rings were blocked on the chains by partial acylation of the amino groups of the polymer. Consequently, a material containing 23 $-NH(CH_2)_{10}-N(H)-(CH_2)_3-$ subunits was shown to entrap 37 CDX rings.

Harada and coworkers studied the threading of cyclodextrins onto long-chain poly(ethyleneglycol) and analogues: poly(propyleneglycol), poly(methylvinylether), etc. The polymers were selectively threaded: poly(ethyleneglycol), only by α-CDX [81] and poly(propyleneglycol), only by *β*-CDX [82]. The γ-CDX forms threaded complexes with poly(methylvinylether) [83]. A polyrotaxane was prepared from a poly(ethyleneglycol) polymer bearing amine end groups [84]. The stoppering reaction involved coupling with 2,4-dinitrofluorobenzene in dmf at room temperature, and the polyrotaxane was obtained in 60% yield. Spectroscopic experiments indicated that *c.* 20–23 α-CDX rings had threaded onto the poly(ethyleneglycol) chain. Consecutive CDXs were shown to be linked by hydrogen bonds, suggesting an alternation of the cone-shaped rings. This tubular assembly was stabilized by covalently linking the CDXs [85]. After cleavage of the stoppers and extrusion of the poly(ethyleneglycol) backbone, a molecular tube constructed from linked cyclodextrins, was produced. Therefore, a poly(ethyleneglycol) chain was used as a template to generate a tubular cyclodextrin polymer.

Aromatic donor-acceptor interactions Stoddart and coworkers discovered that electron-rich *bis*-p-phenylene-34-crown-10 (**25**, Figure 14) formed quite a strong complex with the bis(hexafluorophosphate) salt of electron-poor 4,4′-dimethylbipyridinium (**26**) in acetone and acetonitrile solutions [86]. Conversely, the electron-deficient macrocycle cyclobis(paraquat-p-phenylene) (**27**) was shown to be an ideal receptor for a wide range of aromatic π-electron rich substrates such as 1,4-dimethoxybenzene (**28**) [87]. The crystal structure of this complex shows that

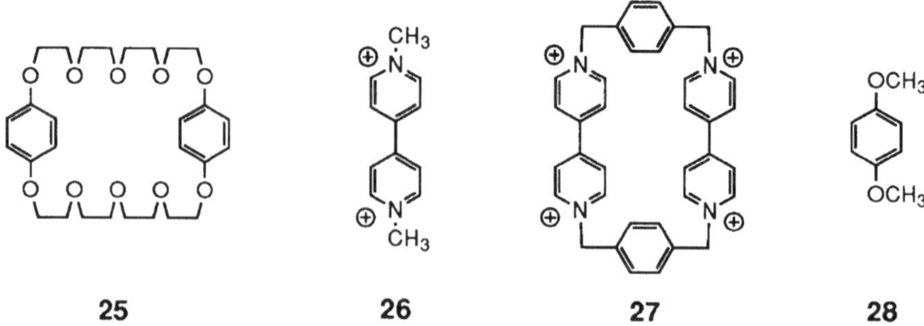

Figure 14 π-electron rich (**25**) and π-electron poor (**27**) macrocycles acting as receptors for the electronically complementary guests (**26**) and (**28**), respectively [86, 87].

the methoxy groups of the guest protrude out of the rims of the tetracationic cyclophane. The threading effect is emphasized when longer guests are utilized. The complex is stabilized by π-donor/π-acceptor interactions between the aromatic rings, as well as H-bonding interactions between the benzylic acid H atoms and the central O atoms of the polyether chains of the macrocycle. These results from host–guest chemistry led to the development of a new family of catenanes and rotaxanes [21]. Rotaxanes were synthesized by threading and stoppering (or slippage), or by clipping. The latter procedure will be discussed in a separate section.

With few exceptions [88–90] rotaxanes involving the tetracationic cyclophane (**27**) were usually prepared by the clipping method [63, 88, 90–97]. The threading and slippage methods were preferably involved for making rotaxanes with 4,4′-bipyridinium-containing threads and electron-rich macrocycles [60–62, 64, 65]. A [3]-rotaxane containing two different rings was synthesized by successive threading, stoppering, and slippage steps [60, 98]. As shown in Figure 15, the functionalized stopper (**29**) and linear component (**30**) were at first allowed to react in the presence of the smaller macrocycle (**25**) in dmf at room temperature, but under 12 kbar pressure. The intermediate two-stations [2]-rotaxane (**31**) was obtained in 19% yield. Next, the larger macrocycle (**32**) was allowed to slip over the triarylmethane stoppers. The reaction was performed in MeCN, at higher temperature than for the threading method (55°C). The [3]-rotaxane (**33**) was obtained in 49% yield. In separate experiments, using bis-aryl crown ethers of variable size, the authors showed that the efficiency of slippage could be controlled by the nature of the R-substituent on the stopper.

Harrison [4, 33, 34] and later, Schill and coworkers [38] had explored temperature-controlled slippage. Another approach was to control the size of the stoppers by complexation. Stoddart and coworkers [90] could thread macrocycle (**27**) onto dumbbells ended by aza crown ethers. The slippage process was inhibited by addition of an alkali cation. Conversely, complexation of the aza-crown after the slippage process inhibited unthreading.

The threading and slippage methodologies were particularly useful for the preparation of the higher order [3]- and [4]-rotaxanes. For example, in the latter case, the reaction produced a mixture of [4]-rotaxane, three-station [3]-rotaxane, and three-station [2]-rotaxane [61]. The yields of the various rotaxanes could be fortunately controlled by the molar ratios of the reactants.

Hydrogen-bonding in ionic rotaxanes In spite of the fact that complexes of ammonium ions with crown ethers have been studied for quite a long time, only very recently was this property of crown ethers used for preparing rotaxanes [99–101]. Actually there were very few reports on the complexation of *dialkyl* ammonium cations and, when the macrocycles used as hosts are small, only face-to-face binding is observed (Figure 16). Increasing the size of the ring could lead to the formation of a threaded species. Accordingly, Stoddart and coworkers [56] took advantage of ionic hydrogen bonds, for making [2]-rotaxane (**37**). After threading of crown ether macrocycle (**34**) onto the secondary ammonium cation-containing

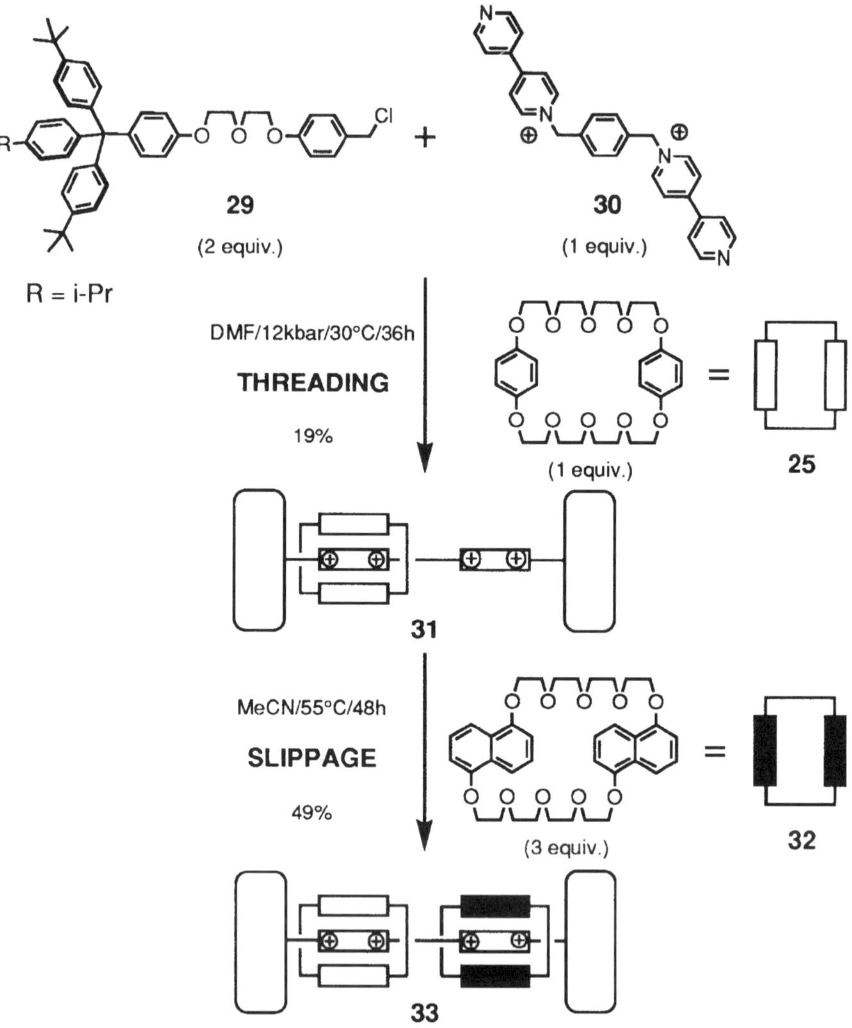

Figure 15 Threading and slippage controlled by π-donor/π-acceptor interactions. Preparation of [3]-rotaxane (**33**) containing two different rings [60, 98].

molecular thread (**35**), the stoppers were constructed *in situ*, by formation of triazole rings from the triaza functions anchored at the extremities of the thread (**35**) and the alkyne di-tert-butylcarboxylate (**36**) (Figure 17). The [2]-rotaxane (**37**) was obtained in 31% yield. The same methodology, adapted to a bis-benzylammonium-containing molecular thread produced a [3]-rotaxane in 10% yield. As shown by a crystal structure analysis, the threaded species are stabilized by NH···O and CH···O hydrogen bonds.

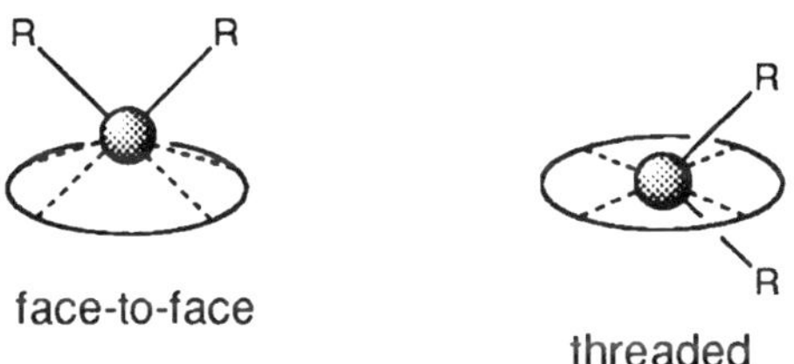

Figure 16 Schematic views of the interaction of a macrocycle with a secondary ammonium cation (sphere). The dashed lines represent hydrogen bonds. Face-to-face (left) and threaded (right) modes of interaction [99–101].

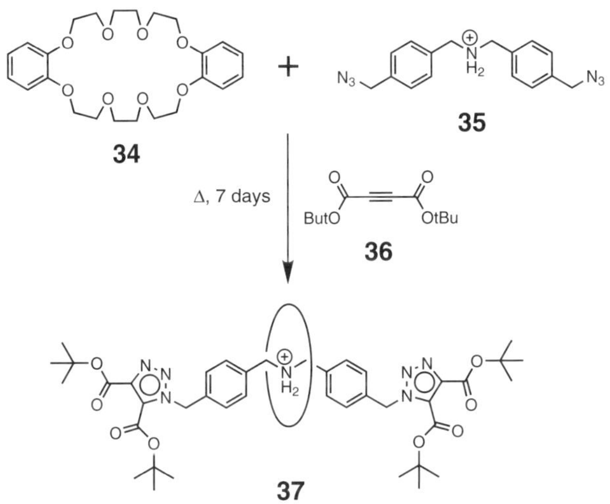

Figure 17 Threading by formation of ionic H bonds. Preparation of [2]-rotaxane (**37**) [56].

Hydrogen-bonding in neutral rotaxanes The independent discovery in 1992 by Hunter [25], and Vögtle and coworkers [26, 102] of catenanes made of interlocked amide cyclophanes led to the development of a novel family of neutral catenanes and rotaxanes. In an example of rotaxane synthesis [103] macrocycle (**38**) and isophthaloyl dichloride (**39**) were mixed in dichloromethane, and the trityl amine (**40**) added to the reaction mixture (Figure 18). The rotaxane (**41**) was obtained in 11% yield. A hypothetical mechanism suggests that the isophthaloyl dichloride (**39**) is incorporated orthogonally into the host macrocycle, owing to van der Waals interactions, $\pi-\pi$ interactions and NH $\cdots$ O hydrogen bonds. Various aromatic acid dichlorides or sulphonyl dichlorides were employed as linear components [104]. The stoppers were usually trityl groups, but porphyrinic units were also introduced [105].

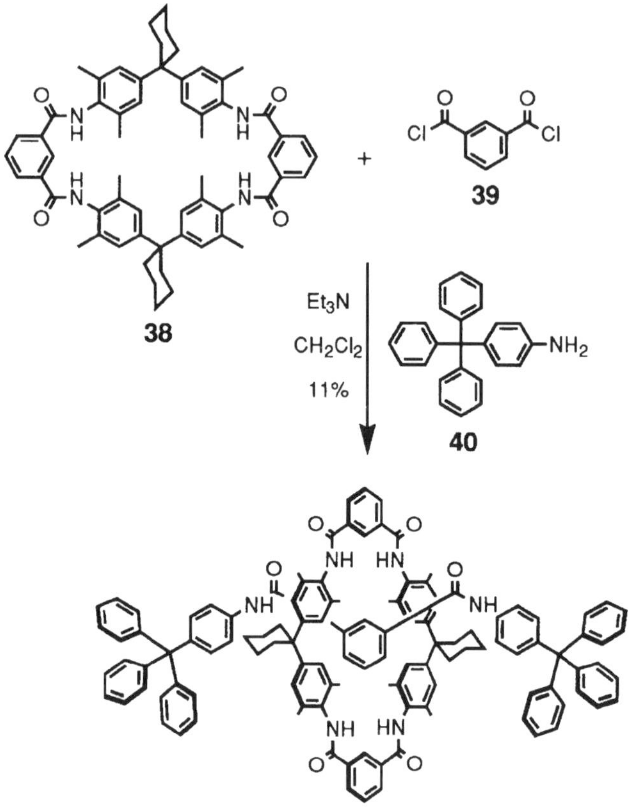

Figure 18 Threading by formation of neutral H bonds. Preparation of [2]-rotaxane (**41**) [103].

The best yield obtained for a [2]-rotaxane was 41%, in the case of a mixed sulphonamide/amide spacer [104]. Combining this non-symmetrical spacer and a macrocycle oriented by alternating sulphonamide and amide linkages, allowed the preparation of chiral rotaxanes, which could be resolved into the separated enantiomers by HPLC on optically active stationary phases [66].

This method also allowed for the preparation of higher order rotaxanes. For example, a [3]-rotaxane could be prepared, in 2% yield, however [57].

Slippage experiments involved extremely hard reaction conditions: the macrocycle was melted at 350°C in the presence of the dumbbell component during very short periods of time (10 min). A [2]-rotaxane was obtained in 3% yield by this method and was shown to be stable at room temperature in dmf solution [106].

3.3 Threading by covalent bond formation

This method was never used for making rotaxanes but it was developed, essentially by Lüttringhaus, Schill, and coworkers, with the aim of synthesizing catenanes. The idea was presented in the 1958 paper [8]. As reported by Schill [42], Lüttringhaus and Schill reacted the macrocyclic ketone (**42**) and the diol (**43**) with the hope of obtaining the threaded species (**45**) by formation of a ketal (Figure 19). The attempt was somewhat risky, because of the possibility of forming the extraannular ketal (**44**) owing to the conformational flexibility of the macrocycle. After a series of chemical transformations, the thread was cyclized and the ketal hydrolyzed. Only macrocycles were obtained, proving that the catenane, and therefore the threaded species (**45**) had not formed. Rigidification of the macrocycle to favourably influence the formation of an intraannular conformation between a ring and a chain attached to it unfortunately did not solve the problem.

Another approach, developed by Lüttringhaus and Isele [52], consisted in constructing the thread inside the ring and was more successful. Accordingly, the inner, primary amine of macrocycle (**46**) was subjected twice to reaction with the long-chain acid chloride derivative (**47**), followed by reduction of the resulting amide to the corresponding amine (Figure 20). Of course, there was still the possibility of forming the threaded (**48**) and unthreaded (**49**) conformers. After a series of chemical transformations, cyclization of the attached thread, followed by the reductive acetylation cleavage

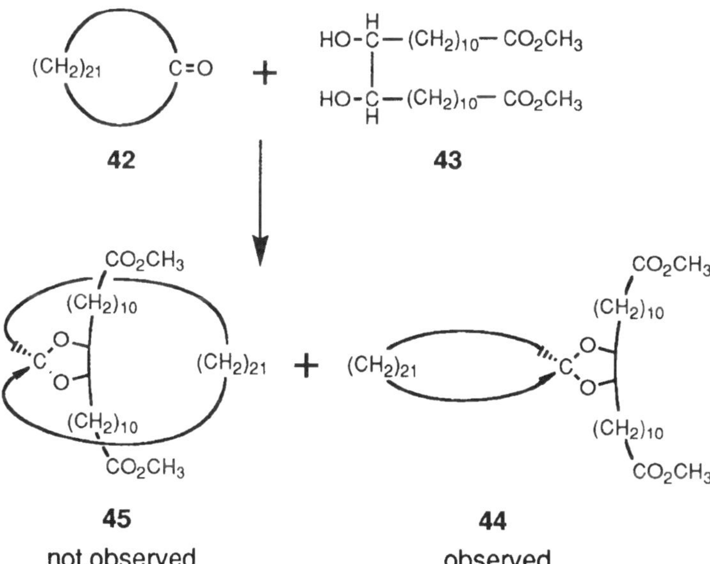

Figure 19 Attempts to threading by covalent bond formation. Only the extraannular conformer (**44**) is obtained. The intraannular conformer (**45**) is not observed [42].

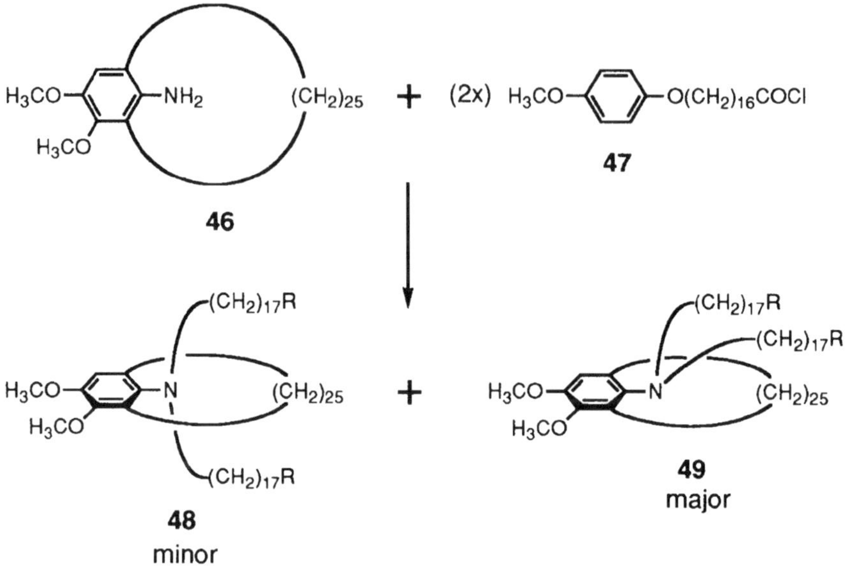

Figure 20 Threading by covalent bond formation. The intraannular conformer (**48**) is observed, but the extraannular conformer (**49**) is the major product [52].

developed by Schill [42], afforded the catenane in 5–8% yield, proving that the threaded species (**48**) had formed, although in low amounts.

4 Clipping

4.1 Control of covalent bonds: the directed synthesis

Threaded species, in which the threaded component is covalently bound to the macrocycle, could also be obtained, in principle, by clipping, that is by cyclization around the thread of a precursor to the macrocycle covalently attached to it. Schill and Tafelmair [42] tried to prepare a rotaxane by intramolecular cyclization of diamine (**50**) (Figure 21). Unfortunately, only the undesired, extraannular compound (**51**), in which the long alkyl chains anchored to the benzene nucleus are lying outside the macrocyclic part of the molecule, was obtained. The formation of the 'threaded' compound (**52**) was not observed.

The 'directed synthesis' controls the selective formation of intraannular species [32, 42, 107–110]. The components are elaborated at a rigid core, an aromatic ketal (Figure 22). The ketal tetrahedral carbon atom in intermediate (**56**), obtained in several steps from guaiacoldialdehyde (**53**) through (**54**) and (**55**) was designed in order to maintain at right angles the precursor to the thread (a benzene ring bearing two functionalized alkyl chains at *meta* positions) and the precursor to the

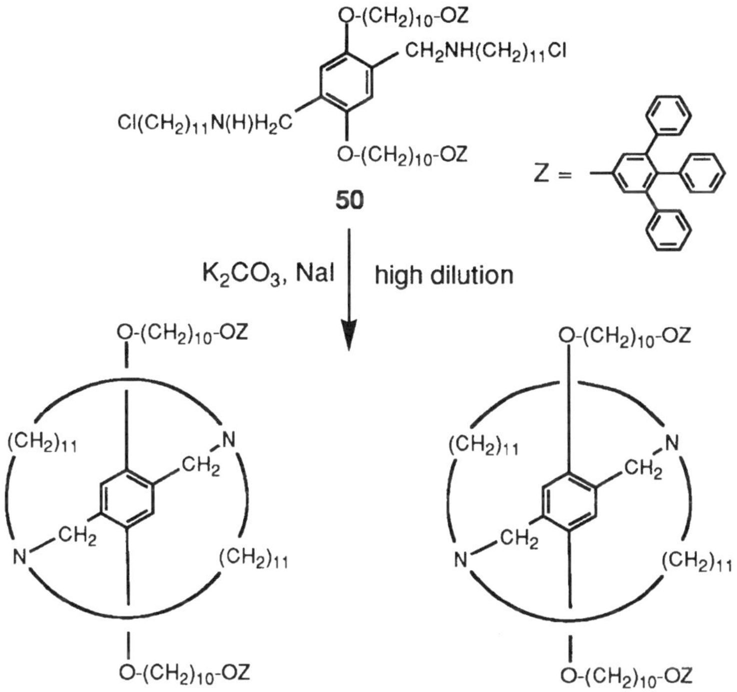

Figure 21 Attempts to clipping by covalent bond formation. Intramolecular cyclization of (**50**) provided only the extraannual isomer (**51**). The intraannular isomer (**52**) was not isolated [42].

macrocycle (two functionalized alkyl chains ready to react with a primary amino group placed in between the precursors of the thread). The cyclization reaction provided the intermediate, 'threaded' compound (**57**) in 26% yield. The stoppers were then grafted, affording (**58**), and the bonds linking the dumbbell to the macrocycle cleaved: hydrolysis of the ketal to provide (**59**) was followed by reductive acetylation, affording the [2]-rotaxane (**60**) [108]. This synthesis was adapted from the first directed synthesis of a catenane by Schill and Lüttringhaus [32].

4.2 Control by non-covalent interactions

Aromatic, donor-acceptor interactions (see section I.3.2) Clipping is the most extensively used method for constructing rotaxanes based on the tetracationic

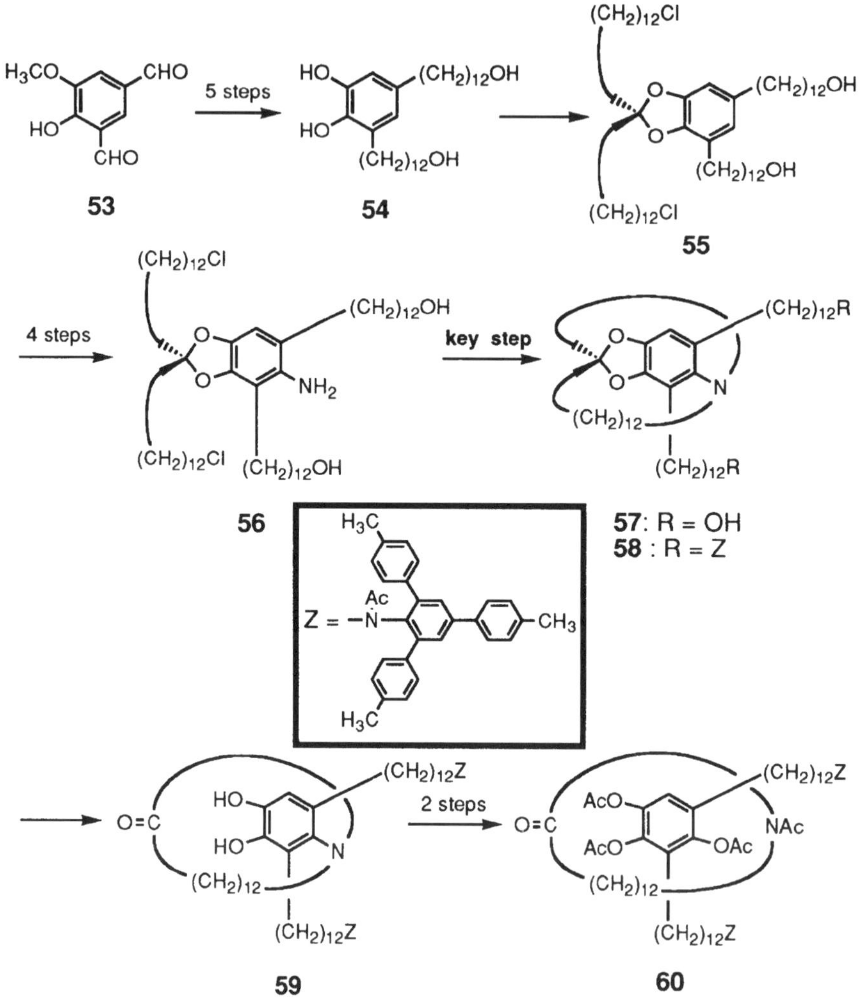

Figure 22 The directed synthesis of [2]-rotaxane (**60**). Cyclization of the key intermediate (**56**) is directed by the very carefully controlled stereochemistry of the ketal carbon atom and provides threaded intermediate (**57**) [108].

cyclophane (**27**) [63, 88, 90–97]. An example is provided in Figure 23 [92]. The precursors to the macrocycle are the bis-pyridinium salt (**61**) and α,α'-dibromo-p-xylene (**62**). The presynthesized dumbbell component (**63**) contains two π-donor sites for donor-acceptor interactions with π-acceptor units. The reaction takes place in mild conditions, i.e. room temperature for a few days. It is believed that α,α'-dibromo-p-xylene alkylates one free pyridine of (**61**) prior to complexation of the

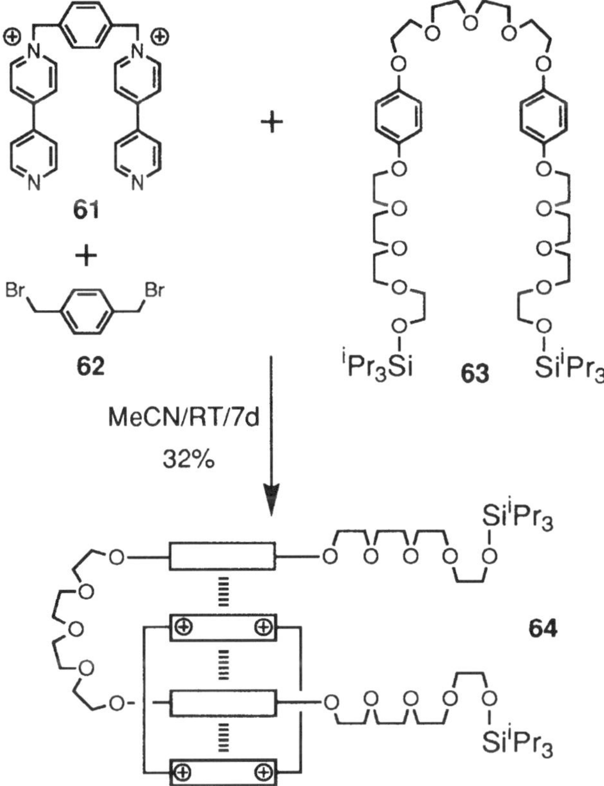

Figure 23 Clipping controlled by π-donor/π-acceptor interactions. Preparation of [2]-rotaxane (**64**) [92].

dumbbell by the precursor to the macrocycle. The reaction of Figure 23 afforded the desired rotaxane (**64**) in 32% yield. Actually the clipping procedure hardly gave yields better than 10–15% [88], except when the dumbbell component contains at least two p-dioxybenzene units, which is the case of the example given here: the combination of these units templates the cyclization of the tricationic precursor of the macrocycle by additional π–π interactions.

The clipping procedure was used for the synthesis of the so-called 'two-stations' [2]-rotaxanes, or 'molecular shuttles' [63, 64, 92–95]. Control of the interaction of the macrocycle with the stations (by photochemical [96, 111] electrochemical, or chemical [28] means) can induce a translational motion of the macrocycle, along the dumbbell, from one station to the other.

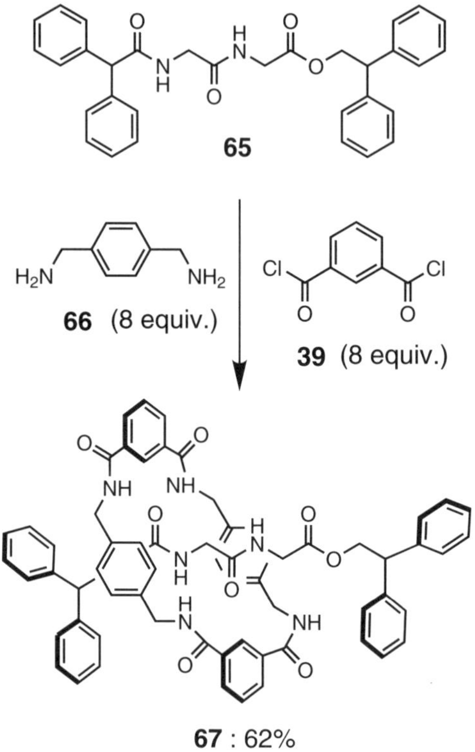

65

66 (8 equiv.)

39 (8 equiv.)

67 : 62%

Figure 24 Clipping by H-bonding interactions. Preparation of [2]-rotaxane (**67**) [113].

Hydrogen-bonding in neutral rotaxanes Rotaxanes in which the macrocyclic and dumbbell components are assembled by hydrogen bonds were usually prepared by the threading and stoppering method (see section I.3.2). Figure 24 shows an example where, on the contrary, the macrocycle is elaborated around the dumbbell, the latter playing the role of a template [112, 113]. When the glycylglycine-containing dumbbell (**65**) was reacted with α,α'-diamino-p-xylene (**66**) and isophthaloyl dichloride (**39**), rotaxane (**67**) was obtained in 62% yield! The adjacent amide groups of the dumbbell adopt transoid conformations which favourably interact via hydrogen bonding with the macrocycle precursors, as shown by a crystal structure analysis of the analogue based on the 2,6-dicarbonylpyridine fragment. The macrocycle dynamics were studied: in a non-polar solvent pirouetting motions around the dipeptide thread are observed, whereas addition of a hydrogen-bonding solvent, such as dmso allows the macrocycle to move freely up and down the peptide chain.

PART II: TRANSITION METAL-TEMPLATED SYNTHESIS OF ROTAXANES

1 Control of Threading by Coordination Bonds

As shown in the previous sections, prerotaxanes, namely the intermediates obtained either by threading or clipping, or the rotaxanes themselves, obtained by slippage, can be stabilized by non-covalent bonding interactions: hydrophobic and aromatic π-donor/π-acceptor interactions, hydrogen bonds involving neutral or ionic species. They can therefore be regarded as host–guest compounds, the guest being the dumbbell component (convex partner), the host being the macrocycle (concave partner). The prerotaxane components could also be assembled around a transition metal, which would therefore play the role of a templating ion in the kinetic sense, as defined by Busch [114].

1.1 Transition metal-templated synthesis of catenanes

In the early 1980s, Sauvage and Dietrich-Buchecker developed the first transition metal-templated synthesis of catenanes [11–15]. It was based on the fact that copper(I), a d^{10} transition metal ion, forms extremely stable complexes with the 2,9-diphenyl-1,10-phenanthroline (dpp) chelate [115]. Two such ligands are entwined around the Cu^{+} cation, the nitrogen atoms of the chelate forming a tetrahedral coordination sphere, as required for the d^{10} electronic configuration. Figure 25 shows the synthetic route that was followed to prepare [2]-catenane (**72**), starting from a functionalized dpp ligand, 2,9-di-p-phenol-1,10-phenanthroline (**68**) [116]. In the intermediate, key complex (**69**), the two chelates (**68**) are perpendicular to each other, the phenol groups of one ligand extending beyond the phenanthroline nucleus of the other. The molecule seems therefore designed to 'ring close with the formation of a catenane' [114]. This was performed by reaction with the diiodo-derivative of pentaoxyethylene glycol (**70**) in high dilution conditions. Copper(I)-complexed [2]-caten*ate* (**71**) was obtained in 27% yield. The metal template was then released by competitive complexation with cyanide, affording the [2]-caten*and* (**72**) quantitatively.

The synthetic route described above corresponds to strategy I of Figure 26. Actually, strategy II was tested at first, since it involves well-controlled steps. It requires the preliminary preparation of one single chelate macrocyclic component. However, once intermediate (F) (called precatenate) is formed, only a single cyclization reaction is needed to afford the interlocked species. Accordingly, yields as high as 42% were observed for this last step.

The clean and selective formation of precatenate (F) deserves some comment. Of course, a homoleptic complex such as (B) of Figure 26 could also have formed. However, as shown in Figure 27, this would imply that the remaining fraction of

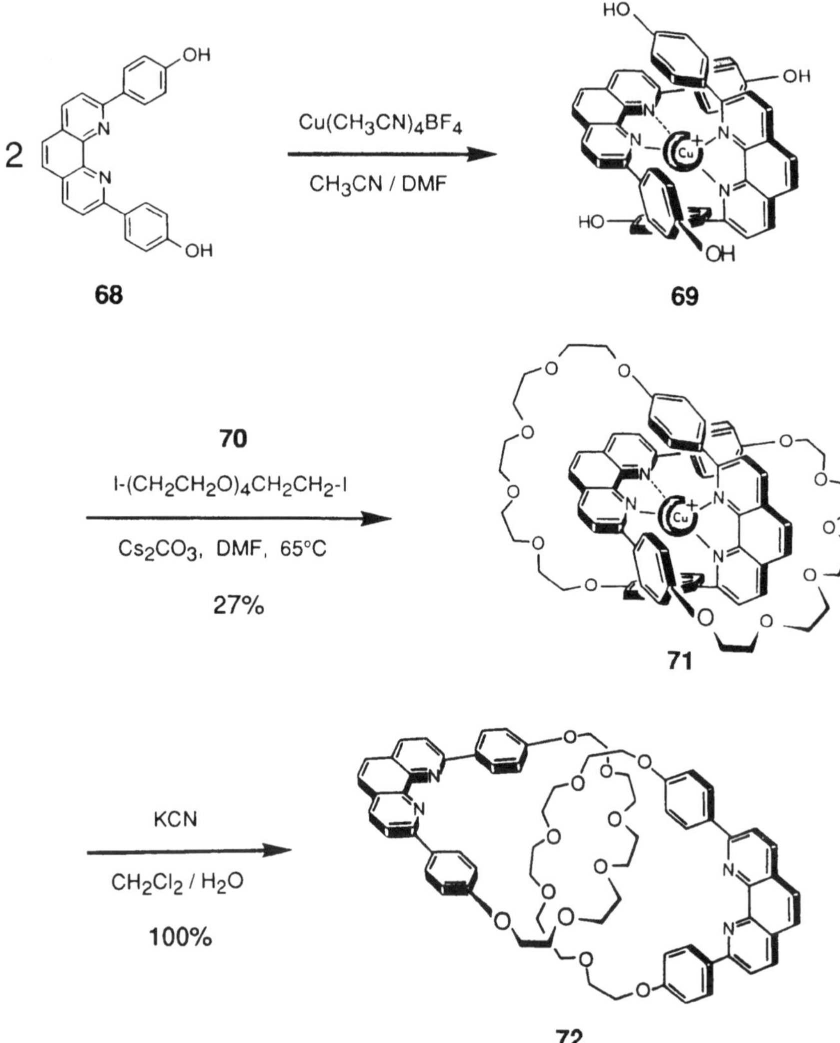

Figure 25 Copper(I)-templated synthesis of [2]-caten*ate* (**71**) and its demetallation to the corresponding [2]-caten*and* (**72**) [116].

copper(I) be complexed by the single chelate macrocycle (E). Owing to obvious steric reasons, two such macrocycles cannot entwine around a Cu^+ cation, and since $Cu(dpp)_2{}^+$ complexes are much more stable than the $Cu(dpp)^+$ homologues [115], bis-chelate complex (F) forms exclusively.

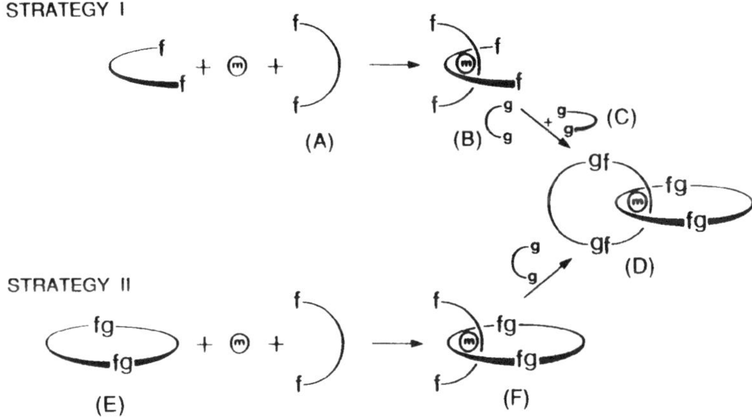

Figure 26 Strategies for the transition metal-templated synthesis of catenanes. The metal (m) predisposes two fragments as open chelates (A) (strategy I) or as a macrocyclic chelate (E) and an open chelate (strategy II) in intermediates (B) and (F) respectively. Cyclization of these intermediate complexes with the chain fragments (C) provides the [2]-catenate complex (D).

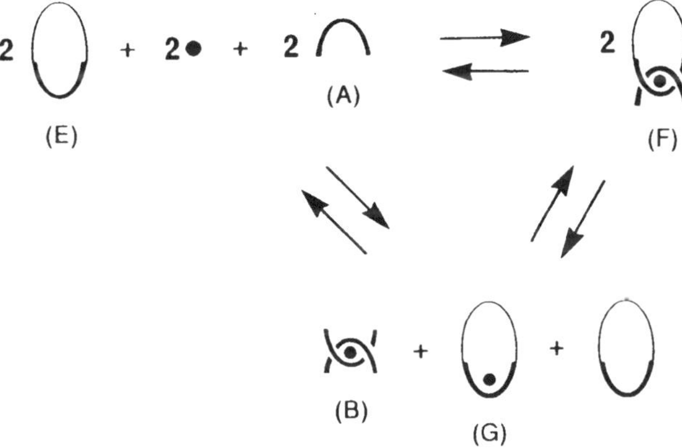

Figure 27 The equilibria shown here involve a mixture of macrocyclic chelate (E), open chelate (A), and metal cation (black disk) on the one hand, and different complexed species on the other hand: the threaded complex (F) or a mixture of homoleptic complex (B), metal-complexed macrocycle (G) and free macrocycle (E).

1.2 Transition metal-controlled threading: a new principle of rotaxane synthesis

The precatenate intermediate can be regarded as a threaded complex, which is thermodynamically stabilized by coordination bonds. From this simple consideration, a transition metal-templated synthesis of rotaxanes was devised, the principle of which is shown in Figure 28 [117]. The threading step (i) is a complexation reaction.

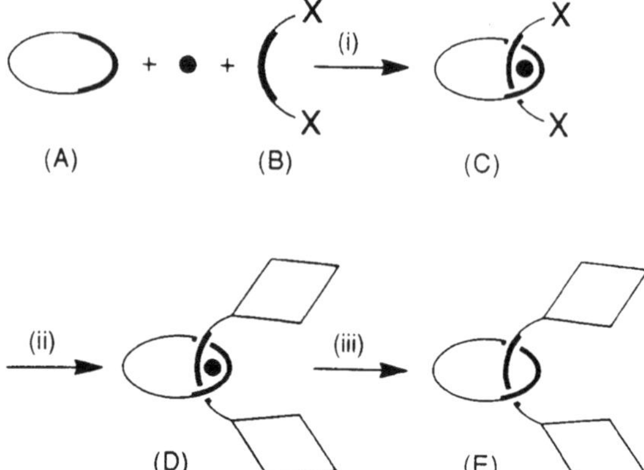

Figure 28 Schematic representation of the principle of transition-metal templated synthesis of rotaxanes from macrocyclic chelate (A), metal cation (black disk) and open chelate (B). The latter bears functions X at its extremities, which will be used for anchoring or constructing the stoppers (represented as diamonds). (i) Threading step; (ii) stoppering step; (iii) removal of the metal template.

It involves a macrocycle incorporating a dpp chelate (A), and a linear fragment incorporating the same dpp chelate (B), and end-functionalized by the appropriate reacting groups X. Thanks to its stereoelectronic requirements, the copper(I) cation templates the assembly of the threaded complex or prerotaxane (C). After stoppering, a copper(I)-complexed [2]-rotaxane (D) is produced. Release of the Cu^{+} cation leaves the template-free [2]-rotaxane (E). Noteworthy, this last step is particular to the transition metal-controlled methodology: neither hydrophobic nor hydrogen-bonding interactions can be cancelled so easily, except by changing the solvent [31, 118]. Aromatic π-donor/π-acceptor interactions can, in principle, be tuned by operating on the redox states of the stacking subunits. However, they are usually not altered, except for controlling the motions of the ring along the dumbbell in the molecular shuttles [28].

Until now only transition metal-controlled threading was considered. As shown in Figure 29, clipping requires that two different open chelates be involved in the complexation step, one being a functionalized dpp chelate (A), the other being a coordinating dumbbell (B). While heteroleptic complexes of kinetically inert transition metals can be formed selectively, this is not the case for copper(I), which forms kinetically labile complexes with dpp-like ligands [119, 120]. The prerotaxane species (C) is therefore in exchange equilibrium with the homoleptic complexes (E) and (D), which are improductive. The clipping route was thus not explored further.

Higher order [n]-rotaxanes could be, in principle, synthesized by the transition metal-controlled threading process, provided that the linear component contains the

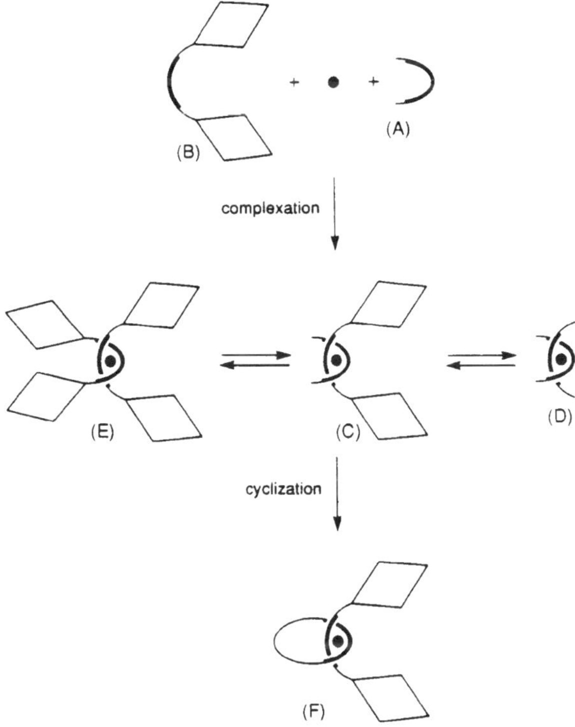

Figure 29 Assembling a dumbbell chelate (B) and an open chelate (A) around a transition metal cation with a kinetically labile coordination sphere will provide a mixture of homoleptic ((D) and (E)) and heteroleptic (C) complexes. Only the latter will be productive for the formation of a [2]-rotaxane.

appropriate number of chelating subunits. This is schematically represented in scheme 8.

Initial multi-threading experiments involved macrocycle (**73**) and the various molecular threads of Figure 30 [121, 122]. These linear components incorporate 2-anisyl, or 2-p-tolyl-1,10-phenanthroline chelates linked by biphenyl (**74**) biphenylether (**75**), $-(CH_2)_4-$ (**76**) or $-(CH_2)_6-$ (**77**) spacers. Mixing (**74**), (**75**), or (**76**) (1 equiv.) with macrocycle (**73**) (2 equiv.) and $Cu(CH_3CN)_4^+$ (2 equiv.) in CH_3CN/CH_2Cl_2 mixtures produced a single complex in all three cases: respectively, (**78**), (**79**), and (**80**). The threaded structure was supported by mass spectroscopy and particularly, ^{1}H-NMR spectroscopy. When the $-(CH_2)_6-$ bridged bis-chelate (**77**) was involved, a mixture of complexes was formed, suggesting that the spacer plays an important role in the control of the complexation properties of the bis-chelate threads. This result was taken into account for the synthesis of [3]-rotaxanes (see below).

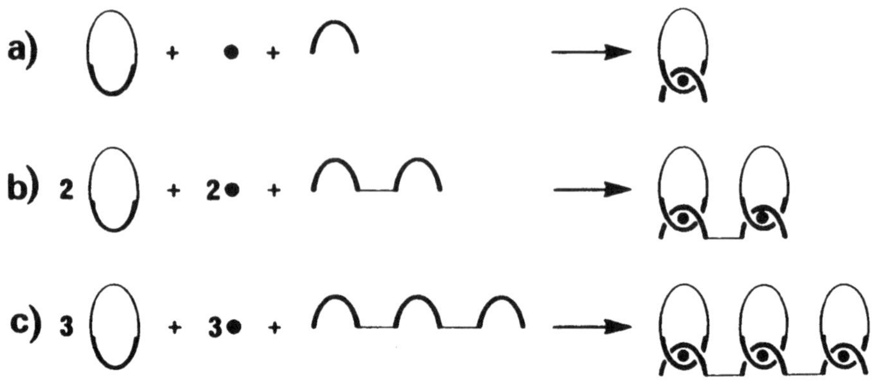

a) + • +

b) 2 + 2• +

c) 3 + 3• +

Scheme 8

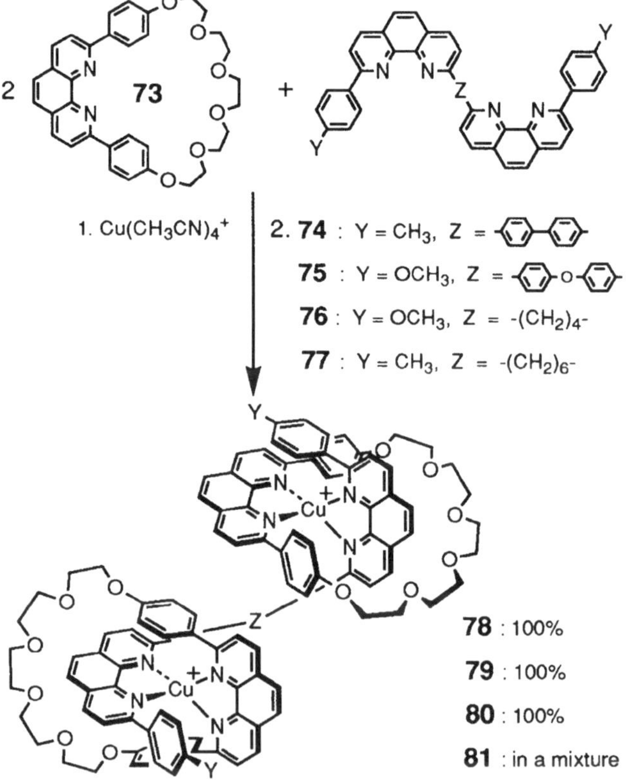

Figure 30 Copper(I)-templated threading of macrocycle (**73**) onto 1,10-phenanthroline-based molecular threads containing two such chelates, (**74**), (**75**), (**76**), and (**77**), to afford threaded complexes (**78**), (**79**), (**80**), and (**81**) respectively [121, 122].

This multi-threading process was put into practice by Amabilino *et al.* for the construction of multiporphyrin assemblies from macrocycles with pendent zinc(II) or gold(III) meso-tetraarylporphyrins [123]. These are shown in Figure 31, together with the linear components used. Mixing (**83**), (**84**), or (**85**) with macrocycle (**82**) in the same conditions as above produced mixtures of complcxes in the case of bis-chelates (**83**) and (**84**), from which the desired threaded compounds could be isolated in 6% (**86**) and 60% (**87**) yields respectively, after chromatography. As noted above, competing complexation reactions, such as formation of a 1 : 1 folded complex in the case of the $-(CH_2)_3-$ bridged thread (**83**), or a dinuclear double helix for the m-xylyl-bridged bis-chelate (**84**) decreased the yields of threaded compound. Nevertheless, when (**85**) was used as linear component, the yield of

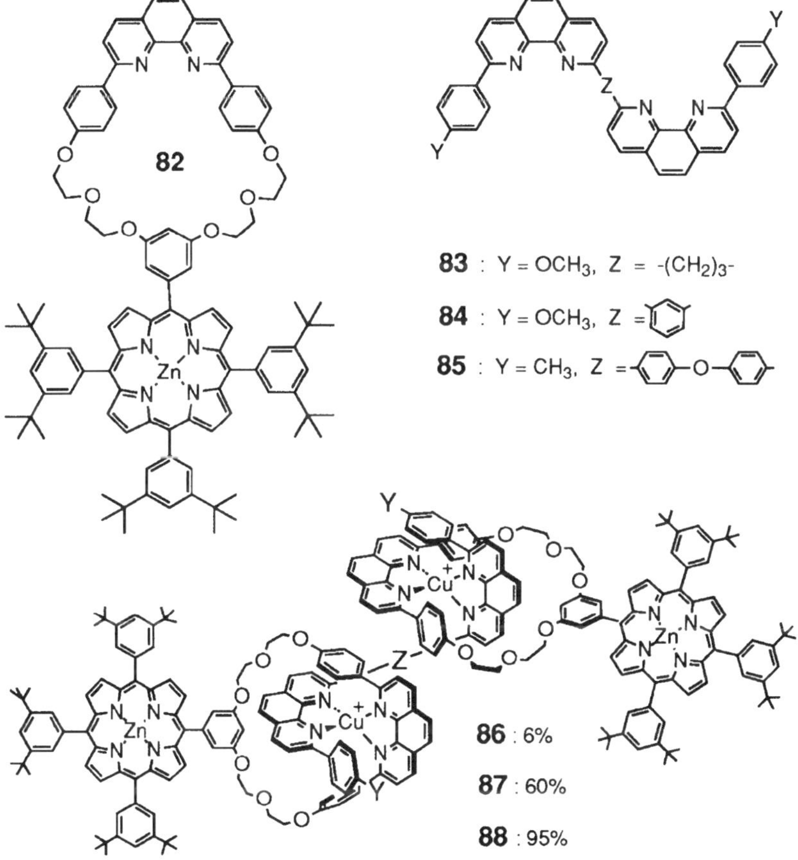

Figure 31 Assembly, by copper(I)-templated threading, of two zinc porphyrins appended to a chelating macrocycle (**82**). Bis-chelate molecular threads (**83**), (**84**), and (**85**) provide threaded complexes (**86**), (**87**), and (**88**) respectively [123].

threaded complex (**88**) was nearly quantitative. Therefore, this compound was used for threading two different porphyrin-containing macrocycles (Figure 32). The macrocycles (**82**) and (**89**) (1 : 1) in CH_2Cl_2, incorporating zinc(II) and gold(III) porphyrins respectively, were combined with a stoichiometric amount of $Cu(CH_3CN)_4^+$ and the required amount of molecular thread (**85**) was added to the solution of complexed macrocycles. As expected, the threaded complexes (**90**), (**91**), and (**92**) had formed quantitatively in 1 : 1 : 2 statistical ratio. They were separated by chromatography and isolated in 20, 10 and 8% yield for (**90**), (**91**) and (**92**) respectively. A remarkable feature of the heteroporphyrinic system (**92**) is its kinetic stability: it does not detectably 'scramble' to either (**90**) or (**91**) when kept in CD_2Cl_2 solution at room temperature for 24 h.

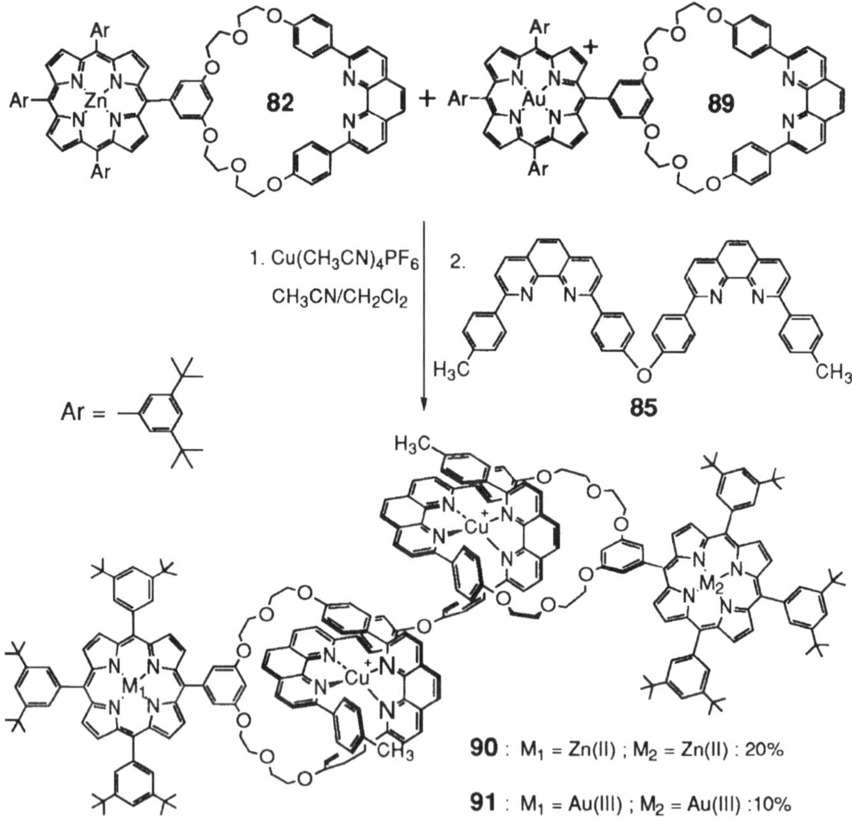

Figure 32 Assembly, by copper(I)-templated threading, of a zinc(II) porphyrin (**82**) and a gold(III) porphyrin (**89**) appended to a chelating macrocycle. The bis-chelate molecular thread (**85**) affords complexes (**90**), (**91**), and (**92**) [123].

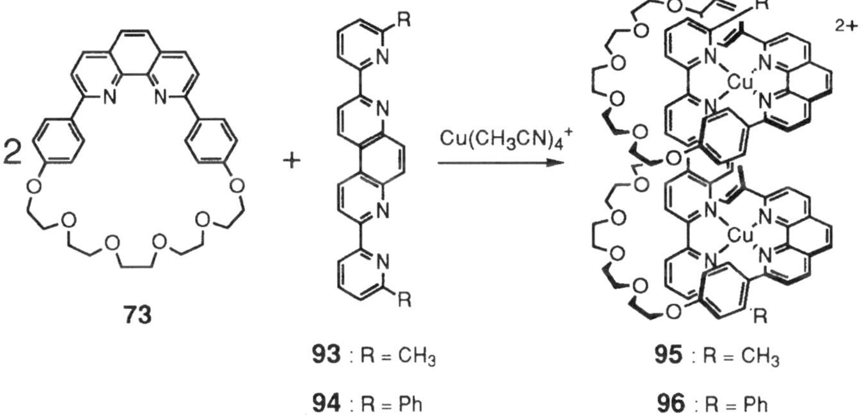

Figure 33 Construction of stacks of macrocycles by the copper(I)-templated threading strategy. Two macrocyclic units (**73**) are threaded onto the rigid bis-chelate rods (**93**) and (**94**), affording complex assemblies (**95**) and (**96**) [124, 125].

Stacks of macrocycles were obtained by Lehn and coworkers, by using rod-shaped bis-bipyridine linear components [124, 125]. An example is shown in Figure 33: copper(I)-templated threading of two molecules of macrocycle (**73**) onto the molecular rods (**93**) or (**94**) produced the rigid rack assemblies (**95**) and (**96**), in which the two threaded macrocycles display the same orientation with respect to the rod component. The use of a sexipyridine ligand allowed for the copper(I)-controlled threading of three macrocycles (equation (c) of scheme 8).

2 [2]-Rotaxanes

In 1991 Gibson and coworkers described the synthesis of a [2]-rotaxane based on macrocycle (**73**) [126]. As shown in Figure 34, copper(I)-templated threading of macrocycle (**73**) onto the dpp derivative (**68**) produced prerotaxane (**97**). This complex was deprotonated with K_2CO_3 and reacted with trityl derivative (**98**), affording the copper(I)-complexed [2]-rotaxane (**99**). Copper(I) was removed by chromatography on an ion exchange resin loaded with CN^-, and the metal-free [2]-rotaxane (**100**) obtained in 42% yield. The free dumbbell component (not represented) was isolated in 9% yield, as side-product, indicating that the prerotaxane was slightly unstable with respect to unthreading in the reaction conditions used. In addition, this experiment established that the ring size of macrocycle (**73**), a dpp-based 30-membered macrocycle, was equivalent to that of the crown ether 30-crown-10 in the presence of the trityl stopper utilized in this synthesis.

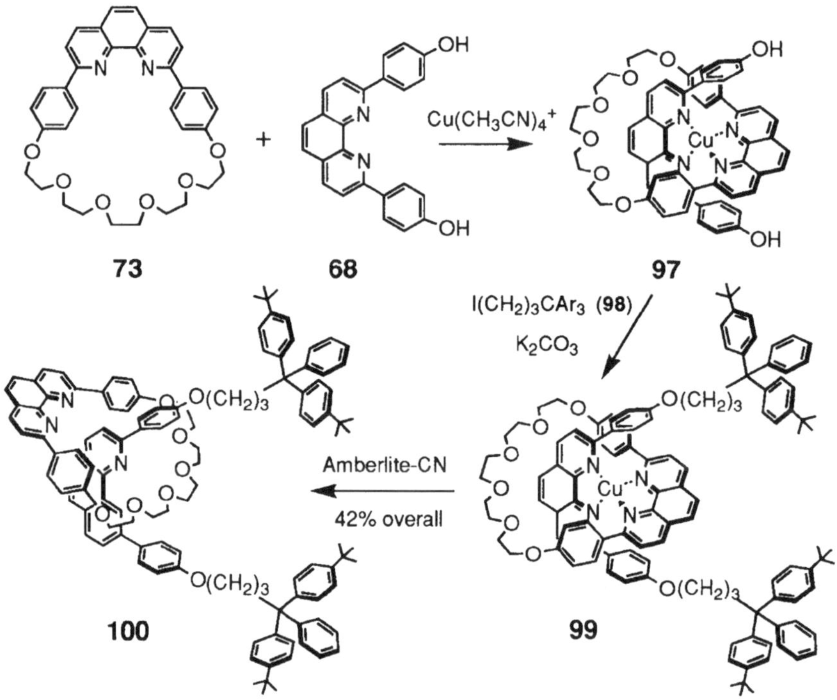

Figure 34 Copper(I)-templated synthesis of [2]-rotaxane (**100**) bearing trityl stoppers [126].

With its spherical shape, the C_{60} moiety is an ideal end group for a dumbbell component (Figure 35). It was used by Diederich *et al.* [127] as stopper for rotaxane (**105**). Macrocycle (**73**) was too large, according to molecular models and a 27-atom-membered homologue (**101**) was chosen. It was threaded onto a dpp chelate bearing alkyne end groups (**102**) [128, 129]. In the resulting prerotaxane (**103**), macrocycle (**101**) wedges the dpp moiety of the thread, as evidenced by the known chirality of the Cu(I) [2]-catenate built on the same macrocycle [130]. Stoppering was performed by Glaser coupling of an excess of alkyne-functionalized C_{60} (**104**) and prerotaxane (**103**). Copper(I)-complexed [2]-rotaxane (**105**) was obtained in 15% yield together with a dumbbell component resulting from the homocoupling of two C_{60} subunits (not shown), which was isolated in 20% yield. The metal-free [2]-rotaxane could not be isolated in pure form, because the nucleophilic CN^- damaged the strained macrocyclic component. Fortunately it was detected by mass spectrometry of the crude reaction mixture, providing a chemical evidence that the macrocycle and the stopper matched well together.

[2]-rotaxane (**110**), prepared by Cárdenas *et al.* involves transition metal-complex fragments as stoppers (Figure 36) [131, 132]. The thread designed for its synthesis

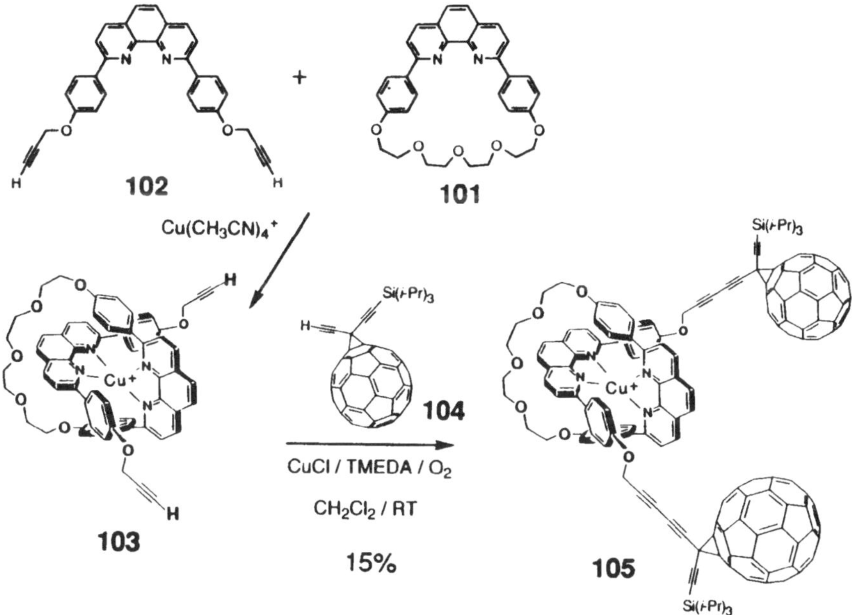

Figure 35 Copper (I)-templated synthesis of metal-complexed [2]-rotaxane (**105**) bearing fullerene stoppers [127].

contains a central dpp chelate bearing two pendent terpyridyl units (**106**). This tris-chelate-containing linear component reacted with stoichiometric amounts of $Cu(CH_3CN)_4^+$ and macrocycle (**73**), producing prerotaxane (**107**) in which Cu(I) is selectively bound to the dpp subunit of the thread. Stoppering was performed by coordinating the pendent terpyridyl subunits to a $[Ru(terpy)]^{2+}$ complex fragment (**108**). The resulting copper(I)-complexed [2]-rotaxane (**109**) was obtained in 33% yield. Removal of the metal template with KCN afforded [2]-rotaxane (**110**) in 59% yield. ^{1}H-NMR experiments established that release of Cu^+ was accompanied by a pirouetting motion of the macrocyclic component, placing the dpp subunit far away from the ruthenium stoppers.

A related molecular thread (**111**), containing a 2-phenyl-1,10-phenanthroline chelate connected to a terpyridyl chelate via a $-(CH_2)_4-$ spacer, was synthesized by Gaviña *et al.* (Figure 37) [133]. This linear component is a semi-dumbbell, since the terpyridyl end is blocked with a trityl derivative. Copper(I)-templated threading of macrocycle (**73**) was shown to take place again selectively at the phenanthroline site. End blocking of the resulting prerotaxane (**112**) with a trityl derivative (**113**) afforded Cu(I)-complexed [2]-rotaxane (**114**) in 40% yield. This metallo-rotaxane differs from those described above in that the dumbbell component entails two chelate subunits: a complexed one (phen), and a free one (terpy). This structural

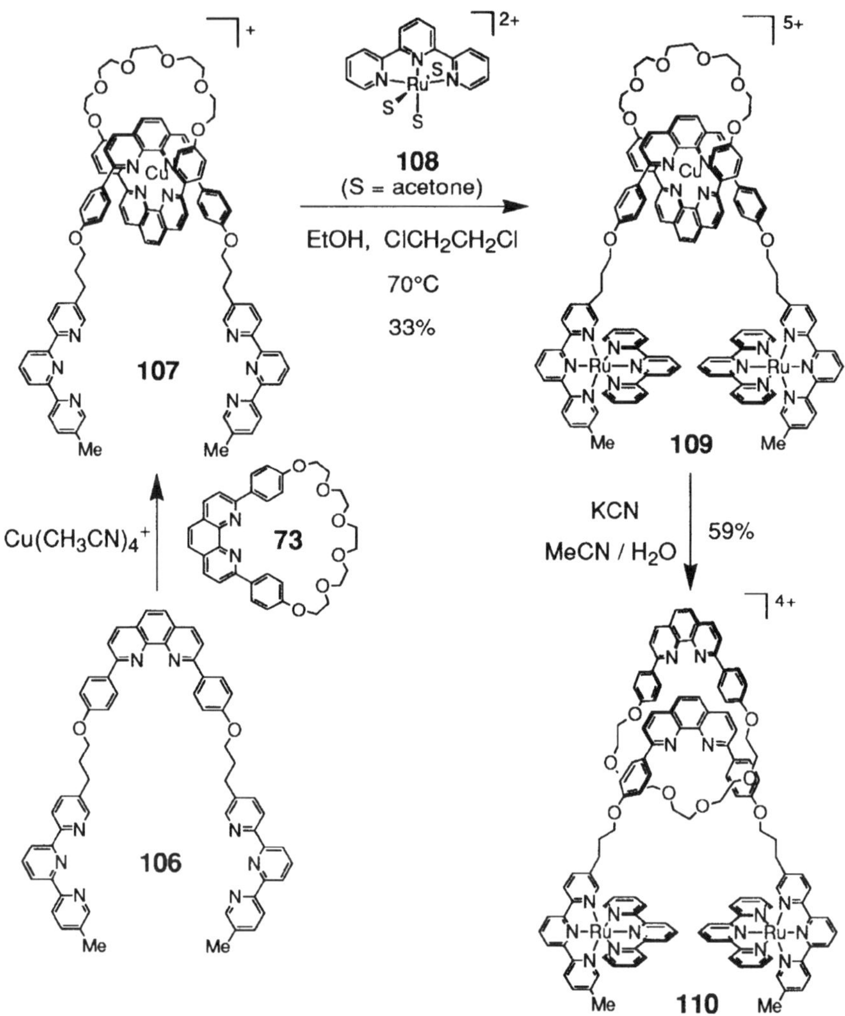

Figure 36 Copper(I)-templated synthesis of [2]-rotaxane (**110**) bearing Ru-complex fragments as stoppers [131, 132].

property, in combination with the redox properties of Cu(I) complexes of 2,9-disubstituted-1,10-phenanthroline ligands [134] was used to study the electrochemically controlled translational motion of the threaded macrocycle between the phen and the terpy chelating sites (Figure 38) [29]. In the initial state (**114**), the macrocycle stays in the phen site, due to the stereoelectronical requirements of Cu(I). Controlled oxidation of Cu(I) to Cu(II) triggers the motion of the complexed ring to the terpy site, since Cu(II) requires higher coordination numbers than Cu(I),

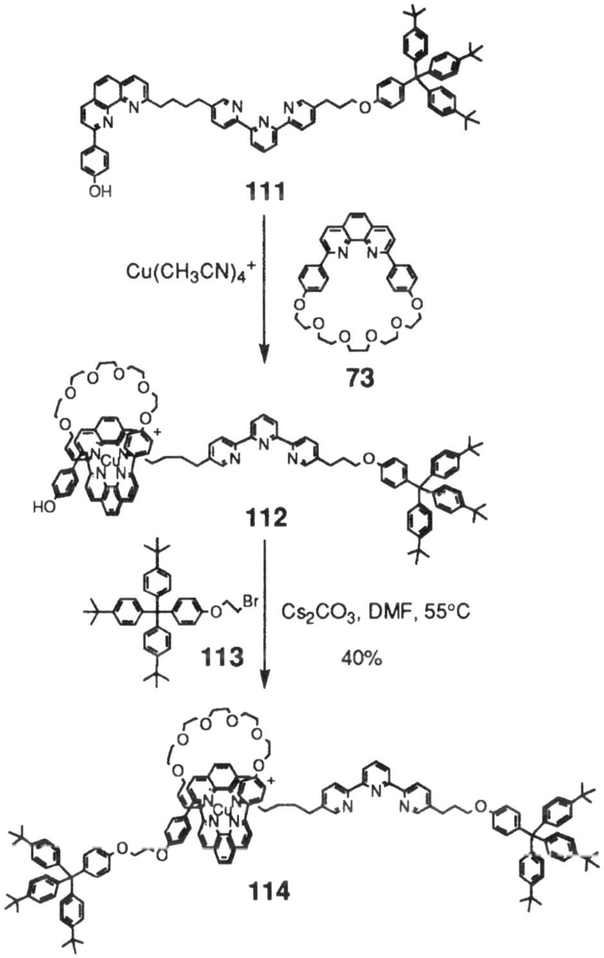

Figure 37 Copper(I)-templated synthesis of metal-complexed [2]-rotaxane (**114**) whose thread contains two different chelates [133].

and affords the Cu(II)-complexed [2]-rotaxane (**115**). This translational motion takes place at a rate of $1.5 \times 10^{-4}\,s^{-1}$, which is much faster than the rate of pirouetting observed in a related molecule, a copper catenate containing macrocycle (**73**) and a macrocycle incorporating dpp and terpy subunits [135, 136]. Interestingly, the motion of the ring back to the phen site, triggered by reduction of Cu(II) to Cu(I), takes place at a slower rate (10^{-4}–$10^{-2}s^{-1}$) than the forward translation motion.

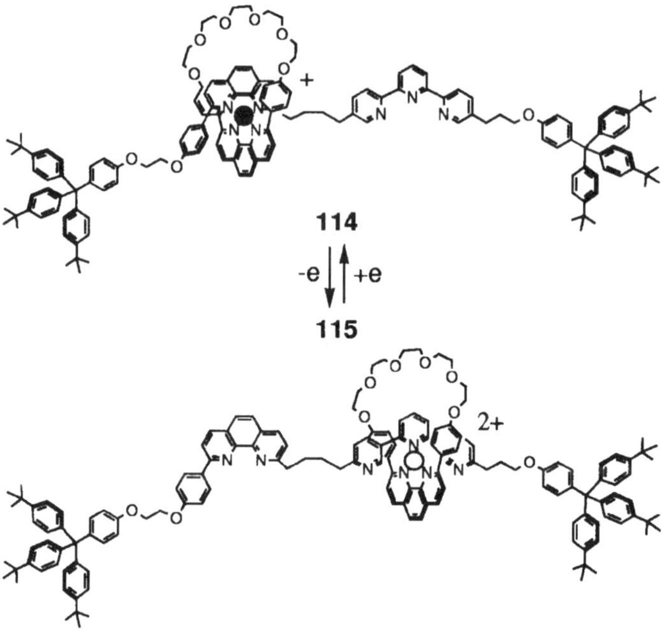

Figure 38 Molecular motions in copper(I)-complexed [2]-rotaxane (**114**) controlled by the redox state of the metal. In the Cu(I) state (**114**) the macrocycle is held at the phenanthroline site of the dumbbell. In the Cu(II) state (**115**) the macrocycle is held at the terpyridine site of the dumbbell. Therefore oxidation of Cu(I) to Cu(II) has triggered a translational motion of the complexed macrocycle along the dumbbell [29].

3 Poly([2]-rotaxanes)

Linear and branched poly([2]-rotaxanes) were synthesized by electro-polymerization. Electroactive films with a poly([2]-rotaxane) backbone structure were produced by Kern *et al.* (Figure 39) [137]. At first, macrocycle (**73**) was threaded onto a dpp ligand terminated with pyrrole subunits (**116**), using either Co(II) or Cu(I) as template, affording prerotaxanes (**117**) and (**118**) respectively. Surprisingly, only in the case of the Co(II)-complexed prerotaxane (**118**) did electropolymerization take place. The branched polymer obtained (**119**) could be easily demetallated using SCN^- as competing complexing agent for Co^{2+}, and the resulting metal-free poly([2]-rotaxane) (**120**) remetallated with Cu^+. Therefore, in spite of the fact that electropolymerization did not take place with Cu(I) as template, the poly([2]-rotaxane) of Cu(I) (**121**) could be prepared using a demetallation-remetallation procedure.

Quite simultaneously, Swager and his coworkers [138] established that macrocycle (**73**) could be threaded onto a 2,2′-*bipyridyl* chelate (**122**) using either Zn(II) or Cu(I) as template metals, affording prerotaxanes (**123**) and (**124**) respectively (Figure 40). The bipy moiety was substituted with bithiophenyl subunits at positions

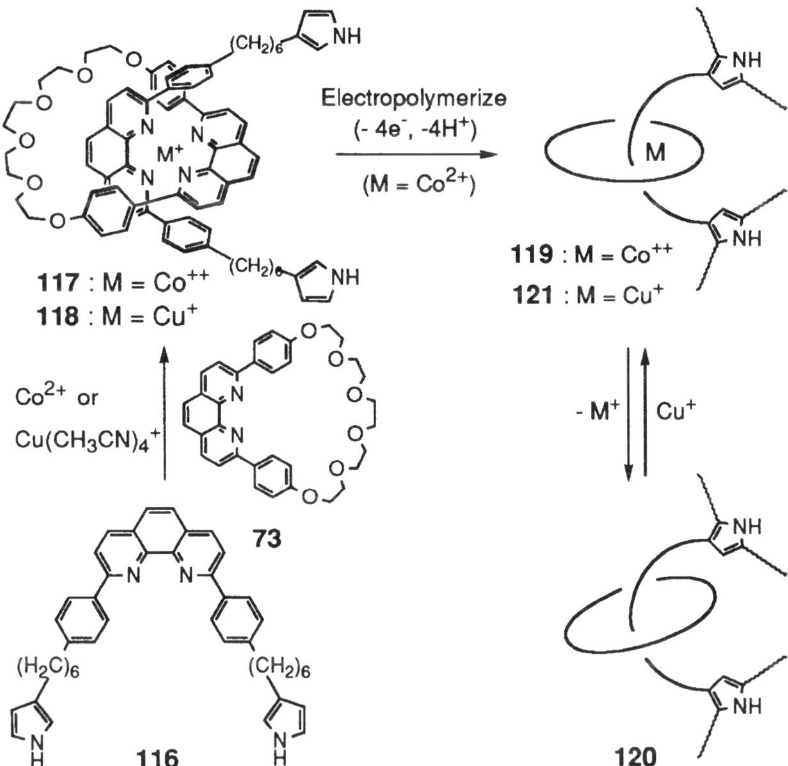

Figure 39 Preparation of poly([2]-rotaxane) (**120**) by electropolymerization of prerotaxane (**117**), formed by cobalt(II)-templated threading of macrocycle (**73**) onto the dpp-based thread (**116**) [137].

5 and 5′ so that the molecular thread was rather a true molecular rod. Electrochemical oxidative coupling of the protruding bithiophenyl substituents afforded linear poly([2]-rotaxanes) (**125**) and (**126**), whose rigid backbones were made of alternating bipyridyl and tetrathiophenyl blocks. Demetallation of the zinc(II)-complexed poly([2]-rotaxane) afforded a metal-free conducting polymer (**127**).

These polymerization experiments also established that other metals than Cu(I) (i.e. Zn(II) or Co(II)) could be used as templating agents for the synthesis of rotaxanes.

4 Rotaxanes with Porphyrins as Stoppers

4.1 [2]-rotaxanes as models of the photosynthetic Reaction Centres

The bis-porphyrin conjugate (**128**) of Figure 41 is composed of zinc(II) and gold(III) mesotetraaryl porphyrins, which are separated by a dpp spacer [139, 140]. This

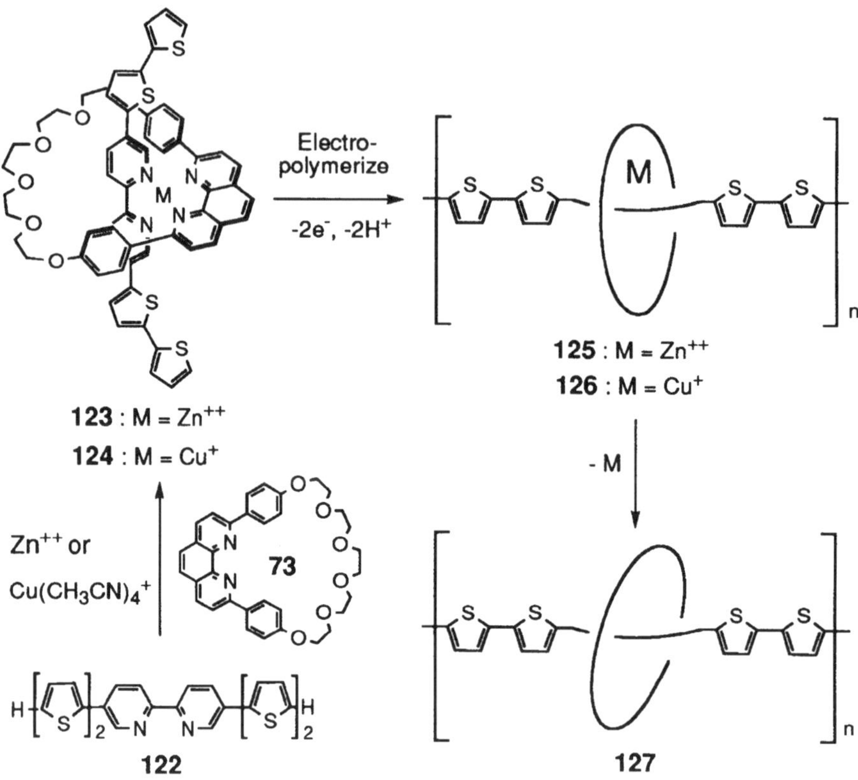

Figure 40 Preparation of poly([2]-rotaxane) (**127**) by electropolymerization of prerotaxane (**123**) or (**124**), formed by transition metal-templated threading of macrocycle (**73**) onto the dpp-based thread (**122**) [138].

molecule was designed to mimic structurally [141] and functionally the association of cofactors SP/BPh, where SP is a dimer of bacteriochlorophylls called 'Special Pair' and BPh is a bacteriopheophytin. These natural porphyrins or, more accurately, chlorins, are buried in the 'Reaction Center', a protein aggregate of the photosynthetic membranes of certain bacteria, where the primary steps of conversion of light energy into chemical energy take place [142–144]. Specifically, light excitation of SP leads to very fast electron transfer to BPh, in 3 ps [145–147]. In bis-porphyrin (**128**), the electron donor is the zinc porphyrin in its singlet excited state, and the electron acceptor is the gold(III) porphyrin in its ground state. The latter is also a very poor energy acceptor. Accordingly, photoinduced intramolecular electron transfer was shown to occur, and to take place in 55 ps [148].

The next step was to take advantage of the coordinating properties of the dpp spacer. Therefore, the complex Cu(**128**)$_2^+$ was prepared and isolated [149]. It is shown schematically in Figure 42. The electron transfer rate between the zinc

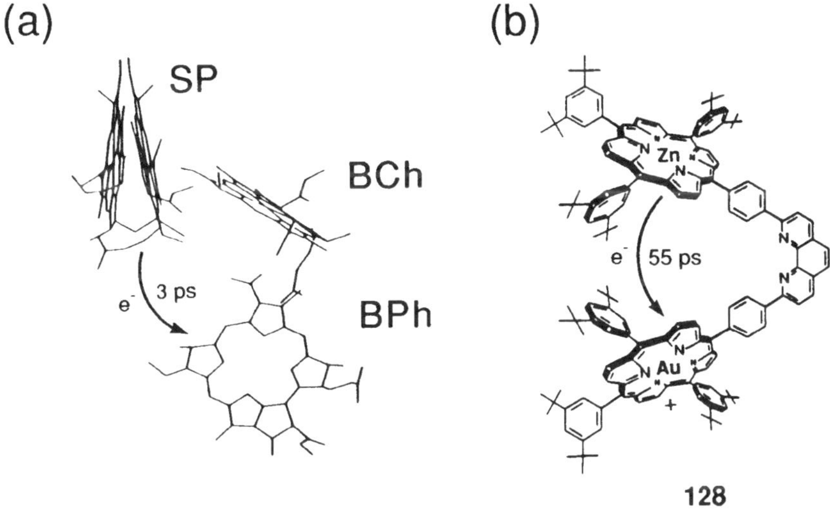

Figure 41 (a) The arrangement of the central cofactors of the Reaction Centre of *Rh. viridis*, according to an X-ray crystal structure analysis. SP, the special pair, is the primary electron donor in its excited state, BCh is an accessory bacterio-chlorophyll, and BPh, a bacterio-pheophtyin, is the primary electron acceptor [142–144]. Photo-induced electron transfer from SP to BPh takes place in 3 ps [145–147]. (b) A synthetic model of the natural system (**128**). [139, 140, 148] SP and BPh are mimicked respectively by a zinc(II) and a gold(III) porphyrin bridged by a dpp spacer.

porphyrin and the gold(III) porphyrin was shown to increase dramatically: a value close to that found in the natural system was obtained! This result was quite exciting, however, as shown in Figure 42, it was not possible to decide which pathway was followed by the electron: was it intraligand (path b) or interligand (path a) electron transfer?

To address this question, a heteroleptic complex, involving bis-porphyrin (**128**) and an auxiliary dpp ligand had to be synthesized. According to the discussion presented in section II.1.2, a rotaxane structure seemed ideally adapted to this particular problem. This was all the more justified as bis-porphyrin (**128**) is a dumbbell-shaped molecule! Therefore a copper(I)-complexed [2]-rotaxane bearing a zinc(II) and a gold(III) porphyrin as stoppers was designed and synthesized, as shown schematically in Figure 43 [117, 150]. Since the end groups are two different porphyrins, they are implemented sequentially. Hence a semi-dumbbell molecule (B) is used for threading of macrocycle (A). The second porphyrin is then attached to the open extremity of the linear component of the prerotaxane (C), affording [2]-rotaxane (D).

The real precursors used and the synthetic route actually followed are represented in Figure 44. Since gold(III) porphyrins are extremely stable square planar metal

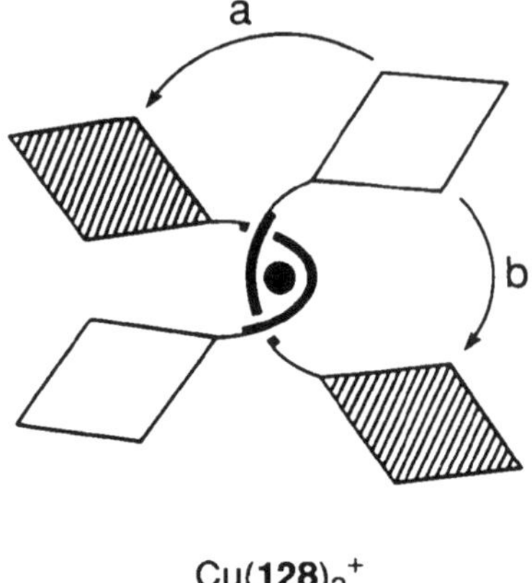

Figure 42 Schematic representation of the complex $Cu(\mathbf{128})_2^+$. The white diamond is a zinc porphyrin and the hatched diamond is a gold porphyrin. The black disk is Cu^+. The thick lines represent the dpp chelating sites of bis-porphyrin (**128**). Paths (a) and (b) represent inter- and intraligand electron transfer processes [149].

complexes, the presynthesized porphyrin was the gold complex. Threading of macrocycle (**73**) onto semi-dumbbell (**129**) in the presence of Cu^+ afforded prerotaxane (**130**) quantitatively. This prerotaxane bears at one extremity an aldehyde function to be used to construct the second porphyrin. This was accomplished by taking advantage of a mild procedure developed by McDonald and

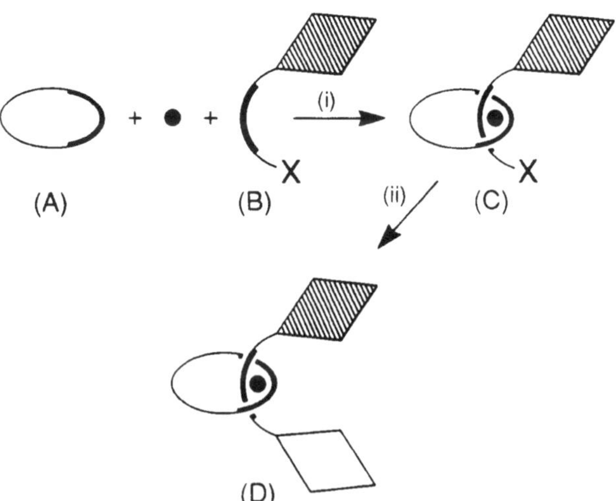

Figure 43 Principle of transition metal-templated construction of a [2]-rotaxane with two different porphyrinic stoppers (same conventions as in Figure 42). (i) Macrocycle (A) is threaded onto chelate (B), end-blocked by a porphyrin at one extremity and functionalized with a reactive group X, which is a precursor to the second porphyrin stopper, affording prerotaxane (C); (ii) construction of the second porphyrin, leading to metal-complexed [2]-rotaxane (D).

Lindsey [151–153]. Accordingly, reaction of (**130**) (1 equiv.), dipyrrylmethane (**131**) (10 equiv.) and 3,5-di-tert-butylbenzaldehyde (**132**) (4 equiv.) afforded, after oxidation of the intermediate porphyrinogen with chloranil (**133**), the free-base rotaxane (**134**) in 25% yield. Metallation with $Zn(OAc)_2{\cdot}2H_2O$ and extensive purification led to copper(I)-complexed [2]-rotaxane (**135**). Noteworthy, the stoppering

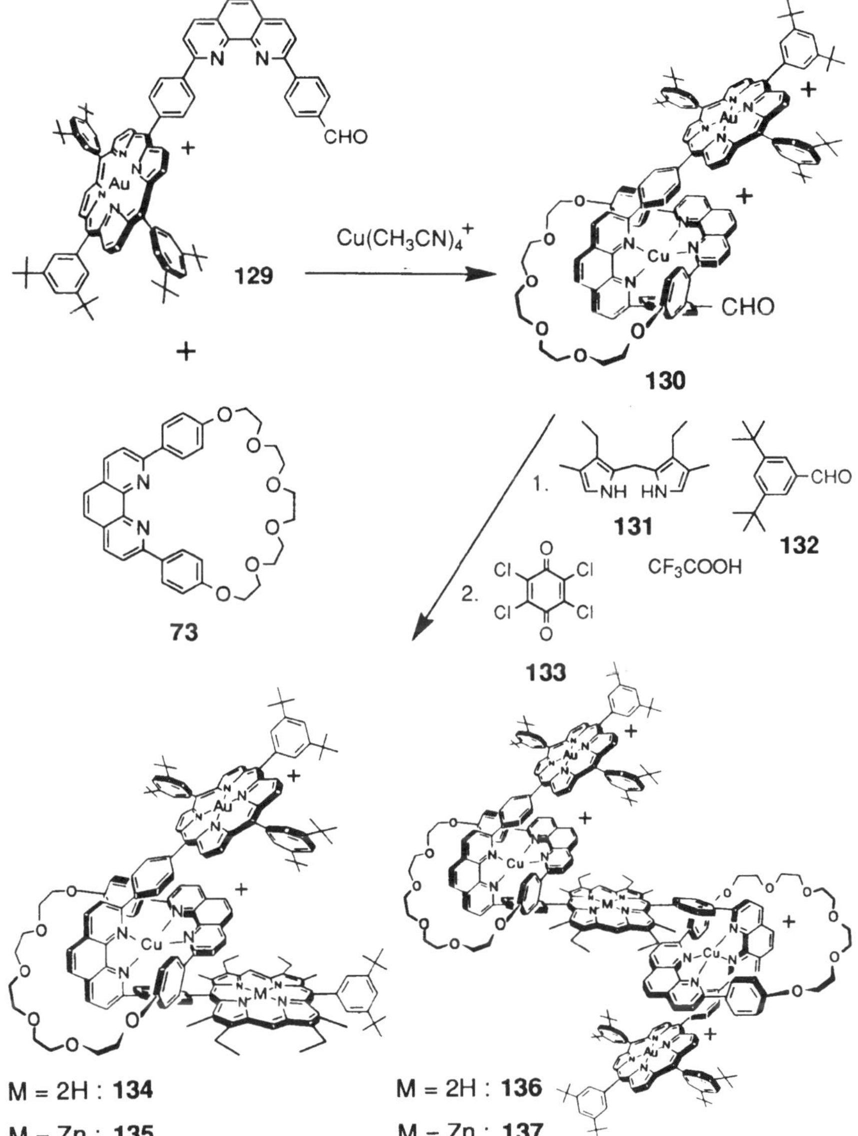

Figure 44 Copper(I)-templated synthesis of Cu(I)-complexed [2]-rotaxane (**134**). The 'compartmental' [3]-rotaxane (**136**) is also obtained [117, 150].

reaction does not involve coupling of a presynthesized stopper to the prerotaxane, as it is usually the case. Rather, it lies in the construction *in situ* of the blocking groups. The reaction also produced a [3]-rotaxane (**136**), which was converted to its zinc complex (**137**). We shall come back in a separate section to the formation of this 'compartmental' rotaxane.

Removal of the Cu^+ template cation of compound (**135**) with KCN produced [2]-rotaxane (**138**) (Figure 45). As observed in some other cases, the demetallation reaction is accompanied with a pirouetting of the macrocycle placing its dpp chelate outside the cleft formed by the porphyrinic stoppers [126, 132].

The rates of photoinduced electron transfer between the zinc porphyrin and the gold(III) porphyrin were 1.7 ps for the Cu(I)-complexed [2]-rotaxane and 36 ps for the free [2]-rotaxane (Figure 46) [154–156]. Therefore, what was initially observed in the case of the homoleptic complex $Cu(\mathbf{128})_2{}^+$ was corroborated by the Cu(I)-complexed [2]-rotaxane, suggesting that in the former molecule, electron transfer takes place between porphyrins of the same ligand (path b of Figure 42). The rate obtained for the free [2]-rotaxane was consistent with that found for the bis-porphyrin conjugate (**128**), the difference arising from a slightly more exergonic process in the case of the [2]-rotaxane.

The increase of the rate of electron transfer by coordination of Cu(I) to the dpp bridging chelate was interpreted in terms of the superexchange effect [157–159]. Speculating that the electron transfer is a through-bond process, coordination of Cu(I) to the dpp spacer lowers the energy of the LUMO of this bridge and therefore increases the electronic coupling between the excited state of the zinc porphyrin component and the ground state of the gold porphyrin electron acceptor [155].

To summarize, the problem that was addressed and issued with the rotaxane structure (**135**) was a specific case of through-bond electron transfer. A rotaxane will also allow for the study of through-space electron transfer, if advantage is taken of

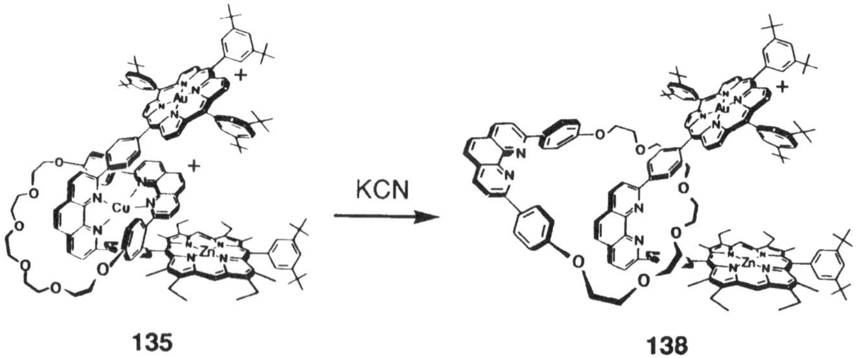

Figure 45 Removal of the metal template, by reaction of Cu(I)-complexed [2]-rotaxane (**135**) with KCN to afford [2]-rotaxane (**138**), bearing metalloporphyrins as stoppers [117].

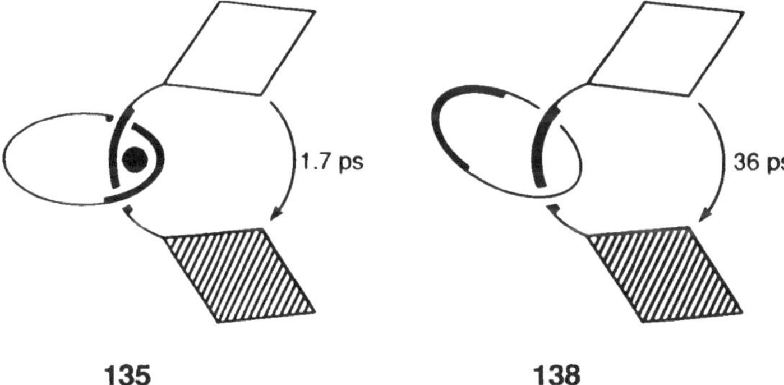

Figure 46 Schematic representation of the electron transfer events occurring in copper(I)-complexed and template-free [2]-rotaxanes (**135**) and (**138**) (same conventions as in Figure 43) [154–155].

the mechanical bond linking the dumbbell and the ring components. Specifically, the electron transfer partners must be incorporated into both components of the rotaxane. Figure 47 shows schematically the [2]-rotaxane that was designed for this purpose [160]. For synthetic reasons, the stoppers were chosen to be the zinc(II) porphyrins, therefore the gold(III) porphyrin was appended to the macrocyclic component.

The precursors and the reaction steps leading to the Cu(I)-complexed [2]-rotaxane (**142**) are represented in Figures 48 and 49. Threading of macrocycle (**89**) onto the dialdehyde derivative of dpp (**139**) afforded prerotaxane (**140**) quantitatively. The two porphyrin stoppers, identical in this case, were therefore constructed in a one-pot procedure, by reacting prerotaxane (**140**) (1 equiv.) with dipyrrylmethane (**141**) (10 equiv.) and 3,5-di-tert-butylbenzaldehyde (**132**) (8 equiv.) in dichloromethane

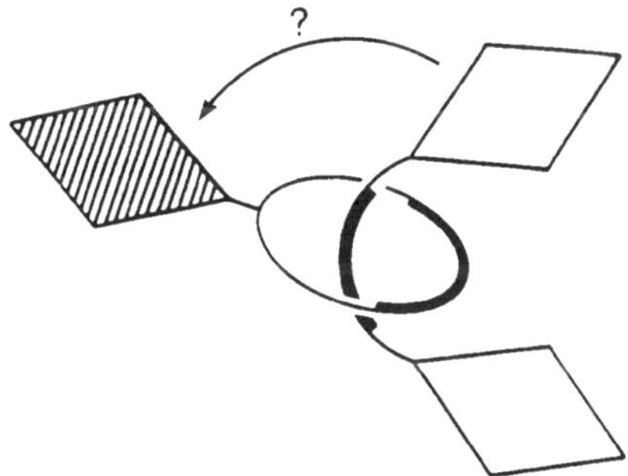

Figure 47 Schematic representation of through-space electron transfer that might occur in a [2]-rotaxane in which the electron donor zinc(II) porphyrin in the excited state and the electron acceptor gold(III) porphyrin belong to different, mechanically linked subunits.

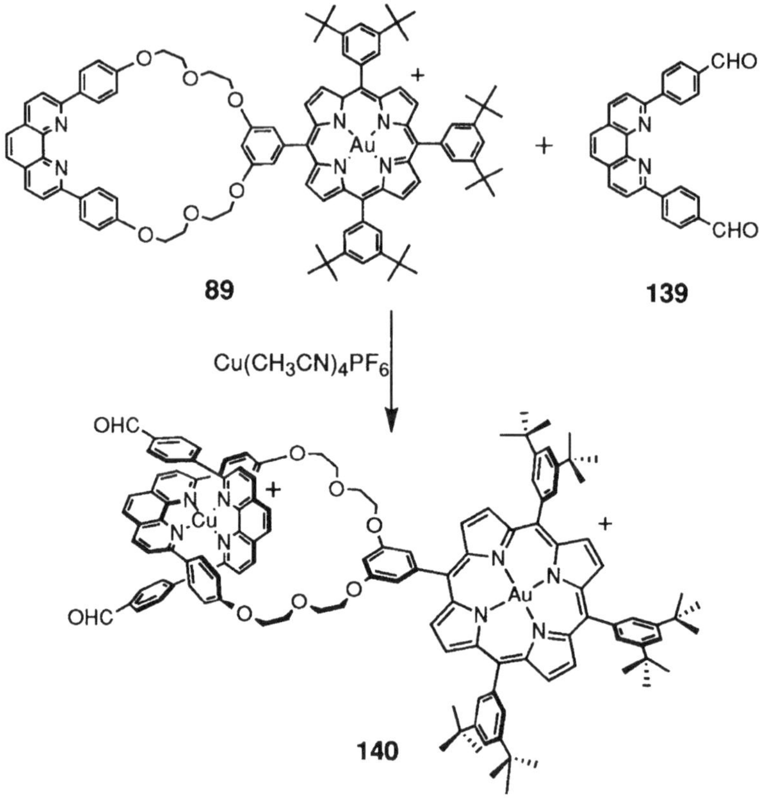

Figure 48 Copper(I)-templated threading of macrocycle (**89**) onto dpp-based dialdehyde (**139**) to afford prerotaxane (**140**) [160].

acidified with trifluoroacetic acid. Cu(I)-complexed [2]-rotaxane (**142**) was isolated in 17% yield, after oxidation of the porphyrinogen intermediates with chloranil (**133**), and metallation of the free-base porphyrinic stoppers with zinc.

Removal of the Cu^+ templating cation was achieved by reaction of copper(I)-complexed [2]-rotaxane (**142**) with KCN, affording free [2]-rotaxane (**143**) (Figure 50). Unlike [2]-rotaxane (**138**) demetallation is not accompanied by a pirouetting motion of the macrocycle. What is rather observed by ^1H NMR spectroscopy is a backward shift of the dumbbell component with respect to the macrocycle, such that the dpp chelate incorporated into the macrocycle is more embedded in the shielding field of the porphyrinic stoppers. The tetrahedral imprint of the copper(I) template could be restored by reacting the free [2]-rotaxane (**143**) with $AgBF_4$ and $LiBF_4$. This produced respectively Ag^+-complexed and Li^+-complexed [2]-rotaxanes (**144**) and (**145**) (Figure 49).

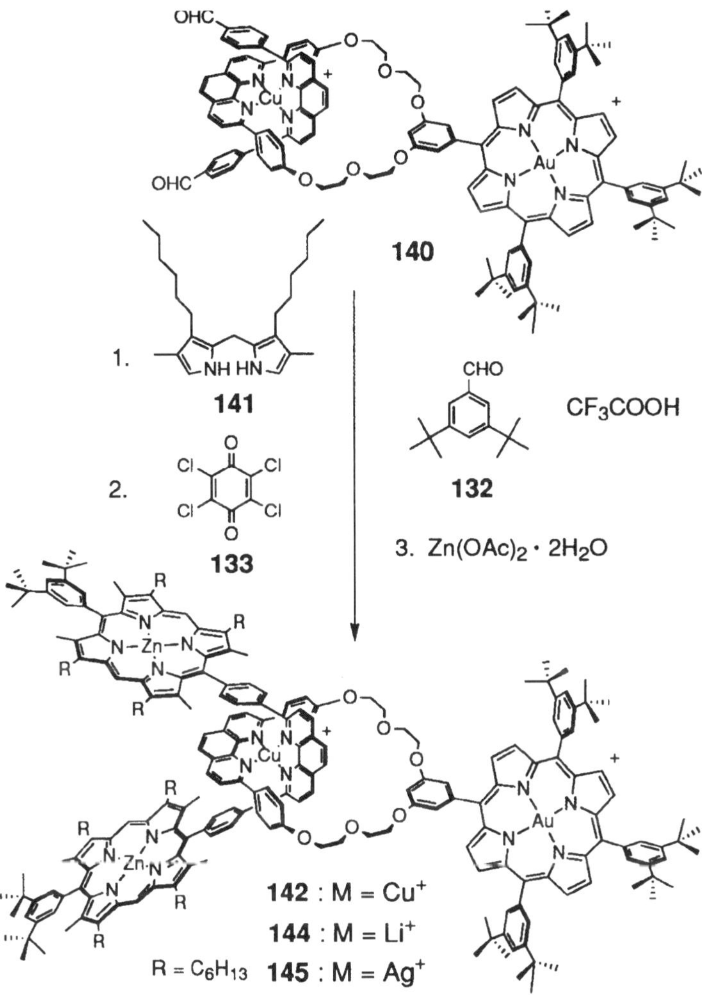

Figure 49 One-pot construction of the two porphyrinic stoppers of copper(I)-complexed [2]-rotaxane (**142**). Also represented are the silver(I)- and lithium-complexed [2]-rotaxanes (**144**) and (**145**) [160].

Preliminary experiments showed that the light emission of the zinc porphyrins was strongly quenched in the case of the metallo-rotaxanes, and, remarkably, also in the case of the free [2]-rotaxane. Whereas for the metallo-rotaxanes, the possibility that the central metal-complex fragment participates in the quenching process cannot be ruled out, for the [2]-rotaxane, it is clear that the only mechanism is through-space electron transfer to the gold(III) porphyrin moiety appended to the macrocycle.

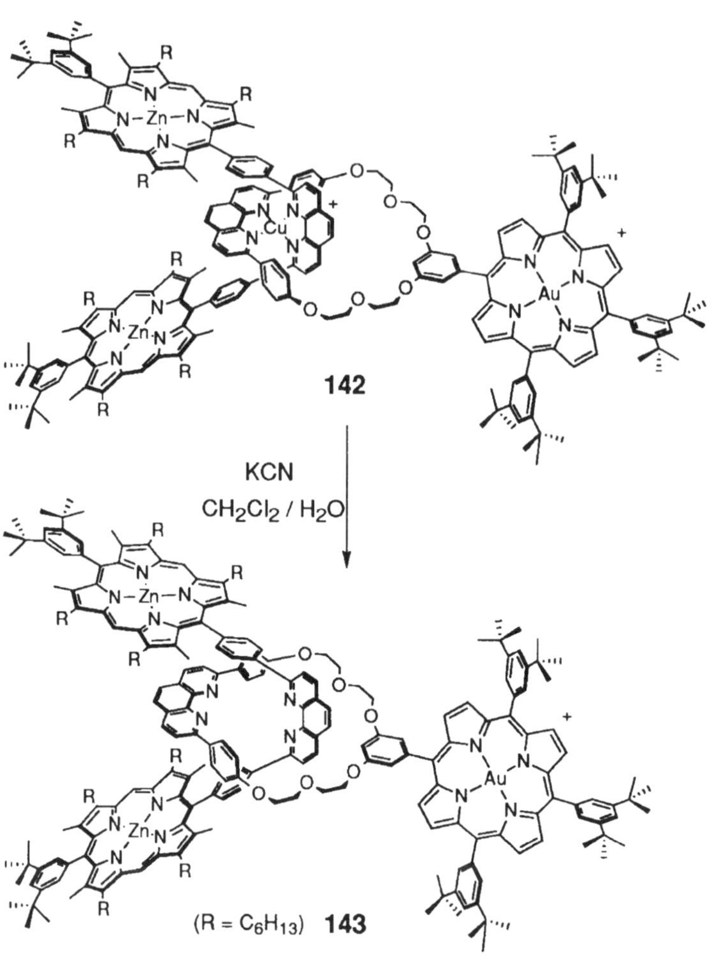

KCN
CH_2Cl_2 / H_2O

142

(R = C_6H_{13}) 143

Figure 50 Removal of the metal template, by reaction of copper(I)-complexed [2]-rotaxane (**142**) with KCN to afford [2]-rotaxane (**143**) [160].

In conclusion, the zinc(II) and gold(III) porphyrin-stoppered [2]-rotaxanes presented in this section allowed the study of two fundamental aspects of electron transfer: the through-bond and the through-space processes.

4.2 [3]-Rotaxanes

It was shown earlier that the Cu(I)-templated threading of two macrocycles onto a bis-phenanthroline-containing thread was possible, provided that the spacer linking the two chelates was well chosen. The principle of metal-templated construction of a

[3]-rotaxane from a molecular thread containing two chelating sites is represented schematically in Figure 51 [59]. At first, two chelating macrocycles (B), complexed with Cu(I), are threaded onto the ditopic molecular thread (A), bearing reactive functions X at its extremities. A prerotaxane (C) is produced (step (i)). The porphyrin stoppers are then constructed, using the terminal functions X of the prerotaxane, affording a bis-copper(I)-complexed [3]-rotaxane (D) (step (ii)).

The precursors used and the synthetic route followed to prepare a bis-copper(I)-complexed [3]-rotaxane fitted with porphyrin stoppers are shown in Figures 52 and 53. The dumbbell precursor is a $-(CH_2)_4-$ bridged bis-phenanthroline bearing two aromatic aldehyde end groups (**146**). Copper(I)-directed threading of macrocycle (**73**) (2 equiv.) was quantitative (Figure 52), as predicted by threading experiments involving the related bis-chelate (**76**).

The prerotaxane obtained (**147**) (1 equiv.) was then reacted with dipyrrylmethane (**141**) (10 equiv.), 3,5-di-tert-butylbenzaldehyde (**132**) (8 equiv.) in dichloromethane solution acidified with trifluoroacetic acid, affording, after oxidation of the porphyrinogen intermediates by excess chloranil (**133**), bis-copper(I)-complexed [3]-rotaxane (**148**) in 35% yield (Figure 53). Remarkably, a copper(I)-complexed [5]-rotaxane (**149**) could be isolated from the reaction mixture, in 8% yield. We shall discuss the formation of this 'compartmental' rotaxane in the last section. The free-base

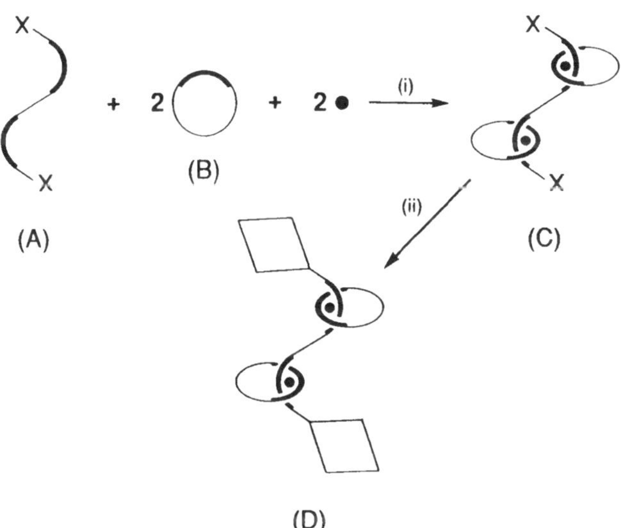

Figure 51 Principle of transition metal-templated synthesis of a [3]-rotaxane, from two chelating macrocycles (B) and a bis-chelate-containing molecular thread (A) functionalized with reactive end groups X (same conventions as in Figure 43). (ii) Threading step, affording prerotaxane (C); construction of the porphyrin stoppers providing copper(I)-complexed [3]-rotaxane (D).

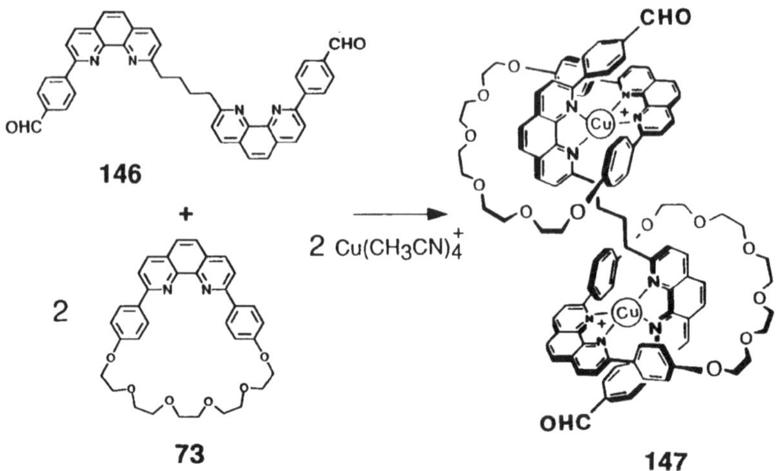

Figure 52 Copper(I)-templated threading of macrocycle (**73**) onto bis-phenathroline-containing molecular thread (**146**) to afford prerotaxane (**147**) [59].

porphyrin components of these rotaxanes were metallated with zinc, affording respectively the corresponding [3]- and [5]-rotaxanes quantitatively.

4.3 'Compartmental' [3]- and [5]-rotaxanes

The condensation of two molecules of prerotaxanes (**130**) or (**147**) with dipyrryl-methanes (**131**) or (**141**) and 3,5-di-tert-butylbenzaldehyde (**132**) produced the [3]- and [5]-rotaxanes (**136**) and (**149**) respectively, represented in Figures 44 and 53 [59, 117].

The formation of [3]-rotaxane (**136**) does not involve any participation of the benzaldehyde derivative (**132**). As shown schematically in Figure 54, two molecules of prerotaxane (A) condense with two molecules of dipyrrylmethane: in structure (C) the central stopper is a free-base porphyrin, which is shared by each threaded macrocycle, and the peripheral stoppers are gold(III) porphyrins. The formation of [5]-rotaxane (**149**) is even more remarkable, because three porphyrins are formed simultaneously, involving the condensation in one molecule of 10 molecular precursors! As before, the formation of the central porphyrin in (D) from two molecules of (B) does not involve the participation of 3,5-di-tert-butylbenzaldehyde (**132**), whereas the latter is involved in the formation of the peripheral porphyrin stoppers. These rotaxanes could be coined 'compartmental rotaxanes', by analogy with the well-known compartmental ligands [161].

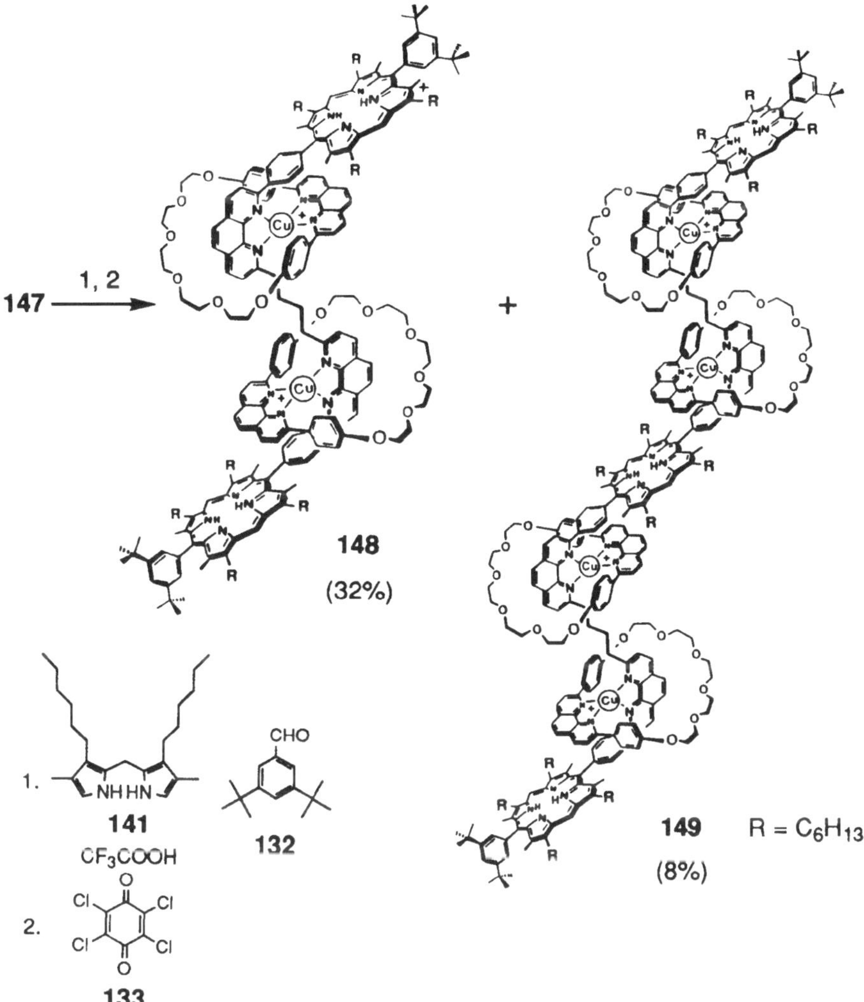

Figure 53 One-pot construction of the two porphyrin stoppers of copper(I)-complexed [3]- and [5]-rotaxanes (**148**) and (**149**) [59].

CONCLUSION AND OUTLOOK

The scope of this chapter was to present a rationale of the synthetic methods used to construct rotaxanes, with particular emphasis on the transition metal-templated route. The key step is to assemble a molecule which can be described as composed of a macrocycle threaded onto a linear molecular component. Therefore, the problem of rotaxane synthesis was mainly the problem of threading at the molecular level.

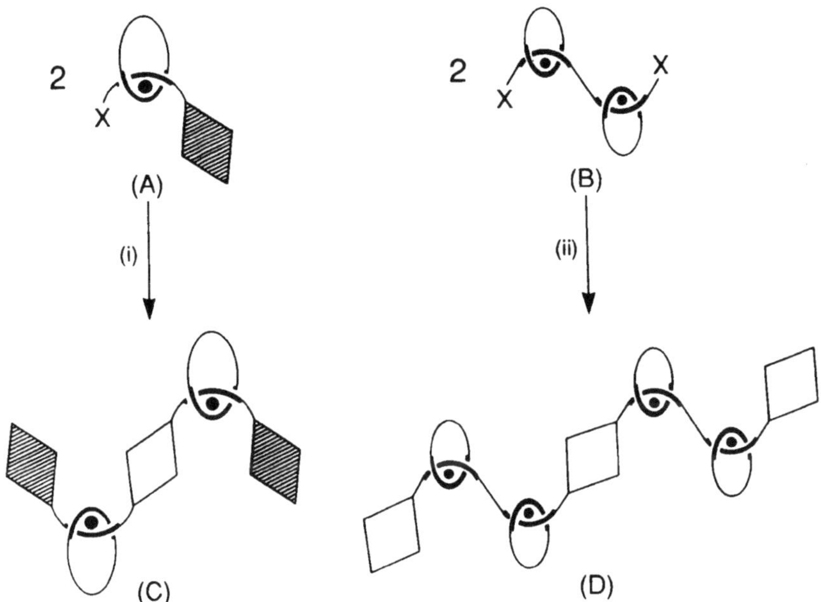

Figure 54 Schematic representation of the formation of 'compartmental' [3]- and [5]-rotaxanes (**136**) and (**149**). (i) Condensation of two pre([2]-rotaxane) units (A) to afford 'compartmental' [3]-rotaxane (C); (iii) condensation of two pre([3]-rotaxane) units (B) to afford 'compartmental' [5]-rotaxane (D) [59, 117].

This challenge has been addressed in very different ways since the very beginning, in the years 1950–60. Efficient template methods involving non-covalent interactions between the components to be assembled have been devised, since then. We have focused on the transition metal-templated threading of coordinating macrocycles onto string- or rod-shaped mono- or bis-chelates, showing that this method was a powerful tool for preparing [2]- and [3]-rotaxanes having a great variety of blocking groups: the classical trityl stopper, and its largest homologue, a dendritic fragment [65], as well as functional groups, such as metallo-porphyrins, fullerene, and transition metal-complex fragments. Polymeric rotaxanes could also be prepared by electropolymerization of threaded complexes bearing electroactive monomeric end units.

Rotaxanes are unique molecules in several respects. In particular, since the ring component is mechanically linked to the linear component, its motions are restricted to one direction of the space: along its holder. In addition, the bonding interactions which were used for elaborating the rotaxane structure, may be used now to control the translational motions of the ring along the thread. External triggers such as light, electrodes, chemical reagents or solvents may be used if the different rotaxane components are endowed with photophysical, electrochemical, chemical or certain physical properties, such as polarity. Control of ring motions in terms of speed and localization is one of the challenges chemists working on rotaxanes are faced with.

ACKNOWLEDGEMENTS

I am grateful to Jean-Pierre Sauvage for his kind invitation to give a contribution to this volume. I also warmly thank Valérie Heitz, Nathalie Solladié and Myriam Linke for their skilful synthetic work on the porphyrin-stoppered rotaxanes, and Anthony Harriman for his expertise in fast kinetics. The CNRS is acknowledged for financial support.

REFERENCES

1. Lehn, J. M. *Angew. Chem. Int. Ed. Engl.* 1988, **27**, 89–112.
2. Lehn, J.-M. *Supramolecular Chemistry, Concepts and Perspectives*, Verlag Chemie: Weinheim, 1995.
3. Schill, G. and Zollenkopf, H. *Nachr. Chem. Techn.* 1967, **15**, 149.
4. Harrison, I. T. and Harrison, S. *J. Am. Chem. Soc.* 1967, **89**, 5723–5724.
5. Frisch, H. L. and Wasserman, E. *J. Am. Chem. Soc.* 1961, **83**, 3789–3795.
6. Sokolov, V. I. *Russian Chem. Rev. (Engl. Transl.)* 1972, **41**, 452–463.
7. Walba, D. M. *Tetrahedron* 1985, **41**, 3161–3212.
8. Lüttringhaus, A., Cramer, F., Prinzbach, H. and Henglein, F. M. *Liebigs Ann. Chem.* 1958, **613**, 185–198.
9. Ogino, H. *J. Am. Chem. Soc.* 1981, **103**, 1303–1304.
10. Armspach, D., Ashton, P. R., Moore, C. P., Spencer, N., Stoddart, J. F., Wear, T. J. and Williams, D. J. *Angew. Chem. Int. Ed. Engl.* 1993, **32**, 854–858.
11. Dietrich-Buchecker, C. O., Sauvage, J.-P. and Kintzinger, J.-P. *Tetrahedron Lett.* 1983, **24**, 5095–5098.
12. Dietrich-Buchecker, C. O., Sauvage, J.-P. and Kern, J.-M. *J. Am. Chem. Soc.* 1984, **106**, 3043–3045.
13. Dietrich-Buchecker, C. O. and Sauvage, J.-P. *Chem. Rev.* 1987, **87**, 795–810.
14. Sauvage, J.-P. *Acc. Chem. Res.* 1990, **23**, 319–327.
15. Dietrich-Buchecker, C. O. and Sauvage, J.-P. *Bioorg. Chem. Frontiers* 1991, **2**, 195–248.
16. Dietrich-Buchecker, C. and Sauvage, J.-P. *Bull. Soc. Chim. Fr.* 1992., **129**, 113–120.
17. Ogino, H. and Ohata, K. *Inorg. Chem.* 1984, **23**, 3312–3316.
18. Ogino, H. *New J. Chem.* 1993, **17**, 603–688.
19. Ashton, P. R., Goodnow, T. T., Kaifer, A. E., Reddington, M. V., Slawin, A. M. Z., Spencer, N., Stoddart, J. F., Vicent, C. and Williams, D. J. *Angew. Chem. Int. Ed. Engl.* 1989, **28**, 1396–1399.
20. Amabilino, D. B. and Stoddart, J. F. *Chem. Rev.* 1995, **95**, 2725–2828.
21. Amabilino, D. B., Raymo, F. M. and Stoddart, J. F. In *Comprehensive Supramolecular Chemistry*; J.-P. Sauvage and M. W. Hosseini, eds., Pergamon: 1996, Vol. 9, pp. 85–130.

22. Wu, C., Bheda, M. C., Lim, C., Shen, Y. X., Sze, J. and Gibson, H. W. *Polym. Commun.* 1991, **32**, 204–207.
23. Gibson, H. W., Bheda, M. C. and Engen, P. T. *Progr. Polym. Sci.* 1994, **19**, 843–945.
24. Gibson, H. W. In *Large Ring Molecules*; J. A. Semlyen, ed., John Wiley and Sons, 1997, pp. 191–262.
25. Hunter, C. A. *J. Am. Chem. Soc.* 1992, **114**, 5303–5311.
26. Vögtle, F., Meier, S. and Hoss, R. *Angew. Chem. Int. Ed. Engl.* 1992, **31**, 1619–1622.
27. Jäger, R. and Vögtle, F. *Angew. Chem. Int. Ed. Engl.* 1997, **36**, 930–944.
28. Bissell, R. A., Córdova, E., Kaifer, A. and Stoddart, J. F. *Nature* 1994, **369**, 133–137.
29. Collin, J.-P., Gaviña, P. and Sauvage, J.-P. *New J. Chem.* 1997, **21**, 525–528.
30. Murakami, H., Kawabuchi, A., Kotoo, K., Kunitake, M. and Nakashima, N. *J. Am. Chem. Soc.* 1997, **119**, 7605–7606.
31. Lane, A. S., Leigh, D. A. and Murphy, A. *J. Am. Chem. Soc.* 1997, **119**, 11092–11093.
32. Schill, G. and Lüttringhaus, A. *Angew. Chem.* 1964, **76**, 567–568.
33. Harrison, I. T. *J. Chem. Soc., Chem. Commun.* 1972, 231–232.
34. Harrison, I. T. *J. Chem. Soc. Perkin Trans. I* 1974, 301–304.
35. Agam, G., Graiver, D. and Zilkha, A. *J. Am. Chem. Soc.* 1976, **98**, 5206–5214.
36. Agam, G. and Zilkha, A. *J. Am. Chem. Soc.* 1976, **98**, 5214–5216.
37. Harrison, I. T. *J. Chem. Soc., Chem. Commun.* 1977, 384–385.
38. Schill, G., Beckmann, W., Schweickert, N. and Fritz, H. *Chem. Ber.* 1986, **119**, 2647–2655.
39. Gilliatt, V. J., Sultany, C. M. and Butcher, J. A., Jr. *Tetrahedron Lett.* 1992, **33**, 6255–6258.
40. Gibson, H. W. and Engen, P. T. *New J. Chem.* 1993, **17**, 723–727.
41. Gibson, H. W., Liu, S., Lecavalier, P., Wu, C. and Shen, Y. X. *J. Am. Chem. Soc.* 1995, **117**, 852–874.
42. Schill, G. *Catenanes, Rotaxanes and Knots*; Academic Press, 1971.
43. Gibson, H. W. and Marand, H. *Adv. Mater.* 1993, **5**, 11–21.
44. Benniston, A. C. *Chem. Soc. Rev.* 1996, **25**, 427–435.
45. Harriman, A. and Sauvage, J.-P. *Chem. Soc. Rev.* 1996, 41–48.
46. Chambron, J.-C, Dietrich-Buchecker, C. O. and Sauvage, J.-P. In *Comprehensive Supramolecular Chemistry*; J.-P. Sauvage and M. W. Hosseini, eds. Pergamon, 1996, Vol. 9, pp. 44–84.
47. The terms 'slippage' and 'clipping' have been introduced by Stoddart and coworkers.
48. Shen, Y. X., Xie, D. and Gibson, H. W. *J. Am. Chem. Soc.* 1994, **116**, 537–548.
49. Anelli, P. L., Ashton, P. R., Spencer, N., Slawin, A. M. Z., Stoddart, J. F. and Williams, D. J. *Angew. Chem. Int. Ed. Engl.* 1991, **30**, 1036–1039.
50. Ashton, P. R., Philp, D., Spencer, N. and Stoddart, J. F. *J. Chem. Soc., Chem. Commun.* 1991, 1677–1679.
51. Ashton, P. R., Philp, D., Spencer, N., Stoddart, J. F. and Williams, D. J. *J. Chem. Soc., Chem. Commun.* 1994, 181–184.
52. Lüttringhaus, A. and Isele, G. *Angew. Chem.* 1967, **79**, 945–946.
53. Wasserman, E. *J. Am. Chem. Soc.* 1960, **82**, 4433–4434.
54. Schill, G., Schweickert, N., Fritz, H. and Vetter, W. *Angew. Chem. Int. Ed. Engl.* 1983, **22**, 889–891.
55. Schill, G., Schweickert, N., Fritz, H. and Vetter, W. *Chem. Ber.* 1988, **121**, 961–970.
56. Ashton, P. R., Glink, P. T., Stoddart, J. F., Tasker, P. A., White, A. J. P. and Williams, D. J. *Chem. Eur. J.* 1996, **2**, 729–736.
57. Vögtle, F., Dünnwald, T., Händel, M., Jäger, R., Meier, S. and Harder, G. *Chem. Eur. J.* 1996, **2**, 640–643.
58. Anderson, S. and Anderson, H. L. *Angew. Chem. Int. Ed. Engl.* 1996, **35**, 1956–1959.

59. Solladié, N., Chambron, J.-C., Dietrich-Buchecker, C. O. and Sauvage, J.-P. *Angew. Chem. Int. Ed. Engl.* 1996, **35**, 906–909.
60. Asakawa, M., Ashton, P. R., Ballardini, R., Balzani, V., Bělohradský, M., Gandolfi, M. T., Kocian, O., Prodi, L., Raymo, F. M., Stoddart, J. F. and Venturi, M. *J. Am. Chem. Soc.* 1997, **119**, 302–310.
61. Ashton, P. R., Ballardini, R., Balzani, V., Bělohradský, M., Gandolfi, M. T., Philp., D., Prodi, L., Raymo, F. M., Reddington, M. V., Spencer, N., Stoddart, J.-F., Venturi, M. and Williams, D. J. *J. Am. Chem. Soc.* 1996, **118**, 4931–4951.
62. Amabilino, D. B., Ashton, P. R., Bělohradský, M., Raymo, F. M. and Stoddart, J. F. *J. Chem. Soc., Chem. Commun.* 1995, 751–753.
63. Ashton, P. R., Johnston, M. R., Stoddart, J. F., Tolley, M. S. and Wheeler, J. W. *J. Chem. Soc., Chem. Commun.* 1992, 1128–1131.
64. Ashton, P. R., Philp, D., Spencer, N. and Stoddart, J. F. *J. Chem. Soc., Chem. Commun.* 1992, 1124–1128.
65. Amabilino, D. B., Ashton, P. R., Balzani, V., Brown, C. L., Credi, A., Fréchet, J. M. J., Leon, J. W., Raymo, F. M., Spencer, N., Stoddart, J. F. and Venturi, M. *J. Am. Chem. Soc.* 1996, **118**, 12012–12020.
66. Yamamoto, C., Okamoto, Y., Schmidt, T., Jäger, R. and Vögtle, F. *J. Am. Chem. Soc.* 1997, **119**, 10547–10548.
67. Pedersen, C. J. *J. Am. Chem. Soc.* 1967, **89**, 2495–2496.
68. Pedersen, C. J. *J. Am. Chem. Soc.* 1967, **89**, 7017–7036.
69. Pedersen, C. J. *Angew. Chem. Int. Ed. Engl.* 1988, **27**, 1021–1027.
70. Saenger, W. *Angew. Chem. Int. Ed. Engl.* 1980, **19**, 344–362.
71. Freudenberg, K. and Cramer, F. *Z. Naturforsch.* 1948, **B3**, 464.
72. Cramer, F. and Henglein, F. M. *Chem. Ber.* 1957, **90**, 2561–2571.
73. Lüttringhaus, A., Cramer, F. and Prinzbach, H. *Angew. Chem.* 1957, **69**, 137.
74. Yamanari, K. and Shimura, Y. *Bull. Chem. Soc. Jpn.* 1983, **56**, 2283–2289.
75. Isnin, R. and Kaifer, A. E. *J. Am. Chem. Soc.* 1991, **113**, 8188–8190.
76. Hannack, R. B., Färber, G., Konrat, R. and Kraütler, B. *J. Am. Chem. Soc.* 1997, **119** 2313–2314.
77. Wenz, G., von der Bey, E. and Schmidt, L. *Angew. Chem. Int. Ed. Engl.* 1992, **31**, 783–785.
78. Wenz, G., Wolf, F., Wagner, M. and Kubik, S. *New. J. Chem.* 1993, **17**, 729–738.
79. Anderson, S., Claridge, T. D. W. and Anderson, H. L. *Angew. Chem. Int. Ed. Engl.* 1997, **36**, 1310–1313.
80. Wenz, G. and Keller, B. *Angew. Chem. Int. Ed. Engl.* 1992, **31**, 197–199.
81. Harada, A. and Kamachi, M. *Macromolecules* 1990, **23**, 2821–2823.
82. Harada, A. and Kamachi, M. *J. Chem. Soc., Chem. Commun.* 1990, 1322–1323.
83. Harada, A., Li, J. and Kamachi, M. *Chem. Lett.* 1993, 237–240.
84. Harada, A., Li, J. and Kamachi, M. *Nature* 1992, **356**, 325–327.
85. Harada, A., Li, J. and Kamachi, M. *Nature*, 1993, **364**, 516–518.
86. Alwood, B. L., Spencer, N., Shariari-Zavareth, H., Stoddart, J. F. and Williams, D. J. *J. Chem. Soc., Chem. Commun.* 1987, 1064–1066.
87. Ashton, P. R., Odell, B., Reddington, M. V., Slawin, A. M. Z., Stoddart, J. F. and Williams, D. J. *Angew. Chem. Int. Ed. Engl.* 1988, **27**, 1550–1553.
88. Anelli, P. L., Ashton, P. R., Ballardini, R., Balzani, V., Delgado, M., Gandolfi, M. T., Goodnow, T. T., Kaifer, A. E., Philp, D., Pietraszkiewicz, M., Prodi, L., Reddington, M. V., Slawin, A. M. Z., Spencer, N., Stoddart., J. F., Vicent, C. and Williams, D. J. *J. Am. Chem. Soc.* 1992, **114**, 193–218.
89. Ashton, P. R., Iqbal, S., Stoddart, J. F. and Tinker, N. D., *Chem. Commun.* 1996, 479–481.

90. Asakawa, M., Ashton, P. R., Iqbal, S., Stoddart, J. F., Tinker, N. D., White, A. J. P. and Williams, D. J. *Chem. Commun.* 1996, 483–486.
91. Ashton, P. R., Grognuz, M., Slawin, A. M. Z., Stoddart, J. F. and Williams, D. J. *Tetrahedron Lett.* 1991, **32**, 6235–6238.
92. Anelli, P. L., Spencer, N. and Stoddart, J. F. *J. Am. Chem. Soc.* 1991, **113**, 5131–5133.
93. Ashton, P. R., Bissell, R. A., Spencer, N., Stoddart, J. F. and Tolley, M. S., *Synlett* 1992, 914–918.
94. Ashton, P. R., Bissell, R. A., Górski, R., Philp, D., Spencer, N., Stoddart, J. F. and Tolley, M. S. *Synlett* 1992, 919–922.
95. Ashton, P. R., Bissell, R. A., Spencer, N., Stoddart, J. F. and Tolley, M. S. *Synlett* 1992, 923–926.
96. Benniston, A. C. and Harriman, A. *Angew. Chem. Int. Ed. Engl.* 1993, **32**, 1459–1461.
97. Ashton, P. R., Huff, J., Menzer, S., Parsons, I. W., Preece, J. A., Stoddart, J. F., Tolley, M. S., White, A. J. P. and Williams, D. J. *Chem. Eur. J.* 1996, **2**, 31–44.
98. Amabilino, D. B., Ashton, P. R., Bělohradský, M., Raymo, F. M. and Stoddart, J. F. *J. Chem. Soc., Chem. Commun.* 1995, 747–750.
99. Ashton, P. R., Campbell, P. J., Chrystal, E. J. T., Glink, P. T., Menzer, S., Philp, D., Spencer, N., Stoddart, J. F., Tasker, P. A and Williams, D. J. *Angew. Chem. Int. Ed. Engl.* 1995, **34**, 1865–1869.
100. Ashton, P. R., Chrystal, E. J. T., Glink, P. T., Menzer, S., Schiavo, C., Stoddart, J. F., Tasker, P. A. and Williams, D. J. *Angew. Chem. Int. Ed. Engl.* 1995, **34**, 1869–1871.
101. Ashton, P. R., Chrystal, E. J. T., Glink, P. T., Menzer, S., Schiavo, C., Spencer, N., Stoddart, J. F., Tasker, P. A, White, A. J. P. and Williams, D. J. *Chem. Eur. J.* 1996, **2**, 709–728.
102. Ottens-Hildebrandt, S., Meier, S., Schmidt, W. and Vögtle, F. *Angew. Chem. Int. Ed. Engl.* 1994, **33**, 1767–1770.
103. Vögtle, F., Händel, M., Meier, S., Ottens-Hildebrandt, S., Ott, F. and Schmidt, T. *Liebigs Ann.* 1995, 739–743.
104. Vögtle, F., Jäger, R., Händel, M., Ottens-Hildebrandt, S. and Schmidt, W. *Synthesis* 1996, 353–356.
105. Vögtle, F., Ahuis, F., Baumann, S. and Sessler, J. L. *Liebigs Ann.* 1996, 921–926.
106. Händel, M., Plevoets, M., Gestermann, S. and Vögtle, F. *Angew. Chem. Int. Ed. Engl.* 1997, **36**, 1199–1201.
107. Schill, G. *Chem. Ber.* 1967, **100**, 2021–2037.
108. Schill, G. and Zollenkopf, H. *Liebigs Ann. Chem.* 1969, **721**, 53–74.
109. Schill, G., Zürcher, C. and Vetter, W. *Chem. Ber.* 1973, **106**, 228–235.
110. Schill, G. and Henschel, R. *Liebigs Ann. Chem.* 1970, **731**, 113–119.
111. Ballardini, R., Balzani, V., Gandolfi, M. T., Prodi, L., Venturi, M., Philp, D., Ricketts, H. G. and Stoddart, J. F. *Angew. Chem. Int. Ed. Engl.* 1993, **32**, 1301–1303.
112. Johnston, A. G., Leigh, D. A., Murphy, A., Smart, J. P. and Deegan, M. D. *J. Am. Chem. Soc.* 1996, **118**, 10662–10663.
113. Leigh, D. A., Murphy, A., Smart, J. P. and Slawin, A. M. Z. *Angew. Chem. Int. Ed. Engl.* 1997, **36**, 729–732.
114. Busch, D. H., Vance, A. L. and Kolchinskii, A. G. In *Comprehensive Supramolecular Chemistry*, J.-P. Sauvage and M. W. Hosseini, eds., Pergamon, 1996, Vol. 9, pp. 1–42.
115. Arnaud-Neu, F., Marques, E., Schwing-Weill, M.-J., Dietrich-Buchecker, C., Sauvage, J.-P. and Weiss, J. *New. J. Chem.* 1988, **12**, 15–19.
116. Dietrich-Buchecker, C. O. and Sauvage, J.-P. *Tetrahedron* 1990, **46**, 503–512.
117. Chambron, J.-C., Heitz, V. and Sauvage, J.-P. *J. Am. Chem. Soc.* 1993, **115**, 12378–12384.
118. Johnston, A., Leigh, D. A., Pritchard, R. J. and Deegan, M. D. *Angew. Chem. Int. Ed. Engl.* 1995, **34**, 1209–1211.

119. Albrecht-Gary, A. M., Saad, Z., Dietrich-Buchecker, C. O., and Sauvage, J.-P. *J. Am. Chem. Soc.* 1985, **107**, 3205–3209.
120. Albrecht-Gary, A. M., Dietrich-Buchecker, C. O., Saad, Z. and Sauvage, J.-P. *J. Am. Chem. Soc.* 1988, **110**, 1467–1472.
121. Chambron, J.-C., Dietrich-Buchecker, C. O., Nierengarten, J.-F. and Sauvage, J.-P. *J. Chem. Soc., Chem. Commun.* 1993, 801–804.
122. Chambron, J.-C., Dietrich-Buchecker, C., Nierengarten, J.-F., Sauvage, J.-P., Solladié, N., Albrecht-Gary, A.-M. and Meyer, M. *New J. Chem.* 1995, **19**, 409–426.
123. Amabilino, D. B., Dietrich-Buchecker, C. O. and Sauvage, J.-P. *J. Am. Chem. Soc.* 1996, **118**, 3285–3286.
124. Sleiman, H., Baxter, P. N. W., Lehn, J.-M. and Rissanen, K. *J. Chem. Soc., Chem. Commun.* 1995, 715–716.
125. Sleiman, H., Baxter, P. N. W., Lehn, J. M., Airola, K. and Rissanen, K. *Inorg. Chem.* 1997, **36**, 4734–4742.
126. Wu, C., Lecavalier, P. R., Shen, Y. X. and Gibson, H. W. *Chem. Mater.* 1991, **3**, 569–572.
127. Diederich, F., Dietrich-Buchecker, C., Nierengarten, J.-F. and Sauvage, J.-P. *J. Chem. Soc., Chem. Commun.* 1995, 781–782.
128. Dietrich-Buchecker, C. O., Khémiss, A. and Sauvage, J.-P. *J. Chem. Soc., Chem. Commun.* 1986, 1376–1378.
129. Dietrich-Buchecker, C. O., Hemmert, C., Khémiss, A.-K. and Sauvage, J.-P. *J. Am. Chem. Soc.* 1990, **112**, 8002–8008.
130. Dietrich-Buchecker, C. O., Edel, A., Kintzinger, J.-P. and Sauvage, J.-P. *Tetrahedron* 1987, **43**, 333–344.
131. Cárdenas, D. J., Gaviña, P. and Sauvage, J.-P. *Chem. Commun.* 1996, 1915–1916.
132. Cárdenas, D. J., Gaviña, P. and Sauvage, J.-P. *J. Am. Chem. Soc.* 1997, **119**, 2656–2664.
133. Gaviña, P. and Sauvage J.-P. *Tetrahedron Lett.* 1997, **38**, 3521–3524.
134. Federlin, P., Kern, J.-M., Rastegar, A., Dietrich-Buchecker, C., Marnot, P. A. and Sauvage, J.-P. *New J. Chem.* 1990. **14**, 9–12.
135. Livoreil, A., Dietrich-Buchecker, C. O. and Sauvage, J.-P. *J. Am. Chem. Soc.* 1994, **116**, 9399–9400.
136. Livoreil, A., Sauvage, J.-P., Armaroli, N., Balzani, V., Flamigni, L. and Ventura, B. *J. Am. Chem. Soc.* 1997, **119**, 12114–12124.
137. Kern, J.-M., Sauvage, J.-P., Bidan, G., Billon, M. and Divisia-Blohorn, B. *Adv. Mater.* 1996, **8**, 580–582.
138. Zhu, S. S., Carroll, P. J. and Swager, T. M. *J. Am. Chem. Soc.* 1996, **118**, 8713–8714.
139. Chardon-Noblat, S. and Sauvage, J.-P. *Tetrahedron* 1991, **47**, 5123–5132.
140. Heitz, V., Chardon-Noblat, S. and Sauvage, J.-P. *Tetrahedron Lett.* 1991, **32**, 197–198.
141. Pascard, C., Guilhem, J., Chardon-Noblat, S. and Sauvage, J.-P. *New J. Chem.* 1993, **17**, 331–335.
142. Deisenhofer, J., Epp, O., Miki, K., Huber, R. and Michel, H. *J. Mol. Biol.* 1984, **180**, 385–398.
143. Huber, R. *Angew. Chem. Int. Ed. Engl.* 1989, **28**, 848–849.
144. Deisenhofer, J. and Michel, H. *Angew. Chem. Int. Ed. Engl.* 1989, **28**, 829–847.
145. Woodbury, N. W., Becker, M., Middendorf, D. and Parson, W. W. *Biochemistry* 1985, **24**, 7516–7521.
146. Breton, J., Martin, J.-L., Migus, A., Antonetti, A. and Orszag, A. *Proc. Natl. Acad. Sci. U.S.A.* 1986, **83**, 5121–5125.
147. Martin, J.-L., Breton, J., Hoff, A. J., Migus, A. and Antonetti, A. *Proc. Natl. Acad. Sci. U.S.A.* 1986, **83**, 957–961.
148. Brun, A. M., Atherton, S., Harriman, A., Heitz, V. and Sauvage, J.-P. *J. Am. Chem. Soc.* 1992, **114**, 4632–4639.

149. Brun, A., Atherton, S. J., Harriman, A., Heitz, V. and Sauvage, J.-P. *J. Am. Chem. Soc.* 1992, **114**, 4632–4639.
150. Chambron, J.-C., Heitz, V. and Sauvage, J.-P. *J. Chem. Soc., Chem. Commun.* 1992, 1131–1133.
151. Arsenault, G. P., Bullock, E. and MacDonald, S. F. *J. Am. Chem. Soc.* 1960, **82**, 4384–4395.
152. Lindsey, J. S., Hsu, H. C. and Schreiman, I. C. *Tetrahedron Lett.* 1986, **27**, 4969–4970.
153. Lindsey, J. S. and Wagner, R. W. *J. Org. Chem.* 1989, **54**, 828–836.
154. Chambron, J.-C., Harriman, A., Heitz, V. and Sauvage, J.-P. *J. Am. Chem. Soc.* 1993, **115**, 6109–6114.
155. Chambron, J.-C., Harriman, A., Heitz, V. and Sauvage, J.-P. *J. Am. Chem. Soc.* 1993, **115**, 7419–7425.
156. Harriman, A., Heitz, V., Chambron and J.-C. Sauvage, J.-P. *Coord. Chem. Rev.* 1994, **132**, 229–234.
157. Marcus, R. A. and Sutin, N. *Biochim. Biophys. Acta* 1985, **811**, 265–322.
158. Hush, N. S. *Coord. Chem. Rev.* 1985, **64**, 135–157.
159. Todd, M. D., Nitzan, A. and Ratner, M. A. *J. Phys. Chem.* 1993, **97**, 29–33.
160. Linke, M., Chambron, J.-C., Heitz, V. and Sauvage, J.-P. *J. Am. Chem. Soc.* 1997, **119**, 11329–11330.
161. Casellato U., Vigato, P. A., Fenton, D. E. and Vidali, M. *Chem. Soc. Rev.* 1979, **8**, 199–220.

Chapter 7

Metallomesogens—Supramolecular Organization of Metal Complexes in Fluid Phases

SIMON R. COLLINSON AND DUNCAN W. BRUCE
University of Exeter, UK

1 INTRODUCTION

Supramolecular chemistry is often thought of as 'chemistry beyond the molecule' [1]. Such a description encompasses the whole field of liquid crystals, where it is non-covalent, intermolecular interactions which determine the molecular organization leading to the various liquid crystalline phases or, in the extreme, to a totally disordered isotropic phase [2]. The science of liquid crystals is really the art of balancing the various intermolecular interactions to achieve a desired liquid crystal phase rather than an ordered crystalline phase. Nature demonstrates this art in its highest form in the self-organization of lipids to produce liposomes and cellular membranes.

In organic liquid crystals the dipole–dipole and dispersion forces are responsible for holding the molecules in anisotropic supramolecular arrangements. The incorporation of metal ions into liquid crystal molecules, so called metallomesogens, further extends the possible intermolecular interactions to include weak metal–metal and metal–ligand effects. A great deal of the early research into metallomesogens involved the coordination of metal ions to suitable donor sites within organic liquid crystals to assess the effect upon the liquid crystalline properties [3]. However, further work has shown that the presence of metal ions can, in favourable circumstances, confer liquid crystalline properties upon molecules exhibiting no

Transition Metals in Supramolecular Chemistry, edited by J. P. Sauvage.

such properties [4]. The incorporation of metal ions also presents opportunities in studying molecules that are paramagnetic [5], redox active [6], highly coloured [7] and highly polarizable [8]. The range of geometries exhibited by metal ions is also far greater than those found in purely organic molecules. Several excellent reviews [9] of metallomesogens have already appeared, however, it is the aim of the current text to try and emphasize certain supramolecular aspects in representative metallomesogen systems.

After an introduction, the review is divided into three main parts looking at three different types of organization leading to mesophase formation. In the first, we consider the mesophases formed by rod-like or calamitic molecules, these tend to be nematic or smectic. Next we look at columnar phases, concentrating primarily on those formed by disc-like molecules, although mentioning briefly polycatenar mesogens. Finally we consider lyotropic liquid-crystalline systems, a newer and expanding area in metallomesogens.

1.1 Terminology

A brief explanation of the terms pertinent to liquid crystal science is undoubtedly warranted for those readers not familiar with the field. The liquid crystal state is found between the solid and liquid states, combining properties from both the crystalline (optical, magnetic and dielectric anisotropy) and the liquid (molecular mobility and fluidity) states. Consequently a liquid crystalline molecule is described as being mesomorphic (from the Greek, *mesos morphe*, i. e. between two states) [9a] and is itself described as a mesogen. The crystal state is usually denoted by Crys or K and melts to form one or more liquid crystal states (mesophases) before being said to clear into an isotropic liquid (I). In thermotropic liquid crystals such transitions are effected by the action of heat, whereas in lyotropic liquid crystals the transitions are observed primarily by the effect of solvent, with temperature as a second variable. Molecules which display both thermotropic and lyotropic mesomorphism are described as being amphotropic.

Thermotropic liquid crystals can then be further subdivided into high molecular mass, main and side-chain polymers [10] and low molecular mass, the latter class of compounds being one of the areas of this review. The phases exhibited by the low molecular mass molecules are then properly described with reference to the symmetry and/or supramolecular geometry of the phases, which are briefly introduced here and are discussed in more detail further below. Thus, the most disordered mesophase is the nematic (N), which is found for calamitic molecules (N), discoidal molecules (N_D) and columnar aggregates (N_C), among others. The more ordered lamellar or smectic phases (S) [11, 12] are commonly shown by calamitic molecules, and there exists a variety of such phases distinguished by a subscripted letter (e. g. S_A, S_B). Columnar phases (often, if incorrectly, referred to as discotic phases) may be formed from stacks of disc-like molecules, or from columnar aggregates of so-called polycatenar mesogens [13] which are discussed below. Although such classifications are valuable, dichotomies exist, for instance where molecules demonstrate nematic and lamellar phases, when they themselves are not of a strictly classical rod shape [14]. A final point to note at this stage is that incorporation of chirality within a mesogenic molecule can lead to chiral modifications of the mesophase, for example the formation of the chiral nematic phase (N*) [15].

Similarly, lyotropic phases may be further subdivided into nematic (N), lamellar (L), micellar cubic (I), bicontinuous cubic (V) and hexagonal (H), with the cubic and hexagonal phases being further classified as normal (oil-in-water, e. g. V_1) or reversed (water-in-oil, e. g. V_2).

1.2 Characterization of liquid crystal phases

The identification of mesophases usually involves three principal techniques namely hot-stage polarizing optical microscopy, differential scanning calorimetry and small angle X-ray scattering.

1.2.1 Polarized optical microscopy [16]

This is usually the initial method of characterization and, in skilled hands, can lead to complete characterization of the mesophase. Typically, the sample (less than 1 mg) is placed between two microscope cover slips and these are in turn placed on a suitably controlled hot-stage through which there is an optical path. The hot-stage itself is mounted on the working stage of a microscope. The technique makes use of the birefringence of mesophases, in that light incident on the samples is first plane polarized, and there is a second polarizer set at 90° to the first between the sample and the objectives (Figure 1). Consequently, in the mesophase two refracted rays are produced which interfere with one another (strictly, the light becomes eliptically polarized) to give a characteristic interference pattern or texture. This contrasts with the situation where the material is in the isotropic state when the sample appears black between the crossed polaroid as only one refractive index is seen. These textures are diagnostic of a particular mesophase and are best developed when they are obtained as part of a cooling sequence, ideally generated from the isotropic phase to obtain what is known as a natural texture [12, 16]. The sequence in which phases are observed can also help with this identification, as mesophases are generally observed in a given thermodynamic order. In certain circumstances the surfaces of the microscope cover slips may be treated to induce surface alignment of the molecules to aid in the phase identification.

Another very useful method often employed in conjunction with microscopy is miscibility. The simplest use of the technique is to bring together on the cover slip an unknown compound with a characterized compound as a contact preparation, with

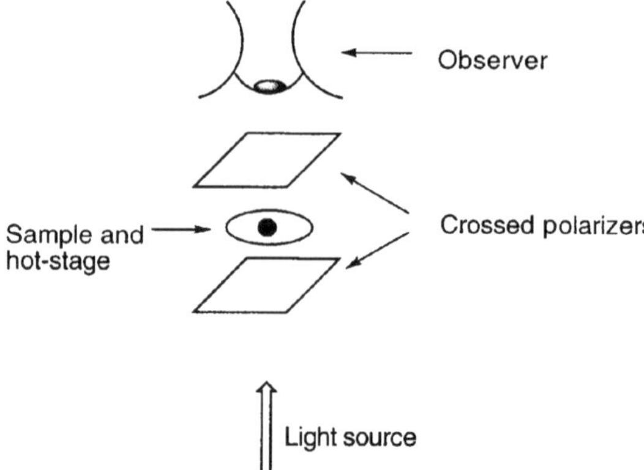

Figure 1 Schematic view of a polarized, hot-stage microscope.

both in their mesophases. If the two materials are co-miscible, then both have the same mesophase (at the temperature in question) and this can be useful conformation of phase identification. Unfortunately, if the two materials are immiscible then no information is obtained as two materials in the same phases are not necessarily miscible (e. g. water and chloroform, both of which are isotropic).

1.2.2 Differential scanning calorimetry (DSC) [17]

In this experiment, the change in the heat capacity of the sample is measured as a function of temperature leading to a measure of the enthalpy exchange accompanying a phase transition. From the thermodynamic point of view, melting transitions are strongly first order while transitions between mesophases are weakly first order or may even be second order (e. g. S_A–N and S_C–S_A can be second order). Information about the phase transition may therefore be derived from the relative magnitudes of the transition enthalpies, so that melting enthalpies are much larger than those found for transitions between mesophases or between a mesophase and an isotropic fluid. While such information is quite useful, it does not allow generalizations to be made and the corresponding entropy changes can be more useful. These can be extracted as it is assumed that at a transition, the system is at equilibrium and, therefore, $\Delta G = 0$ (hence, $\Delta H/T = \Delta S$).

The technique is strictly complementary to optical microscopy, as all changes in optical texture do not necessarily correspond to a change in mesophase type and similarly all phase changes do not always lead to an easily identifiable change in texture. Therefore DSC traces should always be compared with the results of the optical microscopy to be sure of proper correspondence.

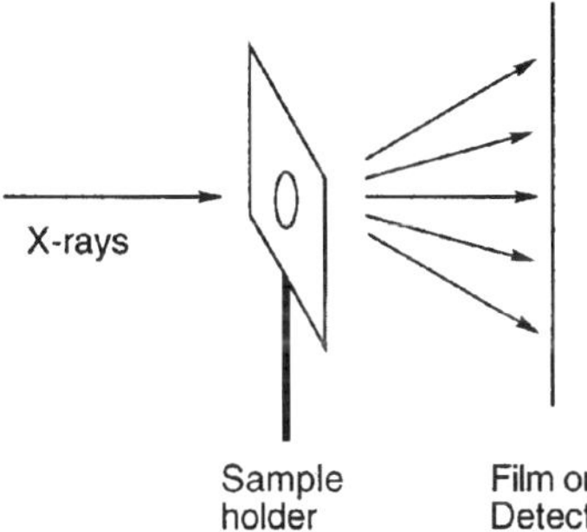

Figure 2 Schematic view of the X-ray scattering experiment.

1.2.3 Small-angle X-ray scattering [18]

X-ray scattering is a very powerful tool for mesophase identification and can provide a great deal of information; it is often the only unequivocal means of phase identification. The experiment is rather simple in principle and relies on the fact that mesomorphic structures are periodic and can therefore diffract; a schematic cartoon of the experiment is shown in Figure 2.

A peak will be observed when the diffracted X-rays interfere constructively. This occurs when the path length difference from two neighbouring regions, that is, $2d \sin \theta$, equals a multiple of the wavelength of the X-rays. Therefore a peak will be seen when the Bragg condition is satisfied:

$$n\lambda = 2d \sin \theta$$

For example in a smectic phase, diffraction lines corresponding to both the layer periodicity and the side-to-side periodicity can be observed; comparison of the observed layer periodicity with the calculated extended molecular length can give information about tilt angles and interdigitation. Similarly, if there is symmetry within the smectic layer this can also be observed as a series of peaks in the wide-angle region. Better data may be obtained if the samples are aligned in a field (typically around 0.6 T for a nematic phase), which in turn allows additional information to be obtained concerning the molecular arrangement.

2 THE STRUCTURE OF NEMATIC AND LAMELLAR (SMECTIC) PHASES

These phases (Figure 3) are commonly thought of as being of calamitic molecules and, for ease of explanation, they will be considered as such in the following section, although as mentioned previously, such a generalization has to be treated with caution. Consequently a nematic phase can be viewed as containing molecules whose long molecular axes are aligned, on average, in the same direction, usually defined by the

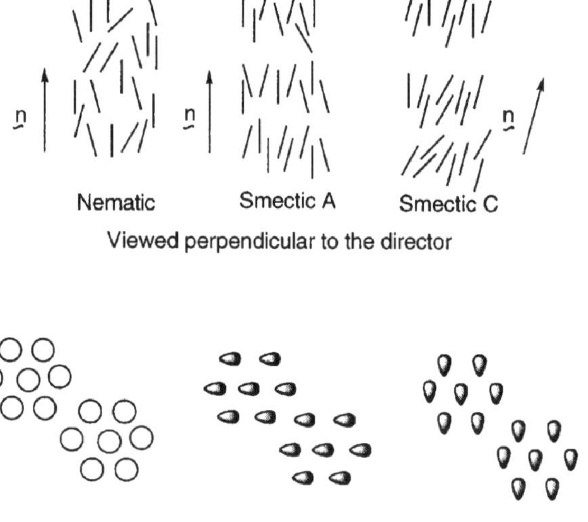

Figure 3 Idealized molecular arrangements within calamatic mesophases.

director, $\underset{\sim}{n}$. There is, however, no positional order between the molecules. The obvious question arises as to why the molecules do not simply adopt a random, isotropic structure. Firstly, the anisotropic molecular structure (in this case, rod-like) is an important factor with respect to the possible packing of the molecules. Furthermore the molecules have an anisotropy of polarizability and often possess a permanent dipole. These properties result in the intermolecular dispersion forces being anisotropic and it is these weak forces, when combined with the anisotropic shape, which stabilize the anisotropic arrangement within the mesophase. The weakness of these forces also helps to account for the highly fluid nature of the nematic phase.

If the intermolecular forces are slightly stronger, then a degree of positional order can be introduced leading to lamellar structures as are found within the smectic mesophases. The simplest arrangement is found in the smectic A phase (S_A, note that the subscript is purely historical in nature). Again, the molecular long axes are orientated in the same direction but the molecules are also loosely associated into layers. This arrangement is still fluid with diffusion between the layers being a facile process and hence, the diagram shows a highly idealized situation. If the molecules are tilted with respect to the layer normally then the smectic C phase (S_C) is generated.

Returning to the smectic A arrangement, then introduction of hexagonal symmetry within the layers produces a more ordered molecular arrangement assigned as the smectic B phase (S_B). Again the phase is fluid and interlayer diffusion occurs, although rotation about the molecular axis is concerted. In a similar manner to the way that the molecules in the smectic C phase are tilted compared to the smectic A, there are tilted (although less ordered) versions of the smectic B phase. These are assigned as the smectic I (S_I) and the smectic F (S_F) corresponding to molecules tilting towards a vertex, or an edge, of the hexagonal molecular arrangement, respectively.

As alluded to earlier, chiral modifications are possible for some of the phases namely the N*, S_C*, S_F* and S_I*. These may be generated either by doping a non-chiral liquid crystal with a chiral additive, or by resolving a racemic material exhibiting one or more of these phases.

Interestingly, introduction of chirality within nematic phases causes the slight and gradual precession of the director, $\underset{\sim}{n}$, producing a helical molecular arrangement. The separation between equivalent molecular alignments, within the helix, is known as the pitch length (p) which is temperature dependent. The helical molecular arrangement has the ability to reflect selectively light of a wavelength equal to np (n is the average refractive index of the material). Consequently the material will exhibit different colours with changes in temperature (thermochroism); this allows their use as, for example thermometers.

In the chiral smectic C phase, a helical molecular arrangement is also observed derived from the tilt direction precessing through the sample. Due to the low symmetry (C_2) within this phase, the molecular dipoles will on average align within a layer, which then becomes ferroelectric. If the pitch length is made large enough, then a preferred macroscopic alignment of the transverse dipoles may be realized. Such materials have demonstrated the ability to be switched between two states on the microsecond timescale, which has generated a great deal of research due to implications for display technology [19].

The final group of lamellar phases to be introduced here are the so-called crystal smectic phases. These are more ordered than the previous smectic examples and are characterized by the appearance of interlayer correlations and, in some instances, by the removal of freedom of molecular rotation. Consequently the (crystal) B, G and J phases are derived from the S_B, S_F and S_I phases, respectively, but with the presence of interlayer correlations. Further the (crystal) E, H and K phases are B, G and J phases which have lost rotational freedom. These phases are still disordered and hence, still intermediate between the solid and liquid states.

3 METALLOMESOGENS EXHIBITING NEMATIC AND LAMELLAR PHASES

3.1. Effects of Complexation

3.1.1 4-Alkyl and 4-alkoxy-4′-cyanobiphenyl ligands

The 4-alkyl-4′-cyanobiphenyl compounds, originally synthesized by Gray and co-workers [20], were responsible for the mass industrial development of liquid crystal

R–C₆H₄–C₆H₄–CN–M(Cl)₂–NC–C₆H₄–C₆H₄–R

M = Pt and Pd

R = $C_9H_{19}O$

Free ligand	Crys • 64 • S_A • 77 • N • 80 • I
M = Pt	Crys • 167 • S_A • 187 • N • 209 • I
M = Pd	Crys • 119 • S_C • 122 • S_A • 146 • I

Figure 4 Palladium(II) and platinum(II) cyanobiphenyl complexes.

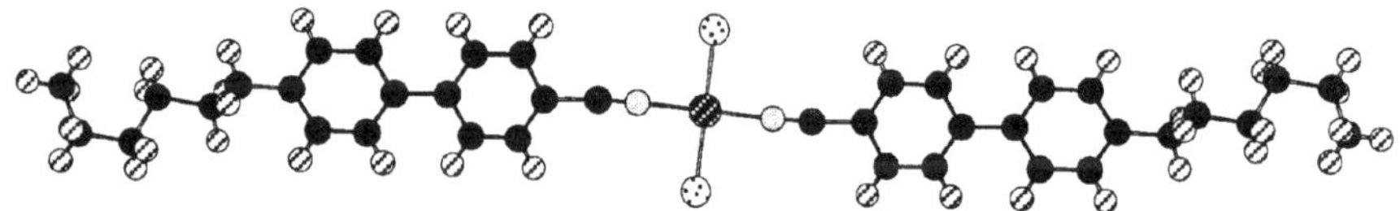

Figure 5 Molecular structure of dichlorobis(4′-octylbiphenyl-4-carbonitrile)platinum(II).

technology and provide an appropriate ligand to begin discussions of metallomesogens. The cyano group, while enhancing the polarizability of the molecule, also offers a donor site at which a metal ion may coordinate [3]. The initial complexes formed were with platinum(II) and palladium(II), yielding *trans*-complexes as shown in Figure 4. These structures were verified by single crystal X-ray structures (e. g. Figure 5), which showed that the metal centres barely perturb the overall rod-like structure of the molecules, with no intermolecular interactions being identified [21]. Complexes of palladium and platinum were found to be essentially isostructural and isomorphous.

The platinum complexes display fairly analogous thermal behaviour to the free ligands although at higher temperatures, while it was found that for certain palladium congeners, a smectic C phase was seen. Therefore, complexation was able to induce the formation of a mesophase which was not present for the free ligand. Physical studies of these materials showed that complexation resulted in a much enhanced polarizability [8] and birefringence [22] over those found for the ligands, consistent with the introduction of a centre of highly mobile electron density.

3.1.2 Complexes of azobenzenes

In the above complexes, mesogens were formed by the complexation of mesogenic ligands to a metal centre. Using entirely different chemistry, Ghedini and co-workers had previously used the same approach in obtaining palladium complexes of

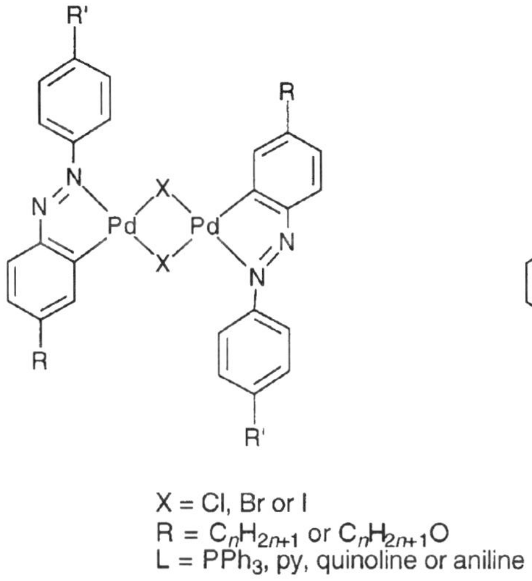

X = Cl, Br or I
R = C_nH_{2n+1} or $C_nH_{2n+1}O$
L = PPh_3, py, quinoline or aniline

Figure 6 Palladium(II) complexes of azobenzenes.

azobenzenes to make complexes of the general type shown in Figure 6. Considering the dinuclear complexes first, it was found from ^{1}H NMR studies that they are present as an approximately 1 : 1 mix of isomers, differing in metallation of the benzene rings [23] and are of an H-like shape depending upon the nature of R and R′.

The free ligands (where R = Me or Et and R′ = hexyloxy or undecyloxy) display monotropic nematic phases, however, upon forming the dinuclear palladium complexes the nematic phases are, perhaps surprisingly, enantiotropic. The melting points are, as expected, raised upon complexation but the increase in the clearing point is somewhat unusual as the structural anisotropy would be expected to be lower in the complexes compared to the free ligands. Indeed, it is somewhat remarkable that these complexes are mesomorphic; certainly they represent a structure which would not readily be observed in purely organic materials.

The effect of the bridging atoms was studied (R = OC_2H_5 and R′ = $C_6H_{13}CO_2$), showing that the melting point increased in the order $Cl_{(Crys\text{-}N)} < Br_{(Crys\text{-}S_A)} < I_{(Crys-S_A)}$ and that the transition temperature to form a nematic phase also increased in this manner [24]. Clearing points, however, followed a different trend, namely Br > I > Cl. The reason for these variations are unclear, however, the appearance of S_A phase for the Br and I complexes may suggest a different geometry of the complex depending upon the nature of the bridging group.

A series of the chloro-bridged complexes have been studied by X-ray diffraction and as a result of the shorter-than-theoretical molecular length, a model was proposed with the molecules containing a rigid core of about 11 Å) with the longer of the two chains tending to fill the free space close to the shorter chain

[25]. The high electronic density associated with the dinuclear centre was suggested to produce electric multipolar interactions and this combined with the sterically rigid core leads to the high transition temperatures in the mesogens.

A single crystal X-ray structure was determined for one of the symmetric complexes ($R = R' = C_6H_{13}$) [26]. This verified the *trans*-arrangement of the metallated ligands with the palladium ions in a square planar geometry, the non-metallated aromatic rings being rotated out of this plane by approximately 35° and 38°, respectively. The molecules are weakly associated into dimers with a palladium–palladium separation of 3.67 Å in the solid state. This particular complex displays unexpected phase behaviour showing a nematic phase between 210 and 220°C and then a crystal E phases up to 235°C; such behaviour, where a mesophase is out of thermodynamic order (the crystal E phase being much more ordered that the nematic), is termed as being re-entrant. It was suggested that this arises due to a transition involving loss of the intermolecular interactions, the monomeric molecules then demonstrate the higher temperature, more ordered phase, although no experimental evidence is presented. The nematic phase is shown to be thermodynamically unstable, as upon standing at 218°C the mesophase changes to that (crystal E) seen at higher temperatures.

In the mononuclear series, for $L = PPh_3$, the structural anisotropy is reduced and the complexes are non-mesomorphic [27]. Similarly with L = aniline, intermolecular hydrogen bonding probably leads to intermolecular hydrogen-bonded interactions which are too strong to allow mesomorphism to be observed. However, with pyridine and quinoline an enhanced nematic range and also smectic phases are seen, although the range is lower for the quinoline due to its greater steric bulk.

Reaction of the dinuclear palladium(II) complexes with thallium(I) cyclopentadienyl yields the mononuclear cyclopentadienyl complexes [28]. When the ligands were simple two-ring azobenzenes as above, the complexes are non-mesomorphic suggesting that the cyclopentadienyl disrupts the structural anisotropy to too large an extent. However, if a further alkoxybenzoate group is incorporated within the ligand framework (Figure 7), then the complexes become sufficiently anisotropic, displaying nematic phases. The melting points are significantly lower than those found in the orthometallated azobenzene palladium complexes, and interestingly, for $n = 10$, the melting point is lower than that demonstrated by the free ligand. This effect is

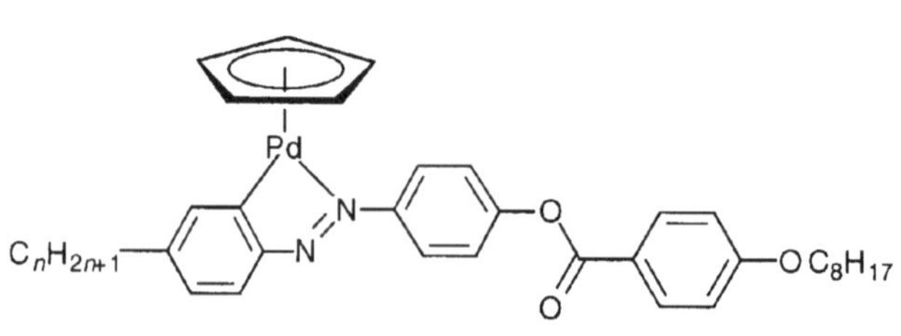

Figure 7 Cyclopentadienyl palladium(II) azobenzene complexes.

likely due to the reduction in structural anisotropy resulting from the introduction of the [CpPd] moiety and is an effect which is discussed again below in relation to high coordination number complexes.

3.1.3 Complexes of imines

Recently Rourke has reported a series of cyclopalladated imine complexes with either acetylacetonate or cyclopentadienyl as the other ligand, as in Figure 8 [29]. Unlike Ghedini's azobenzene complexes these are isomerically pure but still display mesophases at similar temperatures. The square planar $16e^-$ acetylacetonate complexes display both nematic and S_A phases with extended mesophase ranges compared to the free ligands. Not surprisingly, the formally trigonal bipyramidal $18e^-$ cyclopentadienyl complexes show a nematic phase at much lower temperatures and tend to decompose above *c.* 180°C.

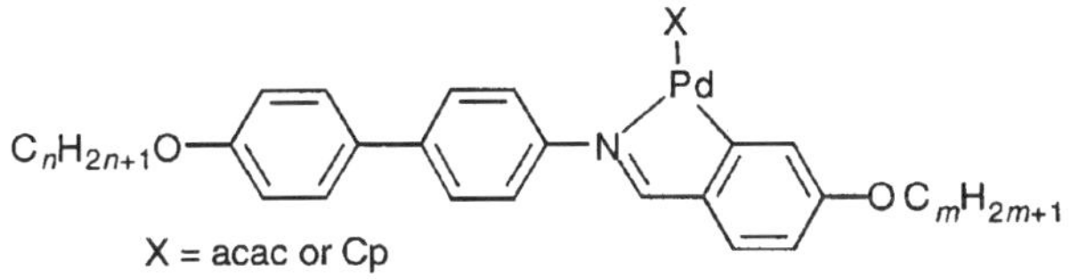

Figure 8 Cyclometallated palladium(II) imine complexes.

3.1.4 Complexes of alkoxydithiobenzoates and alkoxytrithiobenzoates

In the previous examples the ligands were themselves mesomorphic. However, we can now move to the opposite extreme and look at complexes of 4-alkoxydithiobenzoates (abbreviated as n-odtbH, Figure 9) where the ligands are non-mesomorphic [30]. In fact, the lack of mesomorphism of the ligands deserves some comment as the related 4-alkoxybenzoic acids are, in fact, liquid crystals being sufficiently anisotropic due to the formation of hydrogen-bonded dimers. However, with sulphur replacing oxygen, the hydrogen bonding is clearly insufficiently strong to allow the dimer to survive heating, with the monomer being insufficiently

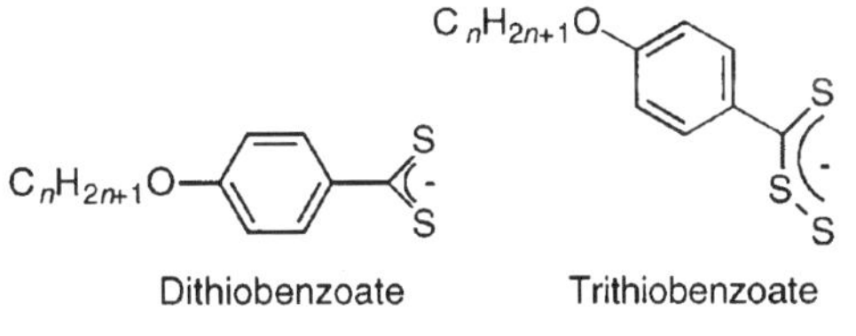

Figure 9 Structures of 4-alkoxydithiobenzoic acid and 4-alkoxytrithiobenzoic acid.

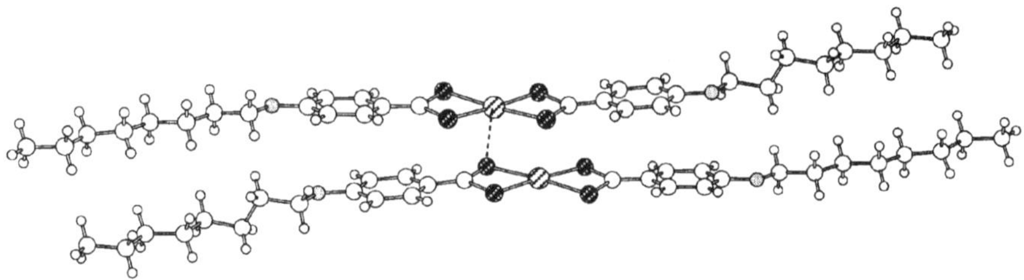

Figure 10 Single crystal X-ray structure of $[Pd(8\text{-odtb})_2]$ showing Pd–S contact.

anisometric to form a liquid crystal mesophase [31]. However, mesomorphism is induced in these molecules upon complexation to gold(III), nickel(II), palladium(II) and zinc(II).

The gold(III) complexes are formulated as $[AuX_2(n\text{-otdb})]$, X = Cl, Br or Me, suggesting it is the enhanced polarizability of the metal centre which is stabilizing the observed S_A mesophase—certainly this is a mesogen of very low structural anistropy. The nickel(II) and palladium(II) complexes, with the formula $[M(n\text{-}odtb)_2]$, have D_{2h} symmetry about the metal and have a more classical, elongated geometry. For shorter chain lengths, the palladium(II) complexes show a nematic phase which is replaced by an S_C at longer chain lengths. An X-ray single crystal structure for the complex $[Pd(8\text{-odtb})_2]$ shows that the molecules are associated into pairs with an intermolecular Pd–S distance of 3.38 Å (Figure 10).

There is debate in the literature as to whether, in general terms, such (supramolecular) intermolecular interactions are important in determining the mesomorphism in metallomesogens where the metal centres are coordinatively unsaturated. In this palladium system, this was investigated using EXAFS [32], which showed that the intermolecular Pd–S contacts could not be observed in the smectic C phase.

The nickel(II) complexes showed a broadly similar mesomorphism at slightly lower temperatures; these materials were also studied independently by Ohta [33]. However, these complexes were especially interesting as they underwent chemical reaction in the mesophase. Thus, at 230°C, the bis(dithiobenzoato)nickel(II) complex rearranges to give a (dithiobenzoato)(trithiobenzoato)nickel(II) complex (Figure 11). Ohta identified these by extraction from a heated sample, while we identified them by independent synthesis, using chemistry described by Fackler [34]. Structurally, the mixed-ligand complex is now much less anisotropic and consequently, its transition temperatures are appreciably lower than those of the symmetric, bis(dithiobenzoato) complex. Further, the bent shape does not readily lend itself to the formation of layered structures and so the nematic phase is the only one observed.

The zinc(II) complexes of the dithiobenzoates were also of some interest as it appeared that different forms could be isolated, depending on whether crystallized or

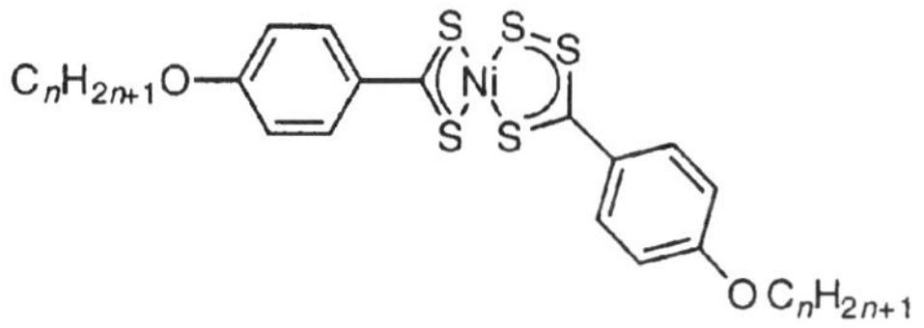

Figure 11 General structure of (dithiobenzoato)(trithiobenzoato)nickel(II) complex.

'as-obtained' complexes were looked at. The most reliable behaviour was found for the crystallized materials which showed nematic and S_C phases, and single-crystal X-ray structures for the butyloxy and octyloxy homologues showed an interesting dimeric arrangement (Figure 12), typical of that found in diethyldithiocarbamate complexes of both zinc and cadmium [35]. However, as determined by osmometry in organic solvents, the complexes were monomeric and so the question arose as to the structure in the mesophase. EXAFS was again able to provide the answer which was the dimer persisted into the mesophases. No definitive structural data were obtained for the as-obtained complexes.

3.1.5 4-Alkoxystilbazoles

Alkoxystilbazoles fall into a sort of 'in-between' scenario as the shorter homologues are non-mesomorphic, while the longer chain homologues show, typically between about 80–90°C, a crystal E and a smectic B phase (the latter over a very narrow range) [36]. However, examples of all chain lengths can be made mesomorphic on complexation to the right metal.

Complexes of Rh(I) and Ir(I) are readily obtained from $[M_2(\eta\text{-COD})_2(\mu^2\text{-Cl})_2]$ (COD = 1,5-cyclooctadiene) and are examples of simple, dipolar mesogens [37]. Complexation led to mesogens which showed nematic and smectic A phases whose stability was very much greater than that of the ligand (e. g. clearing point for the Rh complexes was mostly >130°C), showing how complexation influenced strongly the

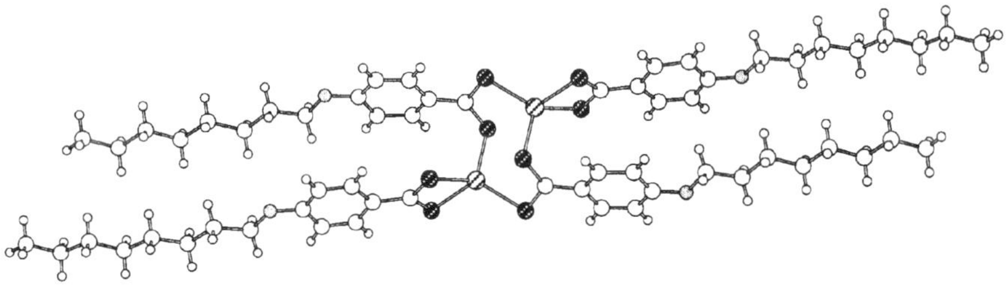

Figure 12 Single-crystal X-ray structure of $[Zn_2(8\text{-odtb})_4]$.

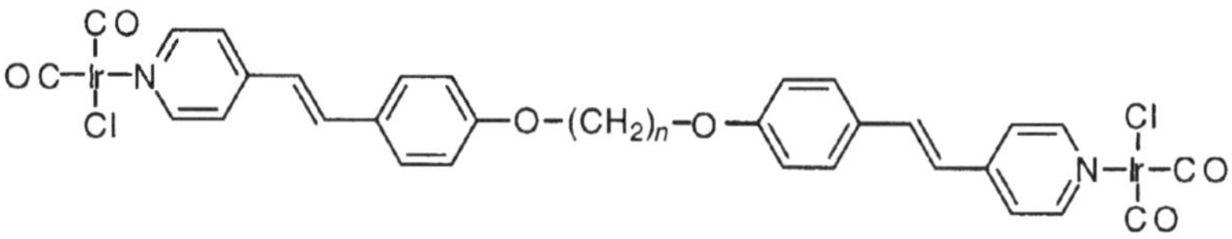

Figure 13 Flexibly linked, dinuclear stilbazole complexes.

liquid crystal properties. However, in the iridium congeners, it was found that two solid forms could be obtained which was interpreted, with the help of infrared spectroscopy, in terms of the well-known stacking of planar Ir(I) complexes. This was initially of passing interest, until dimeric complexes were made [38] using flexibly linked stilbazoles (Figure 13).

It was found that when the flexible chain possessed an even number of atoms then the complexes were not mesomorphic, while when an odd number of atoms was present, mesophases were observed. These observations were interpreted by proposing that an even number of atoms led to an *anti*-arrangement (Figure 14) which encouraged the effective stacking of Ir units, suppressing mesomorphism. However, for odd numbers of atoms in the link, then a kink is effectively introduced developing a *syn*-arrangement between the two Ir-containing ends, frustrating stacking and allowing mesomorphism to be observed.

An even more marked demonstration of the way that complexation can modify the mesomorphism of species such as these is found in the work of Serrano with the related imines shown in Figure 15. Here, none of the ligands was mesomorphic,

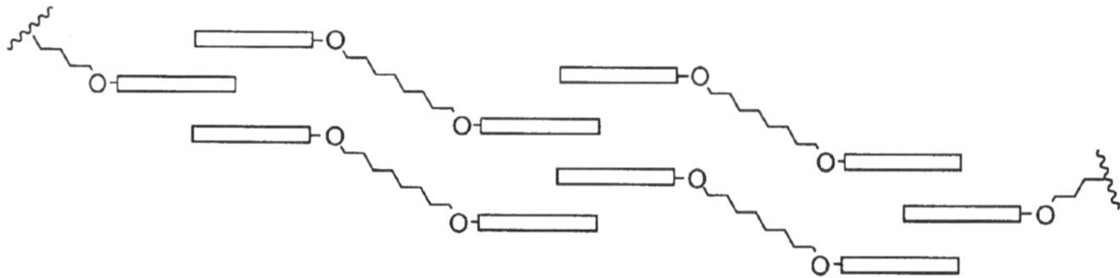

Figure 14 Stacking arrangement possible in dimers possessing even numbers of atoms on the flexible link.

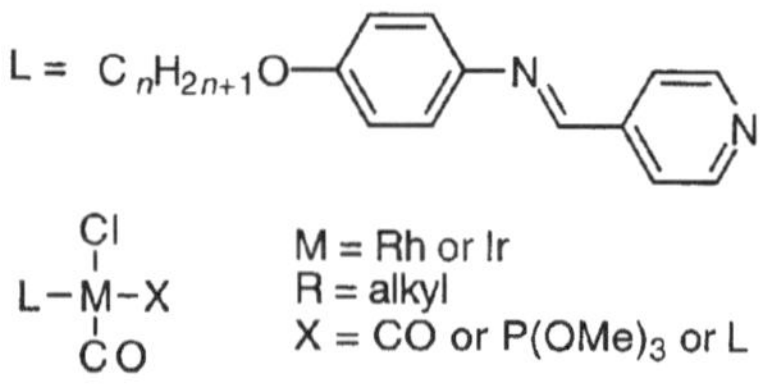

M = Rh or Ir
R = alkyl
X = CO or $P(OMe)_3$ or L

Figure 15 Serrano's iminopyridines and their metal complexes.

while nematic and smectic A phases were observed when bound to *cis*-$[MCl(CO)_2]$ (M = Rh, Ir) [4]. Curiously, the *trans* bis(ligand) complex of rhodium was not mesomorphic, nor is the complex when the CO *trans* to the pyridine ligand is replaced by $P(OMe)_3$ [39].

3.1.6 Isonitrile complexes

Espinet [40] has prepared a series of mesomorphic carbene complexes which are discussed in more detail below, but one series is properly discussed here, these being dinuclear complexes with a bridging octafluorobiphenyl group as shown in Figure 16. It is expected that two rings of the biphenyl ligand will be twisted due to both steric and electrostatic repulsions between the interannular fluorine groups, which leads to a reduction in the π-conjugation. Therefore these complexes are best described as two monomeric molecules linked head to head with no net dipole moment but possessing a complex multipolar structure. Espinet suggests a simple but effective, model for the interactions between two molecules as shown in Figure 17, in which the lamellar arrangement shown is considered unfavourable due to the like interactions required. This therefore rationalizes the disappearance of the smectic A phase in the dimers compared to mononuclear equivalents.

3.2 Effects of a Lateral Substituent on Mesomorphic Properties

In the above section, we have established some simple ideas concerning the way in which a metal may influence liquid crystal properties when bound to mesogenic ligands. The behaviour may also be influenced by lateral substituents which may

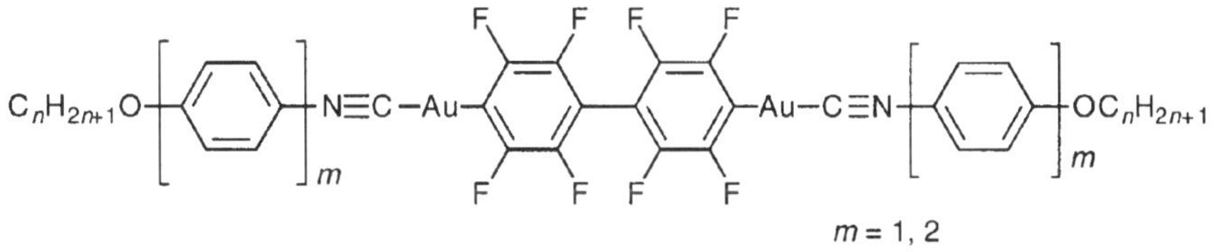

$m = 1, 2$

Figure 16 Espinet's dinuclear Au(I) complexes.

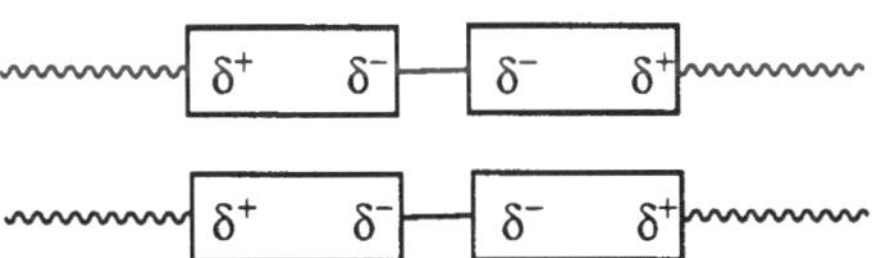

Figure 17 Simple model showing the unfavourable interactions which would be present in the lamellar phase of the dimers.

be incorporated via the metal or via the ligands. In general, lateral substituents divide into two main classes—lateral halo (mainly fluoro) substituents and lateral alkyl(oxy) substituents.

Lateral fluorination [41] has been widely studied in the main in organic systems, and it is found that the effects are to a good degree predictable and rather subtle. Thus, there is a steric effect which is sometimes very pronounced, and there are dipolar effects which often appear more significant in their role, being able to influence nematic and smectic phase stability depending on their position in the molecule. These general effects are largely mirrored in metal-based systems.

Lateral alkylation has, in general, one main effect and that is to promote the nematic phase at the expense of all other phases due to the effective suppression of lateral interactions between the rigid cores.

3.2.1 Alkoxystilbazole complexes

Complexation of 4-alkoxystilbazoles to Pd(II) and Pt(II) leads to the square planar complexes shown in Figure 18. These complexes are, for the main, non-mesomorphic and what little liquid crystallinity is found (S_C phase) is with longer chain derivatives on palladium and at high temperature ($<200°C$). However, analogous complexes may be prepared in which the lateral chloro function is replaced by an alkanoate group; these are obtained by reaction of palladium alkanoate (from palladium acetate and the alkanoic acid) with the stilbazole (Figure 19). Now, the mesomorphism changes and only nematic phases are seen, now at much lower temperatures (e. g. clearing points $<170°C$) [42]. Thus, here the predominant effect is suppression of smectic phases.

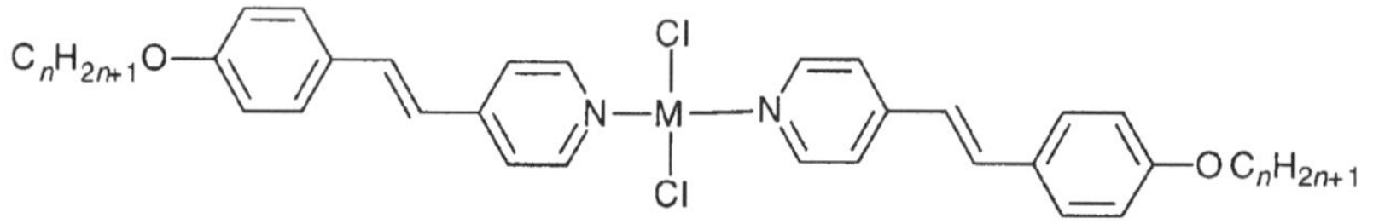

Figure 18 *Trans*-bis(stilbazole)dichlorometal(II) complexes of palladium and platinum.

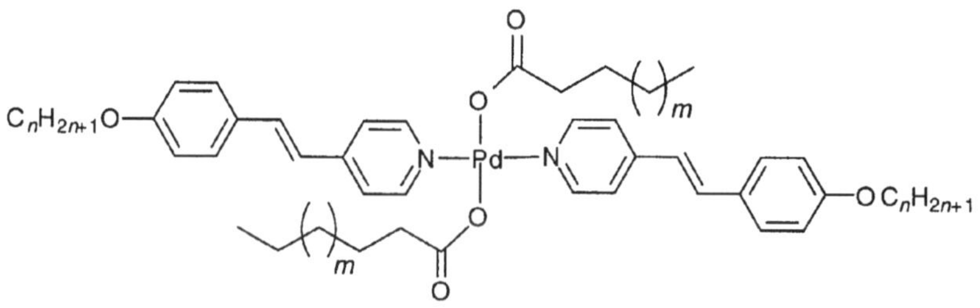

Figure 19 Alkanoate-substituted stilbazole complexes of palladium(II).

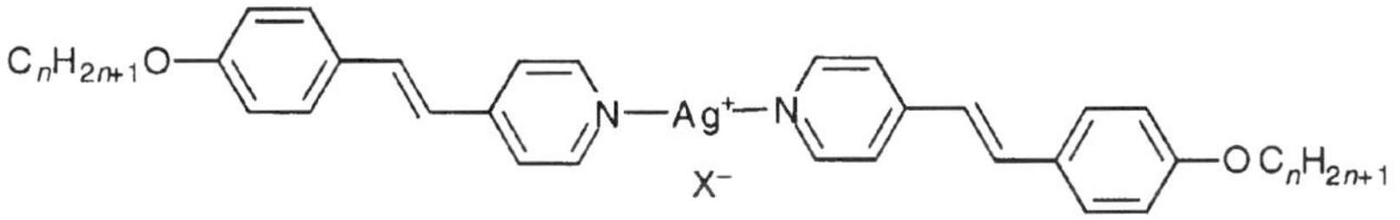

$C_nH_{2n+1}O$ — N—Ag^+—N — OC_nH_{2n+1}
X^-

Figure 20 Stilbazole complexes of silver(I).

Study of the silver(I) stilbazole complexes further develops these ideas. Thus, reaction of AgX ($X = BF_4$, NO_3, CF_3SO_3, $C_mH_{2m+1}OSO_3$) with two equivalents of stilbazole leads to complexes with the general structure shown in Figure 20. Where the anion is a tetrafluoroborate, nitrate or triflate then the clearing temperatures are above 240°C and mostly accompanied by severe decomposition [43–45]. However, use of an alkylsulphate as anion proves beneficial, for instance with dodecylsulphate the clearing temperatures are approximately 180°C [46]. These differences are exemplified by phase diagrams for the triflate and dodecylsulphate complexes, shown in Figures 22 and 23, respectively. A single crystal X-ray structure for an silver(I) octylsulphate complex shows that the octylsulphate anions effectively bridge the silver(I) ions (Ag...O: 2.75, 2.93; and 2.77, 2.89 Å) to produce a dimeric structure (Figure 21) [47].

However, a most significant observation is in the comparison between the two phase diagrams where it is seen that the smectic phases present in the triflate complexes are retained with dodecylsulphate as the anion. This is unusual if the alkylsulphate is considered as a lateral alkyl chain, where it would normally be the case that smectic phases were suppressed. Further, the phase diagram of the dodceylsulphate complexes shows, for seven homologues, a cubic phase between the S_C and S_A phase. Both of these points deserve comment.

A starting point for the discussion rests on the finding that there is no conductivity on the mesophase of the silver complexes of dodecylsulphate, suggesting that the

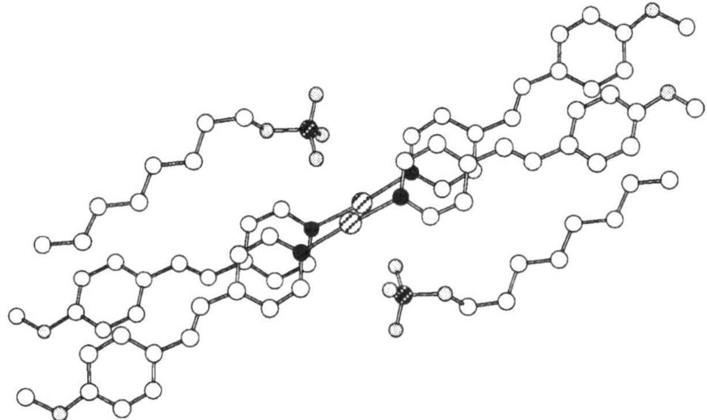

Figure 21 Molecular structure of the methoxystilbazole complex of silver octylsulphate.

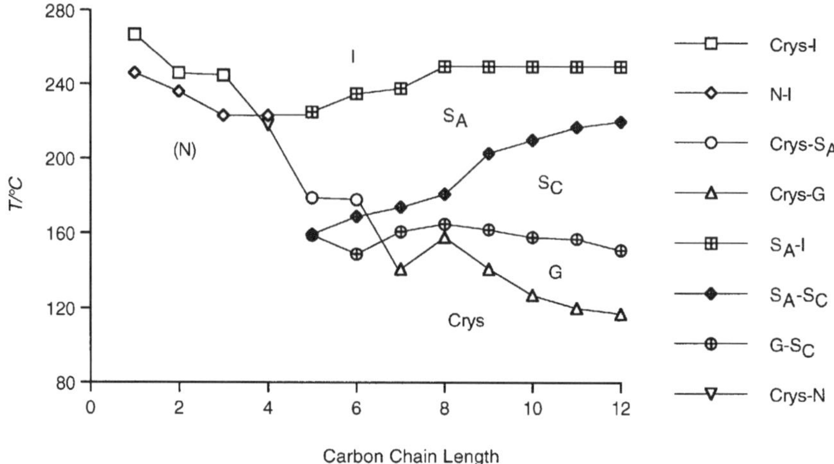

Figure 22 Phase diagram, for the stilbazole complexes of silver triflate.

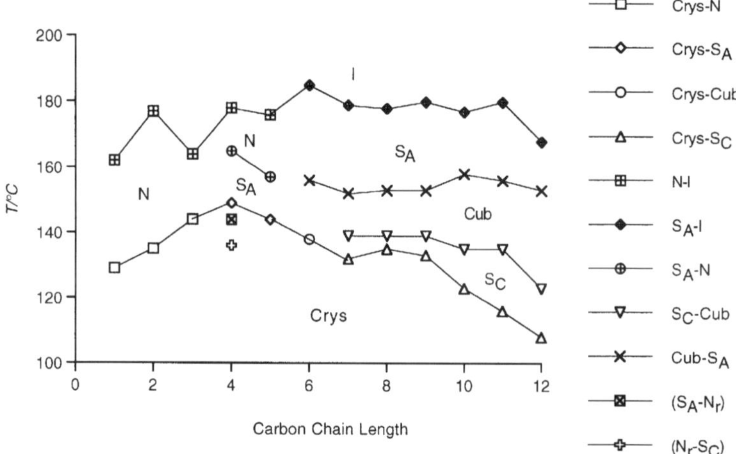

Figure 23 Phase diagram for the stilbazole complexes of silver dodecylsulphate.

species are present as tight ion-pairs. However, the observation of smectic phases suggests that there must be some intermolecular ionic interactions which are able to preserve the necessary layering, although the presence of nematic phases means that these are not too strong that they cannot be broken thermally.

The cubic phase is still a rather rare observation in thermotropic liquid crystal systems, and is found in 'conventional' calamitic materials, as well as in polycatenar materials (*vide infra*). It is a viscous, optically isotropic phase whose kinetics of

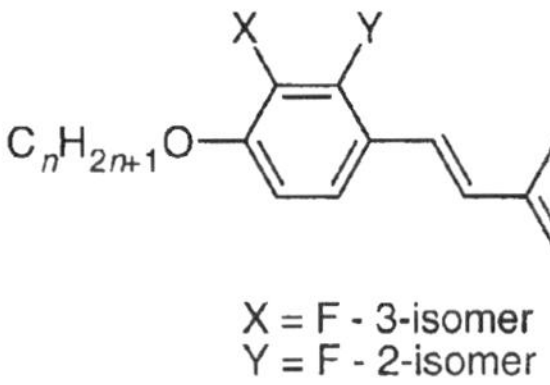

Figure 24 Two isomers of monofluoroalkoxystilbazole.

formation are slow, and is often seen in conjunction with the so-called S_4 phase whose symmetry we have recently found to be tetragonal ($I4_1/acd$) [48]. The symmetry of the cubic phase in this system is $Ia\bar{3}d$ [49], which is the most common symmetry found for thermotropic materials. The formation of this phase is, in general, regarded as arising from a balance between the volume of the rigid core of the molecule and the volume occupied by the terminal chain(s), which can lead to curvature at the aromatic/paraffinic interface when the latter is greater than the former. While, in polycatenar systems, the aromatic/paraffinic curvature appears able to account for cubic phase formation, in calamitic systems, another common feature would seem to be the possibility for intermolecular interactions to occur near the centre of the molecule, and this is a point which requires further study in due course.

Further studies of these silver systems looked [50] at the effects of fluorination using the two monofluorinated stilbazoles indicted in Figure 24 [51]. Thus, where the anion was dodecylsulphate and the fluorination was at the 3-position, the nematic phase is greatly destabilized, the S_A phase stability is increased, the crystal phase is destabilized and the cubic phase is suppressed. Fluorination at the 2-position totally destabilizes the S_A phase, further destabilizes the crystal phase, promotes the nematic phase and retains the cubic phase. Further, this series led to the phase sequence I·N·Cub in one homologue, something which had not previously been seen. The ability to align the nematic phase with an external magnetic field led to a monodomain determination of the space group of the cubic phase as $Ia\bar{3}d$ [49].

This behaviour is rather typical in that 2-fluorination produces a lateral dipolar effect which would be expected to favour a nematic phase and not an S_A phase, while the 'outboard' dipoles resulting from 3-fluorination would be expected to stabilize both crystal and lamellar phases.

3.2.2 Isonitrile complexes

Takahashi has synthesized a series of chloro-gold(I) mesogens incorporating isonitrile ligands in which the unsubstituted free ligands cleared below 100°C and the corresponding complexes cleared with decomposition at approximately 270°C (Figure 25) [52]. However, incorporation of an alkoxy-chain ortho to the isonitrile group destroyed the mesomorphism of the free ligands but now produced complexes

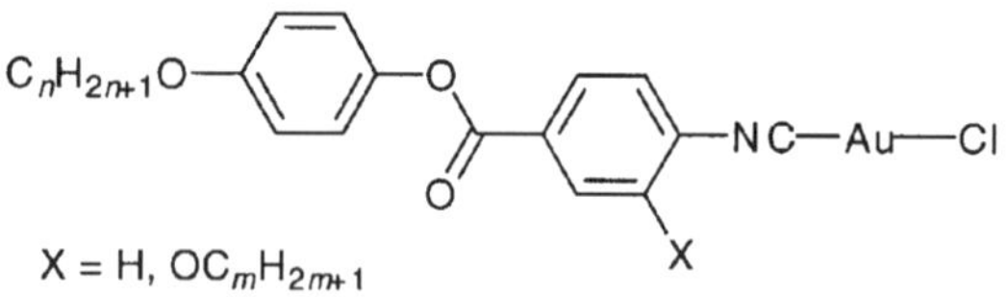

Figure 25 Isonitrilegold(I) mesogens.

clearing without decomposition below 200°C. The alkoxy side chain was proposed to be disrupting gold–gold interactions which are present in many univalent gold complexes [53], which were suggested to be the cause of the high clearing temperatures in the unsubstituted complexes.

He then went on to show that these bound isonitriles could be reacted with, for example, alcohols to give the carbene complexes as two geometric isomers as shown in Figure 26 [54]. Here, the alkoxy chain on the carbene carbon serves as a lateral substituent so that the transition temperatures are lower compared to the parent isonitrile complexes. Aminocarbenes derived from both tertiary and secondary amines were also obtained, and in the former case no mesomorphic materials were found, while in the latter case an S_A phase was seen before the sample decomposed with hydrogen bonding being a possible explanation for the high clearing points observed. Finally, it is noteworthy that a dinuclear carbene complex was formed from *N,N′*-dimethyl-1,6-diaminohexane which displayed an S_A phase before decomposition.

Espinet has also studied gold(I) isonitrile complexes [55]. In the substituted complexes (Figure 27) it is seen that the alkoxy chain length produces little variation in the transition temperatures and the S_A phase dominated. However, for the chlorogold(I) complexes, then the crystal and S_A phase are stabilized in the 3-fluorinated complexes (i. e. $R^2 = F$) compared to the unfluorinated complexes; in the 2-fluorinated complexes ($R^1 = F$), the crystal phase is slightly destabilized over that

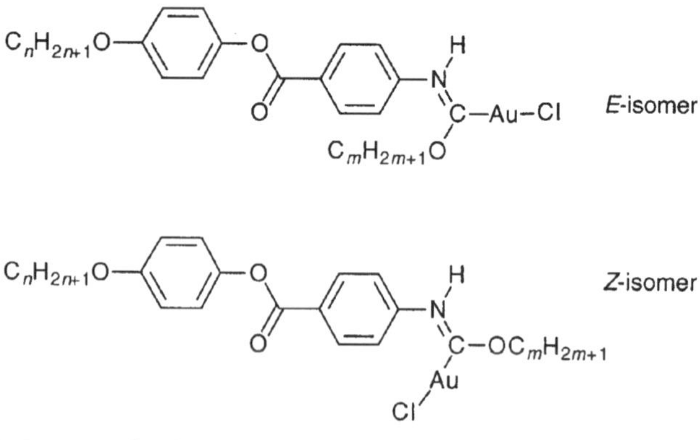

Figure 26 Isomers of gold carbene complexes obtained from coordinated isonitriles.

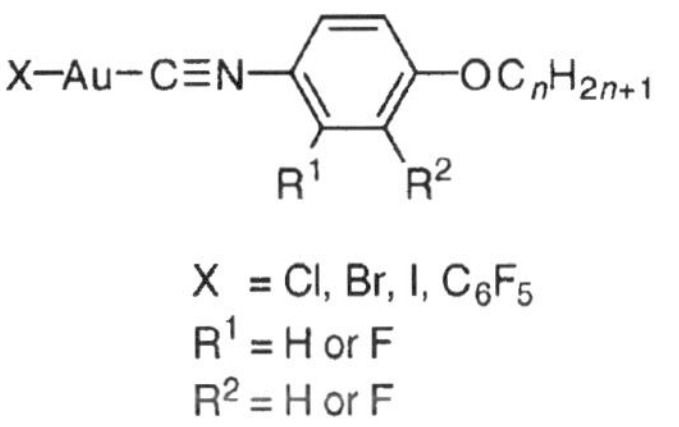

$X = Cl, Br, I, C_6F_5$
$R^1 = H$ or F
$R^2 = H$ or F

Figure 27 Gold(I) isonitrile complexes.

$R^1 = F$ or Br
$R^2 = F$ or Br

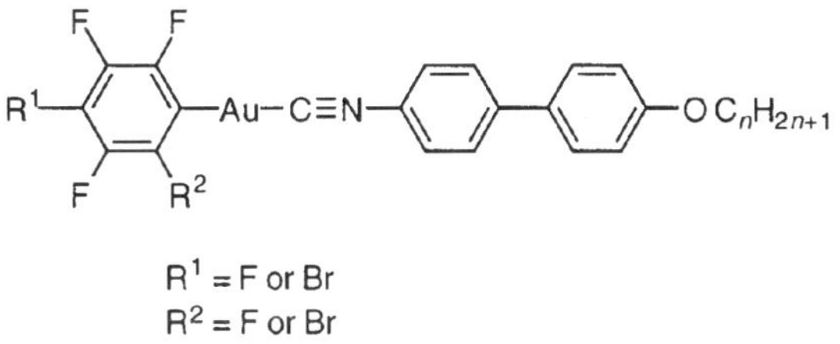

Figure 28 Derivatives of pentafluorophenylgold(I) isonitrile complexes.

of the unsubstituted materials while the S_A phase is massively destabilized [56]. These effects are, once more, in line with what one would expect.

When the halogen attached to gold is replaced by a perfluorophenyl group, the mesomorphism is lost, although it may be regained if the structural anistropy is increased by replacing the 4-alkoxyphenyl isocyanide ligand with a 4-alkoxy*bi*phenyl isocyanide [41]. From this starting point, Espinet went on to investigate the effect of replacing one of the ring fluorines with a bromine atom, bearing in mind the greater steric requirements of the bromine versus its greater polarizability. Thus, the highest transition temperatures were found with a bromine in the 4-position in the ring (Figure 28; $R^1 = Br$). However, for $R^1 = F$ and $R^2 = Br$ (Figure 28), it was found that the clearing points were *higher* than those of the pentafluoro system ($R^1 = R^2 = F$) for $n > 8$, and lower for $n < 8$ which is surprising, especially as S_A mesophases were seen in the bromo-substituted materials suggesting that the steric effects were not at all great. Finally, comparison of the chlorogold(I) complexes and the p-C_6F_4Br analogues shows that the perhalophenyl group produces lower transition temperatures, shorter mesogenic ranges and an enhancement of the nematic phase as a consequence of the greater molecular width.

3.2.3 Salicylaldimato complexes

Complexes of salicylaldimato ligands have been well studied and provide a good system in which to examine the effects of lateral substitution. The parent, unsubstituted complexes (Figure 29; $X^1 = Y^1 = X^2 = Y^2 = H$) show enantiotropic nematic mesophases, with a phase range of 75°C being observed for alkyl chains

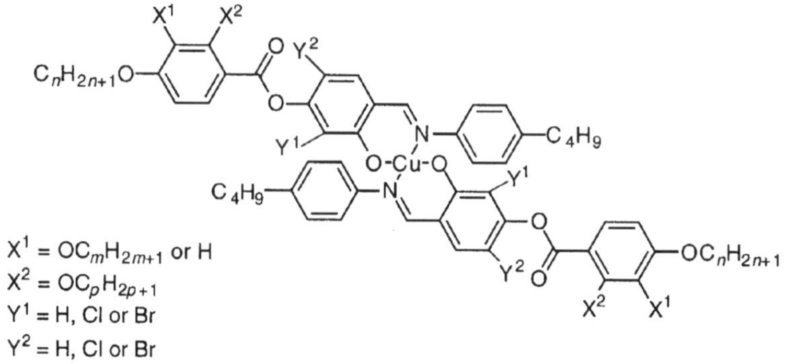

$X^1 = OC_mH_{2m+1}$ or H
$X^2 = OC_pH_{2p+1}$
Y^1 = H, Cl or Br
Y^2 = H, Cl or Br

Figure 29 Substituted copper(II) salicylaldimate complexes.

from dodecyl and higher [57]. Introduction of an additional alkoxy chain [58] of equivalent length at the adjacent 3-position (X^1) produces a lowering of the transition temperatures and stabilizes smectic phases, with the nematic phase no longer being observed. If the alkoxy chains in the 3- and 4-position are inequivalent so that the 4-position is kept as hexyloxy and the 3-position is varied ($n = 6$, vary m), then reductions in the transition temperatures are seen for the lower homologues, while at longer chain lengths, the temperatures are similar to the symmetric case. Keeping a methoxy at the 4-position and varying the chain length at the 3-position fails to generate mesogenic properties due to the chain contributing mainly to an increase in the breadth of the molecule. Substitution at the 3-position, therefore, lowers the transition temperatures and favours layer formation leading to smectic phases.

Conversely substitution in the 2-position (X^2) tend to disfavour intermolecular interactions leading to nematic phases. Thus, an enantiotropic nematic phases was seen for the dodecyloxy complex which cleared at 61.5°C compared to 248°C in the parent complex but with a range of only 9°C, the other members of this series showing monotropic phases. Clearly a fine balance exists between disfavouring intermolecular interactions to lower melting points and destabilizing such interactions with respect to the generation of a mesophase. Further to this point, it is noted that enhancement of the anistropy by incorporation of an azobenzene link in the ligand (Figure 30) gives a nematic range of almost 100°C [59].

Halogenation of these salicylaldimines has also been studied (positions Y^1/Y^2 in Figure 29) [60]. It was found halogenation at Y^1 led to steric repulsion with the aniline-based phenyl ring, disrupting planarity and destabilizing the nematic phase. Introduction of a substituent at Y^2 also causes steric interaction, this time with the carbonyl group of the ester function, again disrupting the planarity of the molecule and reducing the nematic range. If both positions are substituted with chlorine or bromine or a mixture of the two, then only monotropic nematic phases are observed even though the melting points are generally lower than in the parent complexes [61].

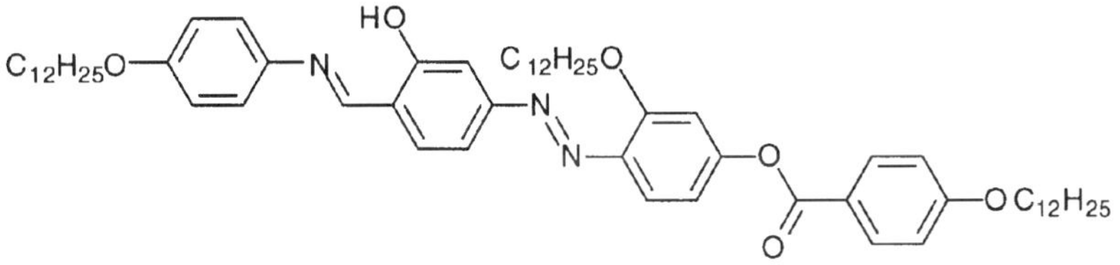

Figure 30 Azobenzene-substituted salicylaldimine.

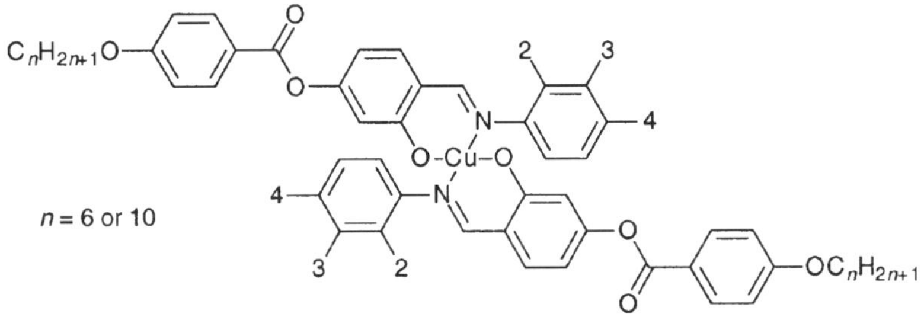

Figure 31 Further examples of substituted salicylaldimines.

Serrano has studied a related series of complexes ($n = 6$ or 10, Figure 31) studying the effects of substituents placed upon the aniline ring [62], compared to the 4-butyl group previously studied. Thus, replacing the butyl chain (4-position) by either a cyano or cyanomethylene group favours S_A phases, while placing a fluorine or a trifluoromethyl in the 3-position reduces the nematic range, to the extent that the phase becomes monotropic with a trifluoromethyl group and $n = 6$. However, there is a beneficial effect upon the nematic phase stability when the ring is fluorinated at both the 2- and 4-positions, for example for $n = 6$ the range is 205–280°C compared to 213–254°C for the 4-butyl complex.

Finally here it is worth noting that substitution by a fluorine atom does not always produce beneficial effects. For example, Chipperfield has synthesized the symmetric salen complexes shown in the Figure 32 containing copper(II), nickel(II) and oxovanadium(IV) [63]. Fluorination reduces the clearing point, as expected, in all but the oxovanadium complex with $n = 6$ where it remains constant. However, the mesophases are severely destabilized either disappearing completely or becoming monotropic.

3.2.4 *Porphyrin complexes*

Through 5,15-substitution about a zinc(II) porphyrin, it has been shown that calamitic molecules may be realized [64, 65]. These complexes (Figure 33) may

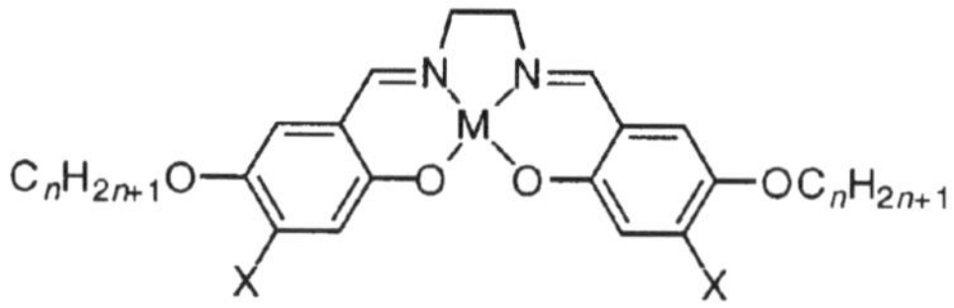

Figure 32 Fluorinated salen complexes.

demonstrate either nematic or S_A phases depending upon the substitution [64], although the mesophases occur at relatively high temperatures, probably due to the aggregation of the porphyrins via electrostatic interactions (e. g. X = Y = H; $n = 7$: Crys·309·N·433·I) [66]. The effects of lateral chains was, therefore, studied with a view to reducing the $\pi-\pi$ interactions by increasing the separations between the molecules [67]. Positioning the alkyl chains on the outer phenyl rings (position Y) removed the mesogenic properties with a reduction in the melting points. However, octyloxy chains attached to the inner phenyl rings (position X) produced the desired effect with the complexes demonstrating nematic phases at temperatures greatly reduced in comparison to the parent zinc(II) porphyrins (e. g. X = C_8H_{17}; Y = H; $n = 7$: Crys·141·N·198·I).

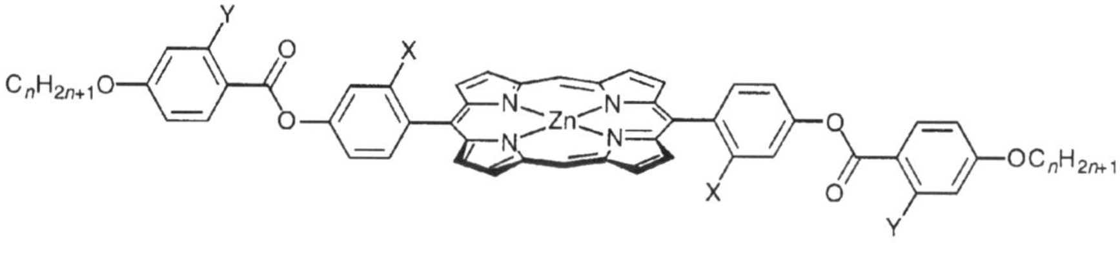

Figure 33 Laterally substituted metalloporphyrins.

4 EFFECTS OF MOLECULAR SHAPE UPON MESOMORPHISM

As discussed earlier, structural anisotropy is paramount in importance when considering the intermolecular interactions necessary to promote mesomorphism. The structural anisotropy is related to the molecular shape and we shall now endeavour to investigate its effect upon the mesomorphic properties of several complexes. In several instances we have discussed N, S_A and S_C phases arising from calamitic molecules, and in later sections we shall discuss disc-like molecules exhibiting columnar mesophases. Unfortunately such rigid definitions of molecular shape in relation to mesomorphism can be misleading as will be demonstrated in the following examples.

4.1 *β*-Diketonate Complexes

The early work of Giroud-Godquin and Billard [68], and later by Ohta [9] dealing with bis(*β*-diketonato) complexes of copper(II) (Figure 34) described the mesophases, not surprisingly, a being discotic (columnar) in nature.

However, in 1986, Chandrasekhar [70] published a derivative (Figure 35) which he claimed to show a biaxial nematic phase. This report was interesting because the biaxial nematic phase (N_b), demonstrated in lyotropic systems [71], had been long sought after in thermotropic materials. Further, the molecules were described as 'bridging the gap' between rod- and disc-like materials (a reference perhaps better reserved for polycatenar liquid crystals—*vide infra*.)

The mesophase was confirmed as one of rod-like molecules due to the discovery of continuous miscibility between 4″-pentyl-4-cyanoterphenyl and the copper(II) complex [73]. Further studies apparently supported the claim that the nematic phase

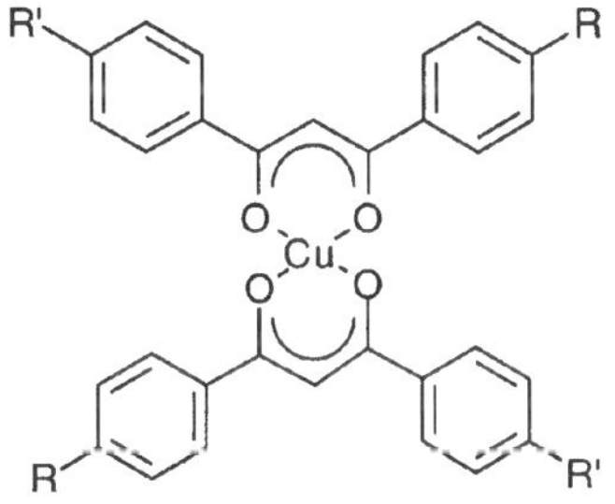

Figure 34 Copper(II) *β*-diketonate complexes which display columnar mesophases.

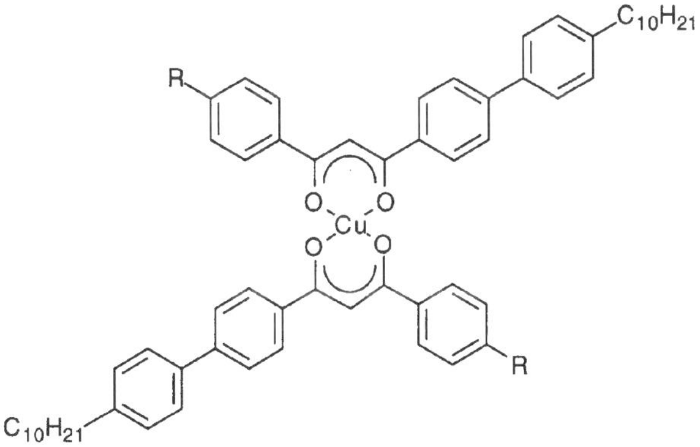

Figure 35 Nematic *β*-diketonate complex.

was in fact biaxial, backed up by conoscopic studies on samples which were held in the correct (homeotropic) alignment by surface coatings and by an external AC field (3 kHz). The conoscopic measurements produced optical figures which gave information concerning the optical axiality of the materials under study. Despite this work, the existence of the N_b phase in these complexes remains to be widely accepted.

However, Chandrasekhar's papers did inspire a great deal of research and, in particular, three groups have studied complexes of a similar molecular shape (Figure 36). Mühlberger and Haase synthesized the complexes where the ring A is cyclohexyl and again these demonstrated a (monotropic) nematic phase [74]. A single-crystal X-ray study of one of these complexes clearly showed the *trans* arrangement of the ligands about the copper(II) ion and the planarity of the molecule. Ohta has reported the related complex where ring A is a phenyl group and $m = 1$, which, despite the molecular shape apparently being rod-like, shows an ordered rectangular columnar phase, apparently through some dimerization [75]. If the methyl group is replaced by a larger group ($m = 2$ or 3) then the molecular interactions necessary for dimerization are reduced due to the additional steric hindrance at the core of the molecule and now a nematic phase is observed; conoscopic measurements determined that this phase was uniaxial [76].

Toyne *et al.* have undertaken a comprehensive structure-property study of a range of related bis(β-diketonate) copper(II) complexes [14b, 77]. It is perhaps surprising in light of Ohta's findings that all such complexes reported by Toyne (Figure 37) for which mesomorphism was reported were nematic or smectic regardless of the fact that the shape of some of the complexes appears to be more disc-like than those of Ohta. Again, the complexes were miscible with 5CT supporting their calamitic structure, providing that there is no direct interaction between the copper(II) ion and the cyano group of the host. A further curious point arises as it would appear that if the R′ substituents are viewed as being lateral substituents, then as the complexes are

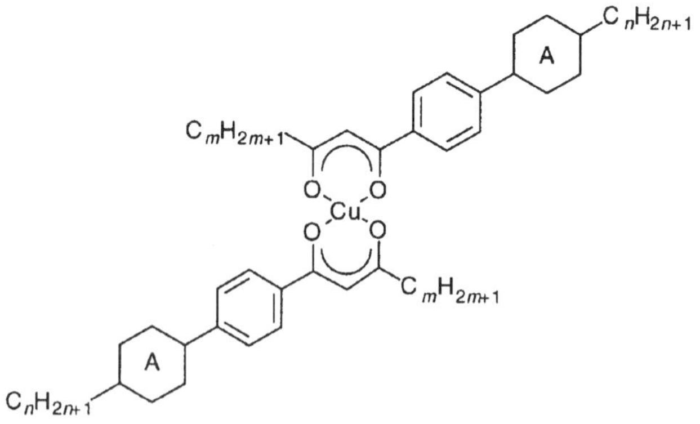

Figure 36 Mesomorphic β-diketonate complexes.

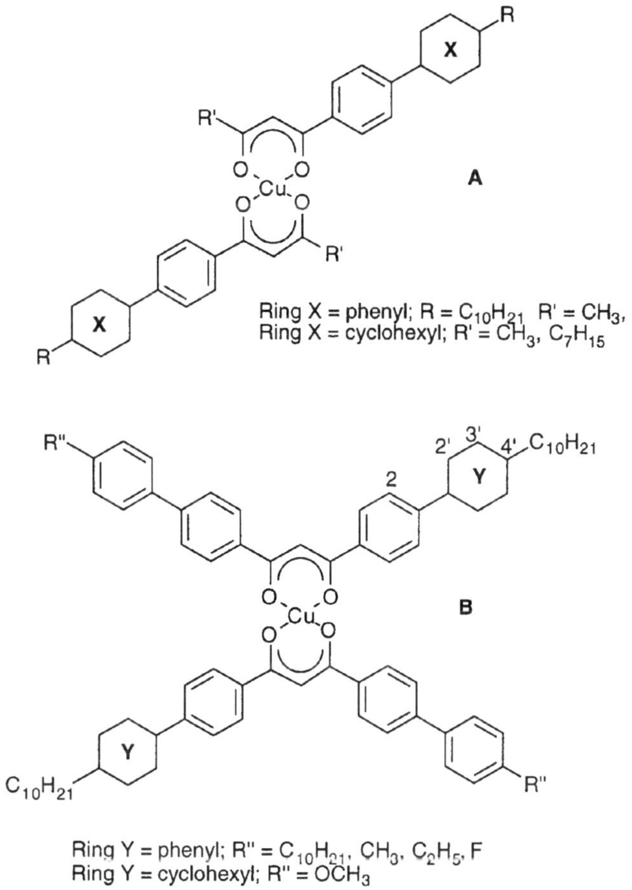

Ring X = phenyl; R = $C_{10}H_{21}$ R' = CH_3,
Ring X = cyclohexyl; R' = CH_3, C_7H_{15}

Ring Y = phenyl; R'' = $C_{10}H_{21}$, CH_3, C_2H_5, F
Ring Y = cyclohexyl; R'' = OCH_3

Figure 37 *β*-Diketonate complexes studied by Toyne and co-workers.

broadened the nematic phase stability apparently increases. Usually, extension of lateral groups reaches a limiting value after which phase stability begins to fall [78].

Various polar substituents on the phenyl or biphenyl 'lateral group' were also investigated (Figure 37; B). Increasing the molecular breadth from phenyl to biphenyl does not result in a reduction in nematic phase stability [77a]; apparently the two opposing effects of steric bulk and polarity with respect to intermolecular interactions are complementary in these systems so that the smectic phase is also stabilized. One apparent anomaly is in the cyano-substituted complexes where the eight-ring complex is purely nematic, while the six-ring complex displays a monotropic S_A phase. The authors suggest molecular associations via antiparallel correlations may occur in the eight-ring complex and such interactions would be less favoured in the six-ring case where the cyano groups are less exposed.

In the series of complexes shown as **B** in Figure 37, the effect of fluorination of the biphenyl ring has been studied [77b]. The effects are similar to those of fluorination discussed above, namely where the fluorine points away from the core, that is substitution in the 2, 3′ or 4′-position then S_A phases are observed whereas fluorination at the 2′-position produces a nematic complex. Fluorines at both the 2 and 2′-position completely destabilizes the mesophases.

4.2 Azobenzene Complexes

Earlier, we considered orthopalladated azobenzene complexes which had a planar, H-shape when the bridging group was a halogen. Alternatively, if the bridging group is a carboxylate then the molecules are best described as open-book dimers (Figure 38), where the alkyl group of the carboxylate is parallel to the spine of the book as is the main axis of the azo-benzene ligand [79]. Such complexes are found to display nematic and lamellar phases, with the open-book molecular shape restricting molecular motion leading to a general preference for the more ordered smectic C phase over the nematic phase.

The effect of length of chain of the alkyl group of the bridging of carboxylate has been investigated [80], and it was found that at both short and long chain length a S_C phase is observed often with a nematic phase, while at medium chain length only nematic phases are seen. This is explained in terms of the medium-length chains acting as lateral substituents and disrupting intermolecular interactions, while the longer chains are able to align along the long molecular axis, favouring the intermolecular interactions necessary for the more ordered phase (Figure 38).

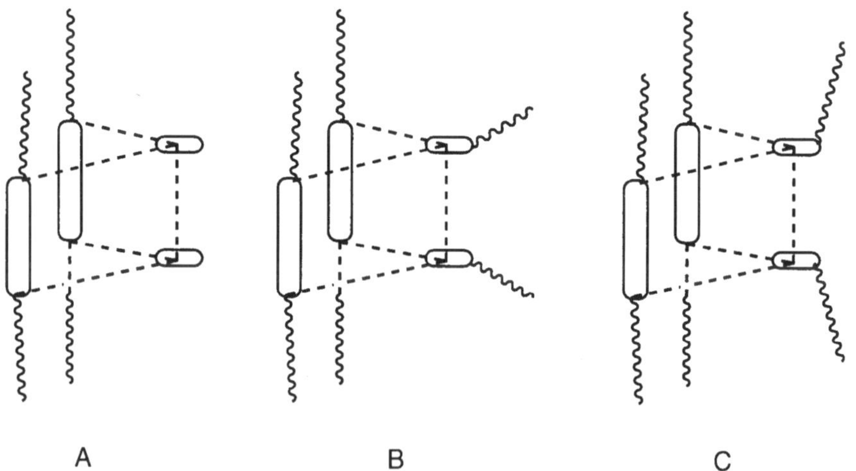

Figure 38 Proposed model for the arrangement of the alkyl bridging carboxylate; A—formate; B—medium chain length; C—long chain length.

4.3 Macrocyclic Complexes

Macrocyclic complexes can be divided into two types, namely those which are largely rigid and those which are flexible. In this latter type, there is interest in seeing how the flexibility of the macrocycle impinges on the mesomorphism and how this can be affected upon complexation.

Thus, in the section on columnar mesophases, several examples of mesomorphic phthalocyanines will be discussed. However, Simon has shown that connecting two phthalocyanines together produces a molecular shape which, from X-ray diffraction

$R = OCH_2CH(C_2H_5)C_4H_9$

M= H or Cu^{II}

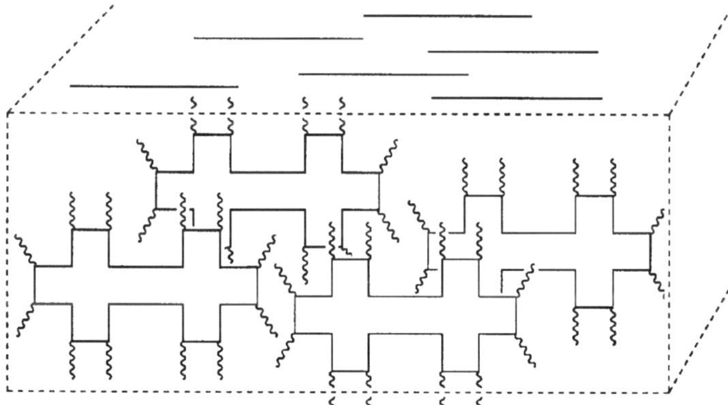

Figure 39 Rigid, dimeric phthalocyanine (above) and a cross-section of lamellar plane proposed from X-ray diffraction studies (below).

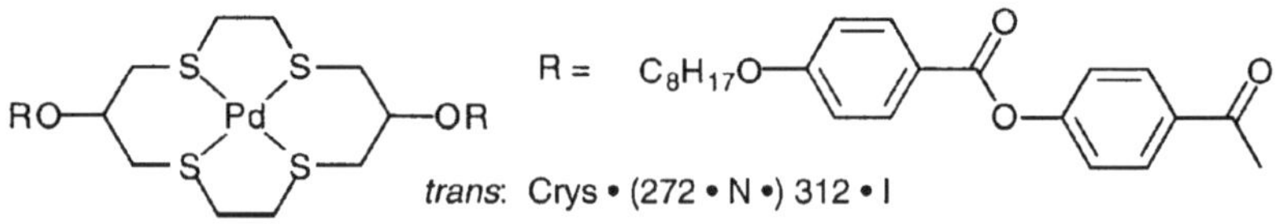

RO
Pd
S
OR
R = $C_8H_{17}O$
O
trans: Crys • (272 • N •) 312 • I

Figure 40 Mesomorphic palladium(II) thioether macrocycle.

studies, produces a smectic-type phase with a nematic order within the lamellar plane (Figure 39) [81]. The fact that the phthalocyanine rings are connected via a rigid phenyl unit leads to a highly anisotropic rotation of the molecule in solution and in the mesophase, which is supported by ESR measurements, and helps to explain the highly viscous nature of the mesophase observed.

In collaboration with Schröder, we have studied a series of macrocyclic thioether metallomesogens, where the functionalization is at ring carbon atoms so that both a *cis* and a *trans* isomer are possible [82]. The free ligands are found to be non-mesomorphic when the pendant groups contains only two ester linkages. However, extension to three-ester pendant groups allows the molecule to function as two independent mesogens linked by a flexible spacer resulting in the appearance of nematic phases. The *trans* isomer of the diester pendant macrocycle becomes mesomorphic upon complexation to Pd(II) due to the enhanced rigidity of the core of the molecule (Figure 40). The molecular shape is more linear for the *trans* complex enhancing the structural anisotropy compared to the *cis* complex, which is consequently non-mesomorphic.

Neve has studied a series of complexes based on non-mesomorphic ligands (Figure 41) obtained by acylation of [18]ane N_2S_4 [83]. A single-crystal X-ray study of one of the silver(I) complexes revealed a tetrahedral geometry for the silver(I) ion and a parallel arrangement for the pendant arms leading to an almost perfect

m+
S
ROC–N
M
N–COR

R = $C_6H_4OC_nH_{2n+1}$

M = Ag, R = $C_{16}H_{33}$, $m = 1$

M = Pd, R = $C_6H_{5-x}(OC_{12}H_{25})_x$: x = 1, 2 or 3; $m = 2$

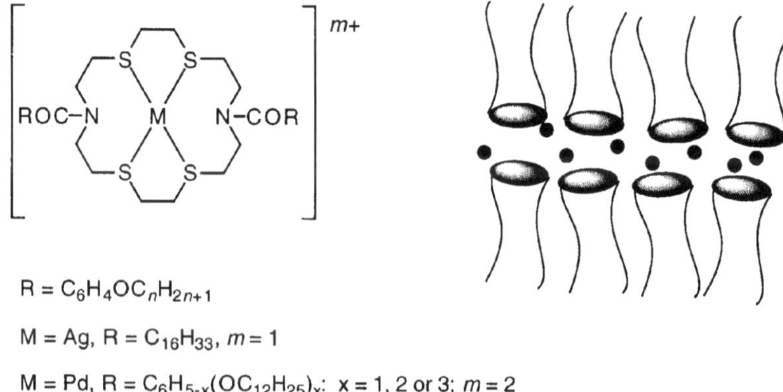

Figure 41 Amphiphilic complexes of Ag(I) with macrocyclic ligands.

U-shape for the molecule [83a]. The cation, therefore, has a strongly amphiphilic character, leading to the formation of a bilayered structure consisting of alternating polar and apolar regions. A monodomain X-ray diffraction study on the silver(I) salts revealed a layer *d*-spacing of 36 Å, and an in-plane modulation (with a period of 60 Å) was detected for the triflate salt. Consequently a bilayered structure with intralayer modulation of the ionic sublayer was proposed as a model for the lamellar phase. Due to the expected structural similarity with the copper(I) complex this model was considered to be representative. The ripple phase suggested is reminiscent of the lamellar $P_{\beta'}$ phase in lyotropic systems, and the amphotropic nature of the silver complexes was demonstrated by swelling experiments in acetonitrile. A stable, lamellar phase was obtained in binary mixtures with up to 20 wt.% of solvent and a slight increase in layer thickness and chain area compared to the anhydrous thermotropic phase was observed.

In the palladium(II) macrocyclic complexes, the metal ions were proposed to be coordinated in a square planar geometry, with a disordered smectic mesophase being observed, even for the polycatenar complexes (i. e. those with more than one chain on each terminal ring). Preliminary X-ray diffraction measurements determined the smectic *d*-spacing to be 53 Å which is comparable to the calculated molecular length of the ligand (50–52 Å) in the fully extended conformation [83b].

Tschierske has reported some novel metallomesogens based on binuclear cyclopalladated cyclophanes (Figure 42) [84]. The free ligands adopt a rod-like shape due to π–π interactions between the aromatic group, rather than forming a typical

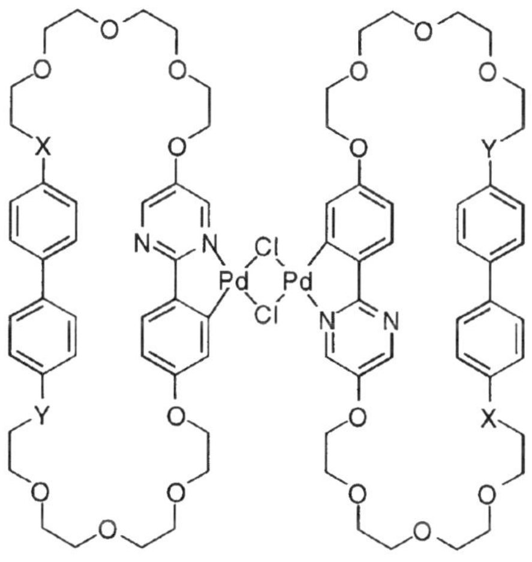

X = Y = O, Crys • 168 • S_A • 208 • N • 226 • I (decomposition)

X = OCH_2, Y = CH_2O, Crys • (91 • N) •118 • I

Figure 42 Novel palladium complexes of macrocyclic pyrimidine ligands.

crown-type conformation [85]. The ligand where X=Y=O, shows a monotropic nematic phase, while that with X=Y=–OCH_2– is non-mesomorphic. However, the complexes show more stable mesophases, although the difference in mesomorphism between the two complexes is not straightforward to understand at first sight.

4.4 Chirality

As mentioned in the introduction, chiral compounds can exhibit chiral mesophases and these are important due to the important physical properties that they may exhibit, including thermochroism, ferroelectric and electroclinic effects [15]. In 1975, Meyer predicted the existence of a spontaneous polarization (P_S) in chiral, tilted smectic phases [86], and the existence of such polar order within a liquid crystal phase has important implications both scientifically and industrially [19]. The asymmetry associated with the chirality may also produce a beneficial lowering of transition temperatures.

In a chiral smectic (S_C*) phase, the tilt angle is the same within a layer, but the tilt direction precesses and traces a helical path through a stack of layers (Figure 43). It has been demonstrated that when such a helix is completely unwound, as in a surface stabilized ferroelectric liquid crystal cell, then changing the tilt of the molecules from $+\theta$ to $-\theta$ by alternating the direction of an applied field results in a substantial electro-optic effect, which has the features of very fast switching ($\approx 1-10\,\mu s$), high contrast and bistability [87]. The smectic A phase of chiral molecules may also exhibit an electro-optic effect, this arises due to molecular tilt fluctuations which occur as the S_A*–S_C* transition is approached, which are combined with a

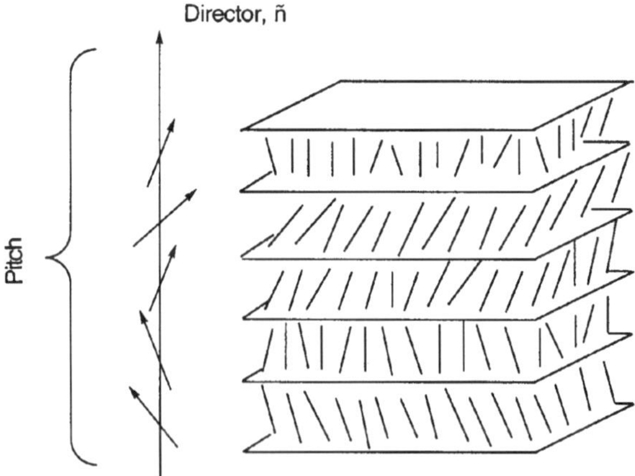

Figure 43 Schematic diagram of an S_C* phase showing the helical precession of the director.

divergence of the smectic layer compressibility, the so-called soft mode (note that the notation S_A* follows a new convention [88] which represents the fact that the phase itself is made up of chiral molecules—i. e. its symmetry is formally different from the same phase of non-chiral molecules—rather than implying that the phase itself is inherently chiral). Such tilt fluctuations in the S_A* phase are connected with local polarization fluctuations along the smectic layers and consequently, application of an external electric field induces an ordering of the molecular lateral dipoles coupled with an induced tilt angle so that a homogeneously aligned S_A* phase is achieved. This effect is termed the electroclinic effect and again has a short switching time and a linear voltage dependence of the induced tilt angle [89].

The first report of a ferroelectric effect by a metallomesogen was for an open-book palladium(II) complexes (Figure 44) of type discussed above, where the chirality was introduced in the bridging carboxylate [14a]. The ^{1}H NMR spectra identified that the complex was a mix of isomers; *trans*-ΔR,R (34%), *trans*-ΛR,R (34%) and *cis*-R,R (32%). Ferroelectric switching was demonstrated, although at a much slower rate (≈1 s) than in organic liquid crystal systems, due to the high viscosity of the material.

A related series of palladium(II) azobenzene complexes with halogen bridging ligands has been prepared by Ghedini, in which the chirality was present in one of

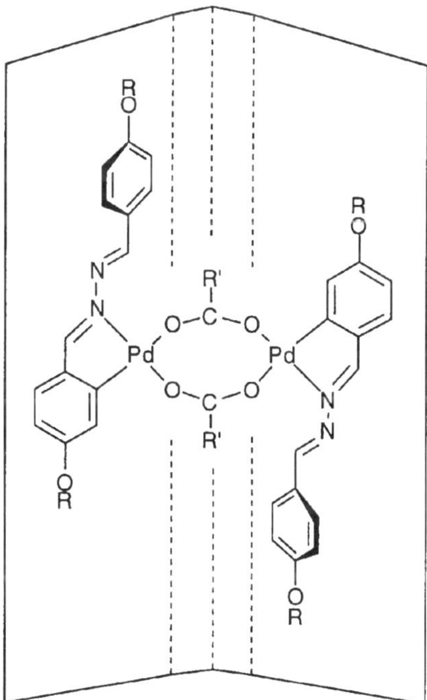

Figure 44 'Book'-shaped palladium complexes showing ferroelectric behaviour.

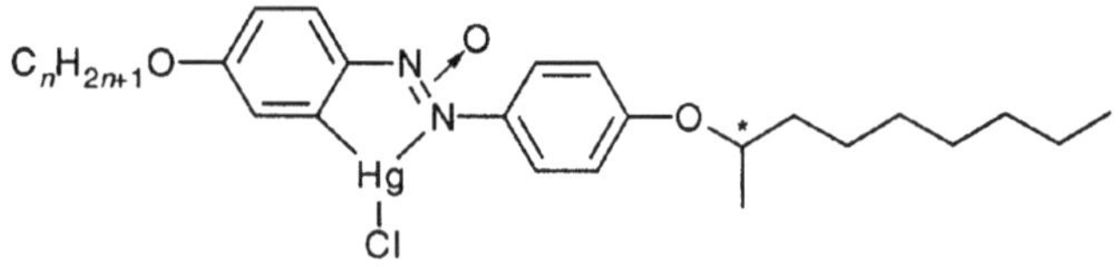

Figure 45 *Ortho*-metallated complexes of Hg(II).

the alkoxy chains of the ligand [90]. This time the complexes were a 1 : 1 mixture of isomers and, in several cases, S_C* phases were reported. Significantly, Ghedini has reported that a related 4,4′-disubstituted azoxybenzene ligand, again with one chiral alkoxy chain, forms complexes with mercury(II) (Figure 45) which exhibit an S_C* phase at room temperature [91].

Espinet has reported similar palladium(II) complexes employing Schiff base ligands with 2-halopropionate as the chiral bridging group [92]. It was shown that the 2-halopropionato bridging group could stabilize mesophases compared to complexes where an acetate bridging group was present. This effect was explained with regard to the new dipole moments that would be present in the 2-halopropionato-bridged complexes, a reduction in the melting point due to the bulkier bridging group and the formation of isomeric mixtures. Exchange of one of the bridging 2-chloropropionate groups for an alkylthiolate produces solely the *cis*-complex and these were reported as the first metallomesogens to display a chiral nematic phase. The complexes displayed an inversion of their helical pitch upon heating, which was explained in terms of the presence of two conformers, which are in temperature-dependent thermodynamic equilibrium with each other, each inducing an opposite helical screw sense in the N* phase.

Espinet and Serrano have previously reported that disrupting the symmetry of a series of dinuclear palladium(II) complexes by forming the mononuclear complexes with a β-diketonate ligand yields a advantageous reduction in the melting point [93]. This chemistry was extended to ferroelectric complexes (Figure 46) to yield complexes that switched much faster than their first reported ferroelectric metallomesogen [94].

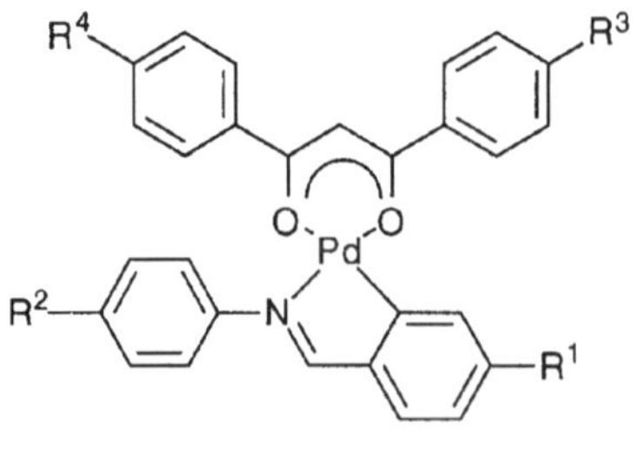

Figure 46 Template for ferroelectric palladium mesogens.

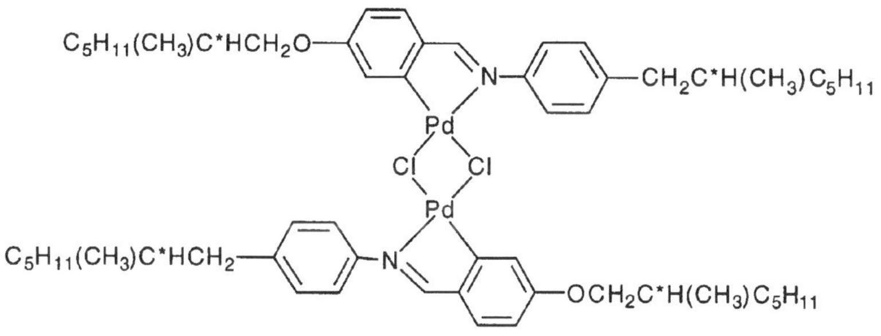

Figure 47 Ferroelectric palladium complexes with a high spontaneous polarization.

In a systematic study Espinet and Serrano have further shown that the value of P_{Smax} in dinuclear palladium(II) complexes is increased if the chiral chain is placed upon the ortho-metallated ring [95]. This effect is due to the chiral chain being fixed to a greater degree in space leading to an increased intermolecular coupling of the molecular dipoles, so that values for P_{Smax} of 44–131 nC cm^{-2} are achieved. Increasing the number of chiral chains to four (Figure 47) yields the largest P_{Smax} of 206 nC cm^{-2}, although again, this complex has a high viscosity and consequent long switching time.

Ghedini has studied a series of salicylaldimato palladium(II) complexes with chiral alkoxy chains on the anilinic portion of the molecule. When the chain is derived from S-(−)-β-citronellol then enantiotopic S_A* and S_C* phases are observed, whereas using *R*-(−)-2-octanol yields a complex displaying a monotropic N* phase [96]. The ferroelectric properties of the former complex have been studied, providing a P_{Smax} of 12 nC cm^{-2} and a response time of 60 ms near the S_A*–S_C* transition [97]. The electroclinic properties of this complex are an induced tilt angle of 12° close to the S_A*–S_C* transition, with a value of 4° near the S_A–I transition but with a shorter response time of only 25 ms.

Serrano has reported the complexation of a chiral salicylaldimato complex with both copper(II) and oxovanadium(IV), with values of 23 and 20 nC cm^{-2} respectively for P_{Smax}, the free ligand having a value of 44 nC cm^{-2} [98]. In all the previous complexes it was found that P_{Smax} was greater in the free ligand when compared to the complex. However, here it was found additionally that the viscosity and the response time was higher for the oxovanadium(IV) complex compared to the copper(II) complex suggesting that there was an additional degree of order present for the former complex, i. e. intermolecular V . . . O interactions. This was supported by the value of the oxovanadium stretching frequency when compared to the value in unassociated oxovanadium(IV) complexes.

Galyametdinov has reported a copper(II) complex with salicylaldimate ligands which bear a chiral alkoxyl chain, the complex displays S_X* (highly ordered tilted smectic), S_C* and N* phases and the P_{Smax} was found to be 25 nC cm^{-2} with a switching time of 10 ms [99]. Magnetic studies showed there was no change in the magnetic behaviour at the phase transitions and no exchange interaction between the copper(II) ions was identified.

4.5 High Coordination Number Metallomesogens

In all the metallomesogens discussed so far, the coordination geometry at the metal centre has been either linear, trigonal planar, square planar, tetrahedral, or square pyramidal. The accessibility of higher coordination numbers is possible through appropriate consideration of the molecular shape and the structural anisotropy. In the complexes considered so far, excluding Neve's tetrahedral silver(I) and copper(I) complexes, the planarity associated with the metal geometry maintains the global anisotropy of the molecule. The oxovanadium bond could be viewed as a lateral substituent, where no intermolecular dipolar interactions are present, perturbing the structural anisotropy and explaining why the mesophase stability is often reduced in comparison to analogous copper(II) complexes.

4.5.1 Nitrogen-containing ligands

In trying to develop a general strategy for designing mesomorphic complexes containing six-coordinate metal complexes, we looked to the work of Deschenaux (*vide infra*) with mesomorphic ferrocenes where it appeared that 1,3-disubstituted ferrocenes required appended groups containing a total of four rings before mesomorphism was observed. This high anisotropy for the ligand is readily understandable if one regards the six-coordinate metal centre as a perturbation to the anisotropy of the ligand. Thus, in order to maintain sufficient anisotropy in the resulting complex, it is necessary to begin with a highly anisotropic ligand.

Consequently, we have successfully produced mesomorphic octahedral manganese(I) and rhenium(I) tetracarbonyl complexes (Figure 48) via orthometallation of highly anisotropic imine ligands [100]. Thus, the free imine ligands show smectic and nematic phases and clear at around 300°C, while the complexes display only a nematic phase; such behaviour is consistent with the metal fragment acting as a lateral substituent reducing intermolecular interactions and so disfavouring smectic phases and lowering the clearing points.

Using a similar approach, we complexed diazabutadiene ligands to bromotricarbonylrhenium(I), again producing mesomorphic complexes (Figure 49), except that this time the reduction in clearing point which was evident in the Mn and Re complexes above, was not observed [101].

The effect of the ligand X (Figure 50) has been studied and shows the expected reduction in clearing temperature with increase in the size of X, to the extent that the triflato derivatives are non-mesomorphic. In the light of our findings with orthometallated imine complexes, it is perhaps surprising that complexation produces higher

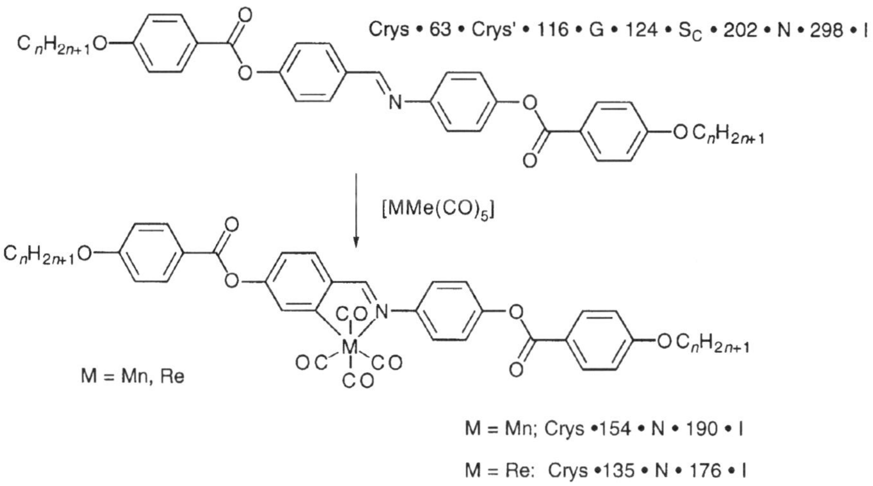

Crys • 63 • Crys' • 116 • G • 124 • S_C • 202 • N • 298 • I

[MMe(CO)$_5$]

M = Mn, Re

M = Mn; Crys •154 • N • 190 • I

M = Re: Crys •135 • N • 176 • I

Figure 48 Mesomorphic complexes of Mn(I) and Re(I).

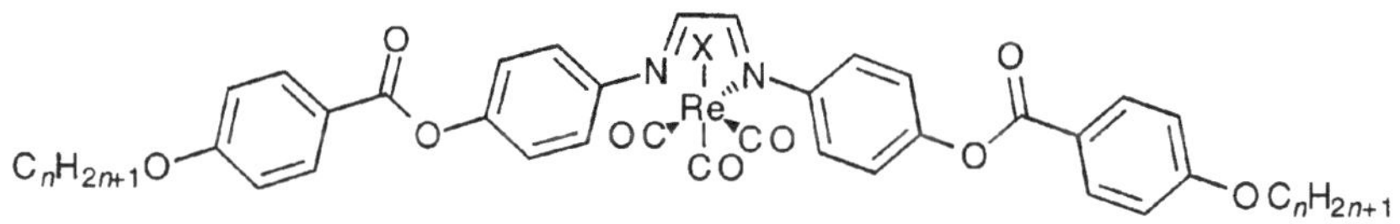

Figure 49 Mesomorphic rhenium complexes of diazabutadienes.

clearing points compared to the free ligands, and we ascribed this to be due to a combination of an increase in rigidity upon complexation as the diazabutadiene will now be held in a fixed cisoid arrangement, coupled with the effects of the dipoles associated with both the cisoid diazabutadiene arrangement and the Re–X bond.

We then synthesized related mesomorphic ligands based upon 2,2′-bipyridines, which, being α-diimines, coordinate in a structurally similar manner to the diazabutadienes. We initially synthesized free ligands [102] containing a total of four rings (Figure 51; $m = 0$) which we found to be mesomorphic but, unexpectedly, none of their complexes (save those of Ag(I)) was found to be mesomorphic [103].

However, by increasing the anisotropy of the ligand (Figure 51; $m = 1$), we found it possible to accomplish the synthesis of complexes of Re(I) which were mesomorphic [104].

Deschenaux and Süss-Fink have recently reported some mesomorphic complexes derived from a $[Ru_2(CO)_4(\mu\text{-}\eta^2\text{-}O_2CR)_2L_2]$ sawhorse unit (Figure 52), showing nematic phases [105]. Again the ligand, L, has to be sufficiently anisotropic to compensate for the bulky bridging carboxylate groups. For example, with L = 4-alkoxypyridine or with a trifluoromethyl substituent on the bridging carboxylate, the

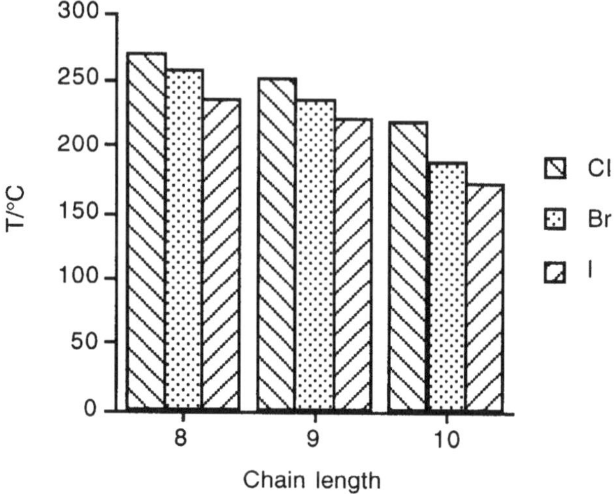

Figure 50 Effect of halide on clearing point for diazabutadiene complexes of Re(I).

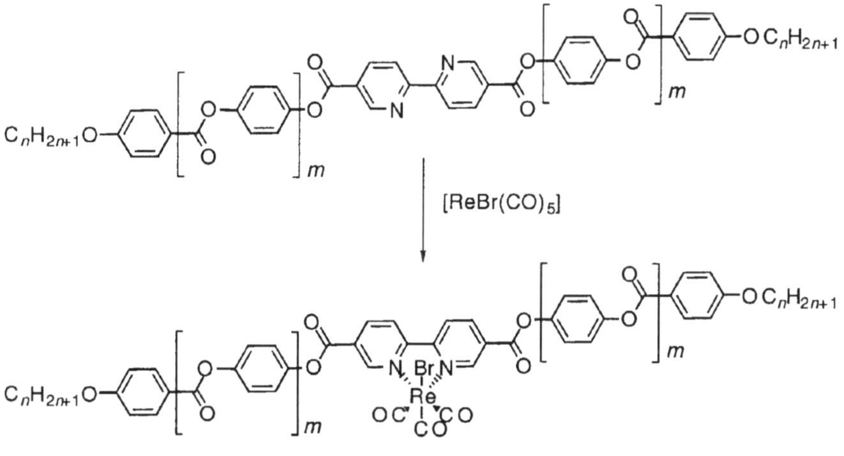

Figure 51 Mesomorphic complexes of 2,2′-bipyridines.

complexes were non-mesomorphic. The authors suggest that $\pi-\pi$ interactions between the aromatic substituent of the carboxylate and the neighbouring pyridine ring are important in organizing the ligands in one direction to favour mesomorphic properties. This is consistent with the observation that groups in the *para* position of these aromatic rings lead to a destabilization of the nematic phase.

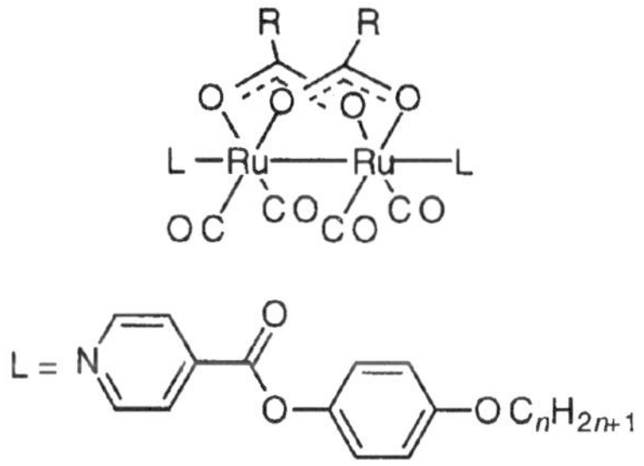

$R = H, CF_3, C_6H_5, p\text{-}C_6H_4OCH_3, p\text{-}C_6H_4CH_3$

Figure 52 Mesomorphic diruthenium complexes.

4.5.2 Iron complexes

Ferrocene, which may formally be considered as containing a six-coordinate iron(II) centre, was first incorporated in a liquid crystal by Mâlthete and Billard (Figure 53) [106]. Clearly the structural anisotropy within the molecule, has to be sufficient to compensate for the large volume associated with the ferrocene unit. Due to the ease of derivatization of ferrocene this problem has been approached in several different directions.

Other monosubstituted ferrocenes have been reported by Imrie and co-workers. For example, the complex shown in Figure 54 is clearly less anisotropic than the original described by Mâlthete and Billard, and shows a nematic phase at just above

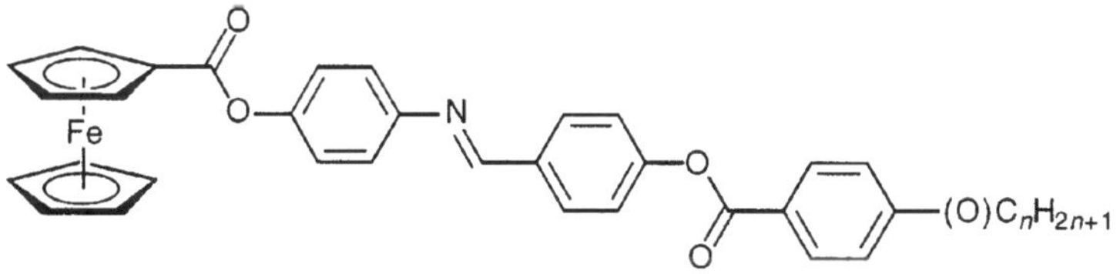

Figure 53 The first mesomorphic ferrocene.

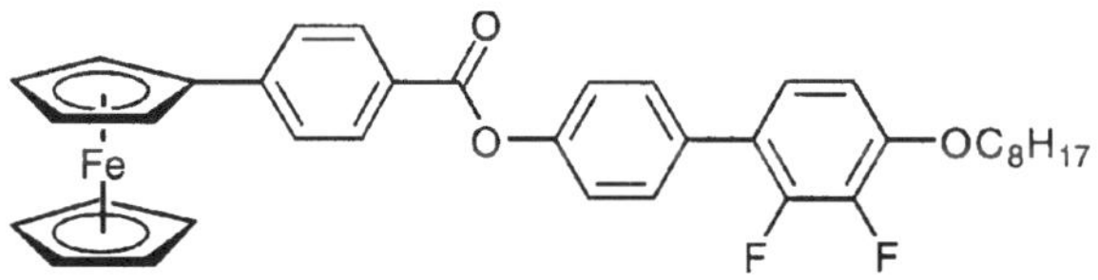

Crys • 115 • N • 129 • I

Figure 54 Mesomorphic, difluorinated ferrocene.

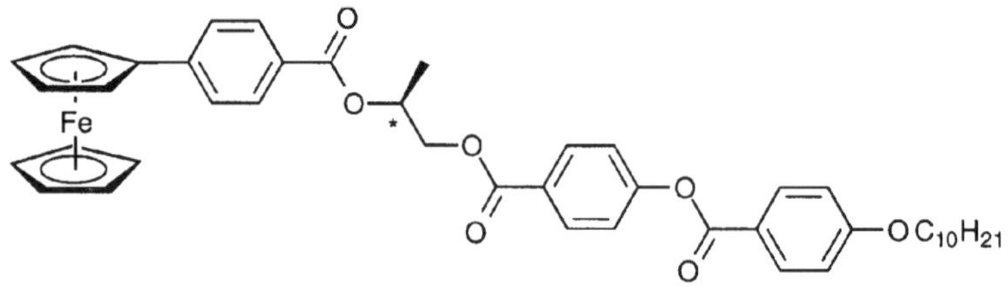

Figure 55 Ferroelectric ferrocene complex.

100°C [107]. Another example from this group (Figure 55) uses a chiral group to generate a complex with a S_C* phase [108].

A great deal of effort has been expended in the synthesis of 1,1′-disubstituted ferrocenes [109], particularly by Toyne *et al.* [110]. The ferrocene complexes were compared to the analogous compounds where the ferrocene was replaced by either a 1,4-disubstituted benzene or cyclohexane, demonstrating that these last two were better at supporting mesomorphism than the ferrocene unit. This is suggested to be due to both the step that the ferrocene unit introduces into the molecule and the low internal energy of rotation of the cyclopentadienyl rings (approximately 4 kJ mol^{-1}) allowing a loss of rigidity and hence, linearity within the molecule. While one might reasonably assume that these mesogens preferentially adopt an extended 'S' conformation in the mesophase, it is interesting to note a mesomorphic [3]-ferrocenophane (Figure 56) with an enantiotropic S_C phase. Interestingly, the analogue [2]-isomer is non-mesomorphic [111].

Deschenaux has shown that the 1,3-disubstituted ferrocene complexes (Figure 57) are far better at supporting mesomorphism when compared to the analogous 1,1′-disubstituted complexes [112]. The former complexes characteristically show enantiotropic S_C, S_A and nematic phases. Asymmetric 1,3-disubstituted ferrocenes have been synthesized and display a rich mesomorphism, for example, S_A, S_C and nematic phases [113].

Not surprisingly it has been demonstrated that a total of three rings plus the ferrocene is insufficient to support mesomorphism in the 1,3-disubstituted

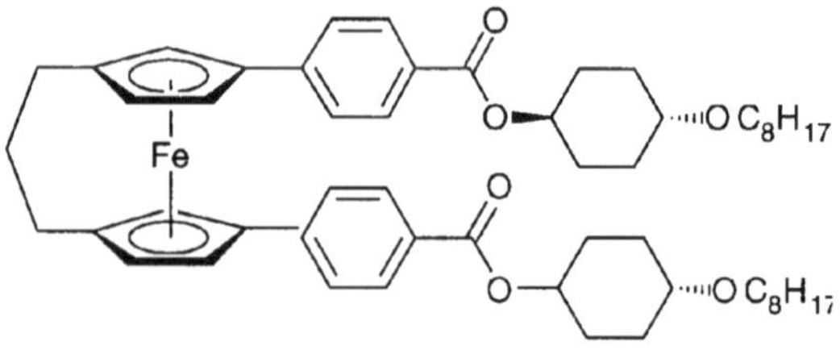

Crys • 100 • S_C • 139.5 • N • 158 • I

Figure 56 Mesomorphic ferrocenophane.

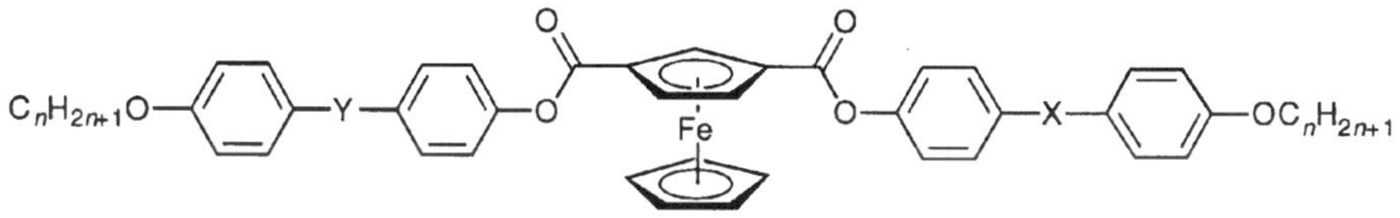

$X = Y$ or $X \neq Y$; typically X,Y = σ-bond, or –COO–

Figure 57 General structure for mesomorphic 1,3-disubstituted ferrocenes.

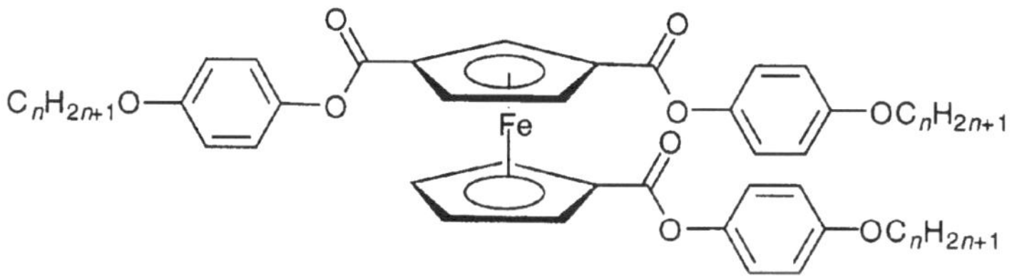

Figure 58 Mesomorphic, trisubstituted ferrocenes.

complexes. However, if the substitution pattern is changed to produce 1,1′,3-trisubstituted ferrocenes (Figure 58), this is found to be sufficient to support monotropic smectic phases [114]. It appears that the three aromatic rings are sufficient to compensate for the unfavourable steric repulsions of the bulky ferrocene core. Incorporation of a further ester and phenyl ring at each position allows enantiotropic S_A and S_C phases to be observed.

Deschenaux has also reported a ferrocene complex (Figure 59) which is non-mesomorphic in the iron(II) oxidation state, but which shows an S_A phase on oxidation to the related ferrocenium ion. X-Ray diffraction studies reveal a *d*-spacing of 39.5 Å, which is comparable to the approximate molecular length of 41 Å [115]. A mesomorphic macromolecule has also been reported by Deschenaux based on a dendritic core substituted with six mesomorphic 1,1′-disubstituted ferrocene units, and displaying an enantiotropic S_A phase [116].

Tschierske has synthesized (Figure 60) some 1,4-disubstituted benzenes and 1,1′-disubstituted ferrocenes which are laterally attached to rod-like 4,4″-disubstituted *p*-terphenyls [117]. The ferrocene complexes display more stable S_A phases in comparison to those derived from the benzene linker, which the authors suggest is due to the flexibility associated with the ferrocene core, allowing an improved arrangement of the individual molecules within the layers compared to the more rigid 1,4-disubstituted benzene. Not surprisingly, when only one terphenyl molecule is attached the ferrocene complex a much smaller S_A range is observed, while the analogous benzene-linked molecule displays S_C, S_A and nematic phases.

Malthête has reported some five-coordinate butadiene iron tricarbonyl complexes (Figure 61) exhibiting a wide range of nematic phase for type A, and nematic, S_A or

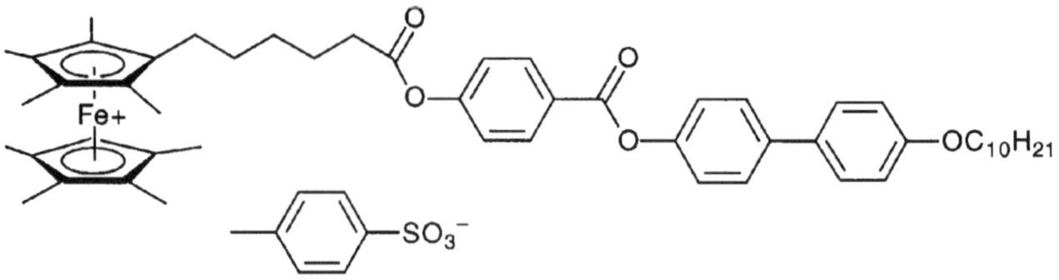

Figure 59 Mesomorphic ferrocenium complex.

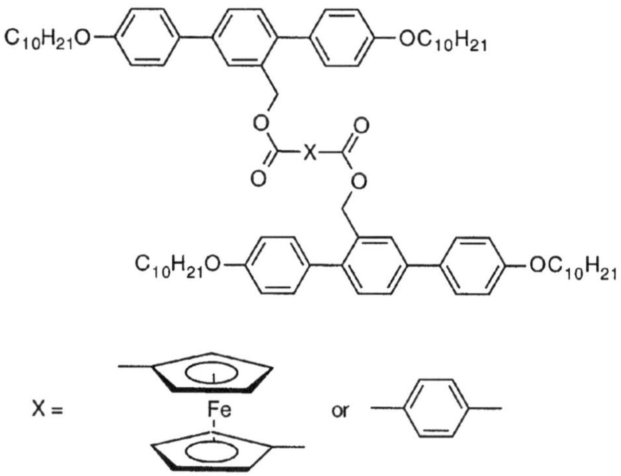

Figure 60 Evaluation of ferrocene as a linker in dimeric mesogens.

both for type B [118]. The asymmetric nature of the complexes helps lower transition temperatures as most exhibit enantiotropic phases for less anisotropic ligands than correlated ferrocene complexes; also the asymmetry means that the complexes are chiral. An enantiomerically pure ferrocene complex of type B ($R = R' = C_{10}H_{21}$) has been synthesized by using an enantiomerically pure starting material, the transition temperatures remain similar to those for the racemic complex but a chiral nematic phase is now seen above the S_A phase [118b].

Deschenaux and Ziessel reported (η^6-arene)tricarbonylchromium complexes (Figure 62) with enantiotropic S_C and nematic phases for the complexes where $n = 8$ or 10, and just a S_C for the higher homologues [119]. When compared to the mesomorphism of the metal-free ligands the complexes exhibit both lower melting and clearing temperatures, and it is possible to view the tricarbonylchromium moiety as a lateral group reducing intermolecular interactions.

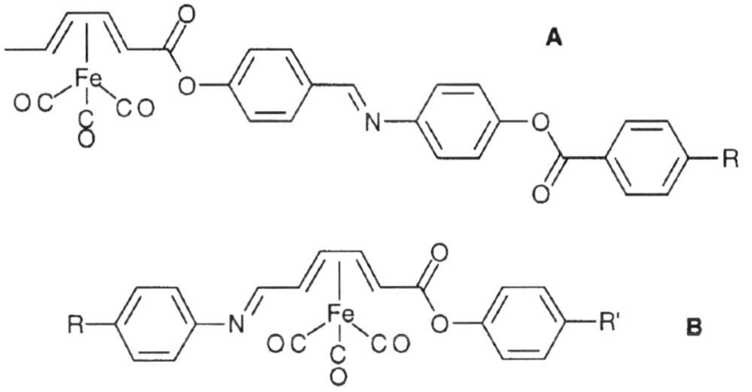

Figure 61 Mesomorphic iron complexes of butadienes.

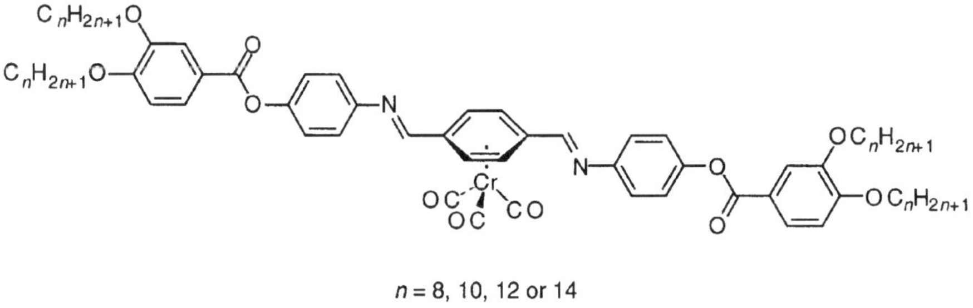

$n = 8, 10, 12$ or 14

Figure 62 Mesomorphic chromarene complexes.

5 MATERIALS FORMING COLUMNAR MESOPHASES

The structure of materials forming columnar mesophases is at first sight somewhat simpler than that of calamitic systems. The disc-like core is often aromatic, and is in general surrounded by six or eight alkyl chains; Figure 63 shows some representative examples of discotic materials.

Traditionally, mesophases formed by such molecules were denoted, for example, D_h or D_r for discotic hexagonal and discotic rectangular, respectively [120]. Such a description actually refers primarily to the structure of the component molecules, rather than the symmetry of the phase. Consequently it has been proposed that the nomenclature be brought more in line with that for smectic and nematic phases, so that the phase be redesignated as Col, describing the columnar nature of the phase, with the specific phase being, for example, Col_h (columnar hexagonal), Col_r (columnar rectangular) and so on (e. g. Figure 64) [121]. It is also common in both the 'D' and 'Col' nomenclature to find an additional subscript 'o' or 'd', meaning *ordered* or *disordered* and referring to the presence or absence, respectively, of liquid-like order within the columns. The order or disorder are usually inferred

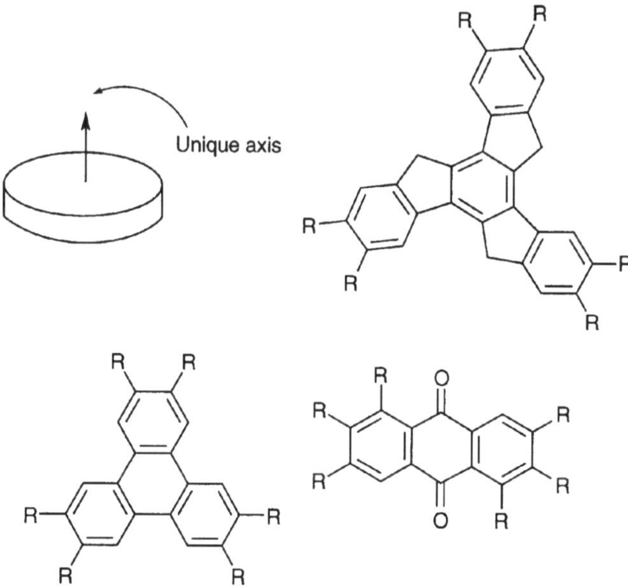

Figure 63 Schematic representation of molecular structure of discotic molecules and three examples of organic discotic molecules.

from the sharpness of the signal in the X-ray diffraction pattern due to the columnar repeat, typically at about 3.4 Å for a columnar phase of planar molecules. However, while it is sometimes the case that a very narrow (ordered) or very diffuse (disordered) signal can be observed, it is true that there exists a spectrum of possibilities, and saying where disordered ends and ordered starts is fraught with difficulty. Therefore, many authors simply do not specify one or the other possibility. However, it is argued by some that a truly ordered columnar phase is akin to a crystal smectic phase, and is not, therefore, regarded as a 'true' liquid crystal mesophase.

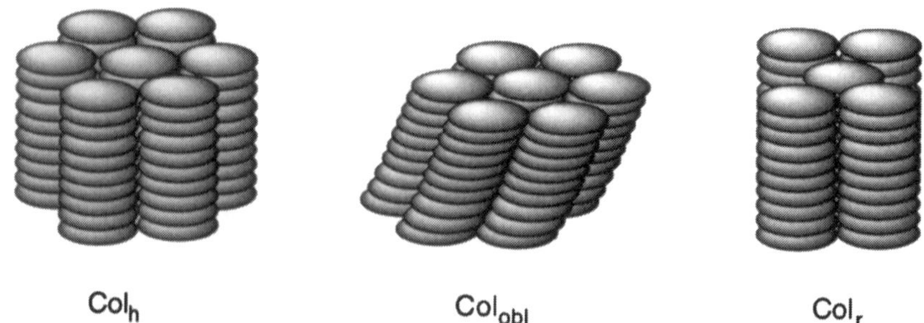

Figure 64 Schematic representation of some columnar phases.

While the lack of consensus is at times frustrating, it is simultaneously quite exciting, pointing to a dynamic and living subject. In general, we will try to follow the 'd' or 'o' if used by the authors, although we shall always use 'Col' in place of 'D'.

5.1 Macrocyclic Complexes

5.1.1 Phthalocyanines

Phthalocyanines represent a class of inherently disc-like molecules and consequently were among some of the first examples of metallomesogens exhibiting columnar mesophases, being first reported by Simon and co-workers in 1982 [122]. Thus, they described an octasubstituted phthalocyanine copper(II) complex (Figure 65) which exhibited a mesophase between 53°C and the start of decomposition at about 300°C; X-ray diffraction studies showed lines with reciprocal spacing of $1:\sqrt{3}:\sqrt{4}:\sqrt{7}$, indicating a hexagonal columnar phase. The distance between the columns was found to be approximately 34 Å, with the stacking distance between the macrocycles within the columns being 3.8 Å, the broad nature of the latter peak suggesting a disordered Col_{hd} phase. The metal-free phthalocyanine displays a slightly smaller

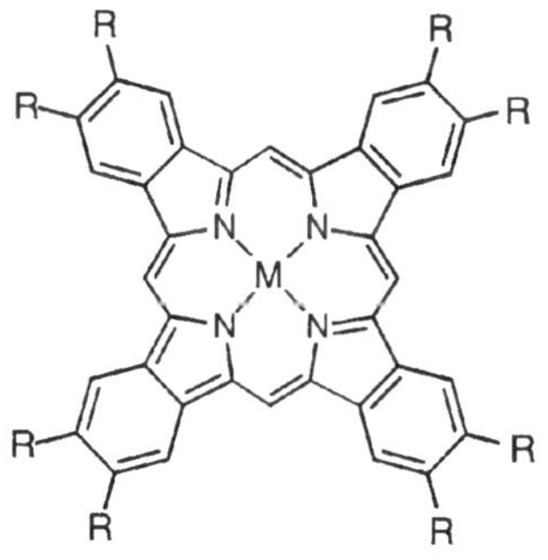

M = Cu:

R = $CH_2OC_{12}H_{25}$ Crys • 53 • Col_{hd} • 300 • I^d
R = C_8H_{17} Crys • 81 • Col_{hd} • 180 • I^d
R = OC_8H_{17} Crys • 112 • Col_{ho} • >345 • I
R = $PhC_{12}H_{25}$ Crys • 67 • K' • 120 • Col_{rd} • 227 • I^d
R = SC_8H_{17} Crys • 75 • Col_h • 320 • I^d
R = SC_8H_{17} & Cl Crys • -5 • Col_h • 320 • I^d

d = decomposition

Figure 65 Mesomorphic phthaloxyanines.

mesophase range of 48 to 264°C; the mesophase which was more ordered and may be assigned as Col_{ho} [123].

More recently Simon has reported a platinum(II) phthalcyanine bearing eight dodecyloxy chains, which also displays a hexagonal columnar phase between 77°C, with the onset of decomposition at about 360°C [124]. Interestingly the intra-columnar periodicity has a value of only 3.29 Å, suggesting some compression due to the overlap of platinum *d*- and/or *p*-orbitals. Not surprisingly the complex demonstrated a tendency to aggregate in organic solvents.

A series of octaalkylphthalocyanine nickel(II) and copper(II) complexes has been studied by Simon, again displaying hexagonal columnar phases, although an extra peak was observed in the X-ray diffraction pattern [125]. This was explained with reference to the D_{4h} symmetry of the molecule disfavouring packing into a regular hexagonal lattice, and to restricted rotation within the columns. Consequently, a model was proposed involving irregular stacking of the molecules within a column composed of alternating pinched areas and areas with larger domains, the periodicity of these areas accounting for the extra peak in the X-ray diffraction pattern (Figure 66). Analogous octaalkoxyphthalocyaninecopper(II) complexes display a Col_{ho} phase, with higher transition temperatures and no transition being detected into the isotropic state upon heating to 345°C [126].

Ohta has studied octaalkoxyphthalocyanines, where the metal-free compounds display Col_{hd}, while the copper(II) complexes display a Col_{rd} phase [127]. The structural rearrangement necessary to change between these two phases is very minor, so that the rectangular symmetry may be obtained from the hexagonal one by simply shifting the layers relative to each other (Figure 67). From the lattice parameters, the plane of the macrocyclic ring is at an angle of 36° to the column axis leading to a shift of the molecular units by 2.5 Å, which can be explained with

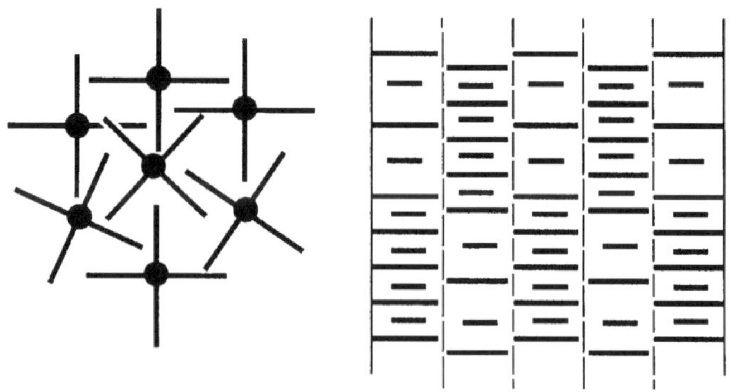

Figure 66 Schematic showing the packing within a cross-section of a column (left) and the alternating pinched and extended areas along a column (right) in Simon's octaalkylphthalocyanines.

Figure 67 Schematic showing the transformation of hexagonal to tetragonal symmetry.

regard to the tendency of copper(II) complexes for axially coordinated ligands, in this case the meso-nitrogen of a neighbouring phthalocyanine ring.

Cook and co-workers have studied complexes with the alkyl groups placed on the non-peripheral positions, to yield 1,4,8,11,15,18,22,25-octaalkylphthalocyanines (Figure 68) [128]. These complexes display an increase in transition temperatures compared to the metal-free phthalocyanine, predominantly in the clearing temperature. The melting points generally increase in the order, $Ni^{2+} < 2H^{+} < Cu^{2+} < Zn^{2+}$, and similarly the clearing points in the order, $Ni^{2+} \approx 2H^{+} < Cu^{2+} < Zn^{2+}$. The main effect of altering the substitution, when compared to the previously discussed phthalocyanine complexes, is that three mesophases are now observed. One was a Col_{rd} phase, while the other two were variants of a Col_{hd} phase, with that at lower temperature having a slightly better developed hexagonal net shown by the presence of a small number of additional reflections; similar behaviour has been reported previously by the same authors [129]. These materials were found to absorb in the near infrared [130].

Replacement of eight alkoxy-groups by eight thioether groups (Figure 69) has been shown to produce a reduction in melting point and an enhanced mesophase range, with recrystallization occurring below room temperature [131]. Phthalocyanine complexes with four thioether groups and chlorine atoms at the other four positions, have been isolated as a mixture of isomers and also display hexagonal columnar phases [132]. Not surprisingly X-ray studies show that the four chain complexes have a smaller inter-columnar distance and an increased intra-columnar long range order as the packing of the alkyl chains is easier when compared to the analogous phthalocyanines bearing eight such chains.

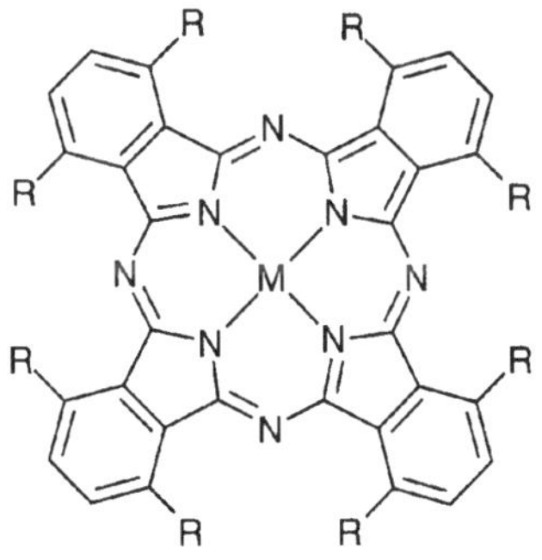

Figure 68 Mesomorphic phthalocyanines.

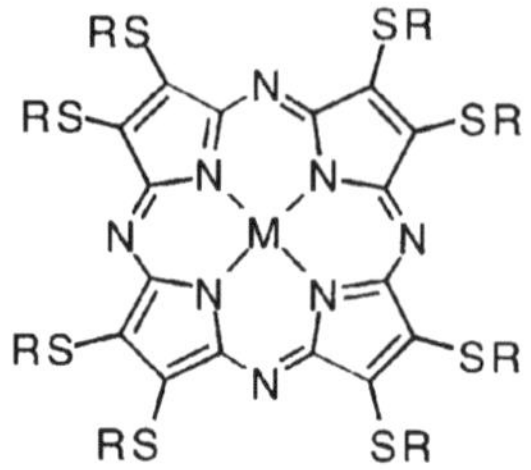

For $n = 8$:

$M = Ni^{2+}$	Crys • 68 • Col$_{hd}$ • 119 • I
$M = Co^{2+}$	Crys • 55 • Col$_{hd}$ • 208 • I
$M = Cu^{2+}$	Crys • 68 • Col$_{hd}$ • 152 • I
$M = Zn^{2+}$	Crys • 41 • Col$_{hd}$ •120 • I

$R = C_nH_{2n+1}$; $M = Ni^{2+}$, Co^{2+}, Cu^{2+} or Zn^{2+}

Figure 69 Thioether-substituted tetraazaporphyrins.

Ohta has also studied triphenylene-based porphyrazinato metal(II) complexes (Figure 70), which have an extended core compared to phthalocyanines [133]. It was found that both the nickel(II) and copper(II) complexes displaced a Col$_{tet.o}$ between $-100°C$ and the onset of decomposition at $300°C$, so that extension of the core size appears to have a beneficial effect with regard to the mesophase range.

Recently, Swager has compared the mesomorphic properties of an octakis(tetradecyloxymethyl)-phthalocyanine copper(II) complex, Crys·80·Col$_{hd}$·280·I, with the analogous octakis(tetradecyloxymethyl)tetra-2,3-thiophenoporphyrazine, Crys·49·Col$_{hd}$·258·I [134]. X-ray diffraction studies show that the lattice constants are 33.5 Å for the phthalocyanine complex and 32.9 Å for the thiophenoporphyr-

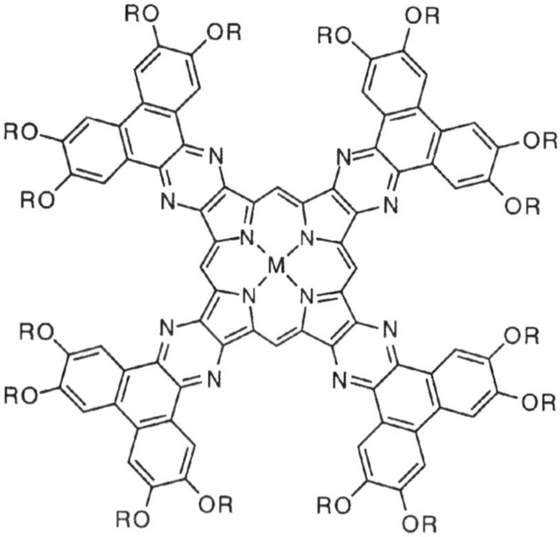

$M = Cu^{2+}$ or Ni^{2+}

Figure 70 Mesomorphic triphenylene-based porphyrazinato complex.

azine, suggesting that the complexes have a similar relative core size. ^{1}H NMR studies show that replacement of the phenyl ring in the phthalocyanine ring with a thiophene ring produces a mixture of four regioisomers and the effect of this will be to induce disorder within the columns leading to a beneficial reduction in the observed transition temperatures.

Simon and co-workers have also studied phthalocyanines with crown-ether moieties attached to the periphery, giving the materials shown in Figure 71 [135]. These compounds were reported as having a monotropic mesophase at 150°C (although melting points were not given, nor was it all clear whether 150°C represented a transition temperature) and X-ray scattering studies, showed the mesophase to be based on a two-dimensional square lattice (lattice parameter 20.8 Å), which stacked to give ion channels described by the crown-ether rings, with a ring–ring spacing of 4.2 Å. Such crown-ether substituted phthalocyanines have been discussed in a recent short review, concerning their applications with regard to ion conduction [136].

Another series of complexes derived from a dihydroxysilylphthalocyanine (Figure 65, M = *trans*-$Si(OH)_2$), showed hexagonal mesophases from −7 up to 300°C [137]. If such complexes were held at around 180°C for several hours, a polycondensation reaction occurred which eliminated water to form polymeric materials with a polysiloxane 'spine' (Figure 72). These polymers displayed a lamellar periodicity of 31 Å from room temperature up to 60°C, where clearing occurred. X-ray diffraction studies in the lamellar phase, showed that the rings were separated by 3.4 Å within a column and that the alkyl chains were molten. Related polymeric

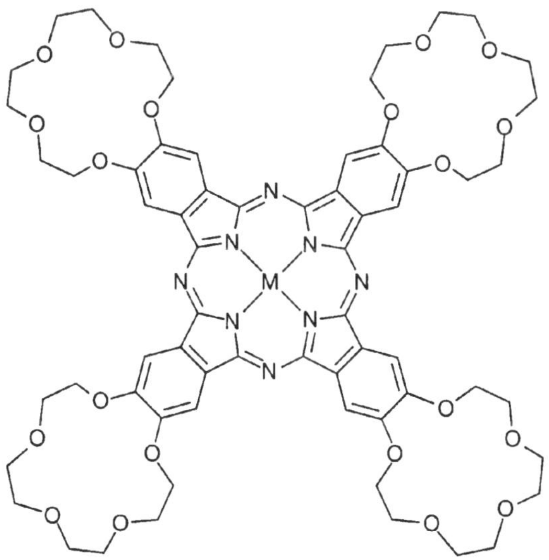

Figure 71 Crown-ether substituted phthalocyanine.

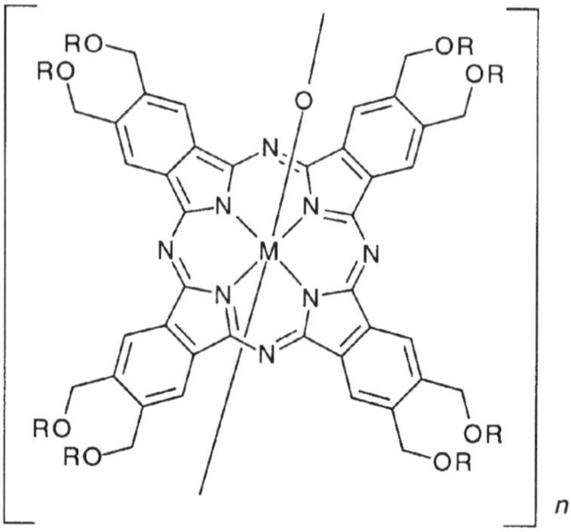

Figure 72 Spinally polymerized phthalocyanines.

materials based on tin have also been reported [138] and again attachment of crown-ether moieties to the periphery of the silylphthalocyanines produced polymeric materials with ion channels, although liquid crystal properties have not been reported [139].

The cavity size of a phthalocyanine is 1.6 Å and consequently metals with larger diameters are required to sit outside of the plane; such is the case with lead(II) ions (ionic radius 2.4 Å). Therefore, the phthalocyanine complex (Figure 65, M = Pb, $R = C_8H_{17}$) was found to exhibit a columnar phase between −45 and 158°C and X-ray diffraction studies showed that the mesophase was again columnar hexagonal, with a stacking repeat distance of 7.4 Å, i. e. twice the thickness of the molecule [140]. This led to the suggestion that the molecules were arranged antiferroelectrically in a tilted stack (Figure 73).

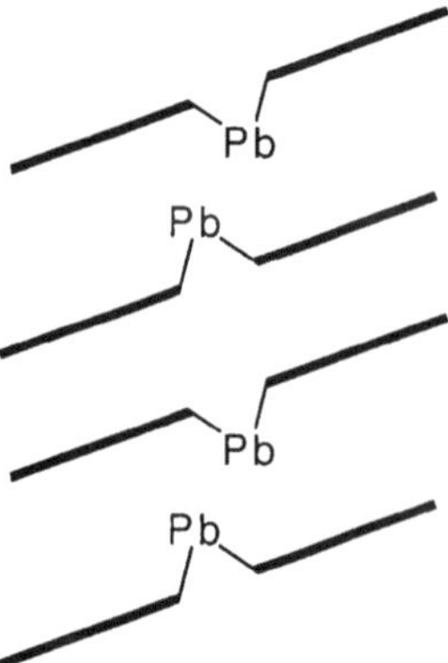

Figure 73 Representation of antiferroelectric structure proposed for a lead phthalocyanine.

Complexes of octasubstituted phthalocyanines with lutetium(III) form sandwich-type complexes with two macrocyclic rings coordinated to one metal ion. When eight alkoxy chains are on the periphery of the molecule, then the columnar hexagonal phases were observed for the dodecyloxy- and the octadecyloxy-substituted complexes with the octyloxy-substituted complex being non-mesomorphic [141]. Ohta has shown that when alkyl chains are on the periphery then the octyl-substituted complex displayed a Col_{hd} phase [142]. When an analogous octadecyloxyphenyl-substituted complex was studied it was found to exhibit $Col_{te.d}$ and Col_{hd} phases, all these studies clearly demonstrate that the nature of the mesophase observed is dependent upon the substitution at the periphery of the phthalocyanine ring [143]. Furthermore, Simon and co-workers demonstrated that such materials have an anisotrophy of conductivity of the order of 10^7 and a conductivity of 3.9×10^{-5} S cm^{-1} at 10 GHz [144].

5.1.2 Porphyrins

Significantly less research has been carried out on metallomesogens with porphyrin ligands compared to those with phthalocyanine ligands. Gregg, Fox and Bard have synthesized a series of octaester and octaalkylether substituted porphyrins (Figure 74) [145]. In the series of octaalkylether compounds, the ligands are generally non-mesomorphic, however, upon complexation mesophases are stabilized. The zinc(II) complex shows a mesophase range of 55°C when the substituent X is either a hydrogen or a cyano group; however, this is reduced to 25°C when a more bulky nitro group is introduced presumably due to buckling of the porphyrin ring. The photophysical properties of these complexes have been studied and the zinc(II) octakis(β-octyloxyethyl)porphyrin, while in the liquid crystal phase, was used to obtain an ordered sample in an electrochemical cell, so that a photovoltaic effect could be observed in the resulting aligned solid state [146].

Shimizu has studied a series of mesomorphic 5,10,15,20-tetrakis(4′-alkyphenyl)-porphyrins (Figure 75) finding that the ligands were non-mesomorphic until the

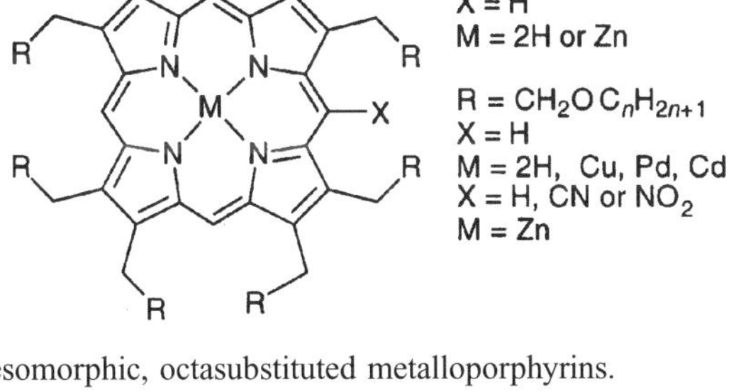

R = $CO_2C_nH_{2n+1}$,
X = H
M = 2H or Zn

R = $CH_2OC_nH_{2n+1}$
X = H
M = 2H, Cu, Pd, Cd
X = H, CN or NO_2
M = Zn

Figure 74 Mesomorphic, octasubstituted metalloporphyrins.

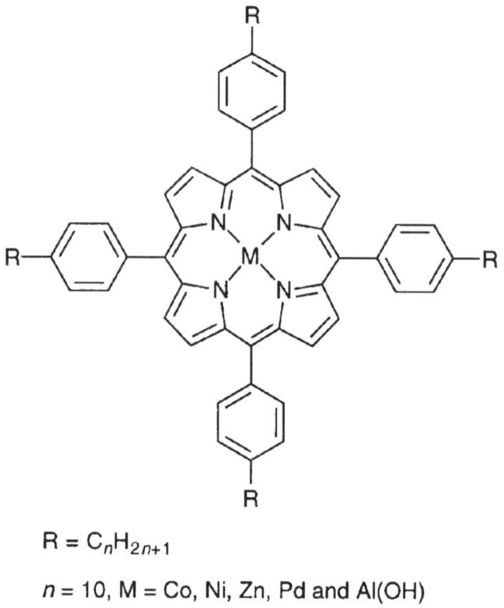

R = C_nH_{2n+1}

n = 10, M = Co, Ni, Zn, Pd and Al(OH)

Figure 75 Mesomorphic metalloporphyrins.

chain length was hexyl [147]. The liquid crystal phase was identified as D_L (the so-called discotic lamellar phase) by X-ray diffraction studies and was also observed when the dodecyl derivative was complexed with cobalt(II), nickel(II), copper(II), zinc(II) or palladium(II). All the metal complexes, except the nickel(II) complex, tend to show an enhanced mesophase range compared to the free ligands, with the largest range of about 168°C seen for the zinc(II) complex. Interestingly, the related porphyrins with *alkoxy*phenyl groups are non-mesomorphic, despite initial claims to the contrary [148].

The hydroxyaluminium(III) porphyrin complex with dodecyl chains displays a Col_{hd} phase upon heating, whereas upon cooling from the isotropic state three mesophases are observed, the higher two were identified as D_L phases and the low temperature phase was not identified [149]. The change in the mesomorphism is ascribed to the formation of a μ-oxo dimer while in the isotropic state, which is then unable to support a columnar arrangement of the molecules.

5.1.3 Tetraaza[14]annulenes

Veber and co-workers have synthesized a series of nickel(II) complexes of tetraaza-[14]annulenes (Figure 76), bearing four, six or eight chains on the periphery of the molecule and displaying a hexagonal columnar phase [150]. Not surprisingly the metal-free, four-chain, dodecyloxy ligand (X = Y = Z = H) has lower transition temperatures

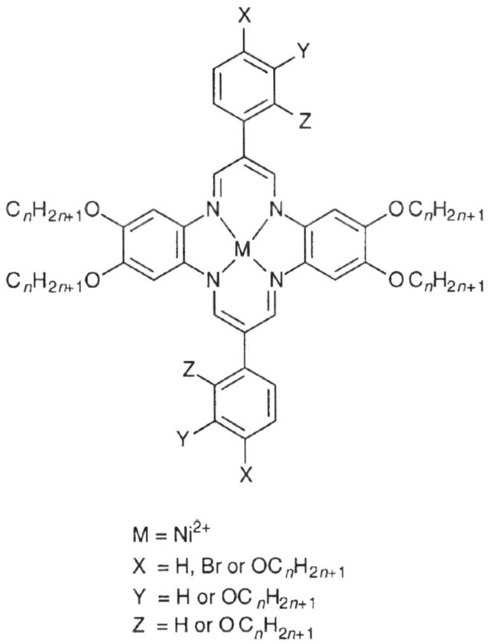
X
Y
Z
$C_nH_{2n+1}O$
$C_nH_{2n+1}O$
OC_nH_{2n+1}
OC_nH_{2n+1}
N
M
Z
Y
X
$M = Ni^{2+}$
$X = H$, Br or OC_nH_{2n+1}
$Y = H$ or OC_nH_{2n+1}
$Z = H$ or OC_nH_{2n+1}

Figure 76 Mesomorphic tetraaza[14]annulenes.

than the complex. X-ray diffraction studies show the separation between the macrocyclic rings within the column remains reasonably constant at ≈3.6 Å, which is quite short suggesting that there are strong $\pi-\pi$ interactions. Attempts at disrupting these interactions in a six-chain complex by placing chains in position Z ($X = Y = H$), destabilizes the liquid crystalline phase completely, although the clearing point is reduced to 122°C, compared to over 300°C when the chain is in the position X ($Z = Y = H$). The complexes with eight chains ($Z = H$, $X = Y = OR$) are best described by a model of the columns as well-defined cylinders of high electron density surrounded by a dense crown of paraffinic chains, such a model is less applicable to the other complexes. Recently, it was shown that exposing a uniformly aligned sample of the eight-chain complexes (with $n = 10$), as either the pure sample or as a polymer disperse liquid crystal, to an argon ion laser results in a reorientation, thereby changing the optical properties of the liquid crystal. This optical storage effect could be easily erased by heating the sample and shearing [151].

5.1.4 Saturated aza-macrocycles

Much of the work in this area concentrates on metal-free macrocycles, often with the nitrogen atoms of the macrocycle functionalized as amides [82]. Reduction of these amides to facilitate complexation to a metal ion destroys the mesomorphism,

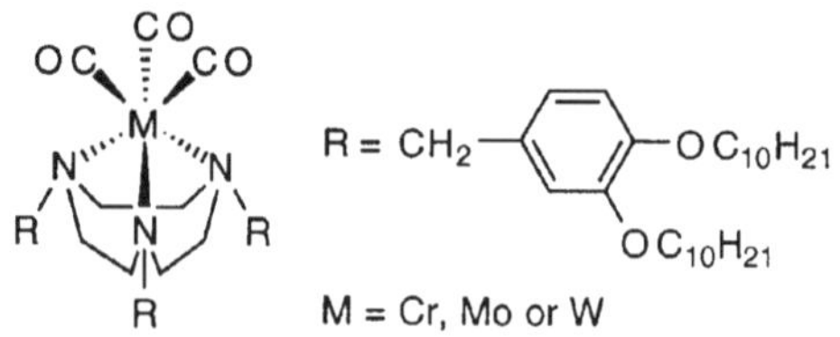

R = CH_2–C₆H₃–$OC_{10}H_{21}$, $OC_{10}H_{21}$

M = Cr, Mo or W

Figure 77 Mesomorphic triazacyclononane complexes.

although it has been shown that upon complexation, columnar mesophases are again observed, for example see Figure 77. Presumably complexation leads to a more rigid conformation of the macrocycle, producing a cone-shaped molecule that may stack to yield a columnar arrangement [152].

5.1.5 Glyoximato complexes

Ohta has studied bis(phenylglyoximato) complexes of nickel(II) and palladium(II) bearing eight alkoxy chains on the periphery of the molecule (Figure 78) [153]. The complexes display a Col_{hd} phase, with the clearing temperatures being higher for the palladium(II) complexes compared to the nickel(II) complexes. Replacement of the alkoxy chain by an alkyl chain was shown to produce a large decrease in both the melting and clearing temperatures.

The complexes exhibit thermochromism as the $3d_{z^2}$-$4p_z$ transition for the nickel(II) and the $4d_{z^2}$-$5p_z$ transition for the palladium(II) ions are sensitive to the intermolecular separation. Therefore, there is a blue shift of these bands as the temperature is increased corresponding to an increase in the internal separation. X-ray studies for a palladium(II) complex, where $n = 12$, supports this premise as the intermolecular separation is 3.3 Å at room temperature compared to 3.5 Å at 150°C.

Solvatochromism is also seen for these complexes and, interestingly, a gel state is observed in hexane solution. When viewed between crossed polarizers, this gel is quite similar to the lyotropic hexagonal phase of poly(γ-benzyl-L-glutamate) in

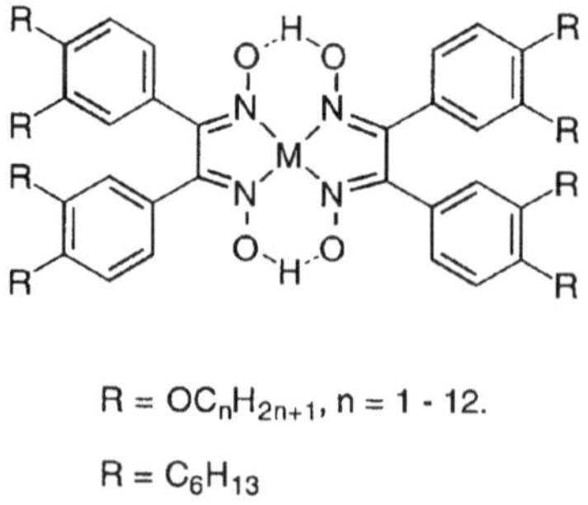

R = OC_nH_{2n+1}, n = 1 - 12.

R = C_6H_{13}

Figure 78 Mesomorphic glyoximato complexes.

dioxane. X-ray studies supported this gel state having a two-dimensional hexagonal structure, with uv studies supporting a similar structure in both a dilute hexane solution and the gel state [154].

5.1.6 Metallocrowns

Serrano has reported a highly novel metallomesogen based upon trimeric gold(I) pyrazolate complexes (Figure 79), displaying a hexagonal columnar mesophases which may be supercooled to $-30°C$ [155]. The lower transition temperatures when X = decyloxy may be explained due to the presence of two isomers as identified by ^{1}H NMR studies. The assignment of the mesophase was supported by X-ray diffraction studies, showing a short range correlation between stacked macrocycles and aliphatic chains of approximately 4.4 Å. In the solid state, X-ray studies show that close-stacked neighbouring molecules are mutually rotated so that the gold ions of one molecule lie on top of the pyrazole ring of the lower neighbour.

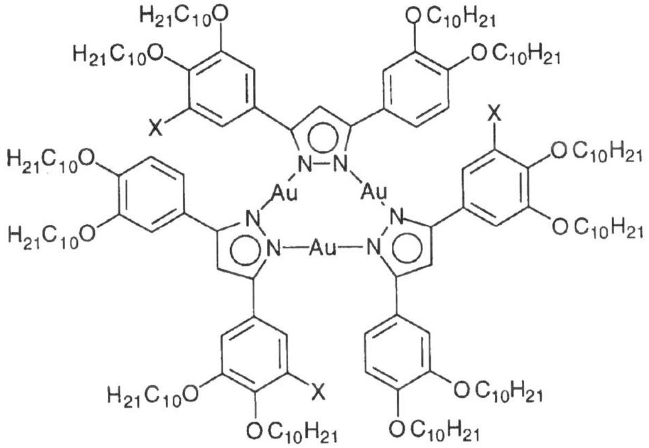

X = H Crys • 59 • Col_{hd} • 64 • I

X = $OC_{10}H_{21}$ Crys • 36 • Col_{hd} • 55 • I

Figure 79 Mesomorphic trinuclear gold(I) materials.

5.2 Carboxylate Complexes

The simplest complexes exhibiting columnar mesophases are the dinuclear metal carboxylates $[M_2(O_2CR)_4]$, derived from simple alkanoic acids which have been synthesized with a range of metals [156]. Early work concentrated upon copper(II) carboxylates, with X-ray studies confirming the phases as hexagonal columnar

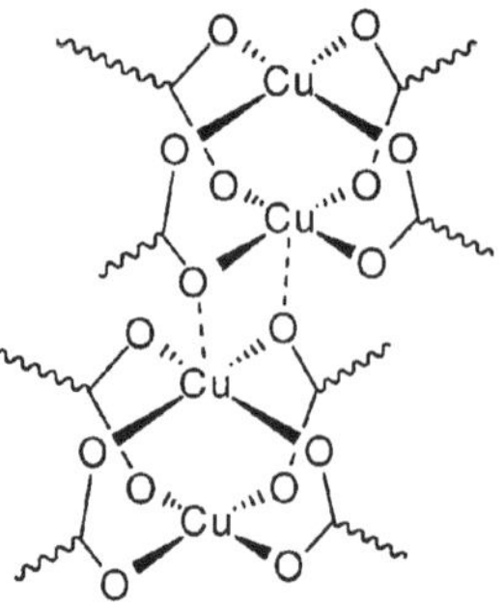

Figure 80 Schematic of the arrangement of copper carboxylates in the Col_h mesophase.

[156a]. EXAFS studies of the crystal identified that the dinuclear units were linked together via axial coordination between a copper(II) ion and a oxygen atom of a neighbouring dimer (Figure 80). Furthermore, these chains are still present within the mesophase, and form the rigid backbone of the columns, with a superposition of binuclear complexes in a four-fold helical fashion. Magnetic studies showed a reduction in the magnetic moment of the copper(II) complexes at the transition to the mesophase, which was attributed to a modification of the relative dispositions of the dimers within the columns [157]. Comparison of the thermal characteristics between a series of such complexes ($[M_2(O_2CR)_4]$, where M = Cr, Mo and W), supports the premise that it is the strength of the intermolecular metal–oxygen bond, that determines the transition temperatures observed [156d].

If branched alkanoic acids are used, then a reduction in the melting point is observed and upon cooling, such copper(II) complexes have been reported to form a glassy state [158]. Interestingly such branched complexes have been shown to exhibit mesophases when pyrazine or 4,4′-bipyridyl are also coordinated to the copper(II) ions to give a polymeric structure (Figure 81). Analogous linear alkanoic acid copper(II) complexes of linear alkanoic acids simply decompose upon heating

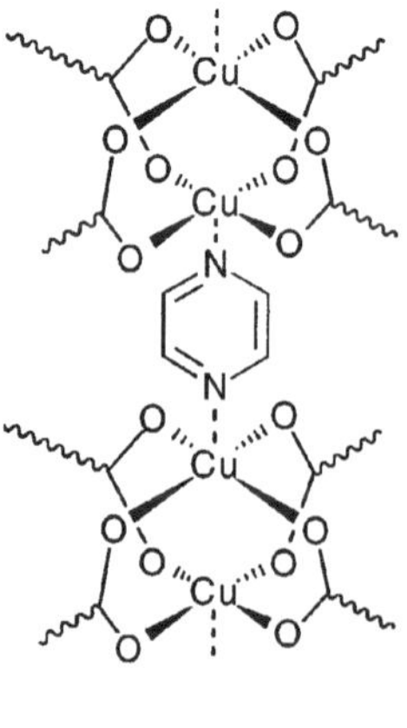

Figure 81 Schematic of the arrangement of copper carboxylates bridged by heterocyclic ligands.

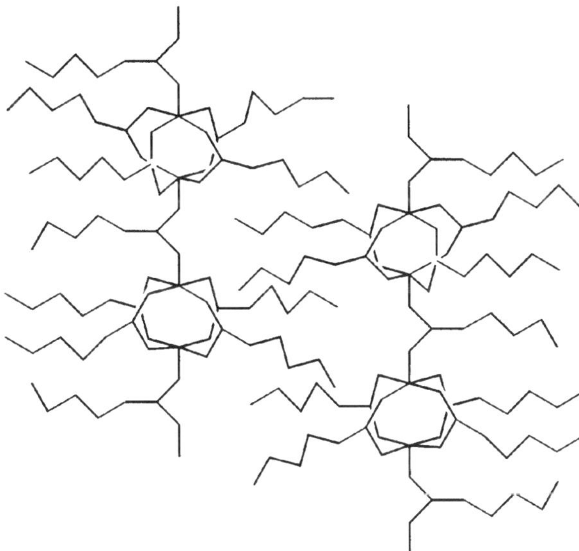

Figure 82 Schematic representation of the two polymeric chains shifted by half a dimer within the mesophase of $[Ru_2(O_2CR)_5]$.

[159], suggesting that the branching allows the alkyl chains to fill the space within the column caused by the coordinated ligand.

Mixed-valence diruthenium(II,III) carboxylates of the formula $[Ru_2(O_2CR)_4X]$ have been studied, and it is found that when the anion X = chloride, the complexes are non-mesomorphic, whereas when X = alkanoic acid or a dodecylsulphate, a mesophase of the hexagonal columnar type is observed; single-crystal X-ray studies show that in the solid state, the anion X bridges the dimeric units [160]. X-ray diffraction studies of the pentakis(carboxylato) complexes suggest that the columns within the mesophase are composed of four dimer chains, with the chains being axially shifted with respect to each other and interdigitated (Figure 82). The chloride-bridged carboxylates are mesomorphic if 3,4-dialkoxybenzoic acids are used rather than simple alkanoic acids, with a lamellar phase being observed below the columnar hexagonal mesophase; again interdigitation is proposed from X-ray diffraction studies and helps to explain the alkyl chains filling the space created by the presence of the bridging chlorides.

Serrano has also reported related mesomorphic rhodium(II) alkanoate complexes, where the 4-alkoxybenzoato complexes display a rectangular columnar phase and a 3,4,5-trialkoxybenzoato complex displays a hexagonal columnar phase. An intermediate situation is seen for a 3,4-dialkoxybenzoato complex, where both phases are observed [161]. Coordination of pyrazine ligands within the 4-alkoxybenzoato complexes completely destabilizes the mesophase.

5.3 β-Diketonate Complexes

The previous discussion of β-diketonate complexes concerned complexes with two or four chains on the periphery of the molecule and how the molecular shape was intermediate between rod- and disc-like. The following discussion will deal with complexes with a greater number of chains on the periphery of the molecule which, in general, behave more like conventional discotics.

Copper(II) complexes with eight alkoxy-chains were synthesized by Giroud-Godquin, and showed a hexagonal columnar phase [162]. In comparison to the four chain copper(II) complexes the melting points are higher and the clearing points are lower leading to a narrower mesophase range.

Serrano reported some thallium(I) complexes with eight or twelve chains on the periphery of the molecule (Figure 83), which display a monotropic Col_{hd} mesophase [163]. A single-crystal X-ray structure of a four-chain derivative shows that the thallium(I) ions are close enough to be regarded as being bonded to form an overall disc-like molecule (although the two 'halves' are not coplanar), with the packing diagram showing that in the solid state the molecules are in a hexagonal columnar stack.

Swager has undertaken a systematic study of a range of metal bis(β-diketonato) complexes, varying the metal ion, the number of chains and the size of the core (Figure 84). Considering first simple bis(β-diketonate) complexes with 10 chains (X = H), the oxovanadium(IV) complexes show a Col_{rd} phase, while the copper(II) and palladium(II) complexes display a Col_{rd} phase, which changes to a Col_{hd} as the chain length is increased [164]. In the series with 12 chains ($X = OC_nH_{2n+1}$), the oxovanadium(IV) complexes are only mesomorphic at longer chain lengths, show-

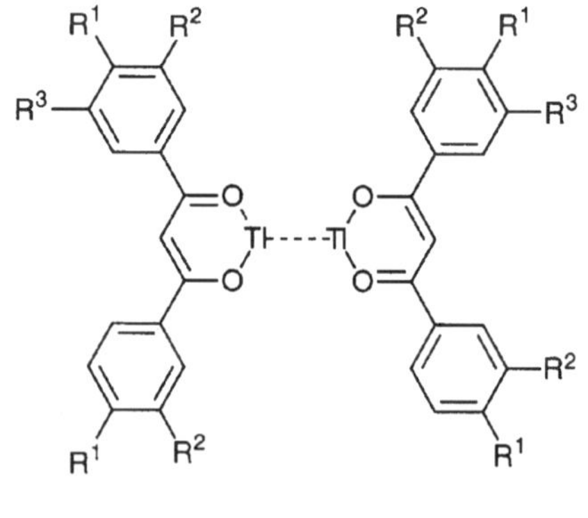

$R^1 = C_nH_{2n+1}$
$R^2 = C_nH_{2n+1}$ or H
$R^3 = C_nH_{2n+1}$ or H

Figure 83 Mesomorphic complexes of Tl(I).

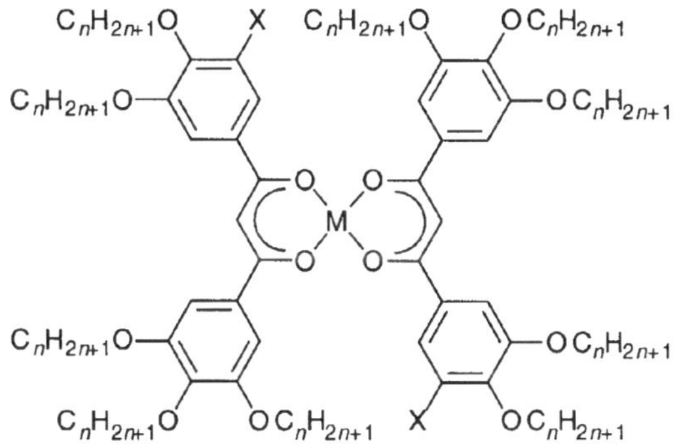

Figure 84 Discotic metal complexes of β-diketonates.

ing a Col_{rd} for the dodecyloxy and tetradecyloxy derivatives, with again the copper(II) and palladium(II) complexes showing a change from Col_{rd} to Col_{hd} as the chain length is increased. These findings are explained in terms of enhanced core–core interactions, accompanying the increase in side-chain density. A single-crystal X-ray structure of an eight-chain oxovanadium(IV) complex, shows that the molecules adopt a tilted columnar structure with antiferroelectric ordering between the oxovanadium(IV) cores.

Swager has also investigated some rhodium(I) and iridium(I) dicarbonyl complexes (Figure 85) of β-diketonate ligands [165]. These display a Col_{hd} phase, which is stable below room temperature at intermediate chain length ($n = 6–10$). As the molecules themselves are not disc-like they are thought to assemble into so-called 'columnar antiphases', however, infrared studies monitoring the carbonyl

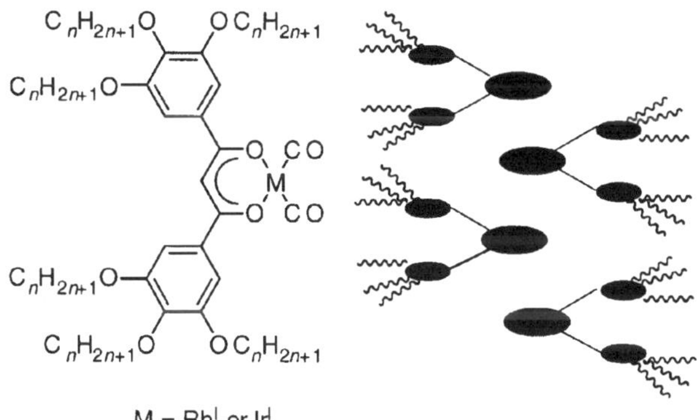

Figure 85 Mesomorphic complexes of Rh(I) and Ir(I) and their proposed arrangement in the mesophase.

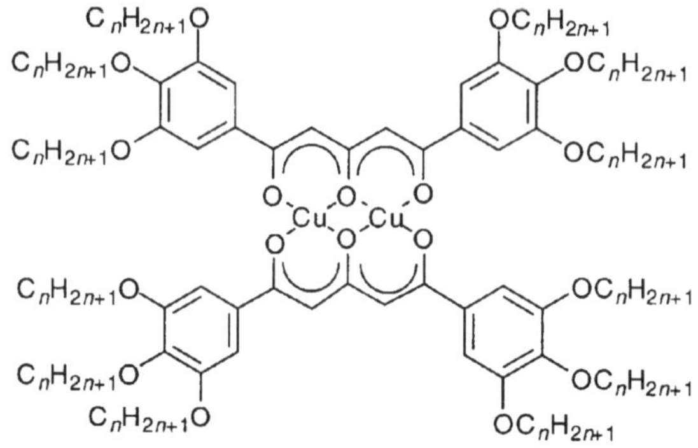

Figure 86 Dinuclear triketonato complexes.

stretching frequency failed to identify any significant intermolecular metal–metal interaction within the mesophase.

Swager has also synthesized a range of 1,3,5-triketonate and 1,3,5,7-tetraketonate ligands, which yield homodinuclear copper(II) complexes (Figure 86). Interestingly with the extended core size compared to the β-diketonate complexes, the complexes now display an enantiotropic Col_{hd} phase with only six chains on the periphery of the molecule [166]. Schiff base derivatives of the 1,3,5-triketonate ligands allow the synthesis of heteronuclear bimetallic liquid crystals of nickel(II) with either copper(II) or palladium(II) ions. A single-crystal X-ray structure of a dicopper(II) Schiff base complex (Figure 87) confirmed that in the solid state, the molecules are correlated via axial interactions between the copper(II) ions and ketonate oxygens on an adjacent complex, so that two neighbouring molecules describe an overall disc-like shape. This is also thought to be the time-averaged situation in the mesophase and that the molecules describe a similar shape is supported by X-ray diffraction

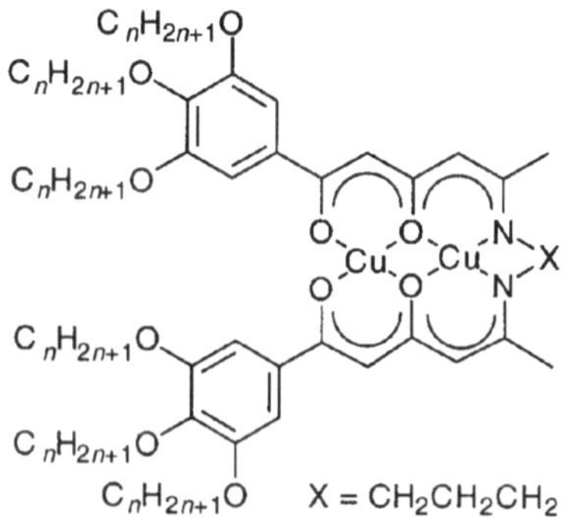

Figure 87 Dicopper(II) Schiff base complex.

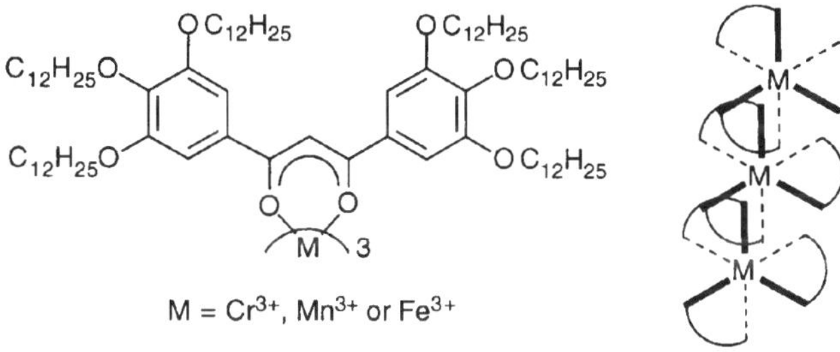

Figure 88 Mesomorphic tris(β-diketonato)metal(III) complexes.

studies, which show that for a given length of side chain the molecules have similar lattice spacings.

It was found that octahedral iron(III), manganese(III) or chromium(III) ions produced mesomorphic complexes when complexed with three β-diketonate ligands, presumably with the molecules associating as in Figure 88 [167].

5.4 Salicylaldimato Complexes

Oxovanadium(IV) complexes of salicylaldimate ligands have been studied and once more, the isolated molecules are not disc-shaped so it is proposed that they are correlated in a time-averaged fashion within the mesophase to produce overall disc-shape (Figure 89), with two possible types of arrangement possible depending upon the substitution at the phenyl ring [168]. The infrared stretch shows that the complexes with the 2,2-dimethylpropyl bridging group are monomeric, while with the propyl bridging group they are weakly associated. However, the strongest interaction is seen with the ethyl bridging group. In support of this it is found that the complex with the substituent in position A and with a 2,2-dimethylpropyl bridging unit is non-mesomorphic as the correlated structure is disfavoured by the steric interaction of the geminal methyl groups with the aromatic ring of the adjacent molecule. However, with substitution at the other position (B = Z), the molecules are able to associate and so Col_{hd} mesophases are observed.

5.5 Enaminoketonate Complexes

Homonuclear dicopper(II) and dinickel(II) complexes (Figure 90) of enaminoketonate ligands have been reported to display an enantiotropic Col_r phase, which could be supercooled to room temperature [169]. Again the molecules are proposed to associate into dimers to produce an overall disc-like shape although if the R-group is t-butyl then the mesomorphism is lost due to steric hindrance preventing the molecular association. EPR studies with the copper(II) complexes showed an

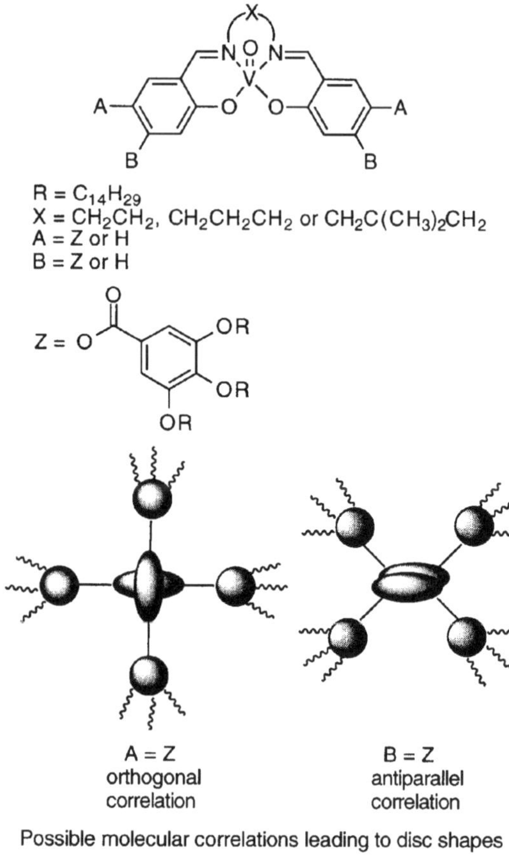

R = $C_{14}H_{29}$
X = CH_2CH_2, $CH_2CH_2CH_2$ or $CH_2C(CH_3)_2CH_2$
A = Z or H
B = Z or H

A = Z
orthogonal
correlation

B = Z
antiparallel
correlation

Possible molecular correlations leading to disc shapes

Figure 89 Mesomorphic salen complexes.

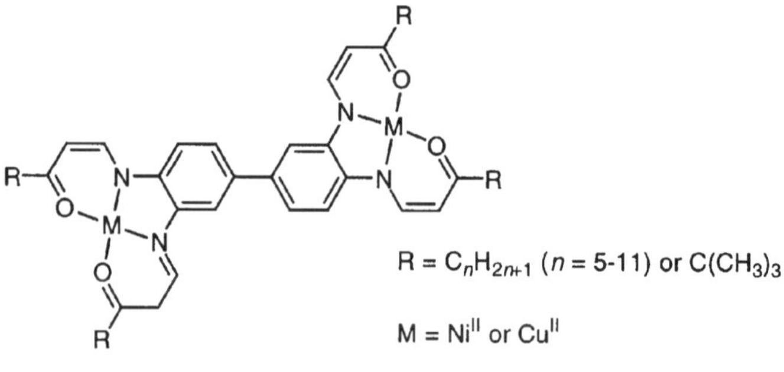

R = C_nH_{2n+1} (n = 5-11) or $C(CH_3)_3$

M = Ni^{II} or Cu^{II}

Figure 90 Homodinuclear enaminoketonates.

exchange-narrowed line for the complexes with the alkyl R groups, and four hyperfine signals for the t-butyl complex, confirming association in the first case and monomeric molecules in the latter. X-ray diffraction studies of the nickel(II) complexes show an intracolumnar distance of about 4.8 Å and a intermolecular separation of about 6 Å in the copper(II) complexes which is still sufficient for the strong spin-coupling interaction observed in the EPR studies.

5.6 Orthometalled Complexes

Earlier, H-shaped orthometallated palladium(II) complexes were discussed which displayed nematic and smectic phases. However, Praefcke has reported related complexes where the number of chains on the periphery of the molecule is increased leading to columnar mesophases. The complex in Figure 91 displays a monotropic nematic discotic phase (N_D) and the structure has been confirmed as that in the diagram by single-crystal X-ray crystallography [170].

Related complexes containing four metal ions and with 12 peripheral chains have also been prepared (Figure 92) and show a $Col_{ob,d}$ mesophase, with unit cell parameters of *ca.* $a = 37$ Å, $b = 23$ Å and $\gamma = 70-75°$, irrespective of the metal ion or the bridging halogen [171]. The similarity between the molecules of this series is supported by two single-crystal X-ray structures of complexes with different bridging halogens [172]. Recently similar complexes with only eight chains in the periphery of the molecule have been reported, which again show columnar phases [173].

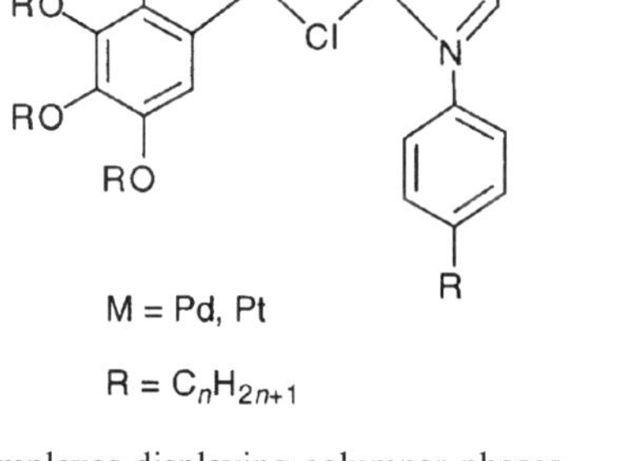

M = Pd, Pt

R = C_nH_{2n+1}

Figure 91 Binuclear complexes displaying columnar phases.

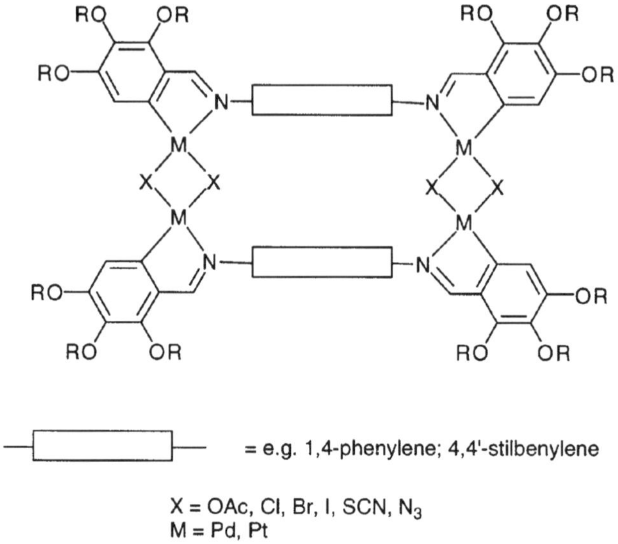

= e.g. 1,4-phenylene; 4,4'-stilbenylene

X = OAc, Cl, Br, I, SCN, N_3
M = Pd, Pt

Figure 92 Discotic tetrametal complexes.

5.7 Polycatenar Complexes

In the section on calamitic materials, we discussed how, by appropriate consideration of the molecular shape of bis(β-diketonato)copper(II) complexes, either nematic and lamellar, or columnar mesophases could be realized. Furthermore, increasing the number of chains on the periphery of calamitic mesogens can lead to molecules which demonstrate both types of mesomorphism, e.g. the phasmids of Malthête [174] and the biforked mesogens of Nguyen [175], which are collectively named as *polycatenar mesogens* [13]. Polycatenar mesogens are normally composed of extended, rigid cores which are terminated, usually but not exclusively symmetrically, by two or three chains bound to the terminal phenyl ring (Figure 93). Thus, the main classes of polycatenar mesogen are the *tetracatenar* and *hexacatenar* cases bearing four and six chains, respectively.

The structure shown in the figure indicates that many substitutional possibilities exist, especially if molecules are not symmetric. However, three main classes tend to predominate, namely tetracatenar materials with terminal 3,4- or 3,5-disubstitution, or hexacatenar materials with terminal 3,4,5-trisubstitution. The mesomorphism of the latter class tends to be dominated by the formation of columnar mesophases, while the former exhibit a progression from nematic and S_C phases to columnar phases with increasing chain length, often through the intermediary of a cubic phase. A 'text-book' phase diagram is shown in Figure 94 for a series of tetracatenar bipyridines [176].

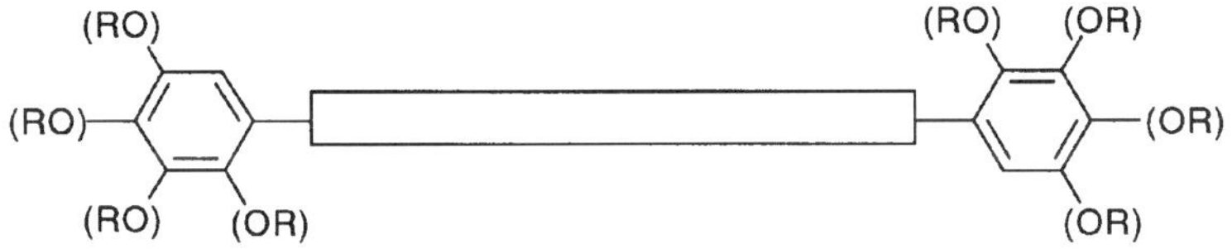

Figure 93 Schematic diagram of a polycatenar mesogen.

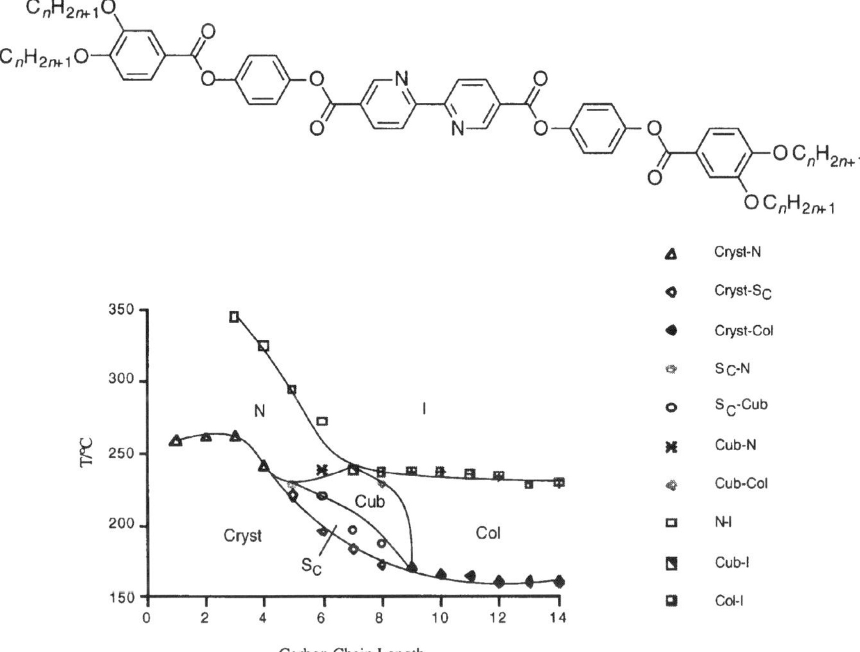

Figure 94 Phase diagram for an homologous series of tetracatenar bipyridines.

This formation of nematic, lamellar and columnar phases within a single class of molecule leads to the idea that they can be considered as a ‘missing link’ between calamitic and discotic mesogens.

5.7.1 Complexes based on polycatenar stilbazoles

We have pursued the study of polycatenar metallomesogens using poly(alkoxy) stilbazoles [177], and have recently undertaken a systematic study of the complexes of series of these ligands with palladium(II), platinum(II) and silver(I) ions.

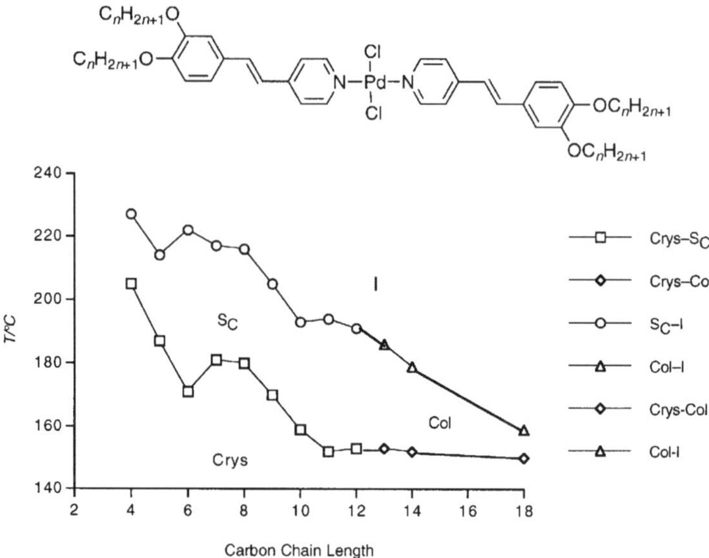

Figure 95 Phase diagram for tetracatenar complexes of Pd(II).

With palladium, we synthesized [178] a full homologous series of complexes based on 3,4-dialkoxystilbazoles and obtained the phase diagram shown in Figure 95.

Thus, it can be seen that the mesomorphism progresses from S_C to columnar with increasing chain length, although there are no complexes which show both mesophases. The mesomorphism of the platinum complexes is almost identical except that columnar phases do not appear until slightly longer chain lengths, which is accounted for by the idea that in polycatenar mesogens, columnar phases are seen when the alkoxyl chains contribute to approximately 60% to the molecular weight of the molecule. With platinum being heavier than palladium, longer chains are, therefore, necessary. For the related hexacatenar bis(trialkoxystilbazole)dichloropalladium(II) complexes, columnar phases were seen for chain lengths above nine, with no other mesophase at shorter chain length.

Bis(3,4-dialkoxystilbazole)silver(I) dodecylsulphate complexes have also been studied, and some of the typical traits of polycatenar mesomorphism are again seen [179]. Thus, the phase diagram in Figure 96 shows both a cubic and hexagonal phase, although note that there is no S_C phase, and that the nematic phase arises from complex behaviour involving the dissociation of one of the stilbazoles. The absence of the S_C phase is explained by the presence of the lateral dodecylsulphate chain which, as was discussed above, can act to suppress lamellar phases, particularly in conjunction with alkoxy chains in the 3-position of the ligand which will tend to broaden the molecule. As part of this study, the cubic phase was identified as having $Ia\bar{3}d$ symmetry using freeze-fracture electron microscopy [180], while a combination

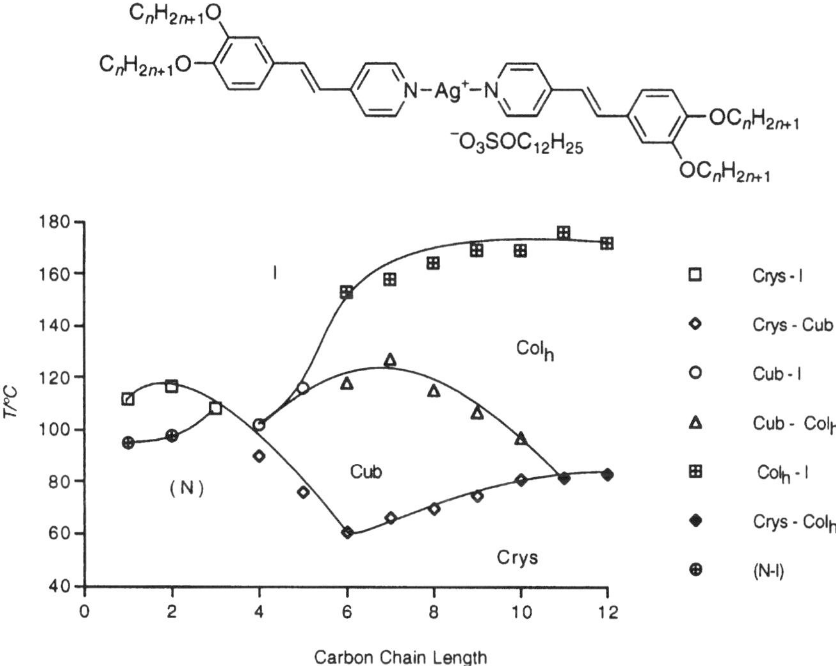

Figure 96 Phase diagram for polycatenar complexes of silver(I).

of dilatometry and X-ray diffraction was used to suggest a model for the cubic-to-columnar phase transition.

5.7.2 Dioxomolybdenum complexes

Swager has described a series of dioxomolybdenum complexes based upon non-mesomorphic substituted pyridinediyl-2,6-dimethanolato ligands [181]. For decyloxy chains and above, the complexes display disordered hexagonal columnar phases as identified by X-ray diffraction studies. Variable-temperature infrared studies looking at the oxomolybdenum symmetric and asymmetric frequencies show that there are weak interactions within the mesophase compared to the isotropic state and that the strongest interactions are seen in the solid state. These interactions are proposed to help stabilize the columnar mesophase observed (Figure 97).

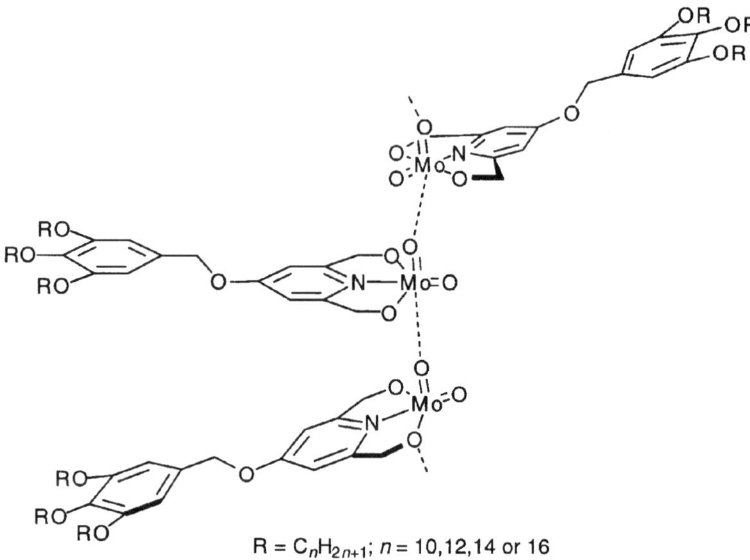

Figure 97 Schematic of the proposed arrangement of dioxomolybdenum complexes in the mesophase.

5.8 Effects of Host–Guest Interactions Upon Mesomorphism

5.8.1 Orthometallated complexes

Praefcke's complexes shown above in Figures 91 and 92 show an N_D phase as single materials. However, charge-transfer complexes may be formed with these complexes as electron donors and 2,4,7-trinitrofluorenone (TNF) as electron acceptor in less than stoichiometric quantities; these complexes are found to display an enantiotropic Col_{ho} phase [171]. The tetranuclear complexes similarly display an enantiotropic Col_{ho} phase upon forming charge transfer complexes with TNF [121].

5.8.2 Calix[4]arenes

Swager has reported some azo-substituted calix[4]arenes, where the free ligands show a mesophase upon the first heating which is then lost upon cooling from the isotropic state [182]. Complexation with an oxotungsten group (Figure 98) locks the conformation of the calix[4]arene in a rigid bowl conformation and columnar mesophases which are stable over a wide temperature range are obtained. Crystallization from dimethylformamide yields a 1 : 1 complex, which surprisingly melts directly into the isotropic state; a similar transition temperature was also noted when a pyridine molecule was complexed. Prolonged heating causes dissociation of the

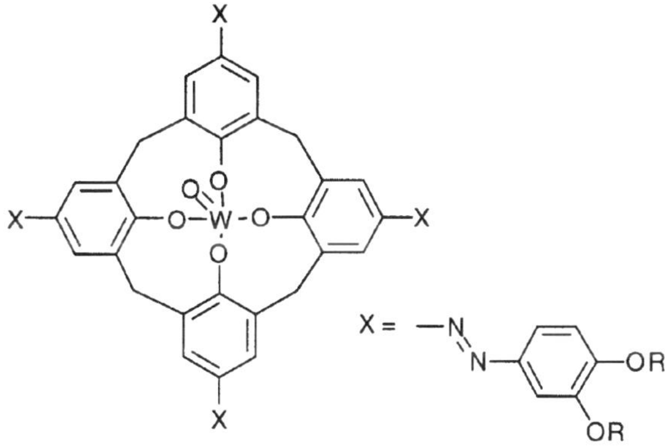

Figure 98 Mesomorphic oxotungsten complexes.

guest molecule and the columnar phase is reformed. The formation of a mesophase when the cavity is empty suggest that within the mesophase the tungsten-oxo group of one molecule protrudes into the cavity of its neighbour.

6 LYOTROPIC METALLOMESOGENS

The third type of organization we wish to consider is one which originates in a manner somewhat different from the first two, in that whereas we have so far considered changes which occur as a result of the action of heat alone, we will now consider systems where the primary means of organization is the action of a solvent.

Consider the situation like this. Suppose we take cetyltrimethylammonium bromide (CTAB) and make a very dilute solution of it in water. CTAB is an amphiphile, that is to say it has one part (its hydrocarbon chain) which is hydro*phobic*, and another part (the ammonium headgroup) which is hydro*philic*. In very dilute solution, CTAB exists as monomers although clearly the solvation is driven by the headgroup. However, at a certain concentration (3.3×10^{-4} mol dm^{-3}) [183], the system forms micelles. This is due to the fact that if the molecules organize themselves so that they form a structure in which the polar headgroups are on the surface and the fatty chains occupy the interior, the unfavourable water/alkane interactions are all but removed (to be replaced by favourable alkane/alkane interactions), while the favourable water/headgroup interactions are retained (Figure 99). The concentration at which this happens is known as the *critical micelle concentration* (cmc). Now suppose that we increase the concentration of CTAB in the solution; what will happen? Well, the concentration of monomeric surfactant does not change, but the concentration of micelles increases. Eventually, a situation can be reached at which the micellar concentration

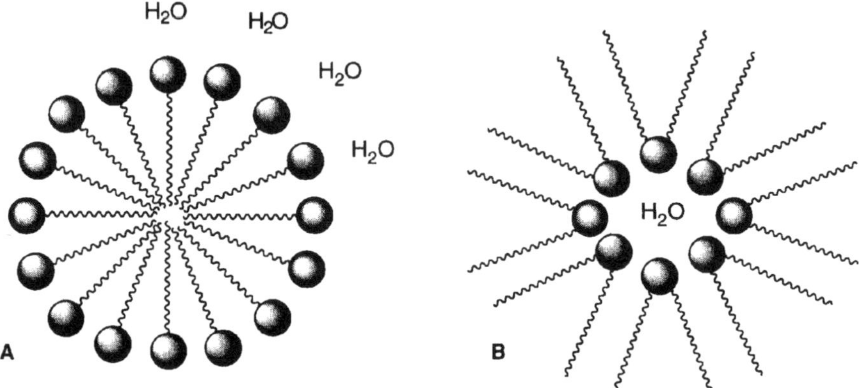

Figure 99 Schematic slice through a normal (A) and a reversed (B) micelle.

is so high that the micelles have no choice but to order themselves. This ordered array of micelles is a *lyotropic liquid crystal phase*.

While thermotropic liquid crystals are known and recognized by all through liquid crystal displays, it is little realized that the largest production of liquid crystals is that of the detergent companies who make tonnes of (surfactant) liquid crystals every year. They pervade our washing and cleaning, our cooking, our bodies, (our cell membranes are liquid crystals, but that's another story), and they are implicated in the fabrication of some of our strongest materials (the fibres of Kevlar are spun from a lyotropic nematic phase in oleum).

Lyotropic liquid crystal phases may conveniently be divided into three types, and there are examples of metallomesogens in each type. We will introduce each type separately and look at how metal-based systems fit in.

6.1 Polymeric Systems

6.1.1 Phase structure

The mesophases formed by polymeric systems are largely nematic or hexagonal phases. Here, we are considering long, normally rigid rods which are dissolved in a solvent and, being polymeric, occur in a variety of lengths. These can then be arranged in a fashion analogous to a thermotropic nematic phase except that the individual rods are solvent separated. If the concentration of these rods is increased, then they can pack to give a more regular structure, namely the hexagonal phase. Conventional examples of polymeric lyotropic systems are the mesophase of Kevlar in oleum (Figure 100) and the chiral nematic and hexagonal phases of DNA in aqueous solution as studied by, for example, Livolant [184].

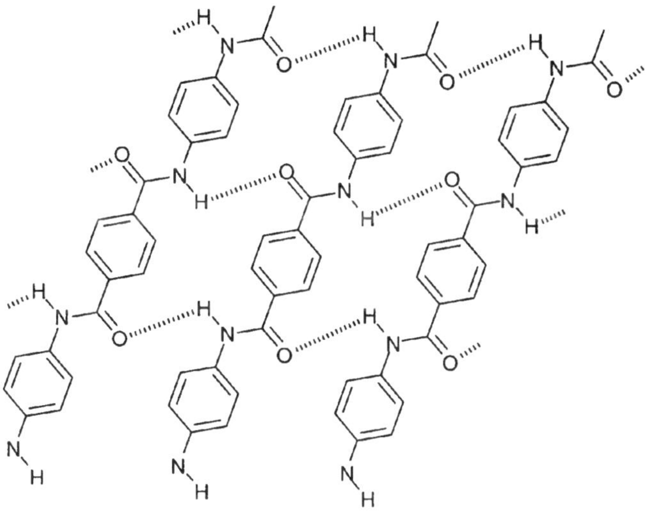

Figure 100 The molecular organization within the aligned nematic phase of Kevlar.

6.1.2 Lyotropic metallomesogenic polymers

In terms of metal-based examples, there are two which are worthy of discussion. The first are some of the earliest metallomesogens studied since the 'revival' of interest in the late 1970s. Thus, Takahashi [185] reported in 1978 that reaction $[PtCl_2(PBu_3)_2]$ with a difunctional acetylene in the presence of copper(I) iodide and base gave polymers (Figure 101) with molecular weight average ($\overline{M}_W$) degrees of polymerization of up to $7 \times 10^4\,g\,mol^{-1}$ (corresponding to a number average ($\overline{M}_W$) of 108). Further, he reported that at about 36% in solution in trichloroethene, the polymers showed nematic liquid crystal phases.

He then extended this study by looking at both mixed-metal copolymers and alternating copolymers. Thus, using ^{31}P NMR measurements he found that various heterometallic and homometallic polymers of 1,4-butadiyne (Figure 102A; M = Pt, M′ = Pd or Ni; M = Pd, M = Ni; M′ = Pd) all had a negative diamagnetic anisotropy, while related homometallic alternating copolymers (Figure 102B) had a positive diamagnetic anisotropy. These rather marked differences were ascribed to a subtle balance between the opposite magnetic anisotropies of benzene and carbon–carbon triple bonds in the complexes [186, 187].

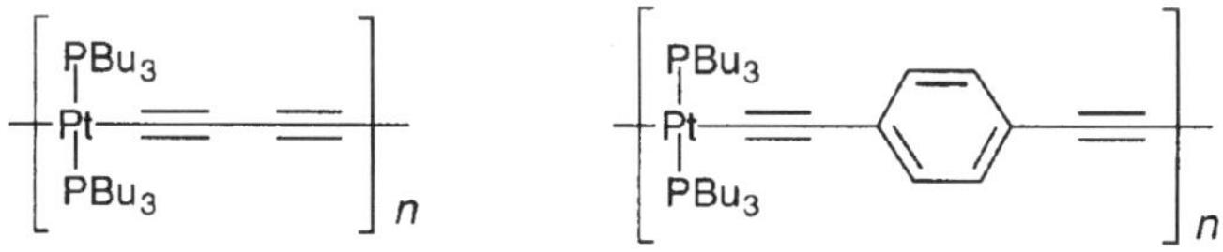

Figure 101 Examples of polymeric Pt acetylide complexes described by Takahashi.

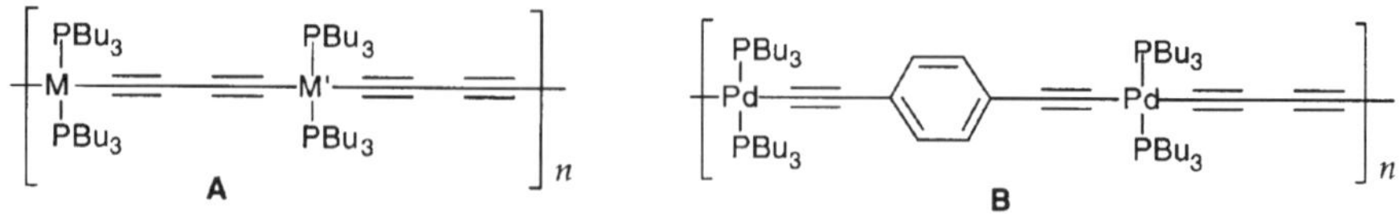

Figure 102 Structure of the mixed-metal polymers (A) and of alternating copolymers (B).

A second noteworthy series is that of the dirhodium tetraalkanaotes (Figure 80 above) which have been extensively studied as thermotropic mesogens (see 5. 2. above). However, in solution in hexadecane, nematic phases are formed as evidenced by the schlieren textures observed [188].

6.2 Amphiphilic Systems

6.2.1 Phase type and structure

The basic mode of mesophase formation is as described above for CTAB. However, as one might expect, things are not quite so straightforward and there are various types of mesophase which can be formed from various types of surfactant. The surfactant structure can be varied so that fluorocarbon chains can be employed in place of the hydrocarbon variety, while anionic (e. g. $-SO_3^-$) and the neutral (e.g. $-(OCH_2CH_2)_n-OH$) polar headgroups are often used. These surfactants can then form a variety of different mesophases as a function of (mainly) concentration in the solvent of choice (normally water). These phases are the lamellar (L_α) phase, a simple bilayer phase, and variations on the cubic (I_1, I_2, V_1, V_2) and hexagonal (H_1, H_2) phases. For these last phases, the subscript '1' implied a 'normal' phase as found in a water-rich system, while the subscript '2' implied a 'reversed' phase as found in an oil-rich system. For the cubic phases, the letter 'I' implied a micellar phase (e. g. I_1 implied a cubic phase of 'normal' micelles), while the letter 'V' implied a bicontinuous phase. Schematic diagrams of some of these phases are shown in Figure 103.

The change between one structure and another can be effected by concentration alone, and an idealized (and fictitious) binary phase diagram formed by a surfactant and water is shown in Figure 104.

In this diagram, L_1 is an isotropic solution of micelles, while L_2 is a solution of reversed micelles in oil. The phases labelled a–d are all cubic phases as follows: a = I_2, b = V_2, c = V_1 and d = I_1 (note that there are no examples known of I_2 phases as yet). What the diagram shows is that as the surfactant concentration increases, the surface curvature of the micelles in 'normal' oil-in-water phases gradually reduces, going through zero surface curvature (L_α phase) and then reverses going into reversed micellar systems. Further, it shows that in principle, one phase is separated

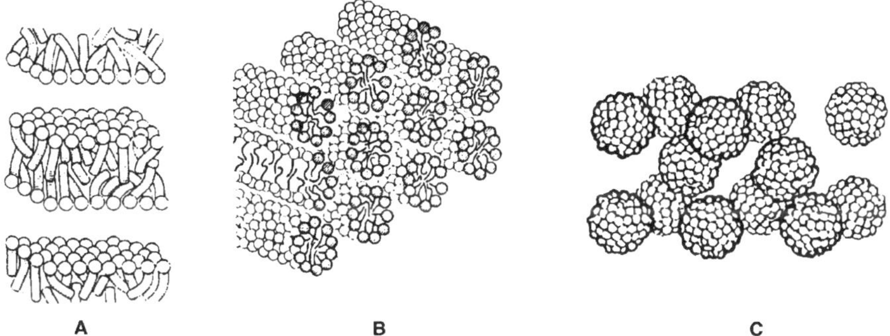

Figure 103 Schematic diagrams of some common lyotropic liquid crystal mesophases: A = lamellar, L_α = hexagonal, H_1; = cubic, I_1.

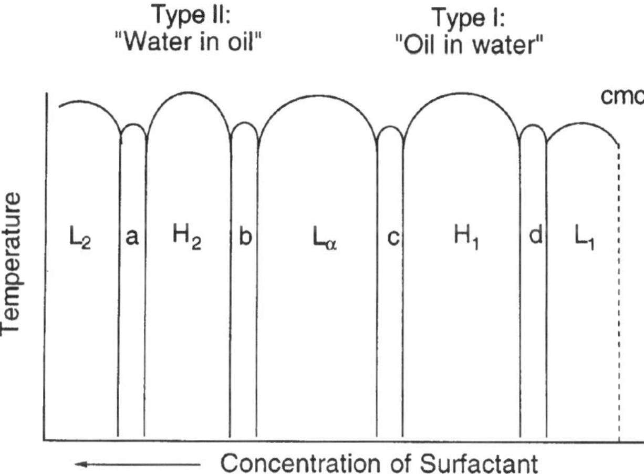

Figure 104 Idealized phase diagram for an imaginary surfactant in water.

from the next by a cubic phase, although in reality, these cubic phases are not always observed.

Techniques for phase identification are similar to those employed for thermotropic liquid crystals, although there are differences. Thus, optical microscopy is the most common technique used, but rather than prepare multiple samples at a range of concentrations it is more common to perform Lawrence Penetration Experiments [189]. In this experiment, some solid surfactant (maybe up to 50 mg) is placed on a microscope slide and a cover slip placed on the top. Water (or which ever solvent is to be used) is placed at the end of the cover slip and proceeds from one end to the other by capillary action, thus, setting up a concentration gradient across the sample. Now, at any given temperature, it is possible to have a snapshot of the whole phase diagram, and clear phase boundaries can often be seen. This experiment can be very useful for quickly establishing the Krafft point (the temperature at which the micelles become soluble).

Having obtained a snapshot of the phase diagram, it is then usual to make up a number of well-defined samples at given concentrations in either water or D_2O. Once these samples have equilibrated (which can take days), they can be investigated by X-ray diffraction methods and, if made up in D_2O, ^2H NMR. The former technique, in conjunction with the microscopy, can identify the phase and give structural parameters, while the, latter technique can given some idea of the phase type and the order within it from the dipolar coupling constant obtained.

Another important idea with amphiphiles relates the physical structure of the surfactant to a prediction of the sort of micelle which it ought to form in water. This idea, developed by Israelachvili [190], looks at the size of the head group in relation to the size of the hydrocarbon chain and considers the surface curvature required to accommodate packing into a three-dimensional structure. As such, the arguments are largely geometric in origin and they can occasionally be found wanting due to unforeseen chemical effects, but nevertheless, they are useful.

The model evaluates a dimensionless quantity, N, given by:

$$N = \frac{AL}{V}$$

where A is the cross-sectional area of the headgroup, L is the length of the hydrophobic chain in the fully *trans* conformation (90% of the chain length is sometimes used), and V is the volume of the hydrocarbon chain.

In their broadest interpretation, values of $N<1$ imply normal micelles, $N=0$ implies a lamellar phase and $N>1$ suggests a reversed phase. However, it is possible to more precise and, for example $0<N<0.25$ suggests a spherical micelle, $0.25<N<0.5$ suggests a rod micelle and $0.5<N$ suggests a disc micelle.

6.2.2 *Amphiphilic metallomesogens forming lyotropic mesophases*

While there are several examples of metal complexes which are amphiphilic in nature, it is in very few cases that lyotropic liquid crystals mesophases have been characterized. Although numerous and strictly classifiable as metallomesogens, in this article we exclude discussion of the amphiphiles with a simple metal ion as the cation (e. g. sodium salts of carboxylic acids), rather concentrating on amphiphiles in which the metal cation is an integral part of the amphiphile.

Probably the first studies of this type were those of Le Moigne and Simon who synthesized what they termed *annelides*, of which they reported an example showing a room-temperature lamellar phase based around Co(III) [191]. Interestingly, these complexes (Figure 105), can be thought of as amphiphilic in two ways, namely because they possess a hydrophobic octadecyl chain and two hydrophilic fragments, namely the cationic cobalt centre *and* the ethyleneoxide chains.

Other amphiphilic cobalt(III) complexes were reported by Yashiro *et al.* [192] (Figure 106A) who measured the cmc to be at very low concentrations (3.3×10^{-5} mol dm^{-3}) in water but did not report any mesophase formation, and by ourselves demonstrating lyotropic mesophases in surfactant ethylenediamine-based complexes including those of chromium (Figure 106B) [193].

We later investigated a series (Figure 107A) of surfactant *tris*(bipyridyl)ruthenium(II) complexes first reported by Seddon [194], which we have shown to be mesomorphic in water [195]. Thus, consistent with the packing constraint ideas of Israelachvili developed above, single-chained derivatives ($m = 1$, $n = 12$–31) showed I_1 cubic mesophases, while double-chained derivatives ($n = 12$, $m = 12$, 19) showed H_1 hexagonal phases. For the mono-chained material with $m = 1$ and

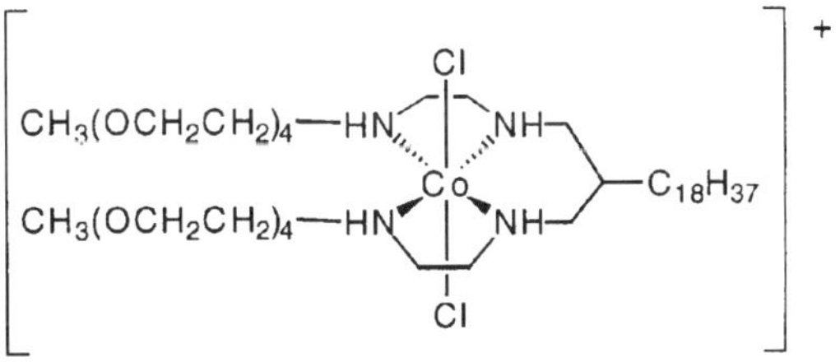

Figure 105 Co(III) annelide complex.

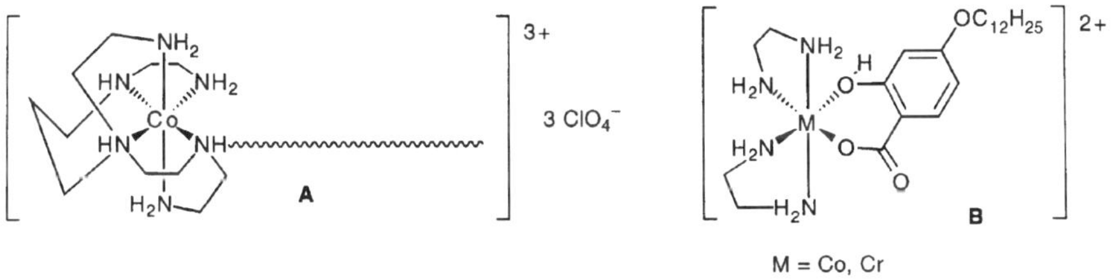

Figure 106 Amphiphilic complexes of Co(III) and Cr(III).

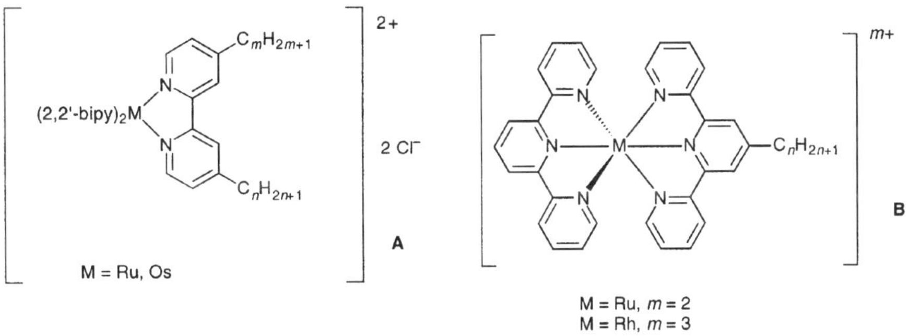

Figure 107 Lyotropic bipyridine and terpyridine complexes.

$n = 12$, we have measured the cmc and found it to be between 5×10^{-3} and 9.5×10^{-3} mol dm^{-3} depending on whether it was measured by NMR or surface tension methods; cmc values for the osmium complexes were consistently lower by a factor of about 2 [196]. It was also found that these complexes would form Y-type Langmuir–Blodgett films and that in alternate layer preparations with long-chain alkanoic acids, pyroelectric response could be generated [197]. Related osmium complexes were subsequently synthesized and were shown to behave in a very similar way [198]. However, related complexes of Rh(III) were not so stable and so in order to generate mesomorphic materials, it was necessary to use terpyridines which bind more strongly (Figure 107B). In this case, mesomorphic materials were obtained in for both Ru(II) and Rh(III), although in some cases it was necessary to use ethylene glycol as the solvent rather than water [199].

Finally in this section, Swager [200] reported amphiphilic complexes of oxovanadium(IV) (Figure 108) which he studied in both binary (surfactant plus water) and ternary (surfactant plus water plus decanol) systems, finding both lamellar (L_α) and hexagonal (H_1) phases.

We can learn several things from the above observations. First, it is amply demonstrated that it is entirely possible to make surfactants based around metal complexes where the complex is an integral part of the polar headgroup and, therefore, the amphiphile itself. Both cationic and anionic surfactants have been realized and in some cases, critical micelle concentrations have been measured; standard packing constraint ideas are also found to be applicable. These observations alone show that surfactant metal complexes are no different in their behaviour from the more traditional organic surfactants. More general remarks will be made at the end of this section.

6.3 Columnar Systems

6.3.1 Phases type and structure

A third type of organization in lyotropic systems is found in mesophases which are strictly analogous to the columnar phases found in thermotropic materials. Also

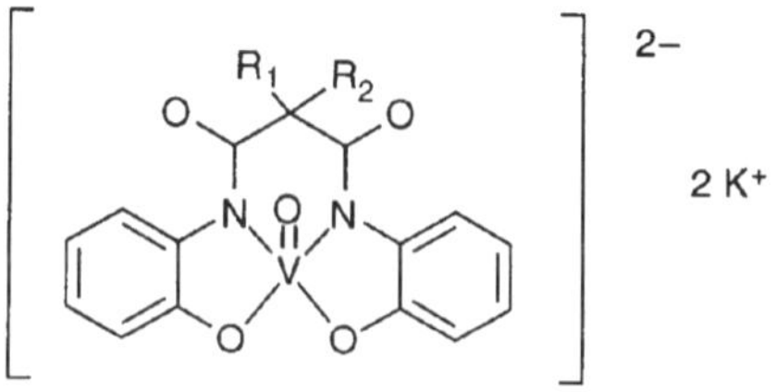

Figure 108 Lyotropic vanadium complexes.

termed *chromonic* [201], these materials differ from the more conventional amphiphilic systems described above in that they have no critical micelle concentration and, therefore, no Krafft point. They were first noticed in rather flat drug and dye molecules (e. g. the anti-asthmatic 'Intal'—disodium chromoglycate; Figure 109) which were substituted by peripheral polar groups.

Aggregation in these systems is by a columnar stacking of the molecules which happens gradually, hence there is no cmc, although it is argued that the driving force for the formation of these columns is not simply the 'attraction' of flat, π-systems [66], but an entropic effect (the hydrophobic effect is entropic) caused by the randomization of water molecules on expulsion from the columns. As the concentration increases and the stacks assemble, the systems show nematic phases followed by a columnar hexagonal phase, previously called the M or 'middle' phase; persistence lengths in these hexagonal phases can be very large—up to 1000 Å. The triphenylenes hexasubstituted by oligo(ethylene oxide) chains and reported by Boden [202] are more conventionally discotic examples of such materials. A schematic diagram of the arrangement in such phases is shown in Figure 110.

6.3.2 *Metallomesogens with lyotropic columnar phases*

The first reports of metallomesogens capable of forming lyotropic columnar phases were by Gaspard, who described mesomorphism in peripherally carboxylated copper(II) phthalocyanines [203], although their phase behaviour was never properly

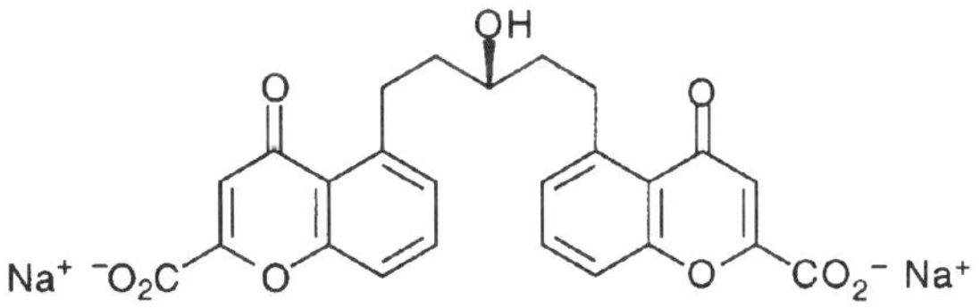

Figure 109 Disodium chromoglycate.

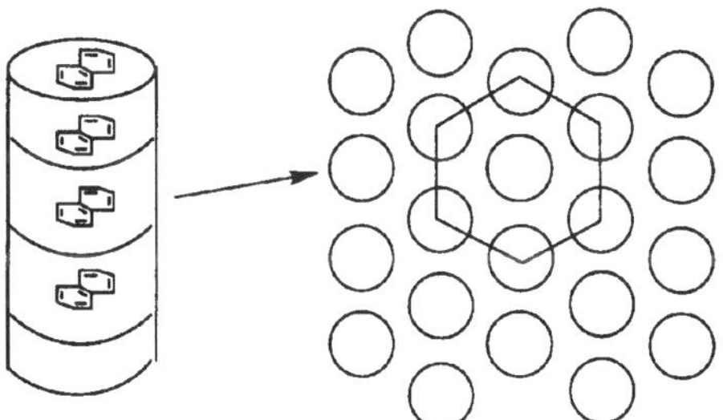

Figure 110 Schematic diagram to show the arrangement of mesogens within the lyotropic columnar phase.

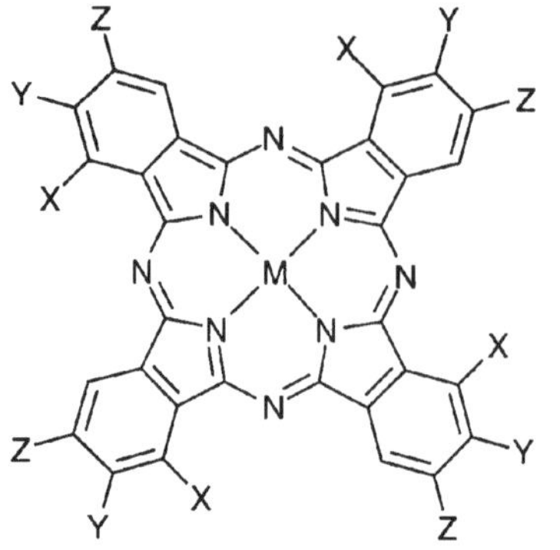

Figure 111 Anionic phthalocyanine liquid crystals.

defined (Figure 111, X, Z = H; Y = CO_2^-, M = Cu). He later reported other derivatives (Figure 111; M = Co, Ni, Pd and Pt) which were not mesomorphic [204].

However, the area was further studied by Usol'tseva [205] who synthesized several peripherally carboxylated phthalocyanines and successfully characterized their phase behaviour as lyotropic columnar. The complexes in Figure 111 with M = 2H, Cu, Zn or Co(II) and X = H, Y = COOH and Z = H or COOH were found to show columnar nematic and hexagonal phases in aqueous ammonia, but this was suppressed when the phthalocyanine ring was substituted according to X = COOH and Y = Z = H or when the central metal ion was Al(III). In this latter case, suppression of the mesomorphism was due to the formation of μ-oxo dimers ([PcAl–O–AlPc]).

However, more remarkable was the discovery that in alkane solvents, large, *ortho*metallated macrocyclic complexes of palladium shown in Figure 92 would form lyotropic columnar phases [171]. These remarkable materials have been shown to form columnar hexagonal phases and, in suitable solvents, lyotropic nematic phases derived from columnar organization. Further, in certain non-mesomorphic examples, mesophases can be induced by the addition of an electron acceptor such as trinitrofluorenone; chiral phases are introduced when the acceptor is resolved 2′-(2,4,5,7-tetranitro-9-fluorenylideneaminoxy)propionic acid (know as TAPA).

REFERENCES

1. (a) J.-M. Lehn, *Angew. Chem. Int. Ed. Engl.*, **29**, 1304 (1990). (b) J-M. Lehn, *Supramolecular Chemistry-concepts and perspectives*, VCH, Weinheim, (1995).
2. H. Ringsdorf, B. Schlarb and J. Venzmer, *Angew. Chem. Int. Ed. Engl.*, **27**, 113 (1988).
3. D. W. Bruce, E. Lalinde, P. Styring, D. A. Dunmur and P. M. Maitlis, *J. Chem. Soc., Chem. Commun.*, 581 (1986).
4. M. A. Esterulas, LA. Oro, E. Sola, M. B. Ros and J. L. Serrano, *J. Chem. Soc., Chem. Commun.*, 55 (1989).

5. (a) M. Marcos, P. Romero and J. L. Serrano, *J. Chem. Soc., Chem. Commun.*, 1641 (1989); (b) M. Marcos, P. Romero and J. L. Serrano, *J. Chem. Soc., Chem. Commun.*, 859 (1990).
6. R. Deschenaux, M. Schweissgarth and A.-M. Levelut, *J. Chem. Soc., Chem. Commun.*, 1275 (1996).
7. K. Ohta, H. Hasabe, M. Moriya, T. Fujimoto and I. Yamamoto, *J. Mater. Chem.*, **1**, 831 (1991).
8. C. Bertram, D. W. Bruce, D. A. Dunmur, S. E. Hunt, P. M. Maitlis and M. McCann, *J. Chem. Soc., Chem. Commun.*, 69 (1991).
9. (a) A. M. Giroud-Godquin and P. M. Maitlis, *Angew. Chem. Int. Ed. Engl.*, **30**, 35 (1991); (b) S. A. Hudson and P. M. Maitlis, *Chem. Rev.*, **93**, 861 (1993); (c) J. L. Serrano, Ed., *Metallomesogens Synthesis, Properties and Applications*, VCH, Weinheim (1995); (d) D. W. Bruce, *Inorganic Materials 2nd Edition*, Ed. D. W. Bruce and D. O'Hare, Wiley, Chichester (1996).
10. L. Oriol and J. L. Serrano, *Adv. Mater.*, **7**, 348 (1995).
11. A. J. Leadbetter, in: *Thermotropic Liquid Crystals*, G. W. Gray, Ed. Wiley, Chichester (1987).
12. (a) G. W. Gray and J. W. Goodby, *Smectic Liquid Crystals; Textures and Structures*, Leonard Hill, Glasgow (1984); (b) D. Demus and L. Richter, *Textures of Smectic Liquid Crystals*, Verlag Chemie, Leipzig (1978).
13. H-T. Nguyen, C. Destrade and J. Mathête, *Adv. Mater.*, **9**, 375 (1997).
14. (a) P. Espinet, J. Etxebarria, M. Marcos, J. Pérez, A. Remón, J. L. Serrano, *Angew. Chem. Int. Ed. Engl.*, **28**, 1065 (1989); (b) N. J. Thompson, G. W. Gray, J. W. Goodby and K. J. Toyne, *Mol. Cryst., Liq. Cryst*, **200** 109 (1991).
15. J. W. Goodby, *J. Mater. Chem.*, **1**, 307 (1991).
16. N. H. Hartshorne, *The Microscopy of Liquid Crystals, The Microscope Series*, **48**, Microscope Publications, London (1974).
17. J. W. Dodd and K. H. Tonge, *Thermal Methods, Analytical Chemistry by Open Learning Series*, Wiley, Chichester (1987).
18. A. de Vries, *Mol. Cryst., Liq. Cryst.*, **131**, 125 (1985).
19. G. W. Gray, *Phil. Trans. R. Soc. Lond. A*, **330**, 73 (1990).
20. G. W. Gray, K. J. Harrison and J. A. Nash, *Electron Lett.*, **9**, 130 (1973).
21. (a) H. Adams, N. A. Bailey, D. W. Bruce, R. Dhillon, D. A. Dunmur, S. E. Hunt, E. Lalinde, A. A. Maggs, R. Orr, P. Styring, M. S. Wragg and P. M. Maitlis, *Polyhedron*, **37**, 1861 (1988); (b) P. Styring, *PhD Thesis*, University of Sheffield (1988).
22. D. W. Bruce, D. A. Dunmur, M. R. Manterfield, P. M. Maitlis and R. Orr, *J. Mater. Chem.*, **1**, 255 (1991).
23. M. Ghedini, S. Armento and F. Neve, *Inorg. Chim. Acta*, **134**, 23 (1987).
24. M. Ghedini, S. Licoccia, S. Armentano and R. Bartolino, *Mol. Cryst. Liq. Cryst.*, **108**, 269 (1984).
25. A. M. Levelut, M. Veber, P. Francescangeli, S. Melone, M. Ghedini, F. Neve, F. D. Nicoletta and R. Bartolino, *Liq. Cryst.*, **19**, 241 (1995).
26. A. Crispini, M. Ghedini, S. Morrone, D. Pucci and O. Fracescangeli, *Liq. Cryst.*, **20**, 67 (1996).
27. M. Ghedini, M. Longeri and R. Bartolino, *Mol. Cryst., Liq. Cryst.*, **84**, 207 (1982).
28. M. Ghedini, D. Pucci and F. Neve, *J. Chem. Soc., Chem. Commun.*, 137 (1996).
29. D. P. Lydon, G. W. V. Cave and J. P. Rourke, *J. Mater. Chem.*, **7**, 403 (1997).
30. H. Adams, A. C. Albeniz, N. A. Bailey, D. W. Bruce, A. S. Cherodian, R. Dhillon, D. A. Dunmur, P. Espinet, J. L. Feijoo, E. Lalinde, P. M. Maitlis, R. M. Richardson and G. Ungar, *J. Mater. Chem.*, **1**, 843 (1991).
31. G. W. Gray and B. Jones, *J. Chem. Soc.*, 4179 (1953).

32. D. Guillon, D. W. Bruce, L. P. Maldivi, M. Ibn-Elhaj and R. Dhillon, *Chem. Mater.*, **6**, 182 (1994).
33. K. Ohta, H. Ema, Y. Morizumi, T. Watanabe, T. Fujimoto and I. Yamamoto, *Liq. Cryst.*, **8**, 311 (1990).
34. (a) J. P. Fackler Jr, J. A. Fetchin and D. V. Fries, *J. Am. Chem. Soc.*, **94**, 7323 (1972); (b) J. P. Fackler Jr, D. Coucouvanis, J. A. Fetchin and W. C. Seidel, *J. Am. Chem. Soc.*, **90**, 2784 (1968).
35. See e. g. M. Bonamico, G. Mazzone, A. Vaciago and L. Zambonelli, *Acta Cryst*, **19**, 898 (1965).
36. D. W. Bruce, D.A Dunmur, E. Lalinde, P. M. Maitlis and P. Styring, *Liq. Cryst.*, **3**, 385 (1988).
37. D. W. Bruce, D. A. Dunmur, M. A. Esteruelas, S. E. Hunt, R. Le Lagadec, P. M. Maitlis, J. R. Marsden, E. Sola and J. M. Stacey, *J. Mater. Chem.*, **1**, 251 (1991).
38. D. W. Bruce and M. D. Hall, *Mol. Cryst. Liq. Cryst.*, **250**, 373 (1994); M. N. Abser and D. W. Bruce, unpublished results.
39. M. A. Esteruelas, E. Sola, L. A. Oro, M. B. Ros, M. Marcos and J. L. Serrano, *J. Organomet. Chem.*, **387**, 103 (1990).
40. R. Bayón, S. Coco, P. Espinet, C. Fernández-Mayordomo and J. M. Martín-Alvarez, *Inorg. Chem.*, **36**, 2329 (1997).
41. M. Hird and K. J. Toyne, *Mol. Cryst., Liq. Cryst.*, in press.
42. J. P. Rourke, F. P. Fanizzi, N. J. S. Salt, D. W. Bruce, D. A. Dunmur and P. M. Maitlis, *J. Chem. Soc., Chem. Commun.*, 229 (1990); N. J. S. Salt, *PhD Thesis*, University of Sheffield (1990).
43. D. W. Bruce, D. A. Dunmur, P. M. Maitlis, P. Styring, M. A. Esteruelas, L. A. Oro, M. B. Ros, J. L. Serrano and E. Sola, *Chem. Mater.*, **1**, 479 (1989); **3**, 378 (1991) (corrigendum)
44. D. W. Bruce, S. C. Davis, D. A. Dunmur, S. A. Hudson, P. M. Maitlis and P. Styring, *Mol. Cryst., Liq. Cryst.*, **215**, 1 (1992).
45. D. W. Bruce, D. A. Dunmur, S. A. Hudson, P. M. Maitlis and P. Styring, *Adv. Mater. Opt. Electron.*, **1**, 37 (1992).
46. D. W. Bruce, D. A. Dunmur, S. A. Hudson, E. Lalinde, P. M. Maitlis, M. P. McDonald, R. Orr, P. Styring, A. S. Cherodian, R. M. Richardson, J. L. Feijoo and G. Ungar, *Mol. Cryst. Liq. Cryst.*, **206**, 79 (1991).
47. H. Adams, N. A. Bailey, D. W. Bruce, S. C. Davis, D. A. Dunmur, P. D. Hempstead, S. A. Hudson and S. Thorpe, *J. Mater. Chem.*, **2**, 295 (1992).
48. A.-M. Levelut, B. Donnio and D. W. Bruce, *Liq. Cryst.*, **22**, 753 (1997).
49. D. W. Bruce, B. Donnio, S. A. Hudson, A-M. Levelut, S. Megtert, D. Petermann and M. Veber, *J. Phys. II France*, **5**, 289 (1995).
50. D. W. Bruce and S. A. Hudson, *J. Mater. Chem.*, **4**, 479 (1994).
51. H. Adams, N. A. Bailey, D. W. Bruce, S. A. Hudson and J. R. Marsden, *Liq. Cryst.*, **16**, 643 (1994).
52. T. Kaharu, R. Ishii and S. Takahashi, *J. Chem. Soc., Chem. Commun.*, 1349 (1994).
53. S. S. Pathanenii and G. R. Desiraju, *J. Chem. Soc.* Dalton Trans., 319 (1993).
54. (a) R. Ishii, T. Kaharu, N. Pirio, S.-W. Zhang and S. Takahashi, *J. Chem. Soc., Chem. Commun.*, 1215 (1995); (b) S.-W. Zhang, R. Ishii and S. Takahashi, *Organometallics*, **16**, 20 (1997).
55. M. Benouazzanne, S. Coco, P. Espinet, and J. M. Martín-Alvarez, *J. Mater. Chem.*, **5**, 441 (1995).
56. S. Coco, P. Espinet, J. M. Martín-Alvarez and A.-M. Levelut, *J. Mater. Chem.*, **7**, 19 (1997).
57. J.-P. Bayle, E. Bui, F. Perez and J. Courtieu, *Bull. Soc. Chim. Fr.*, **4**, 532 (1989).
58. P. Berdagué, F. Perez, J. Courtieu and J.-P. Bayle, *Bull. Soc. Chim. Fr.*, **131**, 35 (1994).

59. P. Berdagué, F. Perez, P. Judeinstein and J.-P. Bayle, *New J. Chem.*, **19**, 293 (1995).
60. E. Bui, J.-P. Bayle, F. Perez, L. Libert and J. Courtieu, *Liq. Cryst.*, **8**, 513 (1990).
61. E. Bui, J.-P. Bayle, F. Perez, and J. Courtieu, *Bull. Soc. Chim. Fr.*, **127**, 61 (1991).
62. E. Campillos, M. Marcos, A. Omenat and J. L. Serrano, *J. Mater. Chem.*, **6**, 349 (1996).
63. A. B. Blake, J. R. Chipperfield, W. Hussain, R. Paschke and E. Sinn, *Inorg. Chem.*, **34**, 125 (1995).
64. D. W. Bruce, D. A. Dunmur, L. S. Santa and M. A. Wali, *J. Mater. Chem.*, **2**, 363 (1992).
65. D. W. Bruce, M. A. Wali and Q. M. Wang, *J. Chem. Soc., Chem. Commun.*, 2089 (1994).
66. C. A. Hunter and J. K. M. Sanders, *J. Am. Chem. Soc.*, **112**, 5525 (1990).
67. (a) Q. M. Wang and D. W. Bruce, *J. Chem. Soc., Chem. Commun.*, 2505 (1996); (b) Q. M. Wang and D. W. Bruce, *Angew. Chem. Int. Ed. Engl.*, **36**, 150 (1997).
68. A. M. Giroud-Godquin and J. Billard, *Mol. Cryst., Liq. Cryst.*, **66**, 147 (1981); (b) A.-M. Levelut, *J. Chim. Physique*, **80**, 149 (1983); (c) A. C. Ribeiro, A. F. Martins and A.-M. Giroud-Godquin, *Mol. Cryst., Liq. Cryst. Lett.*, **5**, 133 (1988).
69. K. Ohta, A. Ishii, I. Yamamoto and K. Matsuzaki, *J. Chem. Soc., Chem. Commun.*, 1099 (1984).
70. S. Chandrasekhar, B. K. Sadashiva, S. Ramesha and B. S. Srikanta, *Pramana-J. Phys.*, **27**, L713 (1986).
71. L. J. Yu and A. Saupe, *Phys. Rev. Lett.*, **45**, 1000 (1980).
72. S. Chandrasekhar, B. K. Sadashiva and B. S. Srikanta, *Mol. Cryst., Liq. Cryst.*, **151**, 93 (1987).
73. (a) S. Chandrasekhar, B. K. Sadashiva B. R. Ratna and V. N. Raja, *Pramana-J. Phys.*, **30**, L491 (1988); (b) S. Chandrasekhar, B. R. Ratna and B. K. Sadashiva, *Mol. Cryst., Liq. Cryst.*, **165**, 123 (1988).
74. B. Mühlberger and W. Haase, *Liq. Cryst.*, **5**, 251 (1989).
75. K. Ohta, O. Takenaka, H. Hasebe, Y. Morizumi, T. Fujimoto and I. Yamamoto, *Mol. Cryst., Liq. Cryst.*, **200**, 109 (1991).
76. K. Ohta, H. Akimoto, T. Fujimoto and I. Yamamoto, *J. Mater. Chem.*, **4**, 61 (1994).
77. (a) N. J. Thompson, J. W. Goodby and K. J. Toyne, *Mol. Cryst., Liq. Cryst.*, **213**, 187 (1992); (b) N. J. Thompson, J. W. Goodby and K. J. Toyne, *Mol. Cryst., Liq. Cryst.*, **214**, 81 (1992).
78. G. W. Gray, *Phil. Trans. R. Soc. Lond. A*, **309**, 77 (1983).
79. P. Espinet, J. Perez, M. Marcos, M. B. Ros, J. L. Serrano, J. Barberá and A.-M. Levelut, *Organometallics*, **9**, 555 (1990).
80. P. Espinet, J. Perez, M. Marcos, M. B. Ros, J. L. Serrano, J. Barberá and A.-M. Levelut, *Organometallics*, **9**, 2028 (1990).
81. D. Lelièvre, L. Bosio, J. Simon, J.-J. André and F. Bensebaa, *J. Am. Chem. Soc.*, **114**, 4475 (1992).
82. A. J. Blake, D. W. Bruce, I. A. Fallis, S. Parsons, H. Rictzenhain, S. Ross and M. Schröder, *Phil. Trans. R. Soc. London. A*, **354**, 395 (1996).
83. (a) F. Neve and M. Ghedini, G. De Munno and A.-M. Levelut, *Chem. Mater.*, **7**, 688 (1995); (b) F. Neve and M. Ghedini, *J. Incl. Phenom. Mol. Recogn. Chem.*, **15**, 259 (1993); (c) F. Neve, M. Ghedini, and O. Francescangeli, *Liq. Cryst.*, **21**, 625 (1996).
84. B. Neumann, T. Hegmann, R. Wolf and C. Tschierske, *J. Chem. Soc., Chem. Commun.*, 105 (1998).
85. P. R. Ashton, D. Joachimi, N Spencer, J. F. Stoddart, C. Tschierske, A. P. White, D. J. Williams and K. Zab, *Angew. Chem. Int. Ed. Engl.*, **33**, 1503 (1994).
86. R. B. Meyer, L. Liebert, L. Strzelecki and P. Keller, *J. Phys.*, Paris, **36**, L69 (1975).
87. N. A. Clark, and S. T. Lagerwall, *Appl. Phys. Lett.*, **36**, 899 (1980).
88. G. W. Gray and J. W. Goodby, personal communications.
89. G. Anderson, D. P. Jeller W. Kuczynski, S. T. Lagerwall, K. Sharp and B. Stebler, *Appl. Phys. Lett.*, **51**, 640 (1987).

90. M. Ghedini, D. Pucci, E. Cesarotti, P. Antogniazza, O. Francescangeli and R. Bartolino, *Chem. Mater.*, **5**, 883 (1993).
91. A. Omenat and M. Ghedini, *J. Chem. Soc., Chem. Commun.*, 1309 (1994).
92. M. J. Baena, J. Buey, P. Espinet, H-S. Kitzerow and G. Heppke, *Angew. Chem. Int. Ed. Engl.*, **32**, 1201 (1993).
93. (a) J. Buey, and P. Espinet, *J. Organomet. Chem.*, **507**, 137 (1996); (b) M. J. Baena, P. Espinet, M. B. Ros and J. L. Serrano, *Angew. Chem. Int. Ed. Engl.*, **30**, 711 (1991).
94. (a) M. J. Baena, P. Espinet, M. B. Ros, J. L. Serrano and A. Ezcurra, *Angew. Chem. Int. Ed. Engl.*, **32**, 1203 (1993); (b) N. J. Thompson, R. Iglesias, J. L. Serrano, M. J. Baena and P. Espinet, *J. Mater. Chem.*, **6**, 1741 (1996); (c) N. J. Thompson, J. L. Serrano, M. J. Baena and P. Espinet, *Chem Eur. J.*, **2**, 214 (1996).
95. (a) M. Marcos, J. L. Serrano, T. Sierra and M. J. Giménez, *Chem. Mater.*, **5**, 1332 (1993); (b) M. J. Baena, J. Barberá, P. Espinet, A. Ezcurra, M. B. Ros and J. L. Serrano, *J. Am. Chem. Soc.*, **116**, 1899 (1994).
96. M. Ghedini, D. Pucci, E. Cesarotti, O. Fracescangeli and R. Bartolino, *Liq. Cryst.*, **15**, 31 (1993).
97. M. Ghedini, D. Pucci, N. Scaramuzza, L. Komitov and S. T. Lagerwall, *Adv. Mater.*, **7**, 659 (1995).
98. M. Marcos, J. L. Serrano, T. Sierra and M. J. Giménez, *Angew. Chem. Int. Ed. Engl.*, **31**, 1471 (1992).
99. M. A. Athanassopoulou, S. Hiller, L. A. Beresnev, Y. G. Galymetdinov, M. Schweissguth and W. Haase, *Mol. Cryst., Liq. Cryst.*, **261**, 29 (1995).
100. (a) D. W. Bruce and X.-H. Liu, *J. Chem. Soc., Chem. Commun.*, 729 (1994); (b) *idem. Liq. Cryst.*, **18**, 165 (1995).
101. (a) S. Morrone, G. Harrison and D. W. Bruce, *Adv. Mater.*, **7**, 665 (1995); (b) S. Morrone, D. Guillon and D. W. Bruce, *Inorg. Chem., 24*, 7041 (1996).
102. D. W. Bruce and K. E. Rowe, *Liq. Cryst.*, **18**, 161 (1995).
103. K. E. Rowe and D. W. Bruce, *Liq. Cryst.*, **20**, 183 (1996).
104. K. E. Rowe and D. W. Bruce, *J. Chem. Soc., Dalton Trans.*, 3913 (1996).
105. R. Deschenaux, B. Donnio, G. Rheinwald, F. Stauffer, G. Suss-Fink and J. Velker, *J. Chem. Soc., Dalton Trans.*, 4351 (1997).
106. J. Malthête and J. Billard, *Mol. Cryst., Liq. Cryst.*, **34**, 117 (1976).
107. P. Loubser, C. Imrie and P. H. van Rooyen, *Adv, Mater.*, **5**, 45 (1995).
108. C. Imrie and C. Loubser, *J. Chem. Soc., Chem. Commun.*, 2159 (1994).
109. (a) P. Singh, M. D. Rausch and R. W. Lenz, *Liq. Cryst.*, **9**, 19 (1991); (b) K. P. Reddy and T. L. Brown, *Liq. Cryst.*, **12**, 369 (1992).
110. N. J. Thompson, J. W. Goodby and K. J. Toyne, *Liq. Cryst.*, **13**, 381 (1993).
111. A. Werner and W. Friedrichsen, *J. Chem. Soc., Chem. Commun.*, 365 (1994).
112. R. Deschenaux and J-L. Marendaz, *J. Chem. Soc., Chem. Commun.*, 909 (1991).
113. R. Deschenaux and J. Santiago, *Tetrahedron Lett.*, **35**, 2169 (1994).
114. R. Deschenaux, I. Kosztics and B. Nicolet, *J. Mater. Chem.*, **5**, 2291 (1995).
115. R. Deschenaux, M. Schweissguth and A.-M. Levelut, *J. Chem. Soc., Chem. Commun.*, 1275 (1996).
116. R. Deschenaux, E. Serrano and A.-M. Levelut, *J. Chem. Soc., Chem. Commun.*, 1577 (1997).
117. J. Andersch, S. Diele and C. Tschierske, *J. Mater. Chem.*, **6**, 1465 (1996).
118. (a) L. Ziminiski and J. Malthête, *J. Chem. Soc., Chem. Commun.*, 1495 (1990); (b) P. Jacq and J. Malthête, *Liq. Cryst.*, **21**, 291 (1996).
119. E. Campillos, R. Deschenaux, A.-M. Levelut and R. Ziessel, *J. Chem. Soc., Dalton Trans.*, 2533 (1996).

120. C. Destrade, N. H. Tinh, H. Gasparoux, J. Malthête and A.-M. Levelut, *Mol. Cryst., Liq. Cryst.*, **71**, 111 (1981).
121. K. Praefcke, J. D. Holbrey, N. Usol'tseva and D. Blunk, *Mol. Cryst., Liq. Cryst.*, **292**, 123 (1997).
122. C. Piechocki, J. Simon, A. Skoulios, D. Guillon and P. Weber, *J. Am. Chem. Soc.*, **104**, 5245 (1982).
123. D. Guillon, A. Skoulios, C. Piechocki, J. Simon and P. Weber, *Mol. Cryst., Liq. Cryst.*, **100**, 275 (1983).
124. J. Vacus, P. Doppelt, J. Simon and G. Memetzidis, *J. Mater. Chem.*, **2**, 1065 (1992).
125. M. K. Engel, P. Bassoul, L. Bosio, H. Lehmann, M. Hanack and J. Simon., *Liq. Cryst.*, **15**, 709 (1993).
126. J. F. Van Der Pol, E. Neeleman, J.,W. Zwikker, R. J. Nolte and W. Drenth, *Liq. Cryst.*, **6**, 577 (1989).
127. K. Ohta, T. Watanabe, S. Tanaka, T. Fujimoto, I. Yamamoto, P. Bassoul, N. Kucharczyk and J. Simon, *Liq. Cryst.*, **10**, 357 (1991).
128. (a) M. J. Cook, M. F. Daniel, K. J. Harrison, N. B. McKeown and A. J. Thompson, *J. Chem. Soc., Chem. Commun.*, 1086 (1987); (b) M. J. Cook, S. J. Cracknell and K. J. Harrison, *J. Mater. Chem.*, **1**, 703 (1991).
129. A. S. Cherodian, A. N. Davis, R. M. Richardson, M. J. Cook, N. B. McKeown, A. J. Thomson, J. Feijoo G. Ungar and K. J. Harrison, *Mol. Cryst., Liq. Cryst.*, **196**, 103 (1991).
130. M. J. Cook, A. J. Dunn, S. D. Howe, A. J. Thompson and K. J. Harrison, *J. Chem. Soc., Perkins Trans I*, 2453 (1988).
131. P. Doppelt and S. Huille, *New J. Chem.*, **14**, 607 (1990).
132. H. Eichhorn, D. Wöhrle and D. Pressner, *Liq. Cryst.*, **22**, 643 (1997).
133. B. Mohr, G. Wegner and K. Ohta, *J. Chem. Soc., Chem. Commun.*, 995 (1995).
134. D. M. Knawby and T. M. Swager, *Chem. Mater.*, **9**, 535 (1997).
135. C. Sirlin, L. Bosio, J. Simon, V. Ahsehn, E. Yilmazer and Ö. Bekâroglu, *Chem. Phys. Lett.*, **139**, 362 (1987).
136. C. F. Van Nostrum and R. J. M. Nolte, *J. Chem. Soc., Chem. Commun.*, 2385 (1996).
137. C. Sirlin, L. Bosio and J. Simon, *Mol. Cryst., Liq. Cryst.*, **155**, 231 (1988).
138. C. Sirlin, L. Bosio and J. Simon, *J. Chem. Soc., Chem. Commun.*, 379 (1987).
139. O. E. Sielcken, L. A. Van de Kuil, W. Drenth, J. Schoonman and R. J. M. Nolte, *J. Am. Chem. Soc.*, **112**, 3086 (1990).
140. P. Weber, D. Guillon and A. Skoulios, *J. Phys. Chem.*, **91**, 2242 (1987).
141. C. Piechocki, J. Simon, J.-J. André, D. Guillon, P. Petit, A. Skoulios and P. Weber, *Chem. Phys. Lett.*, **122**, 124 (1985).
142. T. Komatsu, K. Ohta, T. Fujimoto and I. Yamamoto, *J. Mater. Chem.*, **4**, 533 (1994).
143. T. Komatsu, K. Ohta, T. Watanabe, H. Ikemoto, T. Fujimoto and I. Yamamoto, *J. Mater. Chem.*, **4**, 537 (1994).
144. Z. Belarbi, C. Sirlin, J. Simon and J.-J. André, *J. Phys. Chem.*, **93**, 8105 (1989).
145. (a) B. A. Gregg, M. A. Fox and A. J. Bard, *J. Chem. Soc., Chem. Commun.*, 1134 (1987); (b) B. A. Gregg, M. A. Fox and A. J. Bard, *J. Am. Chem. Soc.*, **111**, 3024 (1989).
146. B. A. Gregg, M. A. Fox and A. J. Bard, *J. Phys. Chem.*, **94**, 1586 (1990).
147. Y. Shimizu, M. Miya, A. Nagata, K. Phta, I. Yamamoto and S. Kusabayashi, *Liq. Cryst.*, **14**, 795 (1993).
148. S. Kugimiya and M. Takemura, *Tetrahedron Lett.*, **31**, 3160 (1990).
149. (a) Y. Shimizu, J. Matsuno, M. Miya and A. Nagata, *J. Chem. Soc., Chem. Commun.*, 2411 (1994); (b) Y. Shimizu, J, Matsuno, K. Nakao, K. Ohta, M. Mioya and A. Nagata, *Mol. Cryst., Liq. Cryst.*, **260**, 491 (1995).
150. (a) S. Forget, M. Veber and H. Strzelecka, Mol. Cryst., *Liq. Cryst.*, **258**, 263 (1995); (b)

S. Forget and M. Veber, *Mol. Cryst., Liq. Cryst.*, **300**, 229 (1997); (c) S. Forget and M. Veber, *Mol. Cryst., Liq. Cryst.*, **308**, 27 (1997).
151. S. Forget and H.-S. Kitzerow, *Liq. Cryst.*, **23**, 919 (1997).
152. (a) A. Liebmann, C. Mertesdorf, T. Plesnivy, H. Ringsdorf and J. H. Wendorff, *Angew. Chem. Int. Ed. Engl.*, **30**, 1375 (1991); (b) G. Lattermann, S. Schmidt, R. Kleppinger and J. H. Wendorff, *Adv. Mater.*, **4**, 30 (1992); (c) S. Schmidt, G. Lattermann, R. Kleppinger and J. H. Wendorff, *Liq. Cryst.*, **16**, 693 (1994).
153. (a) K. Ohta, H. Hasebe, M. Moriya, T. Fjuimoto and I. Yamamoto, *Mol. Cryst., Liq. Cryst.*, **208**, 43 (1991); (b) K. Ohta, H. Hasebe, M. Moriya, T. Fujimoto and I. Yamamoto, *J. Mater. Chem.*, **1**, 831 (1991); (c) K. Ohta, M. Moriya, M. Ikejima, H. Hasebe, T. Fujimoto and I. Yamamoto, *Bull. Chem. Soc. Jpn.*, 3553 **66**, (1993); (d) K. Ohta, M. Moriya, M. Ikejima, H. Hasebe, T. Fujimoto and I. Yamamoto, *Bull. Chem. Soc. Jpn.*, **66**, 3559 (1993).
154. K. Ohta, M. Moriya, M. Ikejima, H. Hasebe, N. Kobayashi and I. Yamamoto, *Bull. Chem. Soc. Jpn.*, **70**, 1199 (1997).
155. J. Barberá, A. Elduque, R. Giménez, L. A. Oro and J. L. Serrano, *Angew. Chem. Int. Ed. Engl.*, **35**, 2832 (1996).
156. (a) A. M. Giroud-Godquin, J-C. Marchon, D. Guillon, A. Skoulios, *J. Phys. Chem.*, **45**, L681 (1984); (b) A.-M. Giroud-Godquin, J.-C. Marchon, D. Guillon, A. Skoulios, *J. Phys. Chem.*, **90**, 5502 (1986); (c) L. Bonnett, F. D. Cukiernik, P. Maldivi, A.-M. Giroud-Godquin, J.-C. Marchon, M. Ibn-Elhaj, D. Guillon and A. Skoulios, *Chem. Mater.*, **6**, 31 (1994); (d) D. V. Baxter, R. H. Clayton, M. H. Chisholm, J. C. Huffmann, E. F. Putilina, S. L. Tagg, J. L. Wesemann, J. W. Zwangiger and F. D. Darrington, *J. Am. Chem. Soc.*, **116**, 4551 (1994).
157. A.-M. Giroud-Godquin, J.-M. Latour and J.-C. Marchon, *Inorg. Chem.*, **24**, 4454 (1985).
158. M. Ibn-Elhaj, D. Guillon, A. Skoulios, A. M. Giroud-Godquin and P. Maldivi, *Liq. Cryst.*, **11**, 731 (1992).
159. G. S. Attard and P. R. Cullum, *Liq. Cryst.*, **8**, 299 (1990).
160. F. D. Cukiernik, M. Ibn-Elhaj, Z. D. Chaia, J-C. Marchon, A-M. Giroud-Godquin,. D. Guillon, A. Skoulios and P. Maldivi, *Chem. Mater.*, **10**, 83 (1998).
161. J. Barberá, M. A. Esteruelas, A. M. Levelut, L. A. Oro, J. L. Serrano and E. Sola, *Inorg. Chem.*, **31**, 732 (1992).
162. A. M. Giroud-Godquin, G. Sigaud, M. F. Achard and F. Hardouin, *J. Phys. Lett.*, **45**, L-387 (1984).
163. (a) J. Barberá, C. Cativiela, J. L. Serrano and M. M. Zurbano, *Adv. Mater.*, **3**, 602 (1991); (b) R. Atencio, J. Barberá, C. Cativiela, F. J. Lahoz, J. L. Serrano and M. M. Zurbano, *J. Am. Chem. Soc.*, **116**, 11558 (1994).
164. (a) H. Zheng, P. J. Carroll and T. M. Swager, *Liq. Cryst.*, **14**, 1421 (1993); (b) H. Zheng, C. K. Lai and T. M. Swager, *Chem. Mater.*, **7**, 2067 (1995).
165. S. T. Trzaska and T. M. Swager, *Chem. Mater.*, **10**, 438 (1998).
166. (a) C. K. Lai, A. G. Serrette and T. M. Swager, *J. Am. Chem. Soc.*, **114**, 7948 (1992); (b) A. G. Serrette, C. K. Lai and T. M. Swager, *Chem. Mater.*, **6**, 2252 (1994).
167. H. Zheng and T. M. Swager, *J. Am. Chem. Soc.*, **116**, 761 (1994).
168. A. G. Serrette and T. M. Swager, *J. Am. Chem. Soc.*, **115**, 8879 (1993).
169. A. Krowczynski, D. Pciecha, J. Szydlowska, P. Przedmojski and E. Górecka, *J. Chem. Soc., Chem. Commun.*, 2371 (1996).
170. (a) D. Singer, A. Liebmann, K. Praefcke and J. H. Wendorff, *Liq. Cryst.*, **14**, 785 (1993); (b) K. Praefcke, B. Bilgin, J. Pickardt and M. Borowski, *Chem. Ber.*, **127**, 1543 (1994).
171. (a) B. Gündoğan and K. Praefcke, *Chem. Ber.*, **126**, 1253 (1993); (b) K. Praefcke, S. Diele, J. Pickardt, B. Gundoğan, U. Nutz and D. Singer, *Liq. Cryst.*, **18**, 857 (1995); (c) K. Praefcke, B. Bilgin, N. Usol'tseva, B. Heinrich and D. Guillon, *J. Mater. Chem.*, **5**, 2257 (1995).

172. J. M. Vila, M. Gayoso, T. Pereira, C. Rodriguez, J. M. Ortigueira and M. Thornton-Pett, *J. Organomet. Chem.*, **426**, 267 (1992).
173. B. Heinrich, K. Praefcke and D. Guillon, *J. Mater. Chem.*, **7**, 1363 (1997).
174. J. Malthête, A. M. Levelut and H. T. Nguyen, *J. Phys. Lett. (France)*, **46**, 875 (1985).
175. H. T. Nguyen, C. Destrade, A. M. Levelut and J. Malthête, *J. Phys. (France)*, **47**, 553 (1986).
176. K. E. Rowe and D. W. Bruce, *J. Mater. Chem.*, **8**, 331 (1998).
177. B. Donnio *PhD Thesis*, University of Sheffield (1996).
178. B. Donnio and D. W. Bruce, *J. Chem. Soc., Dalton Trans.*, 2745 (1997).
179. B. Donnio, B. Heinrich, T. Gulik-Krzywicki, H. Delacroix, D. Guillon and D. W. Bruce, *Chem. Mater.*, **9**, 2951 (1997).
180. B. Donnio, D. W. Bruce, H. Delacroix and T. Gulik-Krzywicki, *Liq. Cryst.*, **23**, 147 (1997).
181. A. G. Serrette and T. M. Swager, *Angew. Chem. Int. Ed. Engl.*, **33**, 2342 (1994).
182. B. Xu and T. M. Swager, *J. Am. Chem. Soc.*, **115**, 1159 (1993).
183. From C. E. Fairhurst, S. Fuller, J. Gray, M. C. Holmes and G. J. T. Tiddy in *Handbook of Liquid Crystals, Vol. 3*, eds: D. Demus, J. Goodby, G. W. Gray, H.-W. Spiess and V. Vill, Wiley-VCH, Weinheim (1998).
184. F. Livolant and A. Leforestier, *Prog. Polym. Sci.*, **21**, 1115 (1996).
185. (a) S. Takahashi, M. Kariya, T. Yakate, K. Sonogashira and N. Jagihara, *Macromolecules*, **11**, 1063 (1978); (b) S. Takahashi, E. Murata, M. Kariya, K. Sonogashira and N. Hagihara, *ibid.*, **12**, 1016 (1979).
186. S. Takahasi, Y. Takai, H. Morimoto, K. Sonohashira and N. Hagihara, *Mol. Cryst., Liq. Cryst.*, **82**, 139 (1982).
187. S. Takahasi, Y. Takai, H. Morimoto and K. Sonogahira, *J. Chem. Soc., Chem. Commun.*, 3 (1984).
188. M. Ibn-Elhaj, D. Guillon, A. Skoulios, A.-M. Giroud-Godquin and J.-C. Marchon, *J. Phys. II France*, **2**, 2197 (1992).
189. A. S. C. Lawrence, *Liquid Crystals 2*, ed. G. H. Brown, Gordon and Breach, London, part 1, p. 1 (1969).
190. J. N. Israelachvili, D. J. Mitchell and B. W. Ninham, *J. Chem. Soc., Faraday Trans. 2*, **72**, 1525 (1976).
191. J. Le Moigne and J. Simon, *J. Phys. Chem.*, **84**, 170 (1980).
192. M. Yashiro, K. Matsumoto and S. Yoshikawa, *Chem. Lett.*, **62**, 985 (1989).
193. D. W. Bruce, I. R. Denby, G. J. T. Tiddy and J. M. Watkins, *J. Mater. Chem.*, **3**, 911 (1993).
194. K. R. Seddon and Y. Z. Yousif, *Tran. Met. Chem.*, **11**, 443 (1986).
195. D. W. Bruce, J. D. Holbrey, A. R. Tajbakhsh and G. J. T. Tiddy, *J. Mater. Chem.*, **3**, 905 (1993).
196. H. B. Jervis, J. D. Holbrey, A. D. Pidwell, G. J. T. Tiddy and D. W. Bruce, in preparation.
197. J. D. Holbrey, M. B. Grieve, T. Richardson and D. W. Bruce, unpublished work.
198. A. D. Pidwell, H. B. Jervis and D. W. Bruce, unpublished work.
199. J. D. Holbrey, G. J. T. Tiddy and D. W. Bruce, *J. Chem. Soc., Dalton Trans.*, 1769 (1995).
200. S. S. Zhu and T. M. Swager, *Adv. Mater.*, **7**, 280 (1995).
201. T. K. Attwood, J. E. Lydon, C. Hall and G. J. T. Tiddy *Liq. Cryst.*, **7**, 657 (1990).
202. N. Boden, R. J. Bushby and C. Hardy, *J. Phys. Lett.*, **46**, L325 (1985).
203. S. Gaspard, A. Hochapfel and R. Viovy, *C. R. Acad. Sci. Ser. C*, **289**, 387 (1979).
204. S. Gaspard, A. Hochapfel and R. Viovy, *Springer Ser. Chem. Phys.*, **11**, 298 (1980).
205. (a) N. V. Usol'tseva, V. E. Maizlish, V. V. Bykova, G. A. Analeva, G. P. Shaposhnikov and N. M. Kormilitsyn, *Russ. J. Phys. Chem.*, **63**, 1610 (1989); (b) N. V. Usol'tseva, V. V. Bykova, N. M. Kormilitsyn, G. A. Ananieva and V. E. Maizlish, *Il Nuovo Cimento*, **12D**, 1237 (1990); (c) N. V. Usol'tseva and V. V. Bykova, *Mol. Cryst., Liq. Cryst.*, **215**, 89 (1992).

Chapter 8

Self-Assembly of Interlocked Structures with Cucurbituril, Metal Ions and Metal Complexes

KIMOON KIM

Pohang University of Science and Technology, South Korea

1 INTRODUCTION

Supramolecular chemistry utilizes relatively weak, noncovalent interactions such as hydrogen bonding, π–π stacking, electrostatic, and van der Waals interactions [1–4]. One of the keywords in supramolecular chemistry is self-assembly; this refers to a process by which a well-defined supramolecular species is produced spontaneously from its component molecules under appropriate conditions. Self-assembly driven by noncovalent interactions has been well recognized in biological systems, as in self-assembly of the DNA double helix from two complementary polynucleotides. In chemical synthesis, self-assembly offers considerable advantages over the stepwise bond formation in the construction of large supramolecular assemblies. Self-assembly processes are (a) economical by virtue of their high convergence and (b) dynamic and reversible, which leads to the thermodynamically most stable product through a self-error correcting mechanism.

Although hydrogen bonding has been most widely used in self-assembly, coordinate bonding has proven to be also very useful in the construction of supramolecular architectures [5–7]. Metal ions, particularly transition metal ions are not only capable of forming relatively strong coordinate bonds but also have preferred coordination geometries, which allows us to construct supramolecular species with unusual structures [8]. For example, Cu(I) ion, which prefers tetrahedral

Transition Metals in Supramolecular Chemistry, edited by J. P. Sauvage.

coordination geometry, has been used in the synthesis of catenanes, double helicates, racks, ladders and grids [1, 8, 9]. Square planar Pd(II) and Pt(II) complexes have been employed in the construction of molecular squares and cages [10–11].

Some of our recent work on the construction of interlocked structures is described in this short review. In the design of these structures, the barrel-shaped molecule cucurbituril is used as a molecular "bead" and metal ions or metal complexes are used as "glue" or "angle connectors". The examples given here demonstrate the efficiency and control in the self assembly of highly organized supramolecular species.

1.1 Interlocked Structures, Catenanes, Rotaxanes, and Molecular Necklaces

Interlocked structures such as catenanes, rotaxanes and knots have intrigued synthetic chemists not only because of their beauty but also because of their potential applications in materials for molecular electronics. A number of excellent reviews on catenanes and rotaxanes are available [12, 13]. The simplest catenane, [2]catenane consists of two mechanically interlocked rings (Figure 1). Many rings are interlocked to each other, e.g. a chain in a polycatenane. In a rotaxane, a circular component is threaded on a linear component having two bulky stoppers at the ends, which prevent dethreading. A pseudorotaxane has a circular component threaded on a string without the bulky stoppers. A polyrotaxane contains many circular components threaded on a long string. A molecular necklace (MN) is a cyclic oligorotaxane in which a number of small rings are threaded onto a large ring [13f, 14]. It differs from [n]catenane where rings are interlocked with each other one by one in a linear fashion. The smallest member of molecular necklaces, [4]MN consists of three small rings threaded onto a large ring. The smaller molecular necklaces [2]- and [3]MN are equivalent to [2]- and [3]catenane, respectively.

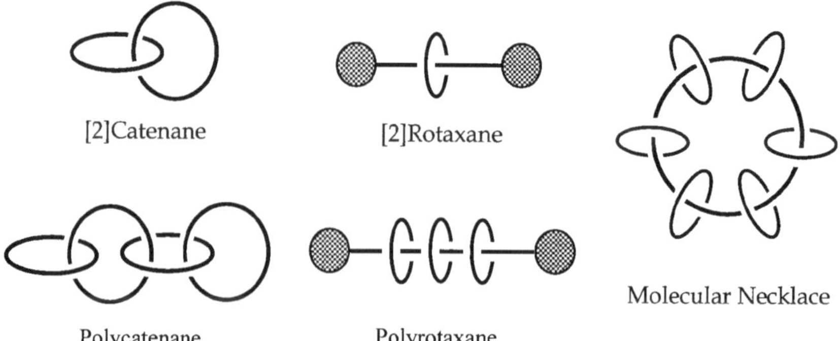

Figure 1 Catenanes, rotaxanes, and molecular necklaces.

1.2 Interlocked Structures Constructed with Cyclodextrins

Cyclodextrins (CDs) are cyclic oligosaccharides composed of six or more α-1,4-linked D-glucopyranose rings (Figure 2) [15]. The α-CD, *β*-CD, and γ-CD contain six, seven and eight glucopyranose rings, respectively. They have rigid, bucket-shaped structures. The open top of the bucket, encircled by secondary hydroxyl groups, has a wider opening than the bottom of the bucket encircled by primary hydroxyl groups. They can bind a wide range of rod-like guest molecules in the rigid, hydrophobic cavity in aqueous solutions. Their propensity to form stable complexes with many different types of guests has led to their development as building blocks in the construction of interlocked structures. Since several comprehensive review articles covering this subject are available [13b, 13c], I shall limit myself here to describe a few examples in this area.

In 1981 Ogino, reported synthesis of [2]rotaxanes incorporating a cyclodextrin as a molecular bead [16]. The [2]rotaxanes (**1**) were synthesized by threading α or *β*-CD with a diaminoalkane to form an inclusion complex followed by attaching *cis*-$[CoCl_2(en)_2]Cl$ (Figure 3). The terminal metal complexes are large enough to prevent the dethreading of the diaminoalkane from the cyclodextrin. More recently, a similar approach was taken by Macartney and coworkers to synthesize [2]rotaxanes (Figure 4). Rotaxane **2** is self-assembled efficiently in solution from α-CD, $[Fe(CN)_5(H_2O)]^{3-}$ and 1,1′-(α,ω-alkanediyl)bis(4,4′-pyridyl-pyridium) ion. The [2]rotaxane is also formed by the addition of α-CD to a solution of the dumbbell component $[(CN)_5Fe(bpy(CH_2)_nbpy)Fe(CN)_5]$. This result infers a slow dissociation of the terminal metal complex, followed by formation of an intermediate pseudorotaxane and reassociation of the metal complex.

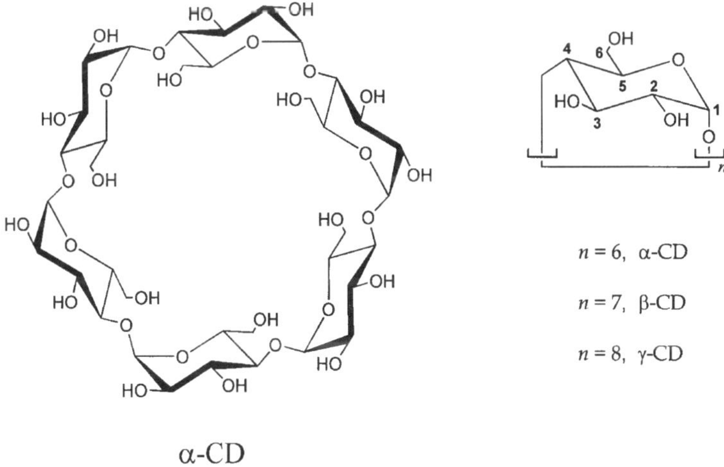

Figure 2 Cyclodextrins

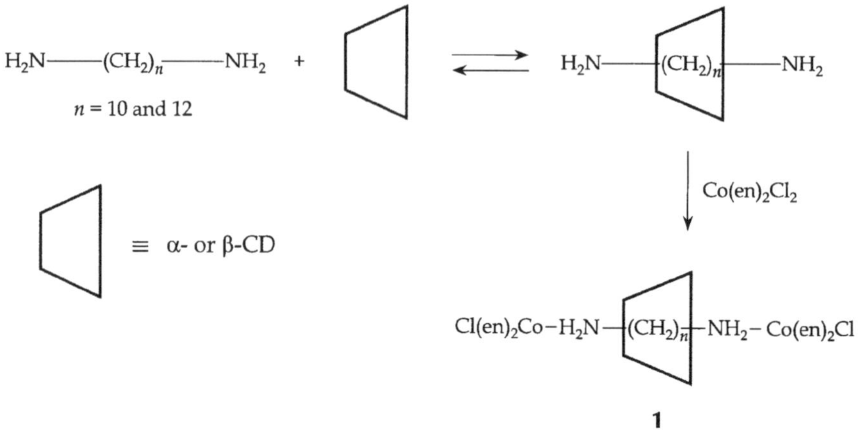

Figure 3 [2]Rotaxanes constructed from α,ω-diaminoalkanes, cyclodextrins and metal complex stoppers reported by Ogino.

Isnin and Kaifer reported the self-assembly of unsymmetrical zwitterionic [2]rotaxane incorporating α-CD [18]. The [2]rotaxane (**3**) was prepared by EDC coupling of potassium 5-amino-2-naphthalenesulfonate with dimethyl(7-oxoalkyl)(ferrocenylmethyl)ammonium cation in the presence of α-CD in water (Figure 5). The unsymmetrical nature of dumbbell components of the [2]rotaxane results in the formation of two orientational isomers (**3a** and **3b**) which were in fact separated.

Several polyrotaxanes containing CDs have been reported [19–21]. Harada and coworkers reported a polyrotaxane in which a number of cyclodextrin beads are

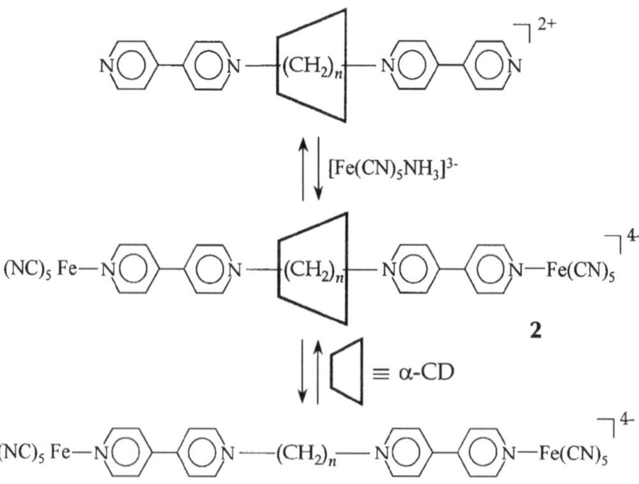

Figure 4 [2]Rotaxane reported by Macartney and coworkers.

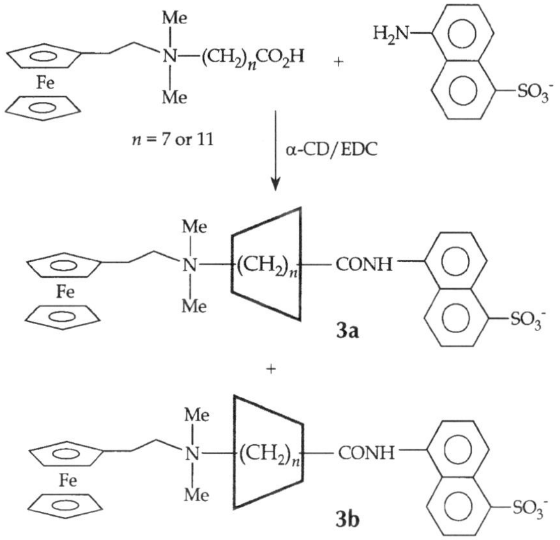

Figure 5 Synthesis of two oriental isomers of a [2]rotaxane reported by Isnin and Kaifer.

thread on a long poly(ethyleneglycol) chain [19]. Reaction of 2,4-dinitrofluorobenzene with poly(ethylene glycol) diamine (PEG-DA) in DMF containing excess α-CD yielded, after purification, the polyrotaxane **4** in 60% yield. It has an average molecular mass of ~25 000 and contains 20–23 threaded CDs. The CD rings are thought to be threaded in a head-to-head/tail-to-tail fashion (Figure 6). More recently, Harada synthesized a tubular polymer using a similar polyrotaxane as a

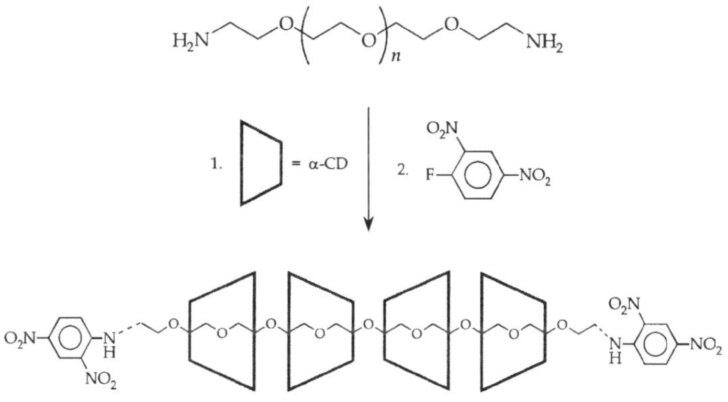

Figure 6 Polyrotaxane containing CDs threaded on PEG-DA reported by Harada and coworkers.

template [19c]. When the hydroxyl groups of α-CD threaded onto a poly(ethylene glycol) backbone were allowed to react with epichlorohydrin, the tori of the CDs became linked together. Cleavage of the stopper groups with strong base resulted in long tube-like structures with an average molecular mass of 20 000 Da.

Wenz and Keller also synthesized similar polyrotaxanes (**5**) (Figure 7) by threading α-CDs onto poly(iminomethylene) chains [20]. Instead of attaching two bulky stoppers at the ends, however, they introduced blocking groups at arbitrary positions on the chain. More specifically, nonspecific reaction of the threaded polymer chain with nicotinyl chloride resulted in trapping of the CD rings located between the two nicotinamide units on the polymer backbone.

Although a number of rotaxanes and polyrotaxanes containing CDs are known, catenanes incorporating CDs are rare. The first example of such catenanes was reported by Stoddart and coworkers [22] 35 years after unsuccessful attempts had been made by Lüttringhaus. Formation of a stable inclusion complex between with *per*-2,6-dimethyl-β-cyclodextrin (DMβ-CD) and the diamine **6** incorporating a biotyl group, followed by reaction of the terephthaloyl chloride produced a mixture of [2]- and [3]catenanes (**7–10**) in low yield (Figure 8). The X-ray crystal structure of **7** reveals that the biotyl unit is located in the cavity of the CD.

1.3 Cucurbituril

Cucurbituril is a macrocyclic cavitand (Figure 9). Several comprehensive reviews on cucurbituril are available [24]. Preparation of cucurbituril first appeared in the literature by Behrend *et al.* in 1905 [25]. Those German chemists reported that glycoluril, which can be obtained from urea and glyoxal, reacts with an excess of

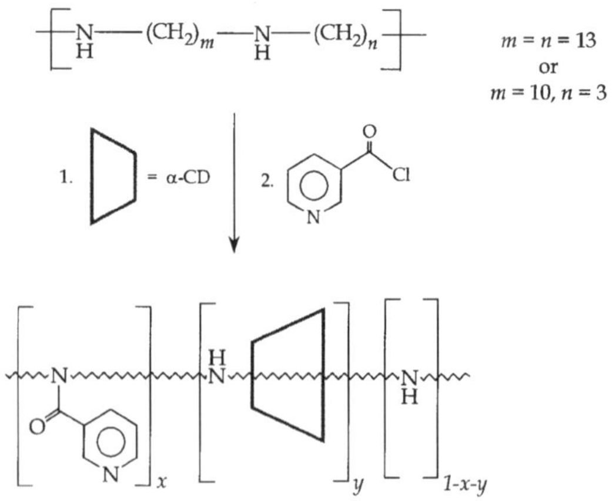

Figure 7 Polyrotaxane reported by Wenz and Keller.

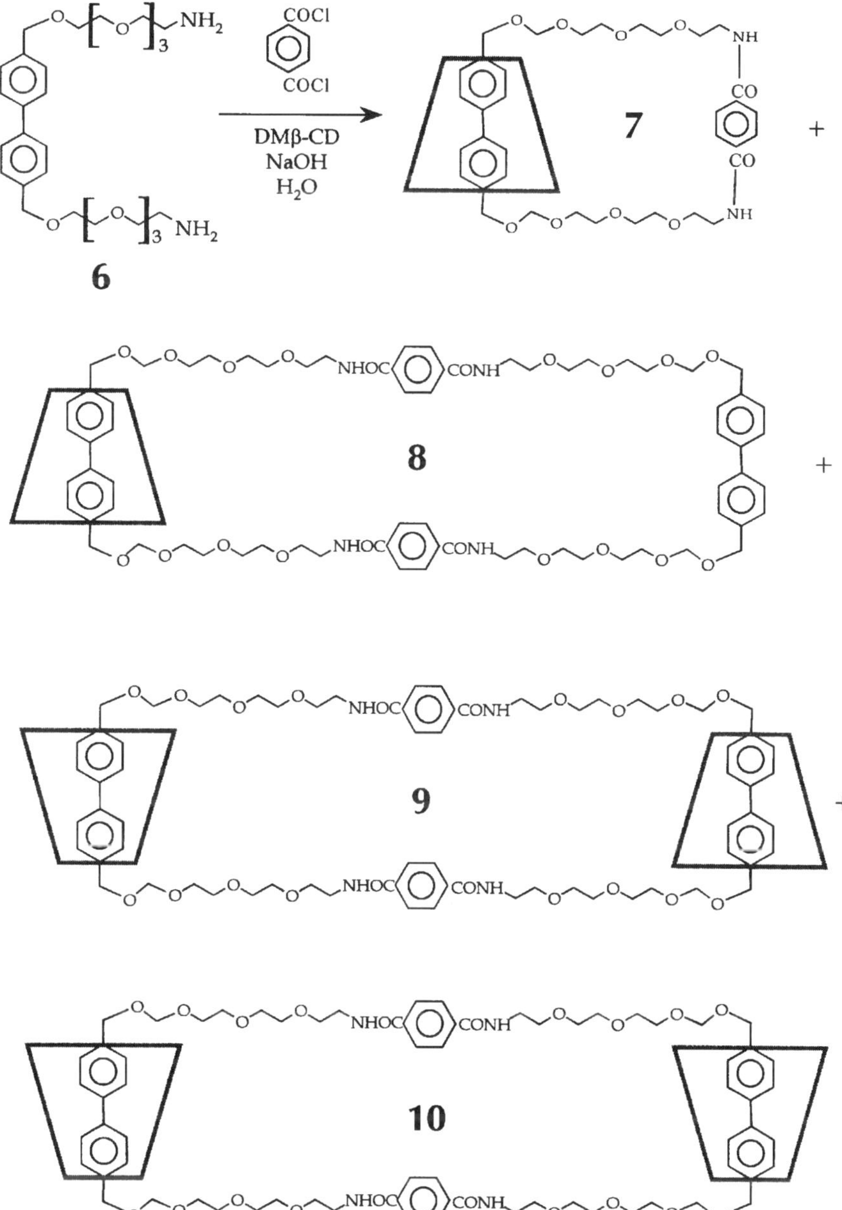

Figure 8 Synthesis of catenanes incorporating CDs reported by Stoddart and coworkers.

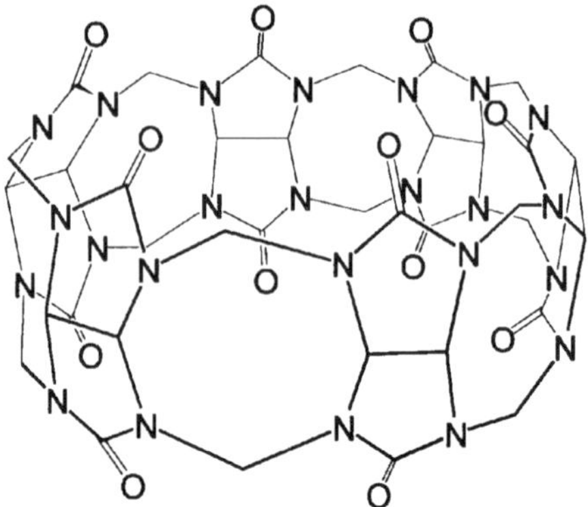

Figure 9 Cucurbituril.

formaldehyde in the presence of acid to yield a gummy precipitate insoluble in all common solvents. They treated the gummy precipitate with hot, concentrated sulfuric acid to dissolve it. The resulting solution was diluted with cold water, and then refluxed before it was allowed to cool to produce a crystalline material. They characterized this compound as $C_{10}H_{11}N_7O_4 \cdot 2H_2O$ solely based on elemental analysis. The substance turned out to be very stable toward various potent reagents including potassium permanganate. A number of crystalline materials were obtained with a variety of metal salts and dyestuffs.

Almost three-quarters of a century later, this substance was rediscovered by Mock and coworkers [26]. With the help of modern instruments they finally characterized it as a hexameric macropolycyclic compound with composition of $C_{36}H_{36}N_{24}O_{12}$. The X-ray crystal structure (Figure 10) of this compound [26] reveals an internal cavity of $\sim$5.5 Å diameter which is accessible from the exterior by two carbonyl-laced portals of $\sim$4 Å diameter. The cavity of cucurbituril has dimensions equivalent to that of α-CD ($\sim$5.7 Å). Similar to cyclodextrins the interior of cucurbituril is hydrophobic and therefore, provides a potential site for inclusion of hydrocarbon molecules. Mock *et al.* proposed a trivial name *cucurbituril* for this compound because of its resemblance to a pumpkin which belongs to the botanical family *Cucurbitaceae*.

Host–guest chemistry using cucurbituril as a host has been studied extensively by Mock and coworkers [27]. One severe drawback of cucurbituril as a synthetic receptor is its extremely poor solubility in virtually any solvents except strongly acidic aqueous solution. Therefore, the host–guest chemistry has been studied only in strongly acidic aqueous solution, typically a 1 : 1 mixture of formic acid and water. Alkyl amines or diamines, which exist as ammonium ions in acidic solutions, form very stable 1 : 1 host–guest complexes (Table 1). Electrostatic interaction as well as hydrogen bonding between the ammonium ions and the carbonyl groups at the entrance of cucurbituril is responsible for the high affinity of these susbtrates. In

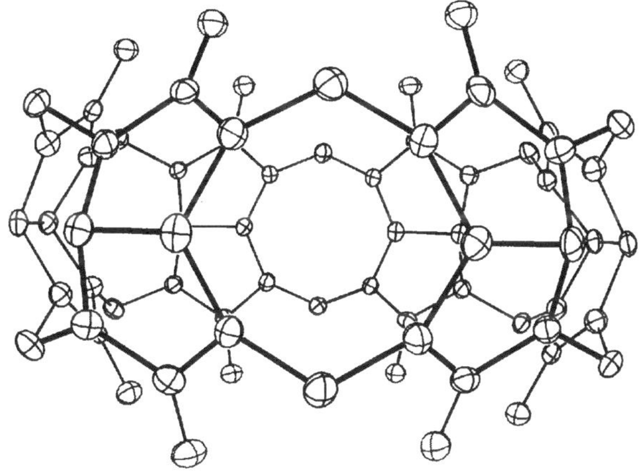

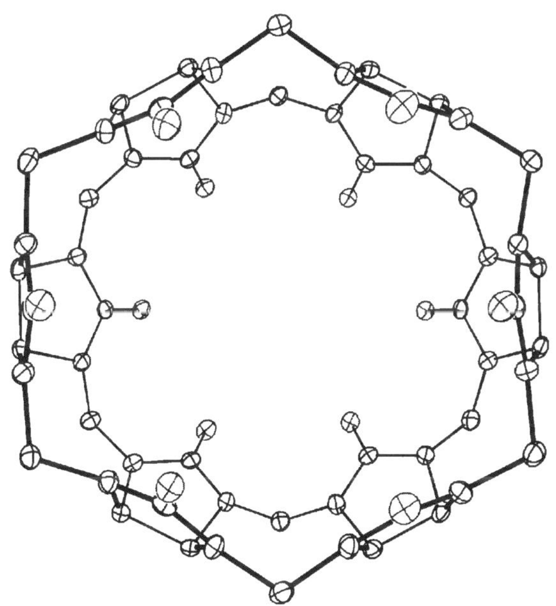

Figure 10 X-ray crystal structure of cucurbituril.

addition, hydrophobic interactions between the alkyl group and the inside wall of the cavity contribute to the stabilities of these host–guest complexes. Shape selectivity associated with the inclusion phenomenon is demonstrated by the fact that *para*-substituted benzene rings are able to adapt to the cavity, whereas *ortho*- and *meta*-substituted counterparts are not included (Table 1).

Although cucurbituril is sparingly soluble in virtually any solvents except strongly acidic aqueous solution, we recently discovered that it dissolves appreciably in aqueous solution of alkali metal salts [28–29]. For example, $\sim 6.6 \times 10^{-2}$ mol of cucurbituril dissolves in 1.0 l of 0.2 M Na_2SO_4 solution [30]. The markedly increased solubility of cucurbituril in neutral aqueous solutions in the presence of alkali metal ions appears to be due to coordination of the metal ions to the carbonyl groups at the portal of cucurbituril. X-ray crystal structure reveals that the two sodium ions and five water molecules bound to the metal ions effectively cover each portal of cucurbituril like a "lid" on a "barrel". Small molecules without ammonium functionality such as THF and benzene can be encapsulated in the sodium ion "lidded" cucurbituril. Furthermore, the encapsulation and release of guest molecules can be controlled reversibly by changing the pH of the medium at ambient temperatures [28]. A similar reversible inclusion phenomenon has been observed in other alkali metal ion complexed cucurbituril [31].

As described above, cyclodextrins have been used as "beads" in the synthesis of rotaxanes and catenanes in recent years. Few of these rotaxanes or oligorotaxanes based on cyclodextrins, however, have been characterized by X-ray diffraction methods presumably due to difficulty in obtaining X-ray quality crystals. Its highly symmetric structure and capability of holding guest molecules make cucurbituril a useful candidate for a molecular "bead" in synthesis of interlocked structures, particularly rotaxanes. Nevertheless, when we initiated this study in early 1990s, no rotaxane or catenane containing cucurbituril had been reported [32]. In the following sections I describe our work on rotaxanes, polyrotaxanes and molecular necklaces incorporating cucurbituril.

Table 1 Affinity data for ligand-receptor complexes of cucurbituril[a]

Ammonium ion	K_f (*rel*)[b]	Ammonium ion	K_f (*rel*)[b]
$Me(CH_2)_2NH_2$	37.6	$NH_2(CH_2)_3NH_2$	2.8
$Me(CH_2)_3NH_2$	307	$NH_2(CH_2)_4NH_2$	480
$Me(CH_2)_4NH_2$	74	$NH_2(CH_2)_5NH_2$	7600
$Me(CH_2)_5NH_2$	7.0	$NH_2(CH_2)_6NH_2$	8600
cyclo-$(CH_2)_3CHCH_2NH_2$	1130	$C_6H_5NH_2$	1130
cyclo-$(CH_2)_4CHCH_2NH_2$	1040	$C_6H_5CH_2NH_2$	58
o- or *m*-$MeC_6H_4CH_2NH_2$			
not bound $NH_2(CH_2)_3NH(CH_2)_4NH(CH_2)_3NH_2$	40 000	*p*-$MeC_6H_4CH_2NH_2$	1.0

[a]From Mock and W. L.; Shih, N.-Y. *J. Org. Chem.* 1986, **51**, 4440.
[b]Formation constant relative to *p*-Me-$C_6H_4CH_2NH_2$, for which the absolute value of $K_f = 320\ M^{-1}$ in 1 : 1 formic acid/water solution at 40°C

2 SIMPLE ROTAXANES

The first simple rotaxane containing cucurbituril as a molecular bead was synthesized by threading cucurbituril with spermine to form a pseudorotaxane and then attaching dinitrophenyl groups to both ends of the spermine unit to prevent dethreading (Scheme 1) [33]. Spermine was chosen as a "string" in this synthesis because it has not only high affinity toward cucurbituril, but also terminal amine groups to which a bulky substituents can be introduced. Although cucurbituril itself is sparingly soluble in water, the resulting pseudorotaxane is quite soluble in water. The drastically improved solubility upon formation of the pseudorotaxane allows the one-pot, high-yield synthesis of the rotaxane **11**.

In the ^{1}H NMR spectrum of **11**, the proton signals of the internal methylene units to the spermine chain, which are now placed inside the cucurbituril cavity, are shifted upfield upon formation of the rotaxane because of the strong shielding effect of cucurbituril. The structure of **11** was confirmed by X-ray crystallography (Figure 11). The cucurbituril molecular "bead" is held tightly at the middle of the string by strong hydrogen bonding between the two inside amine nitrogen atoms (protonated) and the oxygen atoms at the cucurbituril portals. However, it should be noted that the nitrogen atoms form hydrogen bonds with only three of the six portal oxygen atoms. The nitrogen atoms are slightly displaced outside from the mean plane of the six oxygen atoms. The diaminobutane unit of the "string" located inside cucurbituril is lying on a plane with full extension of its length. This structure indicates that the diaminobutane unit has a correct size for inclusion inside cucurbituril. These structural features associated with interaction between the (protonated) diaminobutane unit of the "string" and cucurbituril are also observed in polyrotaxanes described below.

When diethyl ether was layered over the solution containing the inclusion complex of cucurbituril and spermine, the pH of which was adjusted to 8–9 by

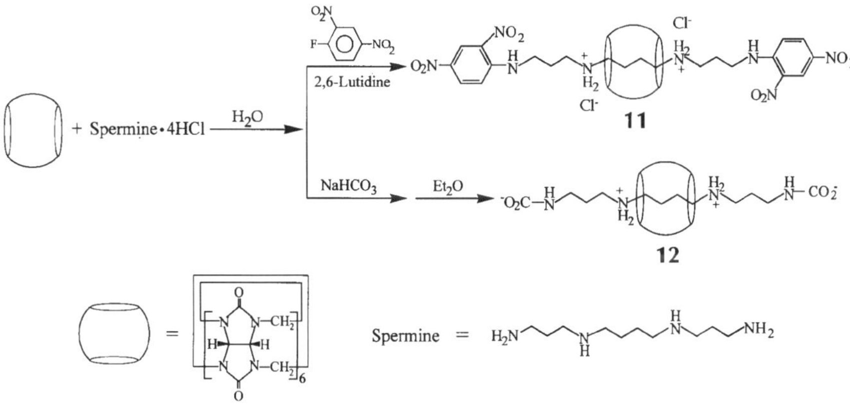

Scheme 1

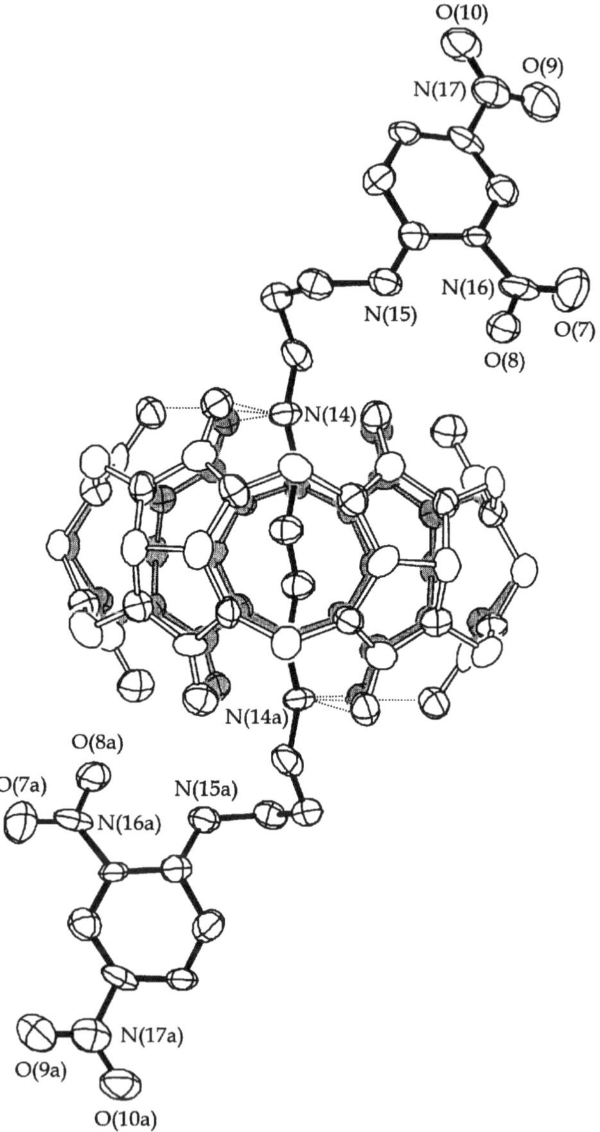

Figure 11 X-ray crystal structure of rotaxane **11**. Counter anions are omitted for clarity.

adding saturated sodium bicarbonate solution, rotaxane **12**, in which both ends of the spermine thread are converted to carbamates groups, was crystallized out [33]. The rotaxane appeared to be produced by the reaction of the terminal amines of the spermine moiety of the inclusion complex with bicarbonate. The structure **12** was determined by X-ray crystallography (Figure 12). Although the size of the terminal carbamate group may not be large enough to be a stopper, its negative charge seems to prevent dethreading effectively. The core structure of **12** is very similar to that of **11**.

The most interesting structural feature of **12** is its solid state structure. The packing diagram (Figure 13) reveals that the terminal carbamate groups of **12** form strong hydrogen bonds with those of neighbor supermolecules. The strong intermolecular hydrogen bonds between the terminal groups of the sperminedicarbamate "strings" link them one by one to form a one-dimensional "polymer" in the solid state. This solid structure, therefore, may be viewed as a "pseudopolyrotaxane" in which cucurbituril beads are threaded by "poly(sperminedicarbamate)". Stoddart and coworkers also reported the syntheses and X-ray crystal structures of "pseudo-

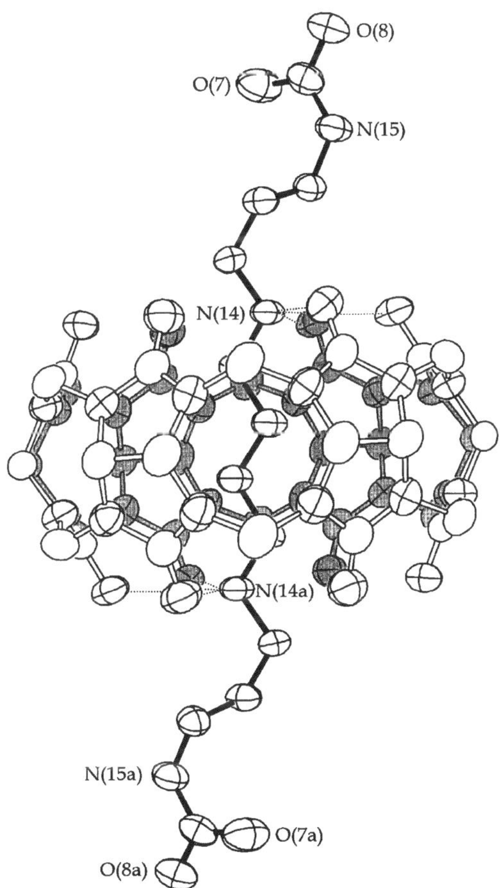

Figure 12 X-ray crystal structure of rotaxane **12**.

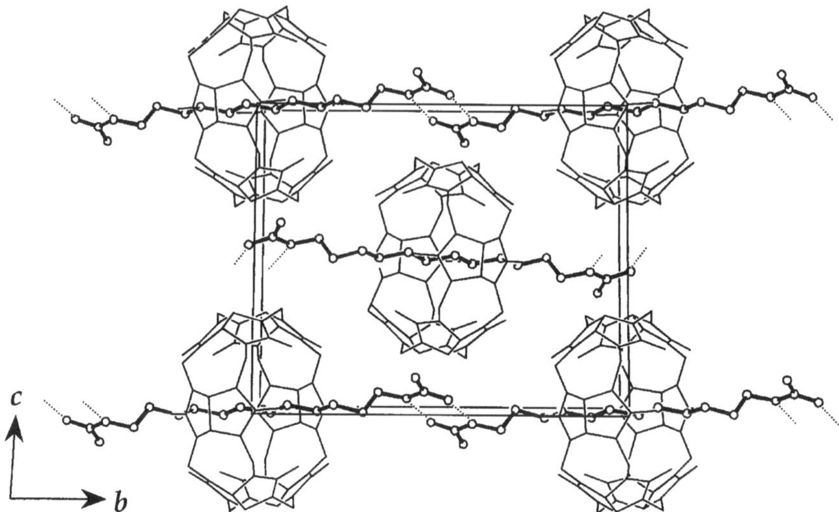

Figure 13 The packing diagram of **12**. Hydrogen bonding interactions are indicted by dotted lines.

polyrotaxanes" in which rotaxane-repeating units are linked by hydrogen bonding in the solid state [34].

3 POLYROTAXANES

3.1. Synthetic Strategy of Polyrotaxanes

Having synthesized the simple rotaxanes using cucurbituril as a molecular "bead" we turned our attention to polyrotaxanes. Polyrotaxanes have been synthesized by threading molecular "beads", particularly cyclodextrins, with organic polymers [19–21, 35]. Typical examples are the ones prepared by Harada *et al.* [19] and Wenz *et al.* [20] described above. Furthermore, no polyrotaxane polymer proven to contain a molecular "bead" in every structural unit of the polymer chain had been reported, although polyrotaxanes with such high structural regularity may posess interesting physical properties not observed in those without such structural regularity. Consequently, we decided to synthesize polyrotaxanes with high structural regularity. Our synthetic strategy has been (1) to thread a molecular "bead" with a short "string" to make a pseudorotaxane, and then (2) to link the pseudorotaxanes by using transition metal ions as "glue" to construct a polyrotaxane on which molecular "beads" are threaded, as shown in Scheme 2 [36]. Prior to our work, neither this approach nor any other synthesis of a polyrotaxane based on a coordination polymer had been reported.

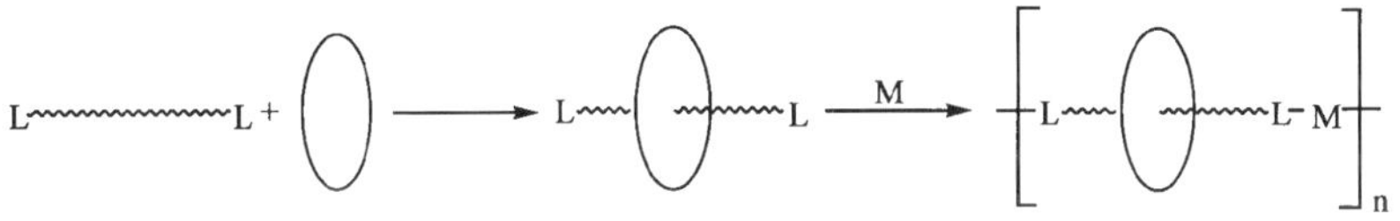

Scheme 2

With this idea in mind, we designed a short "string" in such as way that it can not only form a stable inclusion complex with a cucurbituril "bead", but also bind to transition metal ions using its terminal groups. A 4-pyridylmethyl group was introduced to each amine group of diaminobutane to produce a short "string" (**C4py4**). Indeed, this short "string" forms a stable inclusion complex or a pseudorotaxane with cucurbituril. Moreover, the resulting pseudorotaxane **PR44** is readily soluble in water because of its positive charge (+2). Various transition metal ions such as Cu^{2+}, Ni^{2+}, Co^{2+}, Zn^{2+} and Ag^{+} were then allowed to react with the pseudorotaxane to form polyrotaxanes.

3.2 One-dimensional Polyrotaxanes

The first one-dimensional polyrotaxane **13** was obtained when a solution of $Cu(NO_3)_2$ was allowed to diffuse slowly into a solution of pseudorotaxane **PR44**$^{2+}$ (Scheme 3) [36]. The X-ray crystal structure of **13** (Figure 14) reveals, as expected, cucurbituril "beads" threaded on the coordination polymer, the chain of which is composed of alternating copper ions and **PR44**$^{2+}$. Four nitrate ions are found in the lattice to balance the charge on each repeating unit of the polymer backbone. The cucurbituril "beads" are held tightly on the polymer backbone by strong hydrogen bonds between the protonated amine nitrogen atoms of the "string" and the oxygen atoms at the cucurbituril portals as seen in the above simple rotaxane structures. The coordination geometry of the copper ion is square pyramidal. The two adjacent basal positions are occupied by two pyridine units of two pseudorotaxanes and the remaining basal and apical positions by three water molecules. The

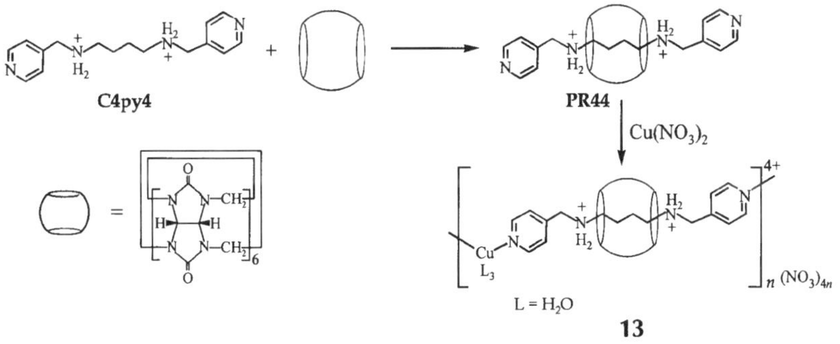

C4py4 PR44 $Cu(NO_3)_2$ $L = H_2O$ **13**

Scheme 3

cis coordination of the two pyridine units to the metal center makes the polymer chain change its direction abruptly at the metal center; consequently, the polymer chain adopts a zigzag shape (Figure 14). This polyrotaxane is unique in several respects, being (1) the first formed on a coordination polymer [37], (2) the first containing a cyclic component in every structural repeating unit, and (3) the first to be structurally characterized by single-crystal X-ray crystallography.

Having successfully synthesized the 1D polyrotaxane **13**, we decided to try other metal ions as "glue" to construct polyrotaxanes using the same strategy. In

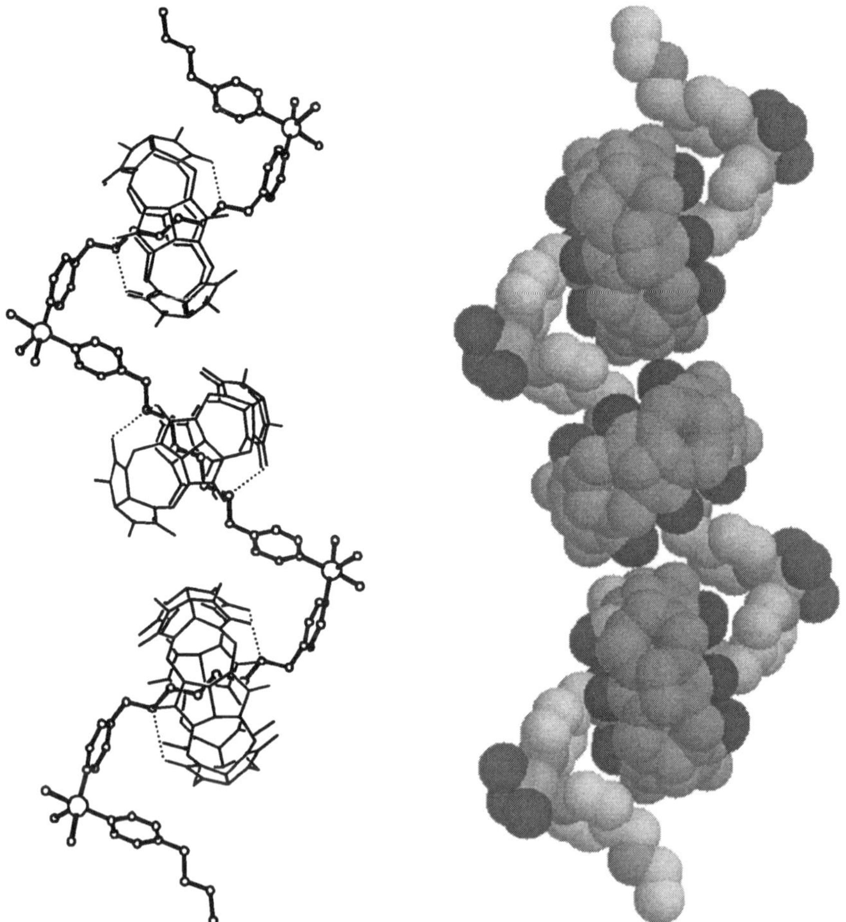

Figure 14 X-ray crystal structure of the 1D-polyrotaxane **13**. Left: a ball and stick representation of **13**: large circles represent Cu^{2+} ions; three small circles linked to each large circle represent water molecules coordinated to the metal ion. Right: space filling representation of **13**.

particular, we chose to use Ag^+ to react with the pseudorotaxane because Ag^+ is well known to form two-coordinate linear complexes. When a solution of silver tosylate is allowed to diffuse into a solution of **PR44**$^{2+}$ (Scheme 4), another one-dimensional polyrotaxane, **14** was obtained as expected [38]. In the structure of **14** (Figure 15), a two-coordinate Ag^+ ion links two molecules of pseudorotaxane **PR44**$^{2+}$ to form a 1D polyrotaxane coordination polymer similar to the one formed with Cu^{2+} ion (**13**). The major structural difference between the two 1D polyrotaxane coordination polymers **14** and **13** is that the two pyridyl units are coordinated to the silver ion in a *trans* geometry in **14** whereas they are bound to the copper ion in a *cis* geometry in **13**. As a result, the former has an almost straight polymer chain whereas the latter exhibits a zigzag shaped polymer chain [36].

Other transition metal ions also work as "glue". When $Zn(NO_3)_2$ and $Ni(NO_3)_2$ are reacted with **PR44**$^{2+}$, the 1D polyrotaxanes **15** and **16**, respectively, are produced (Figure 16 and 17) [39]. In both structures, two pseudorotaxane units are linked by a metal ion which is coordinated by four water molecules and two pyridyl group of the two pseudorotaxanes in an octahedral coordination geometry. The two pyridyl groups occupy *cis* positions in **15** whereas they occupy *trans* positions in **16**. Therefore, the former 1D polyrotaxane has a zigzag shaped polymer chain whereas the latter has a linear structure.

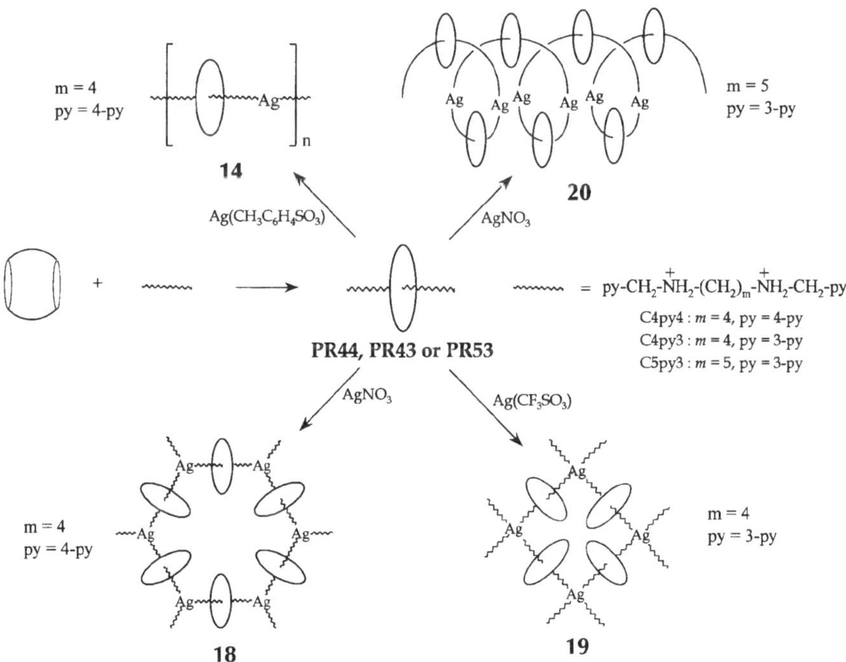

Scheme 4

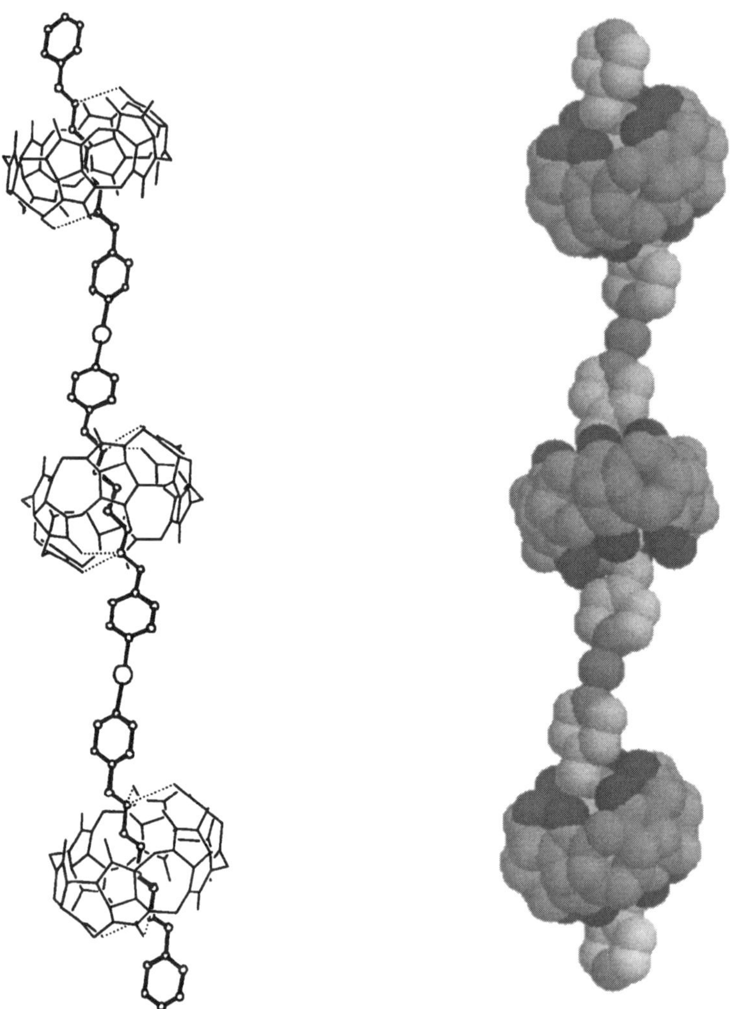

Figure 15 X-ray crystal structure of the 1D-polyrotaxane **14**. Left: a ball and stick representation of **14**: large circles represent AG^+ ions. Right: a space filling representation of **14**.

In this approach not all transition metal salts produce polyrotaxane coordination polymers. For example, when $MnCl_2$ was employed in the polyrotaxane synthesis, a simple rotaxane **17** was produced, in which both ends of the "string" threading a cucurbituril molecule are attached to $MnCl_3(H_2O)$ groups existing in a trigonal bipyramidal coordination geometry with their three Cl^- ions at the equatorial positions (Figure 18) [36]. It would be interesting to study solution properties of

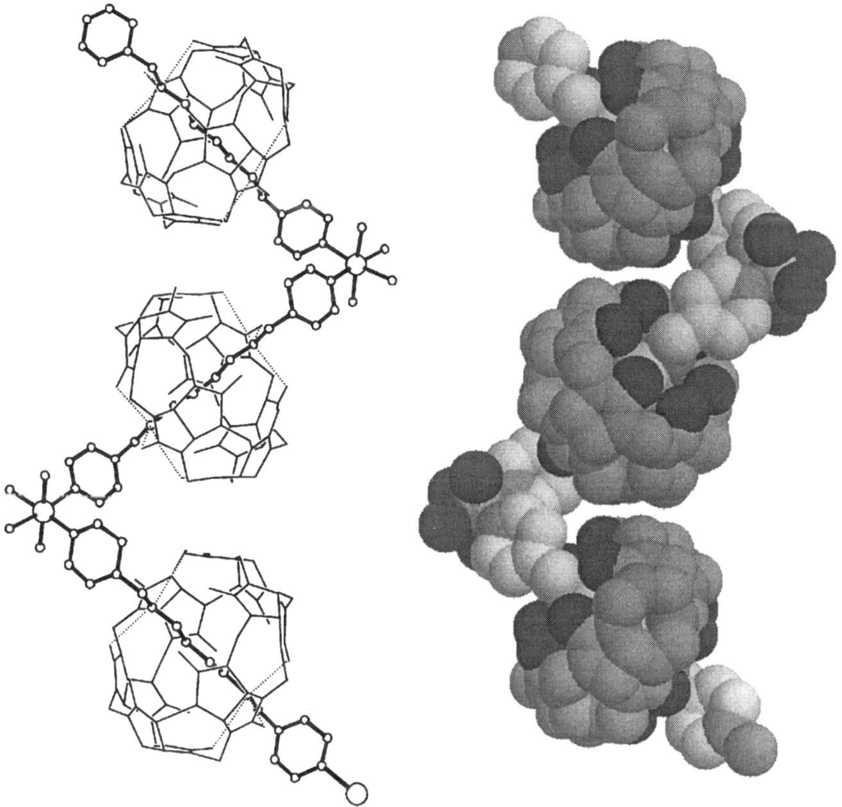

Figure 16 X-ray structure of the 1D polyrotaxane **15**. Left; a ball and stick representation of **15**: large circles represent Zn^{2+} ions; four small circles linked to each large circle represent water molecules coordinated to the metal ion. Right: space filling representation of **15**.

these polyrotaxanes. However, attempts to measure solution properties of these polyrotaxanes have been hampered by their extremely poor solubility in any solvents.

3.3 Two-dimensional Polyrotaxanes

So far I have demonstrated that metal ions play an important role in determining the solid-state structures of polyrotaxanes. Of course, this is not the only important factor. Counter anions also play an important role in determining the solid-state structures. As described above, when silver tosylate is reacted with $\mathbf{PR44}^{2+}$, 1D polyrotaxane **14** is formed. However, when the counter anion is switched to nitrate ion, the same procedure yields a novel 2D polyrotaxane **18**. The X-ray crystal

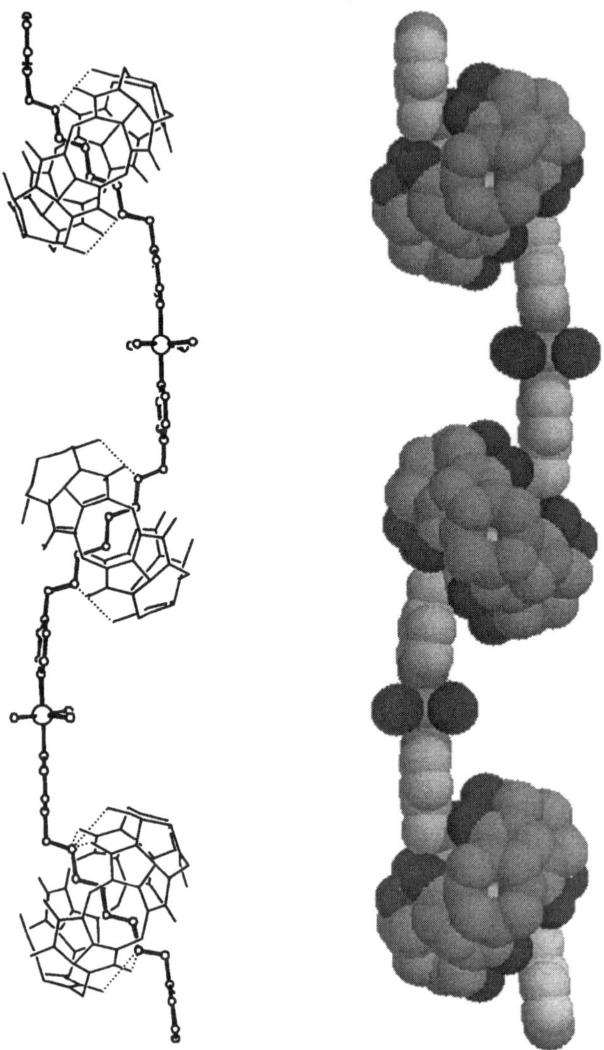

Figure 17 X-ray structure of the 1D polyrotaxane **16**. Left; a ball and stick representation of **16**: large circles represent Ni^{2+} ions; three small circles linked to each large circle represent water molecules coordinated to the metal ion. Right: space filling representation of **16**.

structure of **18** reveals cucurbituril "beads" threaded onto a 2D coordination polymer network (Figure 19) [38].

The 2D network consists of large edge-sharing chair-shaped hexagons with a Ag(I) ion at each corner and a molecule of **PR44**$^{2+}$ at each edge connecting two Ag(I) ions. The edge of the hexagon is 20.9 Å and the separation of the opposite

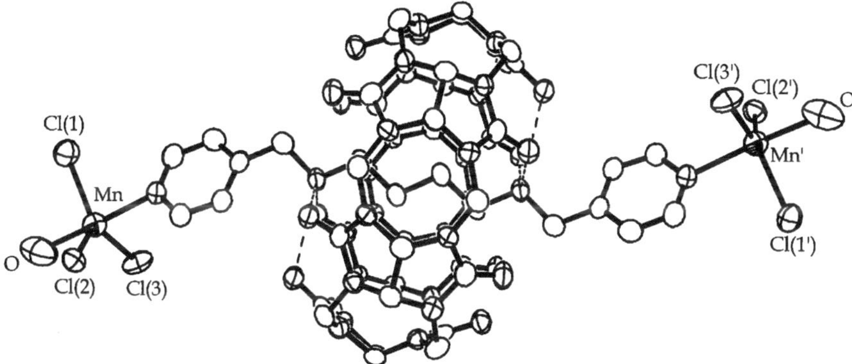

Figure 18 X-ray structure of the simple rotaxane **17**.

corners is 38.0 Å. Each silver ion is coordinated by three $\mathbf{PR44}^{2+}$ and a nitrate ion in a distorted tetrahedral geometry. The 2D polyrotaxane network forms layers stacked on each other with a mean interplane separation of 9.87 Å (Figure 20). There is another 2D polyrotaxane network (denoted **B**) almost perpendicular to the first one (denoted **A**). The dihedral angle between the mean planes of the two networks **A** and **B** is 69.34 °. These networks interpenetrate with full interlocking of the hexagons as illustrated in Figure 21: a hexagon belonging to the network **A** (black) interlocked with four hexagons belonging to **B** (grey) and vice versa. Although interlocking of simple 2D networks has been known (polycatenated 2D nets) [40–47] this structure is the first example of polycatenated 2D polyrotaxane nets [48].

Another 2D polyrotaxane (**19**) was obtained from the reaction of $Ag(CF_3SO_3)$ with **PR43**$(CF_3SO_3)_2$ [39]. In the structure of **19** (Figure 22), cucurbituril beads are threaded onto a square-meshed 2D coordination polymer network. Although the coordination geometry of the silver ion is slightly distorted tetrahedral, the polyrotaxane network has a layer structure in which the 2D networks are stacked along the *c* axis with a mean interplane separation of 12.69 Å. In contrast to **18**, no catenation of the 2D network is observed in this structure. It is interesting to note that a disordered triflate anion is sitting at the middle of the square grid.

3.4 Helical One-dimensional Polyrotaxane

Another unique and interesting twist in this approach to the construction of polyrotaxanes results from changing the "string" from **C4py4** to **C5py3**. The formation of the pseudorotaxane **PR53**$(NO_3)_2$, by threading cucurbituril with *N,N'*-bis(3-pyridylmethyl)-1,5-diaminopentane dihydronitrate (**C5py3**$(NO_3)_2$), followed by the reaction of **PR53**$(NO_3)_2$ with $AgNO_3$ yielded the first helical polyrotaxane **20** (Scheme 4) [49]. In the structure of **20**, cucurbituril "beads" are

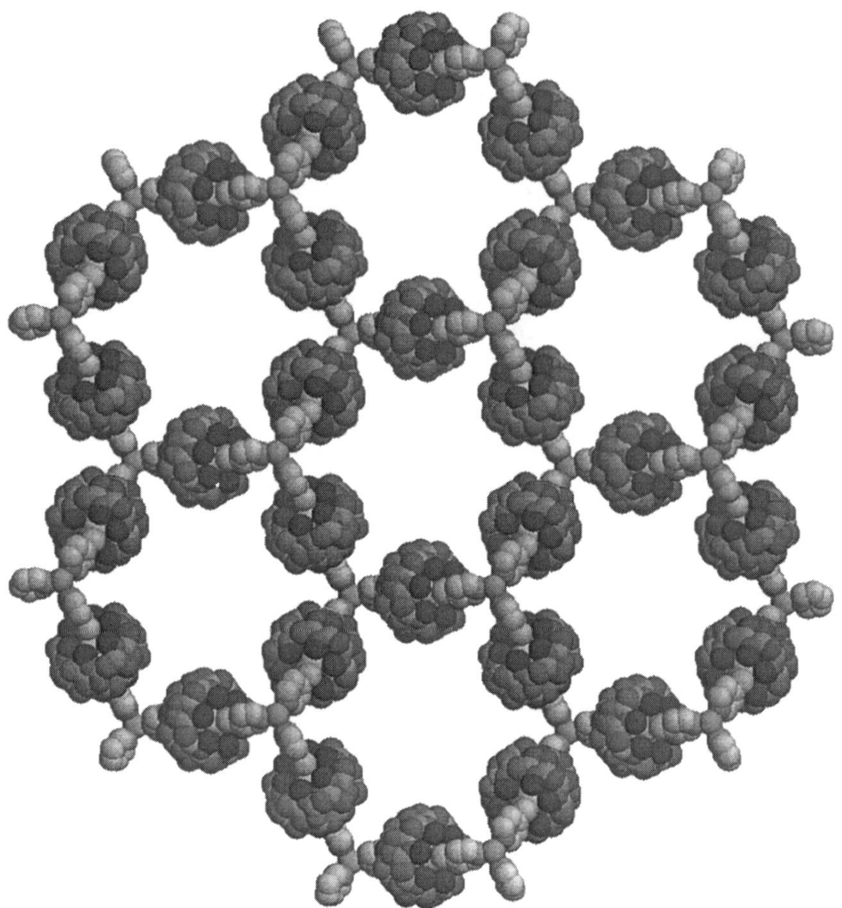

Figure 19 X-ray structure of 2D-polyrotaxane **18**. Each corner of the hexagons is occupied by Ag^{+}. A nitrate ion coordinated to the silver ion is omitted for clarity.

threaded on a helical 1D coordination polymer. The helix is extended along the *b* axis of the crystal with a pitch of 17.9 Å (Figure 23). In each helix, two $\mathbf{PR53}^{2+}$ and two silver ions constitute one turn. Each silver ion is coordinated by two $\mathbf{PR53}^{2+}$ in a linear fashion. It is interesting to note that one of the internal amine nitrogen atom (N(26)) is displaced from the six-oxygen plane of the portal by only 0.037(7) Å, but the other amine nitrogen (N(27)) is displaced by 0.721(8) Å. Moreover, the 3-pyridyl unit attached to N(26) makes a dihedral angle of 61° with the six-oxygen plane whereas that connected to N(27) is nearly parallel to the portal plane (angle = 9°). The parallel conformation of the 3-pyridyl unit requires the polymer chain to change its direction sharply, which eventually leads to a helical structure (Figure 23). There are equal amounts of right-handed and left-handed helices in the crystal; therefore, it

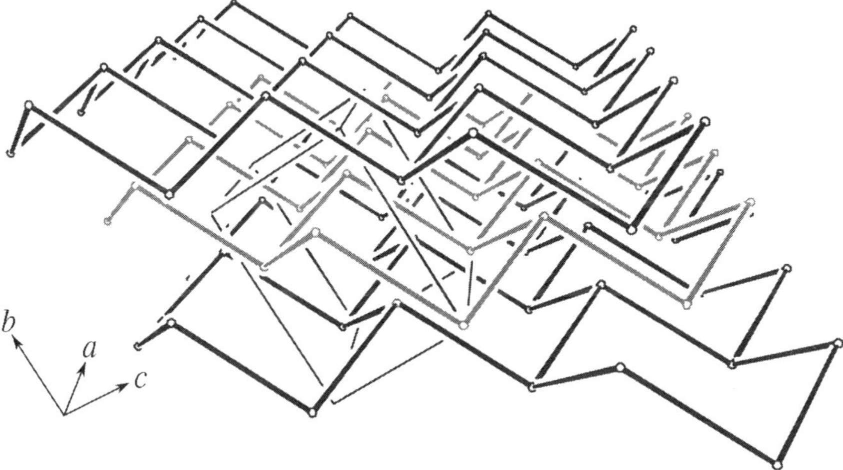

Figure 20 Stacking of the 2D polyrotaxane layers in the structure of **18**.

is a racemic mixture. It still remains to be studied how one enantiomer, either right-handed or left-handed polyrotaxane helices can be enantioselectively constructed. Such chiral helical polyrotaxanes may have interesting applications

4 MOLECULAR NECKLACES

4.1. Molecular Necklace [4]MN

Having synthesized the 1D and 2D polyrotaxanes, we decided to extend our efforts to synthesis of discrete nanometer-size supermolecules. Molecular necklaces were chosen as the first target species. As defined above, a molecular necklace (MN) is a cyclic oligorotaxane in which a number of small rings are threaded onto a large ring. In an elegant synthesis of [3]catenanes, Sauvage and coworkers observed unexpected formation of a mixture of [n]MN ($n = 4$–7), some of which were isolated and characterized by electrospray mass spectrometry [50]. Stoddart and coworkers also isolated and characterized a [4]MN in the synthesis of oligocatenanes [51]. Our strategy to the construction of molecular necklaces is similar to that for polyrotaxanes. In the synthesis of molecular necklaces, however, we use a metal complex with *cis* vacant coordination sites, instead of simple metal salt, as "glue" or an "angle connector" [14]. Metal complexes with *cis* vacant coordination sites have been used as "angle connectors" in the synthesis of molecular squares by Fujita [10, 52], Stang [11, 53] and others [54].

Reaction of **PR44**$(NO_3)_2$ with $Pt(en)(NO_3)_2$ (en = ethylenediamine) in refluxing water for 24 h produces **21** (Scheme 5) in near quantitative yield (by NMR) [14]. The

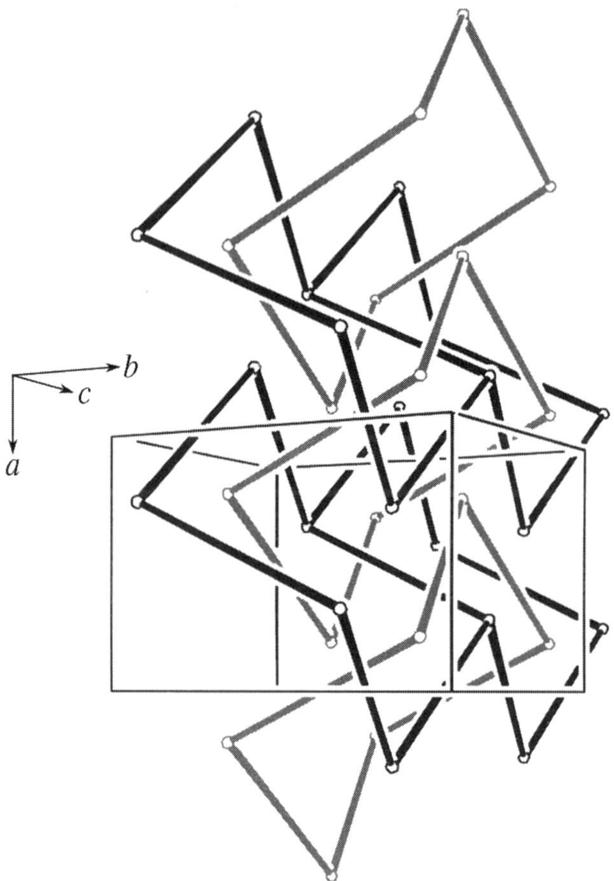

Figure 21 Interlocking of the hexagons in **18**.

same product can be obtained from a 1 : 1 : 1 mixture of cucurbituril, **C4py4**$(NO_3)_2$, and Pt(en)$(NO_3)_2$ under similar reaction conditions. The product was characterized by NMR, electrospray mass spectroscopy and elemental analysis. The ^{1}H NMR spectrum of **21** is very simple, indicating that the supermolecule has a highly symmetric structure. However, these spectroscopic data could not tell us definitely whether it is a [4]MN with a triangular structure or a [5]MN with a square geometry. The structure of **21** was finally determined by X-ray crystallography after over a year's trial of crystallization. Somewhat surprisingly, the X-ray crystal structure of **21** (Figure 24) reveals that three cucurbituril molecular "beads" are threaded on a molecular triangle [55]. Each corner of the triangle is occupied by a Pt(en) moiety and each side by a sigmoidal shaped pseudorotaxane unit **PR44**$^{2+}$, which links the two Pt moieties by coordination at its terminal pyridyl groups. The Pt $\cdots$ Pt

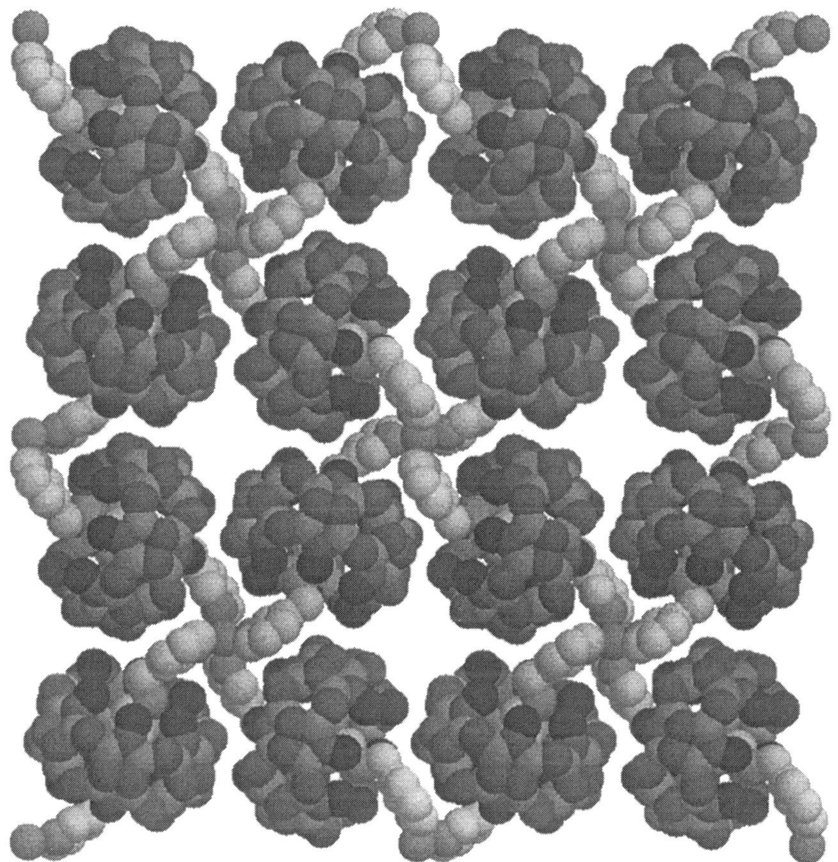

Figure 22 Square-meshed 2D polyrotaxane network of **19**. Each corner of the "square" is occupied by Ag^+. Triflate counter anions are omitted for clarity.

separation is 19.476(1) Å. No unusual bond parameters are observed including the Pt–N distances and the N–Pt–N angles. It is interesting to note that the three cucurbituril molecules in **21** are arranged in such a way that almost no vacant space exists inside the molecular triangle. It appears from the crystal structure that the hydrophobic interactions between the three cucurbituril rings may assist in the efficient formation of the necklace.

4.2 Molecular Necklace [5]MN

We also have constructed a [5]MN in which four cucurbituril beads are threaded onto a molecular square. The synthetic scheme is displayed in Scheme 6. Using an

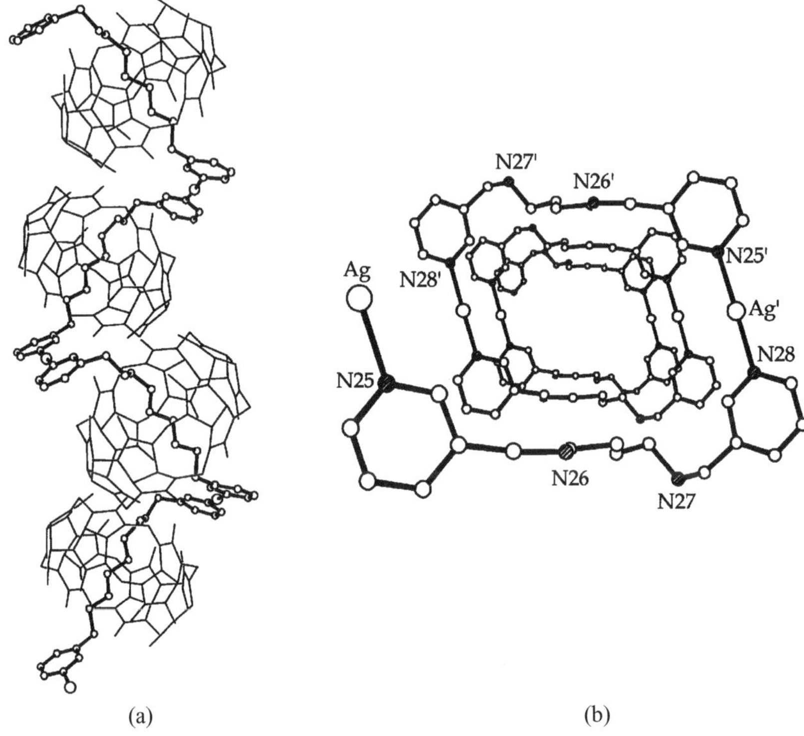

Figure 23 Helical 1D polyrotaxane **20**. (a) Overall Structure of **20** showing cucurbituril beads threaded on a helical 1D coordination polymer. Counter anions are omitted for clarity. (b) The helical 1D coordination polymer threading cucurbituril beads (omitted) in **20**.

L-shape molecule which can hold two cucurbituril beads, we prepared the preorganized pseudorotaxane (**22**) which was then allowed to react with $Pt(en)(NO_3)_2$ to produce the [5]MN **23** [56]. Spectroscopic data of **23** including NMR and mass spectra are consistent with the proposed structure, but we have not been able to grow crystals suitable for X-ray crystallography.

5 CONCLUSION

In this review, I describe our efforts to construct interlocked structures such as rotaxanes, polyrotaxanes and molecular necklaces incorporating cucurbituril as a molecular bead by utilizing the principles of self-assembly and coordination chemistry. A key to the success of this synthesis is the high affinity of cucurbituril toward alkyl diammonium ions, which allows formation of a stable pseudorotaxane

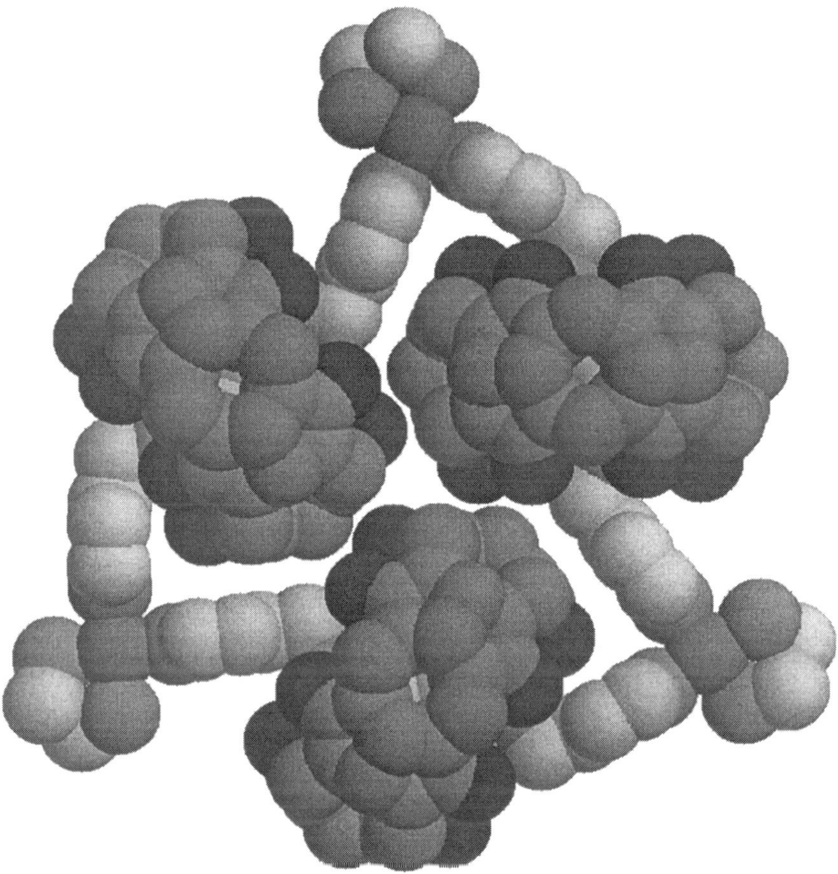

Figure 24 The minimal molecular necklace [4]MN (**21**).

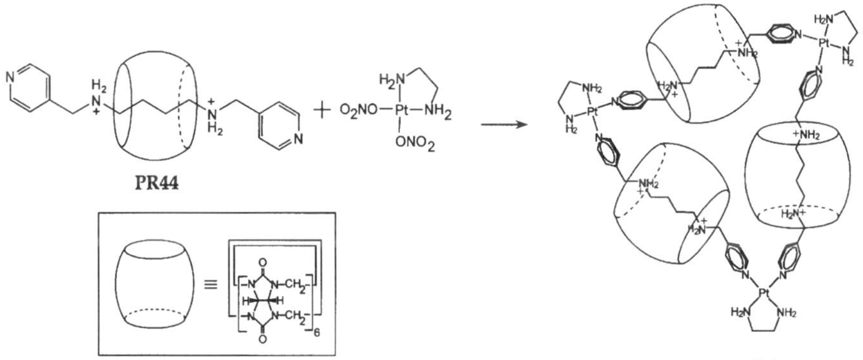

PR44

21

Scheme 5

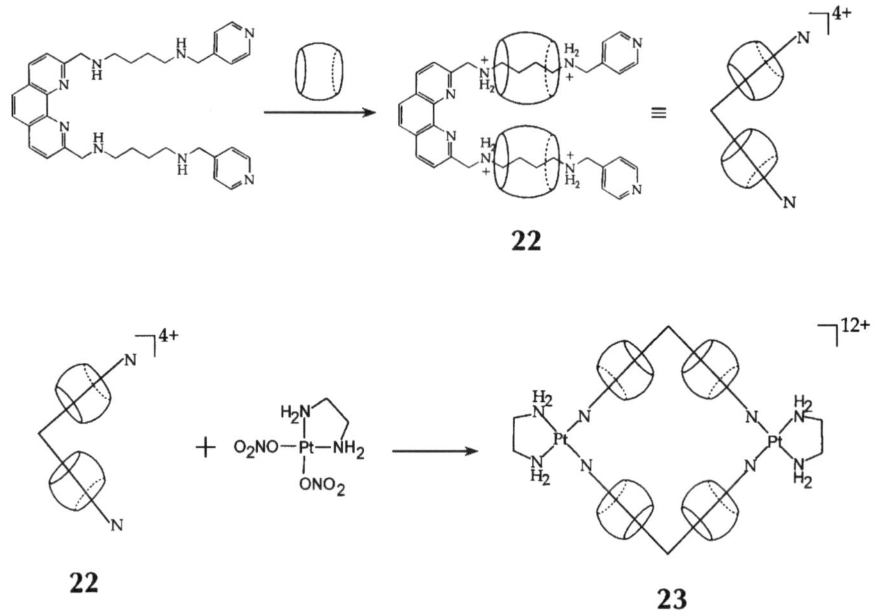

Scheme 6

with a short "string" before polymerization or cyclization occurs. Unlike cyclodextrins, which have been used widely as "beads" in the synthesis of rotaxanes and polyrotaxanes, cucurbituril has a highly symmetrical structure with two identical portals, that allows one to synthesize polyrotaxanes and molecular necklaces with high structural regularity. In the synthesis of polyrotaxanes, however, judicious choice of metal ions, counter ions and "strings" are important in order to construct the desired solid-state structures. In particular, transition metal ions with different preferred coordination geometries lead to polyrotaxanes with different structures. By varying the length and shape of the molecular "string" and the "angel connectors" one should be able to construct a variety of molecular necklaces with different sizes, shapes, and numbers of molecular "beads". Such a highly efficient synthesis of a topologically intriguing supermolecule may provide insights into construction of nanoscale particles with well-defined structures and functions.

ACKNOWLEDGEMENTS

I would like to express my sincere thanks to my coworkers who actually did the work described in this account. Without their talent and hard work this work would have not been possible. Special thanks are due to Dr Ki-Min Park who prepared the figures and schemes in this account. Our work described here was supported by

Korea Science and Engineering Foundation through the Center for Biofunctional Molecules (CBM), by the Ministry of Education through the Basic Science Research Institute (BSRI), Pohang University of Science and Technology and, more recently, by the Creative Research Initiatives Program of the Korean Ministry of Science and Technology through the Center for Smart Supramolecules (CCS). I would like to thank Professors Jik Chin and G. V. Smith for critical reading of the manuscript.

REFERENCES

1. Lehn, J. M. *Supramolecular Chemistry*, VCH: Weinheim, 1995.
2. *Comprehensive Supramolecular Chemistry* (Lehn, J.-M., Chair Ed.; Atwood, J. L.; Davis, J. E. D.; MacNicol, D. D.; Vögtle, F. Exec. Eds., Pergamon Press: Oxford, 1996, Vol. 1–11.
3. *Monographs in Supramolecular Chemistry 1–6*; Stoddart, J. F., Ed.; Royal Society of Chemistry: Cambridge, U.K., 1989, 1991. 1994–1996.
4. *Supramolecular Chemistry*, Balzani, V., DeCola, L., Eds.; Kluwer: Netherlands, 1992.
5. Whitesides, G. M., Simaneck, E. E., Mathias, J. P., Seto, C. T., Chin, D. N., Mammen, M. and Gordon, D. M., *Acc. Chem. Res.* 1995, **28**, 37.
6. Conn, M. M. and Rebek, J., Jr., *Chem. Rev.* 1997, **97**, 1647.
7. Tecilla, P., Jubian, V. and Hamilton, A. D. *Tetrahedron* 1995, **51**, 435.
8. Baxter, P. N. W. in *Comprehensive Supramolecular Chemistry* (Sauvage, J.-P. and Hosseini, M. W. Eds) Vol. 9, Pergamon: Oxford, 1996, 165.
9. Dietrich-Bruchecker, C. and Sauvage, J.-P. *Chem. Rev.* 1987, **87**, 795.
10. Fujita, M. and Ogura, K. *Bull. Chem. Soc. Jpn* 1996, **69**, 1471.
11. Stang, P. J. and Olenyuk, B. *Acc. Chem. Res.* 1997, **30**, 502.
12. Schill, G. *Catenanes, Rotaxanes, and Knots*, Academic Press: New York, 1971.
13. Recent review articles: (a) Sauvage, J.-M. *Acc. Chem. Res.* 1990, **23**, 319. (b) Amabilino, D. B. and Stoddart, J. F. *Chem. Rev.* 1995, **95**, 2725. (c) Philp, D. and Stoddart, J. F. *Angew. Chem., Int. Ed. Engl.* 1996., **35**, 1154. (d) Gibson, H., Bheda, M. C., and Engen, P. T. *Prog. Polym. Sci.* 1994, **19**, 843. (e) Vögtle, F., Dünnwald, T. and Schmidt, T. *Acc. Chem. Res.* 1996, **29**, 451. (f) Chambron, J.-C., Dietrich-Bruchecker, C. and Sauvage, J.-P. in *Comprehensive Supramolecular Chemistry* (Sauvage, J.-P. and Hosseini, M. W. Eds) Vol. 9, Pergamon: Oxford, 1996, 43. (g) Amabilino, D. B., Raymo, F. M. and Stoddart, J. F. in *Comprehensive Supramolecular Chemistry* (Sauvage, J.-P. and Hosseini, M. W. Eds) Vol. 9, Pergamon: Oxford, 1996, 85.
14. Whang, D., Park, K.-M., Heo, J. and Kim, K. *J. Am. Chem. Soc.*, 1998, **120**, 4899.
15. (a) Saneger, W. *Angew. Chem., Int. Ed. Engl.* 1980, **19**, 344. (b) Stoddart, J. F. *Carbohydrate Res.* 1989, **192**, xii. (c) *Comprehensive Supramolecular Chemistry* (Szejtli, J. and Osa, T. Eds) Vol. 3, Pergamon: Oxford, 1996.
16. Ogino, J. *J. Am. Chem. Soc.*, 1981, **103**, 1303; *Inorg. Chem.* 1984, **23**, 3312; *New J. Chem.* 1993, **17**, 683.
17. (a) Wylie, R. S. and Macartney, D. H. *J. Am. Chem. Soc.*, 1992, **114**, 3136. (b) Wylie, R. S. and Macartney, D. H. *Supramol. Chem.* 1993, **3**, 29. (c) Macartney, D. H. and Waddling, C. A. *Inorg. Chem.* 1994, **33**, 5912.
18. (a) Isnin, R. and Kaifer, A. E. *J. Am. Chem. Soc.*, 1991, **113**, 8188. (b) Isnin, R. and Kaifer, A. E. *J. Pure Appl. Chem.* 1993, **65**, 495.

19. (a) Harada, A. and Kamachi, M. *Macromolecules* 1990, **23**, 2821. (b) Harada, A., Li, J. and Kamachi, M. *Nature (London)* 1992., **356**, 325; (c) 1993, **364**, 516; (d) 1994, **370**, 126. (e) Harada, A., Li, J. and Kamachi, M. *Macromolecules* 1994, **27**, 4538.
20. (a) Wenz, G. and Keller, B. *Angew. Chem., Int. Ed. Engl.* 1992, **31**, 197. (b) Wenz, G., Wolf, F., Wagner, M. and Kubik, S. *New J. Chem.* 1993,**17**, 729.
21. (a) Yamaguchi, I., Osakada, K. and Yamamoto, T. *J. Am. Chem. Soc.*, 1996, **118**, 1811. (b) Born, M., Koch, T. and Ritter, H. *Macromol. Chem. Phys.* 1995, **196**, 1761. (c) Born, M. and Ritter, H. *Angew. Chem., Int. Ed. Engl.* 1995, **34**, 309. (d) Born, M., Koch, T. and Ritter, H. *Acta Polym.* 1994, **45**, 68.
22. Armspach, D., Ashton, P. R., Moore, C. P., Spencer, N., Stoddart, J. F., Wear, T. J. and Williams, D. J. *Angew. Chem., Int. Ed. Engl.* 1993, **32**, 854.
23. Lüttringhaus, A., Cramer, F. Prinzbach, H. and Henglein, F. M. *Chem. Ber.* 1958, **613**, 185.
24. Reviews on cucurbituril: (a) Mock, W. L. *Topics in Current Chem.* 1995, **175**, 1. (b) Cintas, P. *J. Incl. Phenom. Molec. Reco. Chem.* 1994, **17**, 205. (c) Mock, W. L. in *Comprehensive Supramolecular Chemistry* (Vögtle, F. Ed.) Vol. 2, Pergamon: Oxford, 1996, 477.
25. Behrend, R., Meyer, E. and Rusche, F. *Liebigs. Ann. Chem.* 1905, **339**, 1.
26. (a) Freeman, W. A., Mock, W. L. and Shih, N.-Y. *J. Am. Chem. Soc.*, 1981, **103**, 7367. (b) Freeman, W. A. *Act Cryst.* 1984, **B40**, 382.
27. (a) Mock, W. L. and Shih, N.-Y. *J. Org. Chem.* 1983, **48**, 3618. (b). Mock, W. L. and Shih, N.-Y. *J. Org. Chem.* 1986, **51**, 4440. (c) Mock, W. L. and Shih, N.-Y. *J. Am. Chem. Soc.*, 1988, **110**, 4706. (d) Mock, W. L. and Shih, N.-Y. *J. Am. Chem. Soc.*, 1989, **111**, 2697.
28. Jeon, Y.-M., Kim, J., Whang, D. and Kim, K. *J. Am. Chem. Soc.*, 1996, **118**, 9790.
29. Metal ion binding phenomenon of cucurbituril was mentioned even in the original report of cucurbituril by Behrend *et al.* [13] Buschmann *et al.* recently studied complexation of alkaline metal ions to cucurbituril by UV-vis spectroscopy: (a) Buschmann, H.-J., Cleve, E. A. Schollmeyer, E. *Inorg. Chim. Acta* 1992, **193**, 93, (b) Hoffmann, R., Knoche, W., Fenn, C. and Buschmann, H.-J. *J. Chem. Soc. Faraday Trans.*, 1994, **90**, 1507. The complex formation constants were measured in the presence of metal salts in the range of 1×10^{-5}–5×10^{-4} M.
30. Solubilities of cucurbituril (mole l^{-1}) in 0.2 M other alkali or alkaline earth metal salts: LiCl, 9.4×10^{-4}; KCl, 3.7×10^{-2}; CsCl, 5.9×10^{-2}; $CaCl_2$, 7.0×10^{-2}.
31. Whang, D., Heo, J., Park, J. H. and Kim, K. *Angew. Chem., Int. Ed. Engl.* 1998, **37**, 78.
32. Mock reported a molecular switch based on a pseudorotaxane. Mock, W. L. and Piepont, J. *J. Chem. Soc., Chem. Commun.*, 1990, 1509.
33. Jeon, Y.-M., Whang, D., Kim, J. and Kim, K. *Chem. Lett.* 1996, 503.
34. Asakawa, M., Ashton, P. R., Brown, G. R., Hayes, W., Menzer, S., Stoddart, J. F., White, A. J. P. and Williams, D. J. *Adv. Mater.* 1996, **8**, 37.
35. (a) Shen, X. and Gibson, H. W. *Macromolecules* 1992, **25**, 2058. (b) Gibson, H. W., Liu, S., Lecavalier, P., Wu, C. and Shen, Y. X. *J. Am. Chem. Soc.*, 1995, **117**, 852. (c) Shen, Y. X., Xie, D. and Gibson, H. W. *J. Am. Chem. Soc.*, 1994, **116**, 537. (d). Loveday, D., Wilkes, G. L., Bheda, M. C., Shen, Y. X. and Gibson, H. W. *J. Macromol. Sci., Pure Appl. Chem.* 1995, **A32**, 1. (e) Marand, E., Hu, Q., Gibson, H. W. and Veytsman, B. *Macromolecules* 1996, **29**, 2555. (f) Gibson, H. W. and Shu, L. *Makromol. Chem. Macromol. Symp.* 1996, **102**, 55.
36. Whang, D., Jeon, Y.-M., Heo, J. and Kim, K. *J. Am. Chem. Soc.*, 1996, **118**, 11333.
37. Recent representative examples of coordination polymers: (a) Munakata, M., Lwu, L. P., Juroda-Sowa, T., Maekawa, M., Suenaga, Y. and Furuichi, K. *J. Am. Chem. Soc.*, 1996, **118**, 3305. (b) Lu, J., Harrison, W. T. A. and Jacobson, A. J. *Angew. Chem., Int. Ed. Engl.* 1995, **34**, 2557. (c) Carlucci, L., Ciani, G., Proserpio, D. M. and Simoni, A. *Angew.*

Chem., Int. Ed. Engl. 1995, **34**, 1895. (d) Khasnis, D. V., Brewer, M., Lee, J., Emge, T. J. and Brennan, J. G. *J. Am. Chem. Soc.*, 1994, **116**, 7129. (e) Decurtins, S., Schmalle, H. W., Schneuwly, P., Ensling, J. and Gütlich, P. *J. Am. Chem. Soc.*, 1994, **116**, 9521. (f) Klüfers, P. and Schuhmacher, J. *Angew. Chem., Int. Ed. Engl.* 1994, **33**, 1742. (g) De Munno, G., Julve, M., Lloret, F., Faus, J., Verdaguer, M. and Caneschi, A. *Angew. Chem., Int. Ed. Engl.* 1993, **32**, 1046. (h) Berardini, M., Emge, T. and Brennan, J. G. *J. Am. Chem. Soc.*, 1993, **115**, 8501. (i) Evans, D. A., Woerpel, K. A. and Scott, M. J. *Angew. Chem., Int. Ed. Engl.* 1992, **31**, 430. (j) Saalfrank, R. W., Lurz, C.-J., Schobert, K., Struck, O., and Bill, E. and Trautwein, A. X. *Angew. Chem., Int. Ed. Engl.* 1991, **30**, 1494.

38. Whang, D. and Kim, L. *J. Am. Chem. Soc.*, 1997, **119**, 451.
39. Park, K.-M., Whang, D., Heo, K. and Kim. K. manuscript in preparation.
40. (a) Abrahams, B. F., Hoskins, B. F. and Robson, R. *J. Am. Chem. Soc.*, 1991, **113**, 3606. (b) Abrahams, B. F., Hoskins, B. F., Michail, D. M. and Robson, R. *Nature*, 1994, **369**, 727. (c) Batten, S. R., Hoskins, B. F. and Robson, R. *Angew. Chem., Int. Ed. Engl.* 1995, **34**, 820. (d) Batten, S. R., Hoskins, B. F., Robson, R. J. *Am. Chem. Soc.* 1995, **117**, 5385.
41. (a) Copp, S. B., Subramanian, S. and Zaworotko, M. J. *J. Am. Chem. Soc.*, 1992, **114**, 8719. (b) Copp, S. B., Subramanian, S. and Zaworotko, M. J. *Angew. Chem., Int. Ed. Engl.* 1993, **32**, 706. (c) MacGillivray, L. R., Subramanian, S. and Zaworotko, M. J. *J. Chem. Soc., Chem. Commun.*, 1994, 1325. (d) Robinson, F. and Zawaorotko, M. J. *J. Chem. Soc., Chem. Commun.*, 1995, 2413.
42. (a) Gardner, G. B., Venkataraman, D., Moore, J. S. and Lee, S. *Nature*, 1995, **374**, 792. (b) Venkataraman, D., Gardner, G. B., Lee, S. and Moore, J. S. *J. Am. Chem. Soc.*, 1995, **117**, 11600.
43. (a) Yaghi, O. M. and Li, G. *Angew. Chem., Int. Ed. Engl.* 1995, **34**, 207. (b) Yaghi, O. M. and Li, H. *J. Am. Chem. Soc.*, 1995, **117**, 10401. (c) Yaghi, O. M., Li. G. and Li, H. *Nature*, 1995., **378**, 703. (d) Yaghi, O. M. and Li, H. *J. Am. Chem. Soc.*, 1996, **118**, 295.
44. (a) Fujita, M., Kwon, Y. J., Washizu, S. and Ogura, K. *J. Am. Chem. Soc.*, 1994, **116**, 1151. (b) Fujita, M., Kwon, Y. J., Kwon, Y. J., Sasaki, O., Yamaguchi, K., Ogura, K. J. *Am. Chem. Soc.*, 1995. **117**, 7287.
45. (a) Soma, T., Yuge, H. and Iqamoto, T. *Angew. Chem., Int. Ed. Engl.* 1994, **33**, 1665 (b) Goodgame, D. M. L., Menzer, S., Smith, A. M. and Williams, D. J. *Angew. Chem., Int. Ed. Engl.* 1995, **34**, 574. (c) Carlucci, L., Ciani, G., Proserpio, D. M. and Sironi, A. *Angew. Chem., Int. Ed. Engl.* 1996., **35**, 1088.
46. Simon, J., André, J.-J. and Skoulios, A. *New. J. Chem.* 1986, 295.
47. (a) Real, J. A., Andrés, E., Muñoz, M. C., Julve, M., Granier, T., Bousseksou, A. and Varret, F. *Science* 1995, **268**, 265. (b) Stumpf, H. O., Ouahab, L., Pei, Y., Grandjean, D. and Kahn, P. *Science* 1993, **261**, 447. (c) Miyasaka, H., Matsumoto, N., Okawa, H., Re, N., Gallo, E. and Floriani, C. *Angew. Chem., Int. Ed. Engl.* 1995, **34**, 1446.
48. After our result was published two other 2D polyrotaxane polymers were reported by Robson and coworkers. (a) Hoskins, B. F., Robson, R. and Slizys, D. A. *J. Am. Chem. Soc.*, 1997, **119**, 2952. (b) Hoskins, B. F., Robson, R. and Slizys, D. A. *Angew. Chem., Int. Ed. Engl.* 1997, **36**, 2336.
49. Whang, D., Heo, J., Kim, C.-A. and Kim, K. *J. Chem. Soc., Chem. Commun.*, 1997, 2361.
50. (a) Bitsch, F., Hegy, G., Dietrich-Bruchecker, C., Leize, E., Sauvage, J.-P. and Dorsselaer, A. V. *J. Am. Chem. Soc.*, 1991, **113**, 4023. (b) Bitsch, F., Dietrich-Bruchecker, C. O., Khemiss, A.-K., Sauvage, J.-P. and Dorsselaer, A. V. *New J. Chem.* 1994, **18**, 801.
51. Amabilino, D. B., Ashton, P. R., Balzani, V., Boyd, S. E., Credi, A., Lee, J. Y., Menzer, S., Stoddart, J. F., Venturi, M. and Williams, D. J. *J. Am. Chem. Soc.*, 1998, **120**, 4295.
52. (a) Fujita, M., Yazaki, J. and Ogura, K. *J. Am. Chem. Soc.*, 1990, **112**, 5645. (b) Fujita, M., Yazaki, J. and Ogura, K. *Tetrahedron Lett.* 1991, **32**, 5589. (c) Fujita, M., Ibukuro, F., Yamaguchi, K. and Ogura, K. *J. Am. Chem. Soc.*, 1995, **117**, 4175. (d) Fujita, M., Sasaki,

O., Mitsuhashi, T., Fujita, T., Tazaki, J., Yamaguchi, K. and Ogura, K. *J. Chem. Soc., Chem. Commun.*, 1996, 1535.

53. Stang, P. J. and Cao, D. J. *J. Am. Chem. Soc.*, 1994, **116**, 4981. (b) Stang, P. J. and Chen, K. *J. Am. Chem. Soc.*, 1995, **117**, 1667. (c) Olenyuk, B., Whiteford, J. A. and Stang, P. J. *J. Am. Chem. Soc.*, 1996, **118**, 8221. (d) Stang, P. J., Cao, D. H., Chen, K., Gray, G. M., Muddiman, D. C. and Smith, R., D. *J. Am. Chem. Soc.*, 1997, **119**, 5563.
54. (a) Slone, R. V., Youn, D. I., Calhoun, R. M. and Hupp, J. T. *J. Am. Chem. Soc.*, 1995, **117**, 4175. (b) Rauter, H., Hillgeris, E. C., Erxleben, A. and Lippert, B. *J. Am. Chem. Soc.*, 1994, **116**, 616.
55. Examples of molecular triangle containing three Pt(en) or Pd(en) units: (a) Schnebeck, R.-D., Randaccio, L., Zangrando, E. and Lippert, B. *Angew. Chem., Int. Ed. Engl.* 1998, **37**, 119. (b) Lee, S. B., Hwang, S., Chung, D. S., Yun, H. and Hong, J.-I. *Tetrahedron Lett.* 1998, **39**, 873.
56. Roh, S.-G., Park, K.-M., Park. G.-J., Sakamoto, S., Yamaguchi, K. and Kim, K. *Angew. Chem.* in press.

Cumulative Author Index

This index comprises the names of contributors to Volumes 1–5 of **Perspectives in Supramolecular Chemistry**.

Cumulative Title Index

This index comprises the titles and authors of all chapters appearing in Volumes 1–5 of **Perspectives in Supramolecular Chemistry.**

Index